Dosimetry in Bioelectromagnetics

Dosimetry in Bioelectromagnetics

Edited by
Marko Markov

CRC Press
Taylor & Francis Group
Boca Raton London New York

CRC Press is an imprint of the
Taylor & Francis Group, an **informa** business

MATLAB® is a trademark of The MathWorks, Inc. and is used with permission. The MathWorks does not warrant the accuracy of the text or exercises in this book. This book's use or discussion of MATLAB® software or related products does not constitute endorsement or sponsorship by The MathWorks of a particular pedagogical approach or particular use of the MATLAB® software.

CRC Press
Taylor & Francis Group
6000 Broken Sound Parkway NW, Suite 300
Boca Raton, FL 33487-2742

First issued in paperback 2019

© 2017 by Taylor & Francis Group, LLC
CRC Press is an imprint of Taylor & Francis Group, an Informa business

No claim to original U.S. Government works

ISBN-13: 978-1-4987-7413-0 (hbk)
ISBN-13: 978-0-367-87882-5 (pbk)

Visit the Taylor & Francis Web site at
http://www.taylorandfrancis.com

and the CRC Press Web site at
http://www.crcpress.com

This book is dedicated to the memory of great scientists Charles Polk, Arthur Pilla, and Nentcho Todorov for their contribution in Bioelectromagnetics, Dosimetry and Magnetic field therapy, as well to my brother Konstantin Markov who passed away during my work on this book

Contents

To the Reader

My work in bioelectromagnetics started way back in 1969 when I joined the Department of Biophysics at Sofia University. My background as a physicist guided me from the first month of my work, with the notion that for an experiment to be scientifically valid it is necessary to have a clear vision for the protocol of the study and exact clarification of the parameters of applied electromagnetic field (EMF) signals.

Over the course of my studies, writing papers, reading the literature, and participating in international meetings, I became more and more convinced that dosimetry is an important factor in bioelectromagnetics.

In the 1980s, I organized in Bulgaria two international conferences, one on electromagnetic fields and another on biomembranes. I planned to entirely dedicate a third one to dosimetry, but because of various circumstances, this did not happen.

Being a member of the Bioelectromagnetics Society, I tried to convince the editors of the two major journals, *Bioelectromagnetics* and *Electromagnetic Fields in Biology and Medicine,* that the instruction to the authors must include strict requirements regarding the dosimetry of the experiments for which manuscripts are submitted, but my efforts failed.

Therefore, I decided to write this book. CRC Press was kind enough to support this venture. I gathered a large international team of scientists from Europe and the United States who joined me in my efforts to describe the basic principles of dosimetry as well as the achievements in this respect in basic science and clinical application.

This team includes physicists, biologists, clinicians, and engineers who recognize the importance of dosimetry in bioelectromagnetics. My list of colleagues and friends includes both senior scientists and young individuals from Zurich, Ljubljana, Modena, Vienna, Niš, Bratislava, Moscow, Yerevan, Wroclaw, and Puschino as well as Americans from Denver, Norfolk, Baltimore, Lexington, Seattle, Kenosha (Wisconsin), and Ft. Lauderdale. I have deliberately not mentioned names, because most of the papers were written by teams of scientists.

I hope that we have been able to cover the physical and engineering principles of dosimetry. This book recognizes the necessity of exact identification of the EMF signals implemented in laboratory studies and clinical trials. At the same time, we have clarified that in bioelectromagnetics, biophysical dosimetry is very important, as it describes the parameters of the exposure to which the target is exposed.

There are several chapters that deal with engineering and physical principles of design and generation of EMF signals with different frequencies and the methods of characterizing the signals. The chapter on dose and exposure in bioelectromagnetics is followed by one dealing with necessary characteristics of quality bioelectromagnetic research as well with the one on physical dynamics as the base for development of biophysical experiments. The chapter detailing electric and magnetic fields as a means of cellular signaling is followed by one explaining the dose and duration of exposure in achieving nonthermal biological effects.

Thermal versus nonthermal mechanisms of EMF interactions with biological systems are discussed in parallel with dosimetry in electroporation. More attention is paid to clinical dosimetry and to accurate reporting of the parameters of exposure that would secure the possibility of reproduction of the laboratory experiments and/or clinical trials.

Last, but not least, we have paid attention to the development of cellular communications, Wi-Fi systems, and devices and also discussed potential hazards of the uncontrolled distribution of these novel achievements of human civilization.

I do believe that the collective efforts of more than 30 scientists from different countries will be a plausible tool for further development of the principles of dosimetry in basic science and clinical application.

Thank you for joining us on this journey.

Marko Markov

Editor

Marko S. Markov received his BS, MS, and PhD from Sofia University, Bulgaria. He was the professor and chairman of the Department of Biophysics and Radiobiology of Sofia University for 22 years. He was an invited professor and lecturer in a number of European and American academic and industry research centers.

Dr. Markov is recognized as one of the world's best experts in clinical applications of electromagnetic fields. He has given 288 invited and platform presentations at more than 70 international meetings. His list of publications includes 196 papers and 18 books.

Dr. Markov has more than 45 years of experience in basic science research and more than 40 years of experience in clinical applications of electromagnetic fields for the treatment of bone and soft tissue pathologies and injuries.

His commercial affiliation started in Bulgaria with a series of contractual appointments and continued in the United States in his capacity as the vice president of three companies involved in the manufacturing and distribution of devices for magnetic field therapy. The spectrum of the signals range from static magnetic fields to 27.12 MHz. The clinical targets are in the area of bone and soft tissue defects, pain control, and innovation of the low-frequency range for the inhibition of angiogenesis and tumor growth. Lately, he has introduced analytical methods for designing signals and devices for bioelectromagnetics research.

In 1981 Dr. Markov wrote his first book, *Professions of the Laser*. In 1988, he coedited *Electromagnetic Fields and Biomembranes* with Martin Blank, which was published by Plenum Press. In 2004, he coedited *Bioelectromagnetic Medicine* with Paul Rosch, president of the American Institute of Stress, and this was published by Marcel Dekker. In 2006, with Sinerk Ayrapetyan, he coedited *Bioelectromagnetics: Current Concepts*. In 2010, he coedited the book *Advanced Electroporation Techniques in Biology and Medicine* with Damijan Miklavcic and Andrei Pakhomov, and this was published by CRC Press. His latest book, also published by CRC Press in 2015, *Electromagnetic Fields in Biology and Medicine*, is currently being translated in China.

Dr. Markov has edited seven special issues for two journals, *The Environmentalist* and *Electromagnetic Biology and Medicine*, that consist of selected papers of biannual International Workshops on Biological Effects of Electromagnetic Fields.

Dr. Markov is a cofounder of the International Society of Bioelectricity, the European Bioelectromagnetic Association (EBEA), and the International Society of Bioelectromagnetism. He has been a member of the board of directors of the Bioelectromagnetics Society and organizer of several NATO research meetings.

Contributors

Marko S. Andjelković
Applied Physics Laboratory
Faculty of Electronic Engineering
University of Niš
Niš, Serbia

Sinerik Ayrapetyan
Life Sciences International Postgraduate
    Educational Center
UNESCO
Yerevan, Armenia
nfo@biophys.am

Frank S. Barnes
Electrical, Computer & Energy
    Engineering
University of Colorado Boulder
Boulder, Colorado
frank.barnes@colorado.edu

Igor Belyaev
Laboratory of Radiobiology
Cancer Research Institute, BMC
Slovak Academy of Sciences
Bratislava, Slovak Republic

and

Laboratory of Radiobiology
General Physics Institute
Russian Academy of Sciences
Moscow, Russia
igor.beliaev@savba.sk

Pawel Bienkowski
EM Environment Protection Lab
Wroclaw University of Technology
Wroclaw, Poland

Carl F. Blackman
Wake Forest Baptist Medical Center
Winston-Salem, North Carolina
carl.blackman@gmail.com

Matteo Cadossi
Orthopaedics and Trauma Clinic I
Rizzoli Orthopaedic Institute
University of Bologna
Bologna, Italy

Ruggero Cadossi
Clinical Biophysics
IGEA
Carpi, Italy
r.cadossi@igeamedical.com

Andrew B. Gapeyev
Institute of Cell Biophysics
Russian Academy of Sciences
Moscow, Russia
a_b_g@gmail.ru

Ben Greenebaum
Department of Physics
University of Wisconsin Parkside
Kenosha, Wisconsin
greeneba@uwp.edu

Yury G. Grigoriev
Russian National Committee on
    Non-Ionizing Radiation Protection
Moscow, Russia
profgrig@gmail.com

Kjell Hansson Mild
Department of Radiation Sciences
Umeå University
Umeå, Sweden
kjell.hansson.mild@radfys.umu.se

Marko Markov
Research International
Williamsville, New York
mamarkov@aol.com

Leo Massari
Orthopaedic and Traumatology
    Department
Cona S. Anna Hospital
University of Ferrara
Ferrara, Italy

Mats-Olof Mattsson
Health & Environment Department,
    Environmental Resources &
    Technologies
Austrian Institute of Technology GmbH
Vienna, Austria

Harvey N. Mayrovitz
Professor of Physiology
College of Medical Sciences
Nova Southeastern University
Fort Lauderdale, Florida
mayrovit@nova.edu

Damijan Miklavčič
Laboratory of Biocybernetics
Faculty of Electrical Engineering
University of Ljubljana
Ljubljana, Slovenia
damijan.miklavcic@fe.uni-lj.si

Wayne Miller
Nura Health, SPC
Vashon, Washington
wayne.miller@nurahealth.com

Andrei G. Pakhomov
Old Dominion University
Norfolk, Virginia
andrei@pakhomov.net

William Pawluk
Medical Editor
Timonium, MD
wpaw@comcast.net

Eva Pirc
Laboratory of Biocybernetics
Faculty of Electrical Engineering
University of Ljubljana
Ljubljana, Slovenia

Lucas A. Portelli
Department of Electrical,
    Computer and Energy Engineering
University of Colorado
Boulder, Colorado

and

Foundation for Research on
    Information Technologies in
    Society
Zurich, Switzerland
lucasportelli@gmail.com

Matej Reberšek
Laboratory of Biocybernetics
Faculty of Electrical Engineering
University of Ljubljana
Ljubljana, Slovenia

Goran S. Ristić
Applied Physics Laboratory
Faculty of Electronic Engineering
University of Niš
Niš, Serbia
goran.ristic@elfak.ni.ac.rs

Stefania Setti
Clinical Biophysics
IGEA
Carpi, Italy

Betty F. Sisken
Department of Biomedical Engineering
   and Department of Anatomy and
   Neurobiology
University of Kentucky
Lexington, Kentucky
bsisken@hotmail.com

Hubert Trzaska
EM Environment Protection Lab
Wroclaw University of Technology
Wroclaw, Poland
hubert.trzaska@pwr.edu.pl

1

Dosimetry in Bioelectromagnetics

Marko Markov
Research International

1.1 Introduction

Despite the increased research on biological effects and clinical use of electromagnetic field (EMF), very little is known about the fact that contemporary magnetobiology and magnetotherapy have begun immediately after World War II by introducing both magnetic and EMFs generated by various wave shapes of the supplying currents. Starting in Japan, this modality quickly moved to Europe, first in Romania and the former Soviet Union. During the period 1960–1985, nearly all European countries designed and manufactured their own magnetotherapeutic systems. Medicine in the 21-century is characterized by an increasing variety of diagnostic and therapeutic devices that implement magnetic and EMFs that utilize time-varying signals. In most cases, however, these devices are manufactured and used without proper knowledge of the principles of planning, conducting research, and magnetotherapy.

The fundamental problem for engineers, physicists, biologists, and clinicians is to identify the biochemical and biophysical conditions under which EMF signals could be recognized by cells in order to further modulate cell and tissue functioning. It is also important for the scientific and medical communities to comprehend that different magnetic fields (MFs) applied to different tissues could cause different effects. What "different MFs" mean is both difficult and easy to define. Easy, because if the second signal has at least one parameter different from the previous one, it is a new signal. The difficulties arise because very few manufacturers, distributors, or authors of published studies exactly describe the parameters of the device. In some cases, it happened because of neglecting the basic principles in biomagnetic technology, but more often, it happens because of lack of knowledge that each and any device to which human being is exposed needs a thorough dosimetry evaluation.

One should not expect that the MF used for superficial wounds might be as beneficial for fracture healing. Once the appropriate signal is engineered, serious attention must be paid to the biophysical dosimetry capable of predicting which EMF signals could be effective and how efficiency could be monitored. This raises the question of using theoretical models and biophysical dosimetry in selection of the appropriate signals and in engineering and clinical application of new EMF therapeutic devices.

Unfortunately, in most cases, the experiment or trial starts with the question, "what signal is available?" Then, the scientist/clinician applies common sense and his/her experience and intuition in selecting the protocol of study. Dosimetry and analysis of the signal most often remain out of consideration. For this reason, many laboratory studies cannot be replicated. Obviously, the situation is more complicated with a clinical trial in which many more variations in the protocol execution in different clinical sites have occurred.

If physical modalities are to be compared with pharmaceuticals, one might see a lot of advantages for magnetotherapy. MF/EMF modalities are usually applied rather than regular pharmaceuticals, offering a plausible alternative with fewer, if any, side effects. This is a tremendous advantage when compared with pharmaceutical treatment in which the administered medication spreads over the entire body, causing adverse effects in different organs, which sometimes might be significant. One should not forget that in order to deliver the necessary medication dose to treat the target tissue/organ, patients should routinely receive doses of medication a hundred times larger than the dose needed by the target. This is exactly the reason that we emphasize the necessity of detailed and accurate physical and biophysical dosimetry.

1.2 Signal Characteristics

Today, magnetic-field-dependent modalities can be categorized into five groups:

- Static MFs
- Low-frequency EMFs
- Pulsed electromagnetic fields (PEMFs)
- Pulsed radiofrequency fields (PRFs)
- Microwave fields (mainly for communications)

Because of the large frequency variations, it is difficult to summarize the signals. An excellent review of the physics and engineering of low-frequency signals was published by Liboff (2004). EMF signals, especially in clinical use, have a variety of designs, which in most cases are selected without any motivation for the choice of the particular waveform, field amplitude, or other physical parameters.

It seems reasonable that the first and widely used waveshape is the sine wave with a frequency of 60 Hz in North America and 50 Hz in the rest of the world. From symmetrical sine waves, engineers moved to an asymmetrical waveform by means of rectification. These types of signals basically flip-flop the negative part of the sine wave to positive, thereby creating a pulsating sine wave. The textbooks usually show the rectified signal as a set of ideal semi sine waves. However, due to the impedance of the particular design, such an ideal waveshape is impossible to achieve. As a result, the ideal form is distorted, and in many cases, a short DC-type component appears between two consecutive semi sine waves. This form of the signals has been tested for treatment of low back pain (Harden et al., 2007) and reflex sympathetic dystrophy (Ericsson et al., 2004). However, the most successful implementation of this signal is shown in animal experiments, causing antiangiogenic effects (Williams et al., 2001; Markov et al., 2004). By investigating a range of amplitudes for a 120 pulses per second (pps) signal, the authors demonstrated that the 15 mT amplitude prevents formation of the blood vessels in growing tumors, thereby depriving the tumor from expanding the blood vessel network, causing tumor starvation and death.

In the mid-1980s, the ion cyclotron theory was proposed by Liboff (1985) and Liboff et al. (1987), and shortly after that, a clinical device was created based on the ion cyclotron resonance (ICR) model. This device is currently in use for recalcitrant bone fractures. The alternating 40 μT sinusoidal MF is at 76.6 Hz (a combination of Ca and Mg^{2+} resonance frequencies). This signal has an oscillating character, but due to the DC MF, it oscillates only as a positive signal.

Using a radio-broadcasting approach, several devices were created to deliver a sine wave signal modulated by another signal. Usually, the sine wave signal is in the high frequency (kHz and MHz) range, while the modulating signal is a low-frequency signal. There are also devices that apply two high-frequency signals, and the interference of both signals results in an interference MF (Todorov, 1982).

In addition to sine-wave-type signals, a set of devices that utilize unipolar or bipolar rectangular signals are available on the market. Probably, for those signals, the most important thing to know is that, due to the electrical characteristics (mostly the impedance) of the unit, these signals could never be rectangular. There should be a short delay both in raising the signal up and in its decaying to zero. The rise time of such a signal could be of extreme importance because the large value of dB/dt could induce significant electric current in the target tissue. Some authors consider that neither the frequency nor the amplitude is so important for the biological response but that the dB/dt rate is the factor responsible for observed beneficial effects. To avoid problems with definition of the rise time, Kotnik and Miklavcic (2006) suggested that the rectangular signals should be replaced by more realistic trapezoid signals. Well, if one accepts this view, why are these signals (easy to characterize) not in common use? Although, it is easy to design such a signal, it is difficult to technically realize this design.

The first clinical signal approved in the United States by the Food and Drug Administration (FDA) for treatment of nonunion or delayed fractures (Bassett et al., 1974, 1977) exploited the pulse burst approach. Having a repetition rate of 15 bursts per second, this asymmetrical signal (with a long positive and very short negative component) has more than 30 years of very successful clinical use for healing nonunion bones. The engineers assume that the cell/tissue would ignore the short negative part of the pulse and would respond only to the envelope of the burst that had a duration of 5 μs, long enough to induce sufficient amplitude in the kHz frequency range. A series of modalities generate signals that consist of single narrow pulses separated by long "signal-off" intervals. This approach allows modification of not only the amplitude of the signal but also the duty cycle (time on/time off).

The pulsed radiofrequency (RF) signal, originally proposed by Ginsburg (1934) and later allowed by the FDA for treatment of pain and edema in superficial soft tissues (Diapulse), utilizes the 27.12 MHz frequency in pulsed mode. Having a short 65 μs burst and 1600 μs pause between pulse bursts, the signal does not generate significant heat during 30 min use.

Which signals and at what conditions could be the most effective? Are certain signal parameters better than others? It should be pointed out that most of EMF signals used in research and as therapeutic modalities have been chosen in some arbitrary manner. Very few studies assessed the biological and clinical effectiveness of different signals by comparing the physical/biophysical dosimetry and biological/clinical outcomes. With the exponential development of the Internet, it is easy to find tens, if not hundreds, of devices that promise to cure each and every medical problem. A careful look at these sites would show that no engineering, biophysical, or clinical evidence is given to substantiate the claims.

Each report of laboratory study or clinical trial should provide information about the following parameters:

- Type of the field (as mentioned earlier)
- Component (electrical or magnetic)
- Intensity or induction
- Gradient (dB/dt)
- Vector (dB/dx)
- Frequency
- Pulse shape
- Localization
- Duration of the exposure
- Depth of penetration

When the amplitude of the applied signal is to be described, it might be the absolute value of the signal in positive direction, peak-to-peak value, or root mean square (RMS). The majority of available instruments are calibrated in RMS, while more advanced solutions allow readings of any value. In very few studies, the Fourier analysis of the signal is performed. In the next section, an example of the importance of the search for harmonics in the signal is shown.

In a systematic review of 56 papers on static MF applications in clinical settings, Colbert et al. (2009) found that only two papers (4%) accurately reported the abovementioned parameters. So, if a publication or any other report does not explain the dosimetry parameters, it is impossible to replicate the study/trial.

It is extremely important for both scientists and clinicians to understand that the availability of many signals/devices on Internet sites is not an advantage, but rather a deficiency. Most of these devices are even not adequately described by the manufacturers and distributors in the sense of their engineering and physical parameters. Some plausible advice could be offered: Before starting the use of a new signal/device, search for detailed characteristics of the new instrument. It is even more important to perform one's own dosimetric evaluation of the device.

Of course, it should be useful to reverse the approach. Instead of empirical design of the signal based on intuition and experience of engineers, let us start with the biological/clinical site. In other words, look what type of signal is necessary for achievement of a biological or clinical response. This requires the exact knowledge of electromagnetic characteristics of normal and injured tissue. Using SQUID magnetometer measurements of normal and injured tissues, we were able to create a therapeutic signal capable of compensation of injury current (Parker and Markov, 2011).

1.3 About Harmonics in the Signal

Every person familiar with time-varying signals knows that in addition to the basic (carrier) frequency, a range of additional frequencies, known as harmonics, is present. Usually, they are multiples of basic frequency. Why do scientific and medical communities neglect the contribution of these harmonic frequencies? The explanation seems to be in the notion that they have small amplitude and thereby are not of significance when the signal's effects are discussed. Is this really true, especially when the target is such a complex medium as a biological system, in which even the same molecules might be in different functional states?

We applied Fourier analysis for the rectified semi sine wave signal discussed earlier and implemented it in another therapeutic system. The frequency spectrum is shown in Figure 1.1. What is seen in this spectrum is that at basic frequency of 120.1 pps, the first

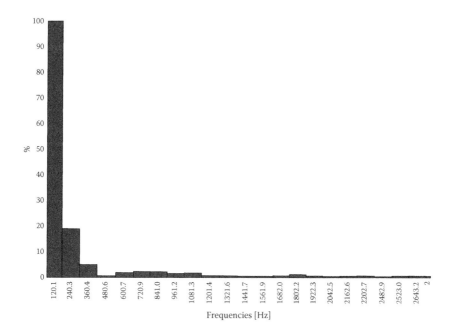

FIGURE 1.1 Spectrum of harmonics at the rectified 60 Hz signal.

harmonic (240.3 pps) has an amplitude equal to 20% of the amplitude of the basic frequency. The third harmonic (360.4 pps) has an amplitude of 5% of the basic amplitude. In addition, the first 20 harmonics, up to 2643 pps, are identified by Fourier analysis (Markov and Hazlewood, 2009).

It remains to be determined what the importance of the impedance for the waveshape and harmonics is for physics and biology. As for many other signals, the biophysical analysis of the rectified signal implemented in this type of therapeutic unit is practically absent. In a study of neurite outgrowth in chick root ganglia, Sisken et al. (2007) described that signal 2 (with larger DC component between semi sine waves) initiated a 20% larger response, as evaluated by the neurite area.

What does this mean? At least for this specific rectified signal, in addition to the basic frequency, the target receives a set of signals with higher frequencies and amplitudes that might be of importance for some processes. It is possible that different generating systems, having different impedances, would generate signals with different spectrum compositions (in respect of harmonics' amplitudes).

All of the above considerations indicate that bioelectromagnetics research needs accurate dosimetry from the viewpoint of both physics and biophysics.

1.4 Physical versus Biophysical Dosimetry

It is important to discuss these two terms because it appears that the issue of dosimetry is not properly interpreted in both the scientific and clinical communities. In our view, physical dosimetry relates to characterization of the device with respect to engineering and physical parameters. This might include such signal parameters as frequency, signal shape, amplitude, repetition rate, rising time, etc. It should be pointed out that these are characteristics of the device and not the actual field distribution inside the target tissue.

Biophysical dosimetry is accounting for the "dose" received by the target tissue. *It should always be remembered that the only valuable information is that which explains the field at the target site.* This goal is easy to achieve because the magnetic properties of air and biological tissues, especially for static and low-frequency fields, are approximately equal. Another important issue is the fact that any physical body attenuates the electric field (EF) component and allows only the magnetic component to penetrate deep into the body (Markov, 1999). Therefore, biophysical dosimetry is mainly dosimetry of the MFs. It should be taken into account that any time-varying MF induces EF, which in some cases might be of importance in analysis and interpretation of the observed effects. In respect of proper dosimetry, the fact that the induced EF generates "back" MFs should be taken into account at any measurements, especially at high-frequency signals (Markov, 2002).

Well, we know the meters that allow us to perform physical dosimetry. How may one perform biophysical dosimetry? There are several methods that might be implemented here. The next section describes the relatively new biophysical method that faciliates not only biophysical dosimetry but also 3-D mapping of the field distribution and the way to predict the "dose" that will be received by the target.

1.5 Calcium/Calmodulin Model as a Plausible Biophysical Dosimeter

In the late 1980s in the laboratory of Lednev, a brilliant experiment was designed in an attempt to search for experimental evidence for his ICR model (Shuvalova et al., 1991). This cell-free calcium/calmodulin model was further developed in my laboratory into a plausible biophysical dosimeter. Both basic science and clinically important data have been obtained during the last decade of the 20th century (Markov, 2004a,b). Let me provide some principal information about this approach: It has been known that the calmodulin molecule has four molecular clefts ready to bind calcium ions. Moreover, calmodulin undergoes conformational changes as each binding site is filled. This binding has a profound effect on cell functioning, since the calmodulin is essential for many biochemical interactions (Markov and Pilla, 1994b). In a series of studies of calcium-calmodulin-dependent myosin phosphorylation, my group demonstrated that specific static MFs, PEMF, and 27.12 MHz PRF could modulate Ca^{2+} binding to CaM in a cell-free enzyme preparation (Markov et al., 1992, 1993, 1994; Markov and Pilla, 1993, 1994a,b; Markov 2004a,b). The ion binding target pathway has been confirmed in other studies using static MFs (Engstrom et al., 2002; Liboff et al., 2003).

The work, conducted within a 10-year period in analyzing various signals, allows us to define and prove the plausibility of using the calcium/calmodulin system as a biophysical dosimeter. The advantages might be seen in the fact that the enzyme, substrate, and calmodulin are extracted from real living tissues and placed in a cell-free solution, securing the homogeneity of the sample. In addition, the small size (100 μL) of the sample provides a nearly perfect opportunity for exact evaluation of the response of the system.

Numerous cellular studies have addressed the effects of EMF on signal transduction pathways. It is well accepted now that the cellular membrane is a primary target for MF action (Adey, 2004; Markov 2004c, 2005). Evidence has been collected that selected MFs are capable of affecting the signal transaction pathways via alteration of ion binding and transport. The calcium ion is recognized as a key player in such alterations.

Clinical experience, as well as results from numerous animal and *in vitro* studies, suggests that the initial conditions of the EMF-sensitive target pathway determine whether a physiologically meaningful bioeffect could be achieved. For example, when a broken bone receives treatment with PEMF, the surrounding soft tissues receive the same dose as the fracture site, but the physiologically important response occurs only in the injured bone tissue, while changes in the soft tissue have not been observed.

This is crucially an important behavior, indicating that MFs are more effective when the tissue is out of equilibrium. Therefore, the experiments with healthy volunteers are not always indicative of the potential response of patients who are victims of injury or disease. The healthy organism has a much larger compensational ability than the diseased organism, which in turn would reduce the manifestation of the response.

Support for this notion comes from a study of Jurkat cells in which the state of the cell was found to be important with regard to the response of tissues to MFs: Normal T-lymphocytes neglect the applied PEMF, while reacting after stimulation with other factors. In other words, it might be approximated by the pendulum effect—the larger the deviation from equilibrium, the stronger is the response (Nindl et al., 2002; Markov, 2007). For example, Nindl et al. demonstrated, in an *in vitro* study, that the initial conditions of lymphocytes are important in terms of the biological effects of those cells in response to MFs.

1.6 Thermal versus Nonthermal

This section deals with the principal difference in the biophysical and engineering approaches to biological mechanisms of EMF-initiated bioeffects. While the biophysical approach is based on experimentally obtained data on biological responses to the applied EMF, the engineering approach strongly relies on specific absorption rate (SAR) values. In most cases, SAR is used to determine the hazardous threshold. The problem of "thermal vs. nonthermal" nature of EMF interactions with biological systems in general follows the level of public interest and funding sources (especially in the United States).

This period started in the late 1950s and continued until the early 1980s and was marked by research on microwave and RF radiation mostly stimulated by military and space research. In the late 1980s and early 1990s, funding and research priorities in the United States shifted toward the low-frequency power line range. During the last 15 years, the research was directed mainly by exponentially increasing cellular communications. Not surprisingly, the availability of funds influenced the directions of research and even the scientific methods applied.

The basic science investigations of EMF interactions developed in two distinct directions searching for thermal and nonthermal mechanisms of action. There is no question for discussion that the science needs to have verifiable and reasonable criteria to evaluate the energy delivered to a given biological body. The problem is, however, what is the driving force in the assumed mechanisms? Classical physics and equilibrium thermodynamics have a simple model: heat. Immediately, an argument comes—the energy an EMF introduces into the biological object is far below the physiological threshold based in equilibrium thermodynamics of kT thermal collision energies. This view is also supported by the notion that any time-varying MF will induce EFs and currents in the tissues subjected to the EMF exposure. For these reasons, the SAR was invented. This "invention" runs through dosimetry and standards in the Western Hemisphere. Even an independent body such as the International Commission on Non-Ionizing Radiation Protection (ICNIRP) based all suggested limits and standards on the thermal mechanisms of action.

In 1953, the late Herman Schwan proposed the value of $100\,W/m^2$ ($10\,mW/cm^2$) as a safe limit for human exposure to microwave energy, based on calculations (Foster, 2005). Even if one accepts the accuracy of the calculation (based on the knowledge of more than six decades ago), there was no biological proof for this value. Since then, many efforts and millions of dollars of funding have been spent to prove this value, all without success.

In an early statement, Becker (1990) pointed out that "based solely on calculations, the magic figure of $10\,mW/cm^2$ was adopted by the Air Force as the standard for safe exposure. Subsequently, the thermal effects concept has dominated policy decisions to the complete exclusion of nonthermal effects. While the $10\,mW/cm^2$ standard was limited to microwave frequencies, the thermal concept was extended to all other parts of the electromagnetic spectrum. This view led to the policy of denying any nonthermal effects from any electromagnetic usage, whether military or civilian."

Here, we are approaching the categories of biological effect, health effect, and health hazard. I take the liberty to claim that these terms are misused. In misusing the words, the scientific community has created havoc in discriminating what is a biological effect, what is a health effect, and what is a hazardous effect. Unfortunately, this was further transferred in the language and terminology of the policy, standard, and regulation bodies.

World Health Organization (WHO) policy is that "not every biological effect is a health effect." Obviously, by saying "health effect," WHO is considering the adverse effects in the sense of diseases, pathologies, and injuries. There is at least one reason for such a statement: Worldwide development of bioelectromagnetic medicine clearly indicates that properly chosen EMF/MF/EF and electric current may be beneficial in treatment of various diseases and injuries, even when all other known medical treatment dramatically fails (Rosch and Markov, 2004; Markov, 2006, 2015). If the action of EMF is to be evaluated, the correct WHO statement should be "not every biological effect initiated by EMF is a health hazard."

There is an abundance of publications pointing out that some biological effects of EMF are reversible and that others are transient. "Transient" indicates biological effects that quickly disappear once the application is terminated. Reversible effects require a longer time to disappear.

So, "hazard" should be kept for irreversible effects caused by short or prolonged exposure to EMF. In the 1990s, the hazard was associated with EMF of power and distribution lines. Now, the power lines are forgotten, and discussions within the scientific community, and among policy makers, the medical establishment, news media, and general public are mostly oriented to cellular communications, mainly cell phones and base stations. There are several international (ICNIRP, ICES) and American (IEEE, ANSI) committees that more or less attempt to direct the world standards.

However, even the simple fact of the existence of several committees indicates the existence of a problem. There should be one recognized and largely accepted standard institution to substitute various national and international standards. Following this idea in the late 1990s, WHO initiated a project involving different laboratories, standard organization, and countries called the "EMF project of harmonization of standards." The problem is that there is no generally agreed upon standard. SAR is favored by the United States, while biological response is the standard favored in Eastern Europe and the former Soviet Union (Markov, 2006; Grigoriev, 2017).

This is a problem with several faces: east vs. west, biophysics vs. engineering, thermal vs. nonthermal. It is curious that all three basically reflect to the last. Why is this so?

Eastern standards are based on biophysics (biological response) and assume nonthermal mechanism(s). Western standards are based on engineering/computation and assume thermal mechanisms. As pointed out earlier, heat-based mechanisms exclude the possibility for occurrence of nonthermal biological effects.

It is obvious that SAR might be a useful criterion, and the only criterion that allows an estimate of the energy absorbed within a biological body. From here to the "thermal monopole" is even less than one step. Every student from high school will tell you that when you have energy absorption, you should expect heating. Yes, but no. In respect of EMF interactions with biological bodies, this approach is rarely helpful. It was already pointed out in the classical thermodynamics dogma "You get energy, you will have heating." Even if one accepts this statement, several questions remain to be answered:

- How does EMF heating occur within complex biological structures?
- Do we have a flow of heat?
- What happens at the interface between tissues with different electrical properties?

These questions, which interpret the physics of interactions, should be complemented with at least two biological questions:

- What are the biological implications of heat generation?
- What is the cascade of events, alterations in the signal transduction, in the enzyme reaction rate?

It is hard to understand why the papers on thermal mechanisms of high-frequency EMF do not consider a set of parameters already in use (Markov, 1994; Valberg, 1995), such as vector, gradient, component, modulation, etc., but emphasize only the SAR values.

Material and ionic fluxes are territories that magnetobiology avoids. Energy interactions are the focus of the research. However, transfer of information is constantly neglected in bioelectromagnetics, even though communication technologies are based on modulation. Nobody is able to estimate the SAR alteration in the human brain that results from voice modulation. This is not, and cannot be, a thermal effect. Here, one should introduce nonequilibrium thermodynamics to search for mechanisms of action, instead of classical, heat-based thermodynamics.

1.7 Thermal Noise

In the 1980s and 1990s, a lot of discussions took place on the problems of thermal noise and "kT." Based on the basic principle of classical thermodynamics, many physicists and physical chemists affirmed that because of the "kT" problem, effects of static and low-frequency EMF cannot be achieved.

In this period, numerous basic science studies had been performed. The experimental data allowed Pilla's group to propose a model that overcomes the problem of thermal noise. In addition, evidence showing that both low-frequency sinusoidal EMF (which induces EFs well below the thermal noise threshold) and weak static MFs, for which there is no induced EF, can have biologically and clinically significant effects has been collected (Shuvalova et al., 1991; Markov et al., 1992, 1993, 1994; Liburdy and Yost, 1993; Markov and Pilla, 1993, 1994a,b; Engstrom et al., 2002; Liboff et al., 2003).

Bianco and Chiabrera (1992) have provided an elegant explanation for the inclusion of thermal noise in the Langevin–Lorentz model: The force applied by an MF on a charge moving outside the binding site is negligible compared to background Brownian motion and therefore has no significant effect on binding or transport at a cell membrane.

1.7.1 SAR: How Important Is This Parameter in Biology?

The technical definition of SAR is directly taken from EMF energy application for dielectric heating, assuming the theoretical estimations of energy absorbed in a heated physical body. The estimations were performed for a flat capacitor in which a homogeneous object was immersed. Note that this case involves a homogeneous structure with linear properties. Can this be applied to a biological body? *I am afraid that the answer is NO.* Biological tissues are not homogenous and are strongly nonlinear systems.

The aim of SAR is to estimate heating of a physical body exposed to EMF energy. It assumes that penetration of this energy does not depend on geometrical size of the body, while it is clear that this process is much more complex. Penetration is a function of the body size and its electric structure, as well as frequency and polarization of the field, reflection and refraction of the energy, and, not to mention, energy dispersion due to radiation and thermoregulation *in vivo*. Thus, the SAR estimation and its spatial distribution within an exposed body by performing external measurement of the EF may be acceptable only in a limited number of cases. A similar conclusion may be drawn for temperature measurements on the surface of the body (Bienkovski and Trzaska, 2017).

Following engineering principles, some international committees such as ICNIRP and IEEE took the liberty to claim that meaningful biological effects might only be thermal. The rise of temperature allows calculation of SAR. The temperature measurements are usually performed on the surface of a body or just below the surface. However, as Bienkovsky and Trzaska (2017) pointed out, even if we assume measurable $\Delta T \approx 0.01\,\mathrm{K}$ the real time of the temperature rise at level of minutes require above megahertz frequencies and E > 100 V/m. While some phantoms allow finding the distribution of temperature within, the accuracy of the procedure is additionally degraded by the heat transfer, cooling, radiation, and the thermoregulation mechanism.

When the temperature issues, especially those related to SAR criteria, are discussed, one question arises: What is the threshold level that determines the effect is thermal? After so many discussions, there is no answer to this question. It reminds me that in the 1990s, I was involved in estimating to what extent the therapeutic device SofPulse that utilizes a 27.12 MHz pulsed RF signal could elevate the temperature at the target. After checking the literature, we created a 2-L saline phantom and found that the prescribed 30-min session with this signal did not elevate temperature more than 0.8°C. At that time, we considered that 1°C could be a plausible threshold level (Markov, 2001). Since that time, I have not seen any "standard" for this level. *Should we consider 1°C as a criterion for thermal effect?* Probably, here it should also be clarified what "room temperature" means. There are a number of publications that state "room temperature," which basically means that no measurements of the temperature in this study were performed. Basic physics states that "room temperature is the temperature in the range between 18.5°C and 19.5°C."

1.7.2 Evidence for Nonthermal Effects

I have made the statement that "Life is a set of electromagnetic events performed in an aqueous medium" (Markov, 2006). This did not happen yesterday; it is a product of the long evolution of physical conditions on our planet and adaptation of the electromagnetic nature of life to these conditions. Take as an example the navigation of birds and fishes along the geomagnetic field, the "suffering" of microorganisms deprived of the usual ambient magnetic and EFs. But, more important is the fact that all of metabolism is in fact transport of electrical charges between different cellular or tissue compartments, or changes in the electrical environment.

It is clear now that the whole biology and physiology of living creature(s) is based on three types of transfer:

- Energy
- Matter
- Information

While the first two processes might be described in terms of classical (equilibrium) thermodynamics, the information transfer obviously needs another approach, which may be found in nonequilibrium thermodynamics. As the late Ross Adey (2004) wrote in his paper (published when he was still alive), "current equilibrium thermodynamics models fail to explain an impressive spectrum of observed bioeffects at nonthermal exposure levels. Much of this signaling within and between cells may be mediated by free radicals of the oxygen and nitrogen species." Cell signaling, signal transduction cascade, and conformational changes are events and processes that may be explained only from the position of nonequilibrium thermodynamics.

For unicellular organisms, the cellular membrane is both the detector and effector of physical and chemical signals. As a sensor, it detects alterations in the conditions in the environment and further provides pathways for signal transduction. As an effector, the membrane may also transmit a variety of electrical (and chemical) signals to neighboring cells with an invitation to "whisper together" as suggested by Adey (2004). It was shown that selected exogenous, weak, low-frequency electric or magnetic fields can modulate certain important biochemical and physiological processes (Todorov, 1982; Detlavs, 1987; Carpenter and Ayrapetyan, 1994; McLean et al., 2003; Rosch and Markov, 2004; Markov, 2006, 2015; Pilla, 2007).

An estimate of detectable EMF exposure can, therefore, only be made if the amplitude and spatial dosimetry of the induced EMF at the target site is evaluated for each exposure system and condition. The electrostatic interactions involving different proteins are assumed to result primarily from electronic polarization, reorientation of dipolar groups, and changes in the concentrations of charged species in the vicinity of charges and dipoles. These effects could be well characterized for interactions in isotropic, homogeneous media. However, biological structures represent complex inhomogeneous systems, and their ionic and dielectric properties are difficult to predict. In these cases, factors such as the shape and composition of the surface and presence or absence of charged or dipolar groups appear to be especially important. The problem of the sensitivity of living cells and tissues to exogenous EMF is principally related to the ratio of signal amplitude to that of thermal noise at the target site. It is clear now that in order for EMF bioeffects to be possible, the applied signal should not only satisfy the dielectric properties of the target, but also induce sufficient voltage to be detectable above thermal noise (Markov and Pilla, 1995). Such an approach relies on conformational changes and transfer of information (Markov, 2004a).

There is a whole series of biologically important modifications appearing under weak static or alternating EMF action that could be explained only from the viewpoint of nonthermal mechanisms. The spectrum includes changes at various levels: alterations in membrane structure and function; changes in a number of subcellular structures such as proteins and nucleic acids; and changes in protein phosphorylation, cell proliferation, free radical formation, ATP synthesis, etc. (Bassett, 1994; Adey, 2004). Other important evidence in favor of the nonthermal character of EMF interaction could be found in the systemic effects (Markov et al., 2004). The wide range of reported beneficial effects of using electric current or EMF/EF/MF therapy worldwide shows that more than 3 million patients received relief from their medical problems: bone unification (Detlavs, 1987; Bassett, 1994), pain (Holcomb et al., 2003; Markov, 2004a), and wounds (Vodovnik and Karba, 1992; Markov and Pilla, 1995) to relatively new applications for victims of multiple sclerosis (Lapin, 2004).

Continuing with the review of nonthermal biological effects, I would point to the fact that EMF effects are better seen within systems out of equilibrium. The observation showed a kind of "pendulum effect"—the larger the deviation from equilibrium, the stronger the response. Such regularity may be seen in changes in the cell cycle, signal transduction, free radical formation and performance, as well as in therapeutic modalities.

The "thermal vs. nonthermal" issue in biology could be clarified in terms of the stress response, the protective reaction of cells to environmental threats that involves activation of DNA and synthesis of stress proteins. Since the stress response is stimulated by a rise in both temperature ("heat shock") and EM fields, it is possible to compare the effects of the two stimuli. Comparing thermal and nonthermal thresholds for the same biological end result (stress protein synthesis), the thermal SAR was ~0.1 W/kg and the nonthermal (EM field) SAR was ~10^{-12} W/kg, many orders of magnitude lower. It is clear that a protective biological response can be activated by nonthermal stimuli at SAR levels orders of magnitude lower than the accepted safety standard (Goodman and Blank, 1998; Blank and Goodman, 2004).

The irrelevance of SAR as a measure of biological response is also apparent when comparing the stress response stimulated by extremely low-frequency and RF fields. Although both fields use the same nonthermal biochemical pathway, the SARs for the responses in the two frequency ranges differ by many orders of magnitude, as shown earlier. Since the same biochemical reactions are stimulated in different frequency ranges at very different SAR levels, SAR cannot be the basis for a biological standard.

The biochemical evidence is even more convincing. From DNA studies, it is clear that thermal and nonthermal stimuli affect different segments of DNA (in the promoter of a stress protein gene) and utilize different biochemical pathways. The specific DNA nucleotide sequence that responds to EM fields does not respond to an increase in temperature. The nucleotide sequences in thermal and EM field domains are different and cannot be interchanged. Finally, when an EM-responsive DNA sequence is inserted into a construct containing a "reporter gene" (CAT or Luciferase), the reporter gene is activated by exposure to EM fields (Blank and Goodman, 1997; Lin et al., 2001).

What about *in vitro* experiments in which isolated organs, tissues, or cells are exposed to EMF? Generally, in these experiments, the temperature is controlled by various regulatory mechanisms. The obvious question arises: In classical thermodynamics, the effect of heat is measured in devices called calorimeters. I cannot recall any paper that reported the calorimetric experiment. Why do the proponents of the thermal mechanisms not conduct such experiments? It would be much more convincing than SAR calculations. *Finally, what is the criteria for "thermal mechanism"—how much should the temperature of the tissue increase for the thermal mechanism to be considered?*

One way or another, dosimetry in bioelectromagnetics is crucially important in both directions: benefit and hazard. Let us do dosimetry following basic science principles and requirements, not with artificially created calculations and criteria. Let us collect actual data that would allow 3-D distribution of the EMF and especially the values of the fields at the *target site.*

1.8 Mobile Phone Hazard

The fast development of satellite communications, followed by wireless communications and, recently, WiFi technology, has dramatically changed the electromagnetic environment. The continuous action of complex and unknown (by sources, amplitudes, frequencies) EMFs is exposed to the entire biosphere and every organism living on this planet. We usually neglect this complex that includes radio and TV transmissions, satellite signals, mobile phones, and base stations and wireless communications.

Speaking of the potential hazard of WiFi technologies, one should not forget that it includes not only mobile phones but also more importantly all means of emitters and distributors of WiFi signals, mainly antennas, base stations, and satellites. In many public locations, own systems are introduced to facilitate the work performance.

As a result, the brains of 7,000,000,000 people are exposed to an unknown spectrum of EMF; yet, there are no criteria for the potential hazard of EMF pollution of the environment: no monitoring, no research, and not even the slightest idea for prevention.

Generally speaking, we do not know if or to what extent WiFi radiation alters the physiology of normal, healthy organisms. The situation becomes more complex when we study the influence on children, on aging adults, or sick individuals, especially on children.

The extensive experience in radiobiology and bioelectromagnetics affirms that biological effects of EMFs are nonthermal and have to be discussed from low energy considerations, such as conformational changes in biological structures, effects on signal-transduction cascade, as well as changes in their manifestations on the level of important biomolecules. More likely, the effects of EMF have informational character, and because of that, they need certain time to manifest and initiate further changes in biochemistry and physiology (Markov, 2012; Markov and Grigoriev, 2013).

In public spaces, such as schools, transportation systems, supermarkets, and hospitals, one may see that nearly any young person is carrying or handling electronic devices from portable electronic games to the latest smart phone. This is what contemporary technology and industry offered as an advance of civilization. The question is, "At what price?"

The cellular telephone delivers a power density of RF radiation about 2 billion times greater than that occurring naturally in the environment. Since mobile phones are designed to operate at the side of the user's head, a large part of the transmitted energy is radiated directly into the person's brain. The absorbed energy could potentially cause dangerous and damaging biological effects within the brain. The small cellular telephones effectively deposit a large amount of energy into small areas of the user's head and brain. At the same time, "smart" phones represent powerful computers that are always in operation—receiving, updating, and transmitting information.

It is reasonable to mention that this problem, which has become aggravated over time, is not new. In 1995, Kane stated that "never in human history has there been such a practice as we now encounter with the marketing and distribution of products hostile to the human biological system by an industry with foreknowledge of those effects."

One of the first papers on the absorption of electromagnetic energy was published by Schwan and Piersol (1954), in which they connected this absorption with tissue composition. It is important to note that tissue composition is very complex and varies from organ to organ and from person to person. From a biophysics point of view, the energy absorption also depends on the depth of penetration for the specific frequency range (for 825–845 MHz, the penetration depth into the brain tissue is from 2 to 3.8 cm) (Polk and Postow, 1986).

Forty years ago, Michaelson (1972) wrote, "It should be understood that a cumulative effect is the accumulation of damage resulting from repeated exposures, each of which is individually capable of producing some small degree of damage. In other words, a single exposure can result in covert thermal injury, but the incurred damage repairs itself within a sufficient time period, for example, hours or days, and therefore is reversible and does not advance to a noticeable permanent or semipermanent state. If a second exposure or several repetitive exposures take place at time intervals shorter than that needed for repair, damage can advance to a noticeable stage."

In other words, the repeated irritation of a particular biological area, such as a small region of the brain, can lead to irreparable damage. Given the existence of energy absorption "hot spots," each damaging exposure to RF radiation provides a new opportunity for the damage to become permanent. Part of the problem is that an exposed person would never know of the penetration and damage.

As seen earlier, the basic science from the 1950s to 1990s reported evidence that high-frequency EMF can have harmful effects on human organisms, especially on the human brain. It has even been detailed with respect to the frequency range: 900–2500 MHz. Numerous studies have pointed out that electromagnetic energy in the 900 MHz region may be more harmful because of its greater penetrating capability compared with 2450 MHz; therefore, more energy in the 900 MHz frequency range is deposited deeply within the biological tissue. In 1976, Lin concluded that 918 MHz energy constitutes a greater health hazard to the human brain than does 2450 MHz energy for a similar incident power density.

Let us remember that studies of diathermy applications consistently show that electromagnetic energy at frequencies near and below 900 MHz is best suited for deep penetration into brain tissue. The depth of penetration is noticeably greater at this frequency range, which includes the portable cellular phone frequencies as compared with higher frequencies. It was also proven that deep-tissue heating is obtained without detecting significant heating in the surface tissues. By their nature, the frequencies that provide the best therapeutic heating would also be the frequencies that could be most hazardous to humans in an uncontrolled situation. High absorption occurs in the inner tissue such as the brain, while fat and bone absorption is many times less.

If tissue damage occurs within a localized region of the brain, it may be completely unnoticed. The threshold for irreversible damage is about 45°C, which is also the temperature at which pain is felt. So, by the time a person exposed to RF radiation feels pain at the skin, that skin is irreversibly damaged, as is the deeper tissue beneath the skin. Similarly, internal heating of the brain tissue would not be sensed as a burning sensation. Likely, there would be no sensation at all. Interest in the ability to "sense" the presence of high levels of RF radiation motivated researchers to determine threshold levels for detecting heat sensations due to RF radiation exposure.

An obvious question arises: If science had developed such knowledge decades ago, why are these questions not at the priority list for research today? We would point here two major reasons: the political power of the industry and the failure of the scientific community.

The major standards and guidelines established by the engineering community, Institute of Electrical and Electronic Engineers (IEEE) in 2005 and International Commission on Non-Ionizing Radiation Protection (ICNIRP) in 2009, provide approach and terminology that are not accepted by physics and biological communities, but nevertheless remain the guiding rules (mainly for the industry). One can only wonder how it is possible to speak about potential "health effects" of RF instead of "health hazard." The misuse of the term "health effect" is probably done on purpose so as to not alarm the general public about the hazard of the use of microwave radiation in close proximity to the human brain. On the other hand, speaking about lack of evidence, the industry experts basically promote the idea that there is no potential hazard from exposure to WiFi and other RF types of radiation.

It has been pointed elsewhere (Markov, 2006) that when the engineering committees state "nonthermal RF biological effects have not been established," basically they are guiding science and society in the wrong direction. Rejecting the possibility of nonthermal effects is not reasonable, but even worse is that they mix "effect" and "hazard." In bioelectromagnetics literature, hundreds of papers discussing the modifications in DNA and other important biological molecules have been published. In the paper of Israel et al. (2013), it was correctly pointed out that even the definition of the thermal effect by these international bodies is not accurate.

The epidemiological teams claim that "there is no consistent evidence for the occurrence of the modification." They also state that "there is no conclusive and consistent evidence that nonionizing radiation emitted by cell phone is associated with cancer risk" (Boice and Tarone, 2011). It is remarkable that this paper was published after International Agency of Research on Cancer (IARC) defined RF as a "possible carcinogen for humans."

For more than half a century, a very serious group of policy-makers and even scientists are playing around with the term SAR. SAR is supposed to provide a measure of absorbed energy in a given tissue: *absorption, not delivery*. However, until today, SAR is more often used to describe the energy delivered by the source of the EMF. One can only wonder how a device may be characterized by the SAR. *To repeat, the SAR identifies the amount of energy that is absorbed in a gram of tissue.*

The inappropriate use of the SAR leads proponents of WiFi technology to affirm that since there is no heating of brain by RF EMF, there is no hazard for the human brain, completely neglecting the fact that most of the biological effects are nonthermal. We are convinced that the safety standards would need to be restated in terms of internal energy absorption.

This is extremely dangerous, especially for children who use mobile phones from an early age without any control from their parents. In evaluation of the mobile phone EMF hazard for children, the main problem is the direct effect of the EMF on the brain of the child, which is the critical organ. One should remember that the child's brain is still in a process of physiological development. We should discuss the basic principles of the hazard of microwave EMF for this part of population who most frequently use contemporary modalities of the wireless communications—children and young adults.

Owing to physiological reasons, the brain of a child exhibits higher rates of specific absorption of EMF compared with an adult's brain. Thus, more compartments of the brain are exposed to high-frequency EMF, including those responsible for intellectual development. Recently, data about unfavorable influence of the EMF on the cognitive functioning of the brain have been published (Grigoriev, 2012).

Taking into account the technological development of wireless communications, it will be fair to say that children at the age of 3–4 will be using mobile communication longer than their parents with respect to their entire life.

I am ready to repeat my statement from the WHO meeting on the harmonization of standards that *neglecting the hazard of high-frequency EMF for children is a crime against humanity* (Markov, 2001).

It should be stressed that while at ionizing radiation, the dose/effect dependence is present in any case of consideration, for nonionizing radiation, the threshold effect is basically absent, and the occurrence of bioeffects needs some time to develop and might be diminished or enlarged by modulation (Grigoriev, 2006a,b).

It is unfortunate that until now, scientists, politicians, and lawmakers underestimate the unfavorable effects on human health that may occur as a result of long, continuous exposure to nonionizing radiation (often exposure from different parameters signals).

We believe that the scientific community must insist more on comprehension of the potential hazard that mobile exchange of information is potentially invoked in human organisms. In that aspect, it is very important that the IARC classified the RF/EMF as a possible carcinogen (2B) (IARC, 2011; IARC WHO, 2011). At the same time, ICNIRP continues to affirm that "the trend in the accumulating evidence is against the hypothesis that mobile phones can cause brain tumors in adults" (Boice and Tarone, 2011).

A question eventually arises: "Why does ICNIRP take this position?" We may offer two different but eventually complementary answers: First, most of the funding for ICNIRP comes from industry and second, eventually more important, most of the members of ICNIRP are engineers for whom only thermal effects may cause biological effects.

It is time that the scientific community in general as well as radiobiologists and magnetobiologists ring the bell that it is time to recognize and estimate the potential hazard for human health of the increasingly elevated background radiation levels.

There is something else that is very important: All the papers that report microwave effects are epidemiological studies, which by definition are long studies requiring a huge cohort of individuals. Well, this is a valuable scientific approach. However, the industry is running much faster than science. The recent reports of the National Toxicology Program deal with a 5-year study of the use of 2G technology. At the same time, the first news about the results came: the U.S. Federal Communication Commission announced that 5G technology is allowed in the United States. But, this is a completely different frequency range of distribution of cellular communications. The only benefit I see from this study is that it confirmed the hazard of 2G phone signals, something which has been known for a long time, but constantly rejected by the industry.

1.9 Conclusion

This chapter:

- Emphasizes the necessity of dosimetry in bioelectromagnetics
- Clarifies the difference between physical and biophysical dosimetry
- Stresses the fact that the only important dosimetry is at the target site
- Demonstrates that nonthermal mechanisms are more likely to cause biological effects
- Clarifies that the SAR cannot be a parameter characterizing the generating system
- Points out the potential hazard of WiFi communications

I would like to finish this chapter with a statement of the group of more than 260 leading scientists delivered to United Nations last year (EMFScientist.org):

The International EMF Scientific Appeals calls upon the United Nations, the WHO, United Nations Environmental Program (UNEP), and the UN member states to:

- *Address the emerging health crisis related to cell phones, wireless devices, wireless utility meters, and wireless infrastructures in neighborhoods*
- *Urge that UNEP initiates an assessment of alternatives to current exposure standards and practices that could substantially lower human exposure to non-ionizing radiation*

Now is the time to ask serious questions about this emerging environmental health crisis.

References

Adey WR (2004) Potential therapeutic application of nonthermal electromagnetic fields: Ensemble organization of cells in tissue as a factor in biological tissue sensing. In: Rosch PJ, Markov MS (eds) *Bioelectromagnetic Medicine*. Marcel Dekker, New York, pp. 1–15.

Bassett CAL (1994) Therapeutic uses of electric and magnetic fields in orthopedics. In: Carpenter D, Ayrapetyan S (eds) *Biological Effects of Electric and Magnetic Fields*. Academic Press, San Diego, pp. 13–48.

Bassett CAL, Pawluk RJ, Pilla AA (1974) Acceleration of fracture repair by electromagnetic fields. A surgically noninvasive method. *Ann NY Acad Sci* 238:242–262. doi:10.1111/j.1749-6632.1974.tb26794.x.

Bassett CAL, Pilla AA, Pawluk RJ (1977) A non-operative salvage of surgically-resistant pseudoarthroses and non-unions by pulsing electromagnetic fields: A preliminary report. *Clin Orthop Relat Res* 124:128–143.

Becker R (1990) *Cross Current.* Jeremy Tarcher Inc., New York, p. 324.

Bianco B, Chiabrera A (1992) From the Langevin–Lorentz to the Zeeman model of electromagnetic effects on ligand-receptor binding. *Bioelectrochem Bioenerg* 28:355–365.

Bienkowski P, Trzaska H (2017) Quantifying in bioelectromagnetics. In: Markov M (ed) *Dosimetry in Bioelectromagnetics.* CRC Press, Boca Raton, FL, pp. 221–235.

Blank M, Goodman R (1997) Do electromagnetic fields interact directly with DNA? *Bioelectromagnetics* 18:111–115.

Blank M, Goodman R (2004) A biological guide for electromagnetic safety: The stress response. *Bioelectromagnetics* 25(8):642–646.

Boice J, Tarone RE (2011) Cell phone, cancer and children. *J Natl Inst Cancer* 103(16):1211–1213.

Carpenter DO, Ayrapetyan S (1994) *Biological Effects of Electric and Magnetic Fields.* Academic Press, New York, vol. 1 (362 p.), vol. 2 (357 p.).

Colbert A, Sauder J, Markov M (2009) Static magnetic field therapy: Methodological challenges to conducting clinical trials. *Environmentalist* (Special issue "Biological effects of electromagnetic fields") 29:177–185.

Detlavs IE (1987) *Electromagnetic Therapy in Traumas and Diseases of the Support-Motor Apparatus.* RMI, Riga, 195 p.

Engstrom S, Markov MS, McLean MJ, Holcomb RR, Markov JM (2002) Effects of non-uniform static magnetic fields on the rate of myosin phosphorylation. *Bioelectromagnetics* 23:475–479.

Ericsson AD, Hazlewood CF, Markov MS, Crawford F (2004) Specific biochemical changes in circulating lymphocytes following acute ablation of symptoms in reflex sympathetic dystrophy (RSD): A pilot study. In: Kostarakis P (ed) *Proceedings of 3rd International Workshop on Biological Effects of EMF*, Kos, Greece, October 4–8, ISBN 960-233-151-8, pp. 683–688.

Foster K (2005) Bioelectromagnetics pioneer Herman Schwan passed away at age 90. *Bioelectromagn Newsl* 2:1–2.

Ginsburg AJ (1934) Ultrashort radio waves as a therapeutic agent. *Med Rec* 19:1–8.

Goodman R, Blank M (1998) Magnetic field stress induces expression of *hsp70*. *Cell Stress Chaperones* 3:79–88.

Grigoriev Y (2006a) Mobile telecommunication: Radiobiological issues and risk assessment. *Proc Latvian Acad Sci B* 60(1):6–10.

Grigoriev Y (2006b) Development of electromagnetic field somatic effects: Role of modulation. *Proc Latvian Acad Sci B* 60(1):11–15.

Grigoriev Y (2012) Mobile communications and health of population: The risk assessment, social and ethical problems. *Environmentalist* 32(2):193–200.

Grigoriev Y (2017) Methodology of standard development for EMF RF in Russia and by international commissions: Distinction in approaches. In: Markov M (ed) *Dosimetry in Bioelectromagnetics.* CRC Press, Boca Raton, FL, pp. 267–288.

Harden RN, Remble TA, Houle TT, Long JF, Markov MS, Gallizzi MA (2007) Prospective, randomized, single-blind, sham treatment controlled study of the safety and efficacy of an electromagnetic field device for the treatment of chronic low back pain: A pilot study. *Pain Practice* 7(3), 248–255. doi: 10.1111/j.1533-2500.2007.00145.x.

Holcomb RR, McLean MJ, Engstrom S, Williams D, Morey J, McCullough B (2003) Treatment of mechanical low back pain with static magnetic fields. In: McLean MJ, Engstrom S, Holcomb RR (eds) *Magnetotherapy: Potential Therapeutic Benefits and Adverse Effects.* TFG Press, New York, pp. 169–190.

IARC (2011) Carcinogenicity of radiofrequency electromagnetic fields. *Lancet Oncol* 12(7):624–625.

IARC WHO (2011) Classifies radiofrequency electromagnetic fields as possibly carcinogenic to humans. Press release no. 208, May 31, 2011, 3 p.

Israel M, Zaryabova V, Ivanova M (2013) Electromagnetic field occupational exposure: Non-thermal vs. thermal effects. *Electromagn Biol Med* 32(2):214–218.

Kane R (1995) *Cellular Telephone Russian Roulette.* Vantage Press, Inc., New York, p. 241.

Kotnik T, Miklavcic D (2006) Theoretical analysis of voltage inducement on organic molecules. In: Kostarakis P (ed) *Proceedings of Fourth International Workshop Biological Effects of Electromagnetic Fields*, Crete October 16–20, 2006, ISBN 960-233-172-0, pp. 217–226.

Lapin M (2004) Noninvasive pulsed electromagnetic therapy for migraine and multiple sclerosis. In: Rosch PJ, Markov MS (eds) *Bioelectromagnetic Medicine.* Marcel Dekker, New York, pp. 277–291.

Lednev VV (1991) Possible mechanism for the influence of weak magnetic field interactions with biological systems. *Bioelectromagnetics* 12:71–75.

Liboff AR (1985) Cyclotron resonance in membrane transport. In: Chiabrera A, Nicolini C, Schwan HP (eds) *Interactions between Electromagnetic Fields and Cells.* Plenum Press, New York, pp. 281–396.

Liboff AR (2004) Signal shapes in electromagnetic therapy. In: Rosch PJ, Markov M (eds) *Bioelectromagnetic Medicine.* Marcel Dekker, New York, pp. 17–38.

Liboff AR, Cherng S, Jenrow KA, Bull A (2003) Calmodulin-dependent cyclic nucleotide phosphodiesterase activity is altered by $20\,\mu T$ magnetostatic fields. *Bioelectromagnetics* 24:32–38. doi: 10.1002/bem.10063.

Liboff AR, Smith SD, McLeod BR (1987) Experimental evidence for ion cyclotron resonance mediation of membrane transport. In: Blank M, Findl E (eds) *Mechanistic Approaches to Interactions of Electric and Electromagnetic Fields with Living Systems.* Plenum Press, New York, p. 109.

Liburdy RP, Yost MG (1993) Time-varying and static magnetic fields act in combination to alter calcium signal transduction in the lymphocyte. In: Blank M (ed) *Electricity and Magnetism in Biology and Medicine.* San Francisco Press, San Francisco, CA, pp. 331–334.

Lin JC (1976) Interaction of two cross-polarized electromagnetic waves with mammalian cranial structures. *IEEE Trans Biomed Eng* 23(5):371–375.

Lin H, Blank M, Rossol-Haseroth K, Goodman R (2001) Regulating genes with electromagnetic response elements. *J Cell Biochem* 81:143–148.

Markov MS (1994) Biological effects of extremely low frequency magnetic fields. In: Ueno S (ed) *Biomagnetic Stimulation*. Plenum Press, New York, pp. 91–102.

Markov MS (1999) Magnetotherapy in the USA. In: *Proceedings of the Second International Conference "Electromagnetic Fields and Human Health"*, Moscow, Russia, pp. 243–244.

Markov MS (2001) Magnetic and electromagnetic field dosimetry: Necessary step in harmonization of standards. In: *Proceedings of WHO Meeting*, Varna, April 2001, ISBN 954911021-4 pp. 169–171.

Markov MS (2002) How to go to magnetic field therapy? In: Kostarakis P (ed) *Proceedings of Second International Workshop of Biological Effects of Electromagnetic Fields*, Rhodes, Greece, October 7–11, ISBN 960-86733-3-X, pp. 5–15.

Markov MS (2004a) Magnetic and electromagnetic field therapy: Basic principles of application for pain relief. In: Rosch PJ, Markov MS (eds) *Bioelectromagnetic Medicine*. Marcel Dekker, New York, pp. 251–264.

Markov MS (2004b) Myosin light chain phosphorylation modification depending on magnetic fields I. Theoretical. *Electromagn Biol Med* 23: 55–74. doi:10.1081/JBC-200026319.

Markov MS (2004c) Myosin phosphorylation: A plausible tool for studying biological windows. Ross Adey Memorial Lecture. In: Kostarakis P (ed) *Proceedings of Third International Workshop on Biological Effects of EMF*, KOS, Greece, October 4–8, ISBN 960233-151-8, pp. 1–9.

Markov MS (2005) Biological windows: A tribute to W. Ross Adey. *Environmentalist* 25(2):67–74.

Markov MS (2006) Thermal vs. nonthermal mechanisms of interactions between electromagnetic fields and biological systems. In: *Bioelectromagnetics: Current Concepts*. Springer, Dodrecht, the Netherlands, pp. 1–16.

Markov MS (2007) Magnetic field therapy: A review. *Electromagnetic Biology and Medicine* 26:1–23.

Markov MS (2012) Cellular phone hazard for children. *Environmentalist* 32(2):201–209.

Markov MS (2015) Benefit and hazard of electromagnetic fields. In: Markov M (Ed) *Electromagnetic Fields in Biology and Medicine*. CRC Press, Boca Raton, FL, pp. 15–29.

Markov M, Grigoriev Y (2013) Wi-Fi technology: An uncontrolled global experiment on the health of mankind. *Electromagn Biol Med* 32(2):200–208.

Markov MS, Hazlewood CF (2009) Electromagnetic field dosimetry for clinical application. *Environmentalist* (Special issue "Biological effects of electromagnetic fields") 29:161–167.

Markov MS, Muehsam DJ, Pilla AA (1994) Modulation of cell-free myosin phosphorylation with pulsed radio frequency electromagnetic fields. In: Allen MJ, Cleary SF, Sowers AE (eds) *Charge and Field Effects in Biosystems 4*. World Scientific, Hackensack, NJ, pp. 274–288.

Markov MS, Pilla AA (1993) Ambient range sinusoidal and DC magnetic fields affect myosin phosphorylation in a cell-free preparation. In: Blank M (ed) *Electricity and Magnetism in Biology and Medicine*. San Francisco Press, San Francisco, CA, pp. 323–327.

Markov MS, Pilla AA (1994a) Static magnetic field modulation of myosin phosphorylation: Calcium dependence in two enzyme preparations. *Bioelectrochem Bioenerg* 35:57–61. doi:10.1016/0302-4598(94)87012-8.

Markov MS, Pilla AA (1994b) Modulation of cell-free myosin light chain phosphorylation with weak low frequency and static magnetic fields. In: Frey A (ed) *On the Nature of Electromagnetic Field Interactions with Biological Systems*. R.G. Landes Company, Austin, TX, pp. 127–141.

Markov MS, Pilla AA (1995) Electromagnetic-field stimulation of soft-tissue-pulsed radio-frequency treatment of postoperative pain and edema. *Wounds* 7(4), 143–151.

Markov MS, Ryaby JT, Kaufman JJ, Pilla AA (1992) Extremely weak AC and DC magnetic field significantly affect myosin phosphorylation. In: Allen MJ, Cleary SF, Sowers AE, Shillady DD (eds) *Charge and Field Effects in Biosystems—3*. Birkhäuser, Boston, MA, pp. 225–230.

Markov MS, Wang S, Pilla AA (1993) Effects of weak low frequency sinusoidal and DC magnetic fields on myosin phosphorylation in a cell-free preparation. *Bioelectrochem Bioenerg* 30:119–125.

Markov MS, Williams CD, Cameron IL, Hardman WE, Salvatore JR (2004) Can magnetic field inhibit angiogenesis and tumor growth? In: Rosch PJ, Markov MS (eds) *Bioelectromagnetic Medicine*. Marcel Dekker, New York, pp. 625–636.

McLean MJ, Engstrom S, Holcomb RR, editors (2003) *Magnetotherapy: Potential Therapeutic Benefits and Adverse Effects*. TFG Press, New York, 279 p.

Michaelson SM (1972) Human exposure to nonionizing radiant energy: Potential hazards and safety standards. *Proceedings of the IEEE* 1:389–421.

Nindl G, Johnson MT, Hughes EF, Markov MS (2002) Therapeutic electromagnetic field effects on normal and activated Jurkat cells. In: *International Workshop of Biological Effects of Electromagnetic Fields*, Rhodes, Greece, October 7–11, ISBN 960-86733-3-X, pp. 167–173.

Parker R, Markov M (2011) An analytical design techniques for magnetic field therapy devices. *Environmentalist* (Special issue "Biological effects of electromagnetic fields") 31(2):155–160.

Pilla AA (2007) Mechanisms and therapeutic applications of time-varying and static magnetic fields. In: Barnes F, Greenebaum B (eds) *Handbook of Biological Effects of Electromagnetic Fields*, 3rd edn. CRC Press, Boca Raton, FL, pp. 351–412.

Polk C, Postow E, editors (1986) *Handbook of Biological Effects of Electromagnetic Fields*. CRC Press, Boca Raton, FL.

Rosch PJ, Markov MS (2004) *Bioelectromagnetic Medicine*. Marcel Dekker, New York, p. 850.

Schwan HP, Piersol GM (1954) The absorption of electromagnetic energy in body tissues. A review and critical analysis. *Int Rev Phys Med Rehabil* 33(6):371–404.

Shuvalova LA, Ostrovskaya MV, Sosunov EA, Lednev VV (1991) Weak magnetic field influence of the speed of calmodulin dependent phosphorylation of myosin in solution. *Dokladi Acad Nauk USSR* 217:227.

Sisken BF, Midkeff P, Tweheus A, Markov MS (2007) Influence of static magnetic fields on nerve regeneration in vitro. *Environmentalist* 27(4):477–481. doi:10.1007/s10669-007-9117-5.

Todorov N (1982) *Magnetotherapy.* Meditzina i Physcultura Publishing House, Sofia, Bulgaria, p. 106.

Valberg P (1995) How to plan EMF experiments. *Bioelectromagnetics* 16:396–401.

Vodovnik L, Karba R (1992) Treatment of chronic wounds by means of electric and electromagnetic fields. *Med Biol Eng Comput* 30:257–266.

Williams CD, Markov MS, Hardman WE, Cameron IL (2001) Therapeutic electromagnetic field effects on angiogenesis and tumor growth. *Anticancer Res* 21:3887–3892.

2

Uncertainty Sources Associated with Low-Frequency Electric and Magnetic Field Experiments on Cell Cultures

Lucas A. Portelli
*University of Colorado
and Foundation for Research
on Information Technologies
in Society*

2.1 Introduction

It is a common occurrence that well-known and established *in vitro* experiments are hard
to reproduce, even in the same laboratory, especially when the effects under scrutiny are
considered "small" in comparison to the expected amount of "biological variability"
associated with the specific experimental outcome (Jayaraman and Hahn, 2009). The
poor consensus within the bioelectromagnetics community regarding the biological
effects of low-level static and extremely low-frequency (ELF) electric and magnetic field
exposures is a typical example, where the signals under scrutiny are below well-known
and established biological effect thresholds (Grosberg, 2003; Mattsson and Simko, 2014;
Lin, 2014). Classical causes for experimental discrepancies range from variations in
traditional factors of chemical, biochemical, biological, and procedural nature (such
as the origin and composition of media, additives, disposables, and reactants) to the
specific techniques used in culture, and even human error (Mather and Barnes, 1998).
However, setting aside these sources of discrepancy, it may still be reasonable to speculate that inherently subtle and selective biological effects that require specific electric
and magnetic field conditions may still be obscured by other sources of deterministic
and avoidable variability in the experimental conditions for three physical parameters:
temperature, electric fields, and magnetic fields. On one hand, some of the proposed
mechanisms of action currently under scrutiny, as well as several experimental reports,
support the possibility both of complex sensitivity of chemical reactions to static and
time-varying magnetic (Buchachenko and Kuznetsov, 2014; Barnes and Greenebaum,
2015; Prato, 2015; Buchachenko, 2016) and electric (Borys, 2012; Hart et al., 2013) fields
of the order of those commonly under scientific scrutiny, which sometimes include
somewhat complex frequency, amplitude, and orientation combinations (Blanchard and
Blackman, 1994; Binhi, 2000), temperature variations (Blackman et al., 1991), and light
sensitivities (Wu and Dickman, 2012; Qin et al., 2016), some of which are reviewed later
in this chapter. On the other hand, inherent nonlinear dose–response effects coupled
with the great capacity for amplification that typically governs biological systems (Vera
et al., 2008) may allow for a small group of cells under specific and localized environmental conditions in culture to be highly influential in generating (or suppressing) the
induced biological effects (or lack thereof) of the entire experimental sample, thereby
concealing its source under the expected variability. Under such premises, reproduction
of the experimental exposure conditions becomes pivotal, as these conditions need to be
emulated with exquisite levels of accuracy for successfully overcoming the indispensable
requirement of the experimental replication (Casadevall and Fang, 2010).

   In view of these premises, this chapter encompasses a review and quantitative perspective for the uncertainty associated with culture temperature and electric and magnetic fields in the context of classical biology experimentation techniques, as well as

customized bioelectromagnetics experimentation. Such environmental parameter assessment, coupled with the potential and known exquisite sensitivity of biological systems, may add to a reasonable explanation for the variability and irreproducibility associated with some *in vitro* ELF electric and magnetic field experimental reports. This chapter will also reiterate some of the simple recommendations and mature techniques to ameliorate such parameters' variability.

2.2 Background Electric, Magnetic, and Temperature Exposure in Biological Experimentation

The electric and magnetic fields as well as the temperature exposure history to which a cell culture is exposed during the course of an experiment include the combination of fields inherent to the laboratory background, inside the dedicated instrumentation (incubators, microscope stages, etc.) and the transients generated by changing the culture location between these environments. Depending on the biological system under consideration and its initial state, the duration and details (magnitude, frequency, spatial distribution) of each of the environmental conditions of the exposure history may or may not be of significance to the biological outcome being studied. However, theoretical, numerical (modeling), as well as experimental assessments (*in situ* measurements and validations) are indispensable for such deductions to be conclusive, and therefore, are scientifically valid. In view of this need, it is a fundamental requirement to have a clear distinction between the factors that make up every environment to which cells are exposed during the course of an experiment, and these are presented in the following sections.

In a typical bioelectromagnetics experimental setup, the total magnetic field to which cultures are exposed is composed of the combination of the *imposed magnetic field* (IMF)—which generation is the main purpose of the exposure system—and the *background magnetic field* (BMF)—which is a combination of the inherent fields in the laboratory. These fields are generated naturally (geomagnetic field (or GMF), space weather, etc.) or artificially via electrical currents or magnetically relevant materials (magnets) within residing instrumentation (shakers, heaters, microscopes, centrifuges, water baths, incubators), the walls or even the surroundings of the laboratory room. Similarly, the total electric field in the culture space is composed of the induced electric field (IEF)—which is introduced by the total changing magnetic field permeating the cultures (IMF + BMF)—and the background electric field (BEF), which is a combination of the fields generated by the accumulated charge in the culture containers and set up in the space by the incubator, the exposure system itself [e.g., parasitic electric fields (PEF) generated in the surroundings of the exposure system coils by the potential drop along its impedances] and even naturally in the atmosphere. Ultimately, the total electric and magnetic field to which a culture is exposed corresponds to the combination of all these fields, with additional unwanted artifacts generated by the interaction of such fields with the exposure system, laboratory equipment, or any other structure with relevant electric or magnetic properties (e.g., ferromagnetic, conductive).

In the case of temperature, the thermal conditions to which a culture is exposed is a combination of *static temperature* (ST)—comprising the "set point" of automatically controlled environments in the laboratory or instrumentation—and a *time-varying temperature* (TVT) component that reflects the unavoidable oscillations introduced by the environmental feedback control systems and perturbations resulting from handling cultures between environments with different ST set points. Such a situation inevitably arises, for example, when a culture is transferred from a biological incubator (e.g., at 37°C) to a laboratory bench or other laboratory equipment, which may be at room temperature (generally recognized as somewhere in the range >18°C–<27°C).

2.3 Electric and Magnetic Fields in Biological Laboratory Environments

BMFs in laboratories exist as a result of the combination of two magnetic field sources: (1) the naturally occurring GMF and (2) the artificially generated (anthropogenic) magnetic field (AMF). The GMF and AMF are individually composed of both a static magnetic field (SMF, 0 Hz) and a time-varying magnetic field (TVMF, >0 Hz) components. In the same way, the BEF is composed of: (1) the naturally occurring electric field (NEF) and (2) the artificially generated electric field (AEF), both also with static (SEF) and time-varying (TVEF) components, the general characteristics of which are known. The magnitude of the SMF component of the GMF currently ranges between ~23 and 65 µT over the planet's surface (Finlay et al., 2010). The amplitude of its TVMF component is typically <0.1 µT, except during the rare cases of large magnetic storms, when it can increase to nearly 1 µT for short periods. It can be composed of frequencies ranging from fractions to several 100 or 1000 Hz, and its direction depends on the location of the source (terrestrial or extraterrestrial), but these frequencies are so large as to be considered homogeneous by the observer (Hansson Mild and Greenebaum, 2006; Zhang et al., 2007). Similarly, the NEF can be on the order of 100–300 V/m, but it can reach 100 kV/m during atmospheric events with frequency components ranging from 0 to 3000 Hz (Krider and Roble, 1986; König et al., 2012). On the other hand, the TVMF component of the AMFs in habitable areas of industrialized countries is primarily a consequence of electrical power distribution and usage and can vary from very localized (i.e., proximate transformers, motors, magnets) to quasi-homogeneous (i.e., distant powerlines), and the spectrum's prime harmonic is typically 50 or 60 Hz (depending on the power distribution convention of each geographic region) and its harmonics (Hansson Mild and Greenebaum, 2006). However, typical signals can be composed of additional frequencies (IEC, 1996). While in most normal habitation environments its amplitude is not >0.4 µT (Ahlbom et al., 2000; Greenland et al., 2000), the BMFs in laboratory environments as well as within controlled environments utilized for cell culture have been shown repeatedly to exceed these baseline values (Hansson Mild et al., 2003, 2009; Moriyama et al., 2005; Úbeda et al., 2011; Schomay et al., 2011; Portelli et al., 2013b; Gresits et al., 2015). Likewise, AEFs in habitational environments were recorded to be as high as 77 V/m, while in close proximity to appliances it can reach the order of 1000 V/m (Hansson Mild and Greenebaum, 2006). The reason for the expectancy of greater fields than in normal habitation environments for biological laboratories stems from the fact that energy consumption is generally greater (5–10 times) and instrumentation and associated energy supplies are more numerous (EPA/DoE, 2008), potentially generating more fields and altering magnitudes, directions, and spectra of existing natural and artificial fields (Portelli et al., 2013b).

2.4 Electric and Magnetic Fields inside Biological Instrumentation

In addition to the BMF in the laboratory, both SMFs and TVMFs are generated by essential instrumentation parts (such as heating elements, fans, and control circuitry) that can ultimately affect the culture designated volumes (CDVs) of incubators, microscope stages, or any other nearby instrumentation-controlled environments, depending on their relative locations and specific architecture, factors that are not standardized in the industry. In this regard, surveys have reported TVMF peak magnitudes of ~1.2 µT for clean benches, 8 µT on inverted microscopes, 11.5 µT in centrifuges, 3.1 µT on cell separators, 20 µT on water baths, 75 µT on shaker tables, 0.9 µT in hemodialysers, and 0.3 µT in deep freezers, all with primary peaks at the same frequencies as their power supply (50 Hz/60 Hz) and harmonics (see Figure 2.1 for a representative example) (Moriyama et al., 2005; Hansson Mild et al., 2009; Gresits et al., 2015). In the case of tissue culture incubators, Figure 2.2 shows how TVMFs can vary as much as nearly 240 µT, both between incubators and within a single incubator. These data include tridimensional measurements of the entire incubator cavity in close proximity to incubator shelves as shown in Figure 2.3 by the use of sensors designed, built, and calibrated to an appropriate culture container scale for such a purpose (Portelli et al., 2013). As a reference, this value exceeds habitation environments by a factor of 600 (55.5 dB) (Ahlbom et al., 2000; Greenland et al., 2000; Hansson Mild et al., 2003, 2009; Hansson Mild and Greenebaum, 2006), while most of the TVMF magnitudes found in habitation environments at 60 Hz typically exceed those found in nature by orders of magnitude. In this regard, Figure 2.4 shows how nearly 93% of all the 567 locations measured in all 21 incubators show magnetic fields that fall above habitation levels. Similarly, Figure 2.5 reveals differences in the SMF nearly as large as 450 µT within the

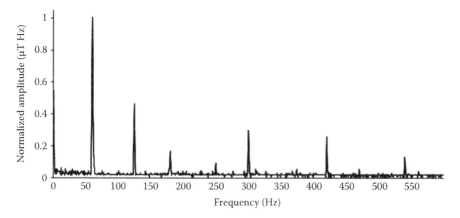

FIGURE 2.1 An example of a typical magnetic field spectrum measured in a CO_2-controlled incubator. Notice how 60 Hz is the dominant frequency while the other prominent peaks correspond to its harmonics. The vertical scale is presented in µT Hz to facilitate the observation of the harmonics. (From Portelli, L.A., Device for controlling the electric, magnetic and electromagnetic fields in biological incubators. PhD dissertation, University of Colorado, 2012; Portelli et al., *Bioelectromagnetics*, 34, 337–48, 2013b.)

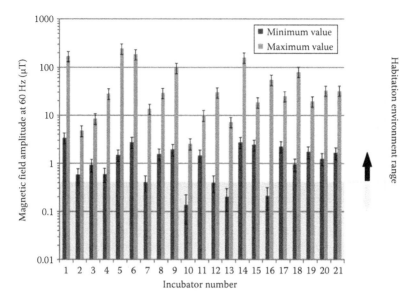

FIGURE 2.2 Maximum and minimum TVMF amplitudes measured (at 60 Hz) in each of 21 CO_2 cell culture incubators. As a comparison, the TVMF in habitation environments is typically $\leq 0.4\,\mu T$ at 50 or 60 Hz as depicted by the shaded area and light arrow. The black arrow represents magnitudes outside of this range. Error bars represent $\pm 0.05\,\mu T\ \pm 24.2\%$ and correspond to error associated with the measurement system and protocol. Notice how all incubators tested had points that exceeded natural and habitation values by at least a factor of three. (From Portelli, L.A., Device for controlling the electric, magnetic and electromagnetic fields in biological incubators. PhD dissertation, University of Colorado, 2012; Portelli et al., *Bioelectromagnetics*, 34, 337–48, 2013b.)

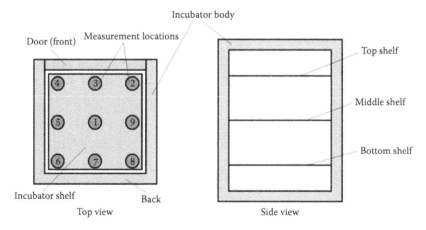

FIGURE 2.3 Diagram of the measurement locations within each incubator. Both SMF and TVMF components were measured for all 3 axes in 27 locations of each incubator. (From Portelli, L.A., Device for controlling the electric, magnetic and electromagnetic fields in biological incubators. PhD dissertation, University of Colorado, 2012; Portelli et al., *Bioelectromagnetics*, 34, 337–48, 2013b.)

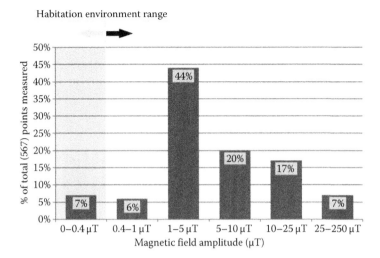

FIGURE 2.4 Distribution of all the TVMF amplitudes measured (at 60 Hz) in all 21 CO_2 cell culture incubators. As a comparison, the TVMF in habitation environments is $\leq 0.4\,\mu T$ at 50 or 60 Hz as depicted by the shaded area and light arrow. The black arrow represents magnitudes outside of this range. (From Portelli, L.A., Device for controlling the electric, magnetic and electromagnetic fields in biological incubators. PhD dissertation, University of Colorado, 2012; Portelli et al., *Bioelectromagnetics*, 34, 337–48, 2013b.)

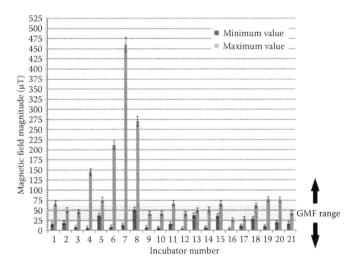

FIGURE 2.5 Maximum and minimum SMF magnitude measured in each of the 21 CO_2 cell culture incubators. As a comparison, the shaded area represents the normal SMF magnitudes found naturally on the earth's surface. Black arrows represent magnitudes outside this range. Error bars represent $\pm 4.0\,\mu T$ ($\pm 2.9\%$) and correspond to errors associated with the measurement system and protocol. (From Portelli, L.A., Device for controlling the electric, magnetic and electromagnetic fields in biological incubators. PhD dissertation, University of Colorado, 2012; Portelli et al., *Bioelectromagnetics*, 34, 337–48, 2013b.)

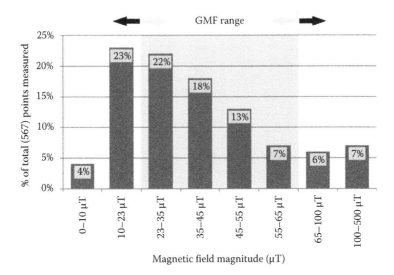

FIGURE 2.6 Distribution of the SMF magnitudes measured in all 21 CO_2 cell culture incubators. As a comparison, the shaded area and light arrow represent the normal and most extreme SMF magnitudes found naturally on the earth's surface. Black arrows represent magnitudes outside of this range. (From Portelli, L.A., Device for controlling the electric, magnetic and electromagnetic fields in biological incubators. PhD dissertation, University of Colorado, 2012; Portelli et al., *Bioelectromagnetics*, 34, 337–48, 2013b.)

same incubator and between incubators, which is 590% (15.5 dB) more than the maximum SMF found naturally on the earth's surface. Furthermore, in that regard, Figure 2.6 shows how nearly 40% of all the points tested in the 21 incubators fall outside of the natural GMF range (23–65 μT). As a comparison, the static component of the earth's magnetic field in Boulder, CO, was ~52.7 μT at the time of the measurements, according to data from the Boulder Geomagnetic Observatory (National Geophysical Data Center at the National Oceanic and Atmospheric Administration, Boulder, CO), which emphasizes the artificial variability introduced by the surrounding instrumentation and building. The same survey revealed inhomogeneities within the same incubator when the TVMF and SMF magnitudes were grouped by shelf and relative position on the shelf (see Figures 2.7 and 2.8). Notice that, for SMFs, locations near the front of the incubator were, on average, the most variable, while those far in the back, especially on the bottom shelf, were the least variable. In contrast, for TVMFs, the smallest and least variable locations were found, on average, at the front of the incubator, which shows that, in the absence of direct measurements, there is little room for educated guesses about the "best" location to perform experiments in a specific incubator, or by analogy, on any other culture-designated laboratory equipment. Additionally, since BMF variations were found in the same brand of incubators but in different locations, recommending any particular brand was not feasible. Instead, these data suggest that BMFs must be monitored and controlled on a per-experiment basis. In addition to the spatial inhomogeneity, BMFs were also found to vary in time, exhibiting multiple magnitudes and frequency components. Figure 2.9 shows a representative

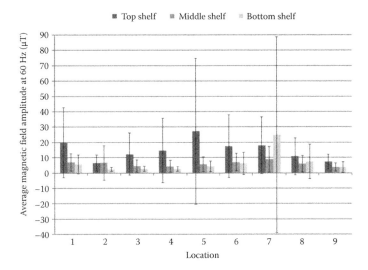

FIGURE 2.7 Average TVMF per location (at 60 Hz). Refer to Figure 2.3 for a diagram of the actual measurement locations within each incubator. The error bars represent ±1 standard deviation. Measurement error is not shown for clarity but was ±0.05 µT ±24.2%. (From Portelli, L.A., Device for controlling the electric, magnetic and electromagnetic fields in biological incubators. PhD dissertation, University of Colorado, 2012; Portelli et al., *Bioelectromagnetics*, 34, 337–48, 2013b.)

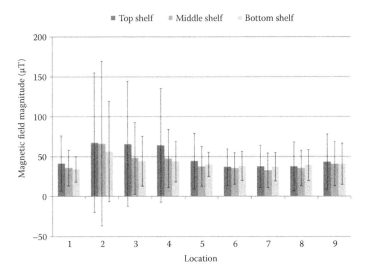

FIGURE 2.8 Average SMF per location. Refer to Figure 2.3 for a diagram of the actual measurement locations within each incubator. The error bars represent ±1 standard deviation. Measurement error is not shown for clarity but was ±4.0 µT ±2.9%. (From Portelli, L.A., Device for controlling the electric, magnetic and electromagnetic fields in biological incubators. PhD dissertation, University of Colorado, 2012; Portelli et al., *Bioelectromagnetics*, 34, 337–48, 2013b.)

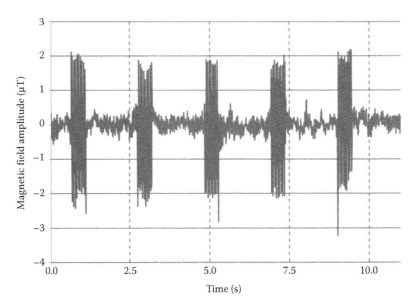

FIGURE 2.9 Example of the intermittence occasionally measured for TVMF in CO_2 incubators. In this case, the 60-Hz TVMF signal observed was modulated by a frequency close to 0.5 Hz. As a result, a substantial change in amplitude is observed. These data were taken on a single axis only (vertical) for one of the incubators measured and is representative of the typical intermittence occasionally observed. Some incubators also exhibited changes in the amplitude of the 60-Hz signal either in a continuous way or in discrete steps. (From Portelli, L.A., Device for controlling the electric, magnetic and electromagnetic fields in biological incubators. PhD dissertation, University of Colorado, 2012; Portelli et al., *Bioelectromagnetics*, 34, 337–48, 2013b.)

example of intermittence in the signal recorded at one of the locations in Figure 2.3 that was generated by one of the incubator heaters. As a result, the values reported for the peak amplitude of the 60-Hz spectrum component in this study represent a lower limit of the existing TVMF magnitudes. Also, all the measurements made for TVMF and SMF components correspond to a spatial discretization of the actual BMF in the incubator, which leaves room for higher or lower SMF or TVMF values to exist elsewhere. Additionally, while all measurements reported here were made with the door closed, substantially different SMF and TVMF magnitudes were observed depending on the position of the incubator door. Therefore, cell cultures will experience variable BMFs during transportation that depend on the opening and closing of the incubator, especially for the cultures placed near the door. Finally, incubators presenting the worst-case SMF and TVMF values were dismantled to seek the locations and natures of the sources. Fans and resistive heaters were found to be the main causes for the extreme TVMF magnitudes, while permanent magnets (door, solenoids) were the main sources of high SMFs. The SMFs and TVMFs measured inside the incubators were generated outside of the culture space, as all these components were found in the incubator instrumentation cavities. The shelves, metallic or not, of the incubators surveyed were mostly nonmagnetic and were found to contribute insignificantly to these fields (Portelli et al., 2013).

In the case of AEF, accumulated charge is expected to be evenly distributed on the conductive metallic surfaces of the interior of the incubator, resulting in negligible or small electric fields. Although direct measurements in incubators are scarce, some measurements inside magnetic open shields (conductive) have reported levels as small as 0.1 V/m (Schuderer et al., 2004). Therefore, the only fields remaining to which cultures are exposed correspond to the combination of fields due to the charge accumulated in the culture containers themselves—for which there have been no assessments to date—and of the PEF. In this regard, PEF may present amplitudes similar to or greater than the induced fields in culture containers. Some measurements have shown fields as high as 300 V/m from exposure systems utilized in bioelectromagnetics research with frequency components that will depend on the signal injected into the system (Schuderer et al., 2004).

All these factors combined, create a very complex set of electric and magnetic signals to which cell cultures are exposed, which is hard to predict without direct measurements.

2.5 Electric and Magnetic Fields in Culture-Designated Containers

Electric and magnetic field conditions in the culture space depend on culture container size and shape, liquid volume (and shape), liquid electrical properties (Bassen et al., 1992), and the location of the culture container within the imposed electric (BEF + AEF) and magnetic (BMF + IMF) fields. Typically, in bioelectromagnetics research, culture containers are magnetically coupled with experimental setups (i.e., based on coil systems) and, ideally, should be sufficiently decoupled from the surrounding fields for the imposed signal effects to be studied. In such cases, the electric and magnetic fields to which cells are ultimately exposed correspond to the IMF and the IEF due to the IMF. In terms of the spatial quality of the imposed fields, it is imperative to have a clear definition of the homogeneity on the CDV as IMF inhomogeneities outside of the desired tolerance may appear depending on their relative location. For example, in the case of flat culture containers (e.g., petri dish) that are axially aligned with single-loop coil systems, the IMF will increase as the radius of the dish approaches that of the single loop [i.e., variability >20% if CDV radius >40% single round coils radius (Frix et al., 1994)]. However, displacing the dish from the geometrical center of the single loop while still aligned axially will eventually induce the opposite effect. Therefore, while the reported values at the centers of both configurations may be the same, the decay will be in opposite directions when moving out in the radial plane, which is of particular importance when the size of the dish (or other CDV) is comparable to that of the coil system. Similarly, in the case of IEFs, the orientation of the containers inside the exposure system (and its tolerance) is sometimes an overlooked factor of possible significance (McLeod et al., 1983). For example, when the IMF is homogeneous and perpendicular to the dish culture plane (and in the case of adherent cultures), the IEC vortex will be at its center and its magnitude will increase linearly as a function of the radius. Thus, the electric field exposure of cultures is variable, [except in the case of annular culture containers, where the gradient is significantly minimized (Bassen et al., 1992; Liburdy, 1992)], and about 50% of the cells are exposed to electric fields that range from 70% to 0% of the maximum IEF. On the other hand, when a homogeneous IMF is parallel to the culture plane, the IEF vortex will be located on a line parallel to the culture surface and separated from it by about half of the height of the culture liquid. In this case, the IEF gradient is significantly diminished from the perpendicular case, as most of the cells on the culture plane are exposed to reasonably homogeneous magnitudes (Bassen et al., 1992).

Although the IEF can be roughly estimated with simplified forms of the Maxwell–Faraday equations that combine the rate of change (dB/dt) of the IMF, its magnitude, and the area of the container, these estimations may differ from reality in a number of factors. For example, aside from the fact that it is usually regarded as a quantitative variable which is assigned qualitatively, cell morphology or confluency (aggregation, stacking) can also be a factor of significance (Wang et al., 1993). The complexity of current flow in close proximity to cell clusters can introduce significant gradients in the vicinity of cell surfaces (~±50% for spherical cells) from the nominally induced or applied electric field under homogenous medium conditions. Deviations of the same order are also expected for the introduction of dissimilar cellular shapes (spheres vs. rods) (Hart, 1996; Hassan et al., 2003), and the possibility of electrical coupling between cells (gap junctions) can significantly increase the IEF inside cells (inherently small ~>60 dB for $r = 10\,\mu m$) depending on the specific type of culture and the aggregation (Fitzsimmons et al., 1994).

In the case of the temporal quality of the magnetic fields, unsatisfactory measurements or incomplete descriptions of the IMF signal spectrum may compromise replication of the magnetic conditions and largely contribute to IEF uncertainty. For example, the frequency components of the TVMF introduced by current intermittence depend on the specific characteristics of the switching circuitry that feeds the devices generating the magnetic fields, such as heaters, electromechanical valves, etc., and on the electrical characteristics of the devices themselves and surrounding magnetically relevant material. Furthermore, since these devices are usually part of a feedback loop, the frequencies at which the switching occurs also depend on the current state of the parameter being modified by the magnetic-field-generating device (like temperature). If the device is a switched heater, for example, it will be in the "on" state an increasing amount of time as the incubator's temperature falls below the nominal temperature. Subsequently, the time it will take for the temperature to recover will depend on the thermal load imposed by culture containers, perturbations introduced by the researcher, and other variable circumstances. Therefore, special attention must be paid to measurements that encompass signals with rich harmonic content, such as intermittent and powerline signals, since unaccounted harmonic components can yield significant IEF magnitude differentials in wet cultures (Hansson Mild et al., 2003, 2009; Schuderer et al., 2004; Portelli et al., 2013b).

As mentioned before, IMF and IEF artifacts have been shown to be caused by interaction of the imposed fields with nearby ferromagnetic structures. Such materials (e.g., nickel–chrome based coatings and iron) are commonplace, for example, in microscope objectives and other components (springs, screws, studs, etc.). Numerical modelling has shown the appearance of large spatial gradients (>200% or >6 dB) in microscopes CDVs. Models generated from the interaction of the IMF and these parts, have shown to introduce gradients so sharp, that conventionally-sized (~0.5–1 cm) measurement equipment would be able to detect them only partially (Chaterjee et al., 2001). Along the same lines, numerical modeling has also shown that slight (~1°) objective/coil misalignment introduced an IMF perturbation of up to 193% (5.7 dB) several millimeters away from the focal plane, amounting to 136% (2.7 dB) in IEF distortions (Publicover et al., 1999).

Additionally, the introduction of additional spectral components might be in order, depending on the saturation of these metallic materials (Lekner, 2013). These factors are to be taken into consideration, for example, in the case of single-cell homogeneous magnetic field exposures using microscopes, as the IEF is generally assumed to be null, as the CDV of relevance generally resides in the proximity of the center of the culture plane (field of view) where simplified approximations predict a vortex under homogeneous (and centered to the objective) IMF exposure.

2.6 Electric and Magnetic Field Environment Reproducibility in Contemporary Bioelectromagnetics Experimentation

Independently from sources of electric and magnetic field exposure uncertainty associated with exposure system and experimental design itself, additional uncertainty stems from unavoidable assumptions and speculation needed to reconstruct the exposure conditions as a result of incomplete or ambiguous exposure system physical and numerical design and validation, monitoring, and reporting, indispensable in subsequent instances of study replication. In this regard, Golbach et al. carefully reconstructed (estimated) 52 electric and magnetic exposure conditions of 31 previously selected publications regarding *in vitro* ELFMF exposures, based on the exposure system and experimental setup description details (published or clarified through author correspondence) from a study replication perspective, finding significant uncertainty potential (Golbach et al., 2015). After some extra processing, this assessment (see Figure 2.10) shows how 55% and 73% of the IMF and IEF exposures, respectively, could potentially exceed the maximum reported fields by at least 3 dB (41%). Such uncertainty is rooted in the fact that reconstruction often required the use of assumptions as responsiveness from authors was mostly lacking (<20% response rate) and detailed experimental system descriptions were rare. In this regard, only 53% of publications provided a magnitude qualifier ("RMS," "peak," "peak-to-peak," etc.), 16% of the exposures had the potential to have variations >10% from the intended IMF nominal (and homogeneous) value, 35% of exposures were ambiguous, and only 54% and 48% described the CDV and culture container size and shape, respectively. Additionally, 86% did not provide measurements or estimates of the spectrum or time-domain description of the IMF signal or direct IEF measurements on the wet culture space (leading to the contemplation of worst-case harmonic content). Also, 39% of the exposures were performed under microscopes for which objective materials and geometries were undescribed, and the IMF homogeneity assessments reported were most of the times made based on measurements along only one axis, where the diameter of the sensors were of comparable size to that of the culture container (few mm to cm), and with unspecified survey resolutions that could partially obscure the presence of potentially significant spatial gradients. In the case of time-varying AMFs, only 45% reported any kind of residual time-varying AMF magnitude measurement, and PEF shielding attempts were reported to be made in only 22% of the exposures by partially surrounding the generating coils with conductive material, and only 6% reported residual time-varying PEF measurements and calculations, leaving

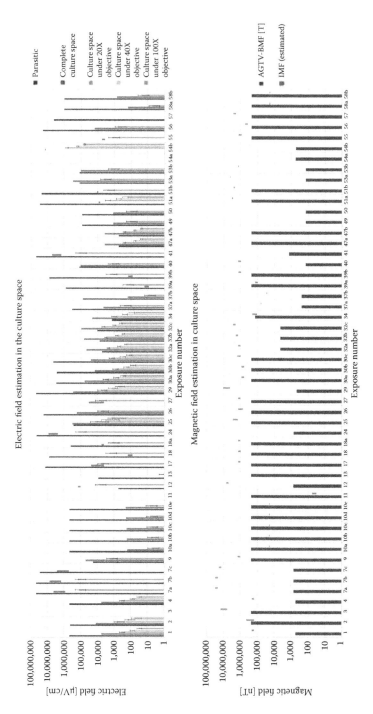

FIGURE 2.10 Estimation of the possible electric and magnetic field exposure parameters on exposures in Golbach et al. (2015). IMF: imposed magnetic field; TV-BMF: time-varying background magnetic field (50 Hz/60 Hz only). Bars denote the range (uncertainty) over which the electric and magnetic exposure values could be found for all portions of the culture space considered based on the reported exposure system and culture container characteristics. Dark bars represent the maximum measurement (or, in absence of measurement, estimation) for PEF and TV-BMF in the culture space. Calculations are made for all three microscope objectives for reference in the cases where the precise objective used was not specified. Notice how artifacts have the potential to dominate over the desired exposure parameters for many cases. (From Golbach, L.A. et al., *Environ. Int.*, 92–93, 695–706, 2015.)

ample room for speculation. In this regard, 22% and 82% of the exposures could have AMF and PEF magnitudes greater than the maximum reported IMF and IEF (due to the maximum IMF reported). Similarly, only 63% of exposures reported a static AMF measurement or calculation. Of this group, only 12% of the exposures were performed within GMF conditions, 45% in environments with static fields lower than those found in nature below normal conditions (hypogeomagnetic), and 6% were performed in higher fields (hypergeomagnetic). Although the IMF and IEF spatial distributions were implicitly or explicitly described for all exposures with respect to the wet culture liquid surface [e.g., for 80% of the exposures, the IMF was perpendicular to it (subsequently generating perpendicular IEFs) while being parallel for the rest], the spatial distribution of both AMF and PEF artifacts remains largely unknown for most of the exposures, while for the SMF it is sometimes (44%) described. As a result, the extent of their contribution to the total electric and magnetic field vector to which cells in culture are ultimately exposed is uncertain, but in the worst case, would correspond to the sum of the worst-case magnitudes plotted in Figure 2.10.

Additionally, simple errors in the reporting of the physical parameters were detected on 10% of the exposures (diameter reported as radius, µT as mT, switching RMS values for peak values or peak-to-peak value to calculate induced field, etc.). Some of these discrepancies were easily detectable and conciliation was possible by cross-checking with other reported parameters, diagrams, and known relations between them, highlighting the importance of thorough experimental setup description in uncertainty reduction. Nevertheless, estimates for 12% of the exposure parameters studied were not possible, as these were incomplete beyond reasonable estimation [e.g., square wave without measured (or even presumed) slope rate of change, etc.], rendering the exposures irreproducible.

Other complementary factors that are unaccountable without a thorough experimental description could significantly contribute to the total electric and magnetic fields to which individual cells were exposed in each experiment. For example, only 12% of the cases where cultures could be found in close proximity to ferromagnetic or conductive materials during exposure (metallic microscope parts, shelves, instrumentation, etc.) were complemented with numerical estimations of the localized perturbations introduced by such materials, and only 14% of exposure descriptions were complemented with direct signal spectrum or time-domain measurements on the CDV. Additionally, in the case of perpendicular IMF exposures on adherent and nonadherent cultures (71% of exposures), small magnitudes on the induced gradient are increasingly susceptible to perturbation as proximity to the IEF vortex increases. Coincidently, measures to generate quasi-homogeneous IEFs (i.e., annular rings) were used in only 10% of the exposures. Similarly, only 12% of exposures have included explicit details of the cellular confluency at the time of exposure, leaving ample room for speculation in the IEF as to which individual cells may have been exposed. Finally, factors like greater signal magnitude and frequency components, distortions due to generation or measurement (discrepancies in equipment calibration, linearity, etc.), differentials in the amounts of liquid in the culture containers (Freshney, 2015), and the effects of the meniscus (Greenhill, 1984; Schuderer and Kuster, 2003) could have a significant impact.

Finally, the lack of explicit descriptions of the management and measurement of other physical parameters that may entail significant biological relevance unavoidably introduce uncertainty, which could be of significance to the observation of small biological effects due to ELF electric and magnetic fields. For example, parameters inherent to electrically driven systems (coils, heated stages) like vibration/sound (Tanaka et al., 2003; Shikata et al., 2007; Ito et al., 2011; Muehsam and Ventura, 2014) and temperature gradients and differentials (Blackman et al., 1991; Weaver et al., 1999) seldom presented real-time and *in situ* measurements. In this regard, only 15% of exposures described reproducible sham exposures by utilizing bifilar coils introducing much similar vibration and heating profiles to the exposed samples. Perhaps more importantly, light irradiance (W/cm²) was not described in any of the experiments in which a microscope was used during exposure, although it has been repeatedly pointed out that the presence of light, whether alone (Grzelak et al., 2001) or in combination with electric and magnetic fields (Gegear et al., 2008; Okano, 2008; Wu and Dickman, 2012; Barnes and Greenebaum, 2015; Prato, 2015; Qin et al., 2016), is likely to have biological relevance.

2.7 Potential Biological Relevance of Electric and Magnetic Fields Differentials

There are strong classical and fundamental reasons why magnetic fields should not affect cellular biological systems (Weaver et al., 1999; Foster, 2000; Adair, 2003; Vaughan and Weaver, 2005; Sheppard et al., 2008), leaving the molecular mechanism of action that explains these effects currently open for debate. However, in view of the multiple and repeatedly observed behavioral effects on complex biological systems (insects, reptiles, fish, rodents, birds, and ruminants) for small electric (<1 nT/cm) (Adair et al. 1998; Kajiura and Holland, 2002; Huveneers et al., 2013) and magnetic (0.01–430 µT) fields (Burda et al., 2009; Wu and Dickman, 2012; Prato, 2015; Qin et al., 2016), leaving aside pure scientific curiosity and as yet inconclusive health effects (Savitz et al., 1988; IARC, 2002; Schreier et al., 2006; Davanipour and Sobel, 2009; ICNIRP, 2010; Grellier et al., 2014; Huss et al. 2014; Pedersen et al., 2014; Redmayne and Johansson, 2014; Slusky et al., 2014; Schüz et al., 2016), several other possibilities have been explored theoretically and experimentally. These include the "ion cyclotron resonance model," "ion parametric resonance model" (Adair, 1998; Belova and Panchelyuga, 2010), "molecular gyroscope model" (Binhi et al., 2001), "Lorentz models" (Muehsam and Pilla, 2009a,b), "DNA antenna model" (Blank and Goodman, 2011), "radical pair model" (Timmel and Henbest, 2004), and other modifications and combinations of these (Prato, 2015). Currently, the radical pair model is a promising theory gaining increasing attention, and several aspects are being explored both experimentally (Müller and Ahmad, 2011; Usselman et al. 2016) and numerically (Solov'yov and Schulten, 2009). In this regard, for example, it is still not clear what structures would allow such effects to happen or be amplified within standard environmental conditions (Engström, 2007; Liedvogel and Mouritsen, 2010; Bounds and Kuster, 2015). However, it is encouraging in the sense that it supports the notion that the concentration of free radicals can be both decreased and increased by changing the magnitude of the magnetic field (Rodgers et al., 2007; Rodgers and Hore, 2009). This, in turn, can be expected to both inhibit and accelerate biological processes such as growth rates with dependency on cell type and initial conditions (like cell-type specific redox status) (Simko, 2007; Buchachenko, 2016) providing a possible explanation for some of the conflicting results in the literature (Barnes and Greenebaum, 2015).

In practice, some experiments have shown that TVMF magnitudes, such as the ones reported here, have been able to affect cell cultures when compared to cells cultured in conditions similar to those found in nature (TVMF < 0.1µT). A brief review of the most recent data is presented here. In the case of human breast cancer cells (MCF-7), effects have been found to occur for magnitudes as small as 1.2µT, with exposures as short as 24h at 60 Hz having been linked to growth (Blackman et al., 2001; Girgert et al., 2005) and several other cellular functions (Girgert et al., 2008, 2009, 2010). Greater TVMF magnitudes were shown to induce heat shock proteins (HSP70) in eukaryotes at 60–80µT (at 50 Hz for 30 min) for HL60 cells (Tokalov and Gutzeit, 2004), at 25–100µT (at 50 Hz for 1h) for K562 cells (Mannerling et al., 2010), and other protective proteins at 38µT (at 50 Hz for 5h) for M14 and A375 cells (Basile et al., 2011). Similarly, *E. coli* morphotype, viability, and transposition activity were shown to be affected by 100µT (at 60 Hz) for exposures as little as 20 min (Del Re et al., 2003; Cellini et al., 2008). Similarly, cells of various types have been shown to behave differently when exposed to SMFs outside the natural GMF range, especially below 23µT. These environments have been referred to as "magnetic vacuum," "zero fields," "near-zero fields," "weakened GMF," "perturbed geomagnetic fields," and "low-level fields," to name a few. For example, chromatin formation as well as tubulin self-assembly were shown to be affected for free cell assays and VH-10 cells, respectively, in short time-frame exposures (~20 min) (Belyaev et al., 1997; Wang et al., 2008). Human primary Human Umbilical Vein Endothelial Cells (HUVEC), animal primary Pulmonary Artery Endothelial Cells (PAEC), human cancer (HT1080, HCT116, SH-SY5Y, AsPC1), and primary E180 mice skeletal muscle cells growth rates were shown to be affected by exposures as short as 24h, while other multiple cell functions were shown to respond in the order of minutes to a few hours, affecting in some cases H_2O_2 and oxygen radicals (Martino et al., 2010a,b; Martino, 2011; Martino and Castello, 2011; Mo et al., 2014; Fu et al., 2016). Similarly, pathogenic bacteria (*Pseudomonas, Enterobacter,* and *E. coli*) resistance to antibiotics (Poiata et al., 2003; Creanga et al., 2004) as well as changes to the genome (Binhi et al., 2001) were shown to be influenced by such exposures. Reports from more extensive studies have also revealed effects on proliferation, differentiation, and other cellular processes after exposure of human leukemia, astrocytoma, and lymphoma cells to altered BMF conditions (both for the SMF and TVMF components), exhibiting dose-, time-, and cell-type-dependent responses on the order of hours (Chen et al., 2000; Wei et al., 2000; Mangiacasale et al., 2001; Robison et al., 2002), while similar observations were made for mouse lymphocytes (Zmyslony et al., 2004). Finally, experiments at other frequencies different from 50 Hz/60 Hz have suggested the possibility for biological relevance in very selective ways. For example, 2.5µT at 7 but not 18 Hz for 3–5 days has been shown to modulate intracellular calcium concentration, metabolic activity, proliferation, and differentiation of human adult cardiac stem cells (Gaetani et al., 2009). Also, the concentrations of oxygen radicals and Nitric Oxide (NO) in human neutrophils, macrophages, and HT1080 cancer cells were found to be sensitive to electric fields of small magnitudes (10^{-4} V/m at sub-Hertz frequencies) to the point of DNA damage in under 5 min of resonance with metabolic oscillations (Rosenspire et al., 2001, 2005), which underscores the potential relevance of signal intermittence (as shown in Figure 2.9).

2.8 Temperature in Biological Laboratory Environments

The standard temperature (ST) of biological laboratories generally defined as ranging between 20°C and 25°C. However, environmental temperature controller's guidelines, typically set based on the human "zone of thermal comfort," are expected to perform with an uncertainty of 0.5°C–2°C depending on the class. As a result, in principle, the expected ST can be extended to anywhere between >18°C and <27°C. Additionally, for the case of TVT, it is recommended that the rate of the variation does not exceed 2.2°C/h, while if the peak of the variation is <1.1°C there is no rate limit set, allowing for multiple oscillation frequencies within that band. Additionally, even if the laboratory is compliant with these recommendations under standard conditions, thermal pockets and gradients (>2°C) can be generated due to unavoidable differential air circulation within the laboratory itself and remain unaccounted due to the discrete location of sensors and actuators (ASHRAE, 1997; Riederer, 2002; Riederer et al., 2002). In this regard, previous observations have found a clear relationship between the moderate shifts in room temperature and temperature regulation of standard laboratory equipment. For example, shifts of +6°C/–3°C (from 20°C) were shown to elicit changes in heated microscope stages (2.1°C/–2.5°C), slide warmers (3.4°C/–2.4°C), and incubators (0.9°C/–0.5°C) (Butler et al., 2013). Besides the earlier facts, such variations can be also introduced by unconventional atmospheric circumstances and the environmental control capacity of the buildings. Because of this, laboratories handling sensitive cultures are causally recommended to be set away from the facilities' top floors, where typical fluctuations may tend to be greater. Additionally, although for most locations worldwide, power distribution is fairly predictable, brownouts and blackouts may occur unexpectedly (and in extraordinary situations, frequently), impacting directly the ability of laboratory equipment to regulate their environments within manufacturer specifications. In this regard, incubator set points have been observed to shift significantly (up to 1°C) in situations where "backup" generators are available due to differentials in the power quality of the emergency systems, markedly affecting embryonic cultures (Gardner et al., 2012). Contingency for such events, as well as thermal monitoring logs, is of particular interest to the researcher, especially with regard to overnight cultures where human reaction can be quite delayed.

2.9 Temperature in Instrumentation and Its Effect on Culture-Designated Containers

A reasonable and a generally accepted assumption is that, under any practical and standard utilization regime, the ST of any modern instrumentation with biological incubation systems at any point inside its controlled environment (if properly calibrated) may not have a difference between any point and the absolute set point would be >±0.25°C (±0.5°C in more extreme cases). Additionally, since the modern incubator is an automatic feedback controlled system, its temperature (and every other controlled environmental variable) is expected to have an oscillatory behavior (TVT) around this set point, but in general, if it is even acknowledged, its magnitude is expected to be not greater than

±0.15°C (Weaver et al., 1999; Walker et al., 2013), resulting in a maximum uncertainty on a contemporary biological incubator of ~0.6°C by the sum of the squares. Contemporary experienced researchers typically resort to further verification of the incubator's temperature display validity by performing direct measurements using glass thermometers, thermocouples, thermistors, RTDs, and so on directly inside the chamber or in water baths, thus usually confirming this assumption (Patching and Rose, 1970; Freshney, 2015). However, although such equipment would be appropriate for measurements in incubation systems where temperature has already stabilized (after several hours to days), they are inherently insufficient to satisfactorily capture the existence of thermal perturbations of shorter time frame. The reason stems from the fact that the physical properties of the chosen thermometry systems itself are usually many times different from the culture containers and culture medium that will be utilized in real experiments, making its thermal constant inherently different, and in principle, inadequate to capture transient thermal perturbations. Nevertheless, some early observational studies have reported transients of up to 1°C/min when inserting culture containers from the incubator to the heated microscope stage with differences from the incubation temperature of up to 5°C (Cooke et al., 2002). In the case of culture incubators, early observations have shown that tissue culture container temperatures may take 1–2 h to reach steady state after insertion (37°C) from room temperature (Bender and Brewen, 1969). However, adequate measurements to capture the real extent of thermal transients or oscillations were challenging as they required nonconventional instrumentation and procedures. More recently, Portelli et al. designed, built, and calibrated a set of temperature probes to mimic normal culture container conditions in a way that the temperature in such places was minimally perturbed. These sensors were placed inside an incubator and perturbations were introduced in a controlled way, revealing greater variations than previously contemplated (Portelli et al., 2013a). Figure 2.11 shows inhomogeneity upon insertion to the incubator and differences from the incubator set point of up to 5°C and 12°C, respectively, with equilibration (recovery) reached after only ~135 min. Additionally, homogeneity was also shown to be significantly affected by the size and shape of containers, their location within a stack, the specific incubator (size, shape, and type), and container distribution within its interior. Figure 2.12 shows examples of the variabilities in the order of 6.9°C (max. rec. ~87 min) for the type of container, while differences as big as 4.9°C (max. rec. ~376 min) between commonly utilized plastic petri dish sizes with recovery times in the order of hours spanning approximately fivefold. Moreover, the temperature to which cultures are exposed is also sensitive to the position within the dish as well as to the liquid amount variations. In this regard, Figure 2.13 shows examples where the differences during insertion were as big as 3.1°C (max. rec. ~58 min), showing potentially relevant heterogeneous conditions even within the same culture container in standard laboratory practice. Similarly, perturbations during incubation (door openings) show equally variable thermal histories depending on the factors previously described, with the addition of the duration and modulation of the perturbation. In this regard, early observations have pointed out that a 1–2 h period was needed for thermal equilibration after a 1–2°C drop was introduced by opening the incubator door (Bender and Brewen, 1969), and more recent observations have proposed door openings and location within the incubator as potential factors of concern (Higdon et al., 2008). In Figure 2.11, notice how under single and consecutive

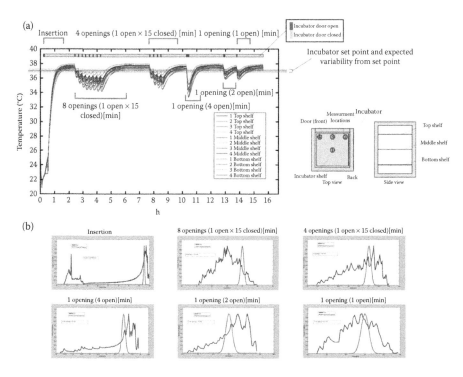

FIGURE 2.11 Assessment of culture container (30-mm petri dish, liquid volume = 2.4 mL) temperature during thermal perturbations on CO_2-humidified ~160 L incubator. (a) Testing protocol involves six different scenarios: (1) insertion (MI = 5°C, MDS = 12°C, TE = 135 min), (2) 8 consecutive 1-min openings followed by 15 of closed door (MI = 2.7°C, MDS = 3°C, TE = 90 min), (3) 4 consecutive 1-min openings followed by 15 of closed door (MI = 2.2°C, MDS = 2.6°C, TE = 65 min) (4) 1–4 min openings (MI = 2.0°C, MDS = 2.5°C, TE = 60 min), (5) 1–2 min openings (MI = 0.9°C, MDS = 1.1°C, TE = 50 min), and (6) 1–1 min opening (MI = 0.8°C, MDS = 0.9°C, TE = 45 min). The incubator was allowed to reach thermal equilibrium without perturbation for at least 24 h prior to the experiment. Insertion was done as fast as possible, but in <2 min. The door(s) was held open to its limit for the remainder time until reaching the 2-min mark and temperature was recorded after insertion. Each subsequent door opening was made in the same fashion. Diagram (right) shows the approximate location of the culture containers within the incubator. Red lines denote the incubator set point (programmed) as well as ±0.25°C traditionally expected variability for incubation systems. Abbreviations: MI = Max inhomogeneity; MDS = max deviation from set point; TE = time to equilibrium. (b) Histograms showing thermal dosage in comparison with the incubator set point and traditionally expected variability from the time of the perturbation until thermal equilibrium. Notice how for most exposures the thermal dosage depends on the perturbation type. (From Portelli et al., *Temperature variations in biological samples is a potential confounder for experimental variability and reproducibility.* The Bioelectromagnetics Society (BEMS) Annual Meeting, Thessaloniki, Greece, 2013a.)

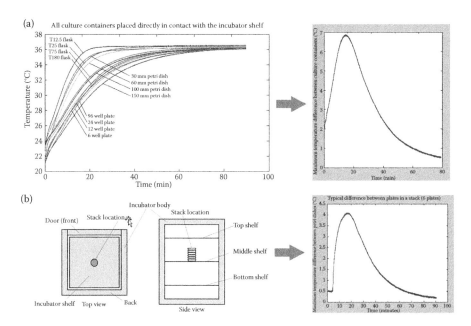

FIGURE 2.12 Assessment of culture container temperature as a function of type and size (a) and location within a stack (b) during thermal insertions on CO_2-humidified ~160 L incubator. Insertion was done as fast as possible, but in <2 min. The door(s) was held open to its limit for the remainder time until reaching the 2-min mark and temperature was recorded after insertion. Curves shown on the right denote examples of the worst-case difference between all containers involved in the same study. For (a) (MI = 6.9°C, TE = 87 min). For (b) experiments were carried out for petri dishes sized 30 mm (MI = 4.9°C, TE = 87 min), 60 mm (MI = 4.7°C, TE = 126 min), 100 mm (MI = 4.1°C, TE = 208 min), 150 mm (MI = 3.4°C, TE = 376 min). Abbreviations: MI = Max inhomogeneity; TE = time to equilibrium. (From Portelli et al., *Temperature variations in biological samples is a potential confounder for experimental variability and reproducibility.* The *Bioelectromagnetics* Society (BEMS) Annual Meeting, Thessaloniki, Greece., 2013a.)

perturbations (1 min open/15 min closed), the maximum inhomogeneity was recorded to be up to 2.0°C and 2.7°C, while the difference from the incubator set point could reach 2.5°C and 3°C (max. rec. ~90 and 45 min, respectively). Therefore, in essence, smaller containers have the ability to heat and cool quicker than their bigger counterparts, making them better for minimizing variation during insertion, but not optimal for withstanding perturbations (door opening) and vice versa. Also, in the case of the common practice of stacking culture containers, the greatest temperature difference always appears to be present between the containers at the extremes of the stack (top and bottom) and the containers at the center of the stack where they are more thermally isolated. More importantly, Figure 2.14 shows an example evidencing the inaccuracy of the temperature reported by the incubator integrated display [>3°C peak, ~1.5°C (average)] during and after thermal perturbation (consecutive door openings). The same investigation showed that although the incubator display reported the average recovery time to be <15 min after insertion (for 10 different incubators), this value was approximately ninefold shorter from that recorded inside the culture containers, evidencing significantly greater uncertainty. Because of this, as the sensors reporting the incubators display temperature (typically just one) are placed in discrete locations behind or on the incubator internal chamber and not within the culture containers where the cells are actually cultured, they can only be intended to provide what can be called a "reference temperature" and should be expected to differ significantly from the real culture temperature, especially during thermal perturbation recovery for which *in situ* and real-time measurements (as the ones shown in Figures 2.11 through 2.14) are needed.

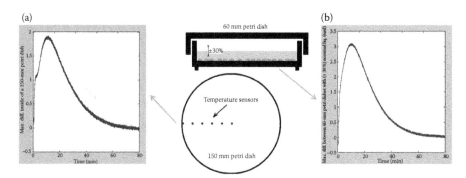

FIGURE 2.13 Assessment of (a) position within the culture container and (b) liquid variation temperature after insertion in a CO_2-humidified ~160 L incubator for containers in contact with the middle shelf (center). In this example, the assessments were made in (a) Petri dish (150 mm) (MI = 1.8°C, TE = 58 min) and (b) Petri Dishes (30 mm) with variations of ±30% of the nominal liquid volume (2.4 mL). Abbreviations: MI = Max inhomogeneity; TE = time to equilibrium. (From Portelli et al., *Temperature variations in biological samples is a potential confounder for experimental variability and reproducibility.* The *Bioelectromagnetics* Society (BEMS) Annual Meeting, Thessaloniki, Greece., 2013a.)

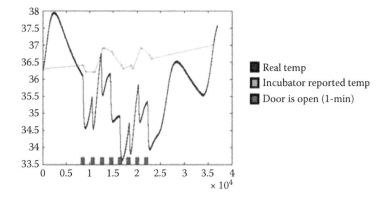

FIGURE 2.14 Culture container vs. reported incubator display temperature (8 consecutive 1-min openings followed by 15 of closed door) in ~160 L incubator. (From Portelli et al., *Temperature variations in biological samples is a potential confounder for experimental variability and reproducibility.* The Bioelectromagnetics Society (BEMS) Annual Meeting, Thessaloniki, Greece., 2013a.)

Additionally, evidence of potentially significant thermal discrepancies due to transient perturbations between incubators has also been reported. For example, temperature recovery for a 5-s opening of incubators was shown to last six times more (~5 vs. ~30 min) for two incubators of different sizes and architectures, and simple thermistor measurements have also suggested significant differences (>1°C) between the incubator interior and the culture container during this period (Fujiwara et al., 2007). Similarly, in another report, a factor >4 was observed when two incubators of different sizes and architectures were recovered from a 2°C culture container cooling (Cooke et al., 2002). In addition, the common practice to share most of the laboratory equipment (including biological incubators) between several researchers may expose culture containers to further thermal perturbations in a "passive" way, inevitably multiplying the thermal perturbations to which individual cultures should be minimally exposed. Additionally, perturbations by insertion and door openings may not be homogeneous to all cultures as different locations within the same incubator (i.e., near the door on the top shelf vs. near the back on the bottom shelf) will have different exposures to the perturbation (i.e., during insertion or a temporary door opening), leading to fundamentally different thermal transients to reach thermal equilibrium (see Figure 2.15).

Finally, thermal inhomogeneity and susceptibility to perturbations have been observed on a variety of laboratory instrumentation. For example, thermal baths water surface temperature has been observed to drop as much as 2°C upon removing the lid allowing for evaporation and creation of thermal gradients inside the liquid (Butler et al., 2013). Similarly, heated microscope stages (tools widely used in assisted fertilization)

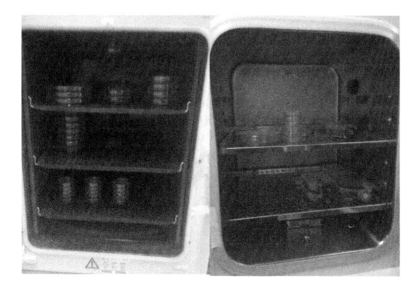

FIGURE 2.15 Example of common practice in a typical shared laboratory incubator. (From Portelli et al., *Temperature variations in biological samples is a potential confounder for experimental variability and reproducibility.* The Bioelectromagnetics Society (BEMS) Annual Meeting, Thessaloniki, Greece., 2013a).

have been shown to present localized thermal gradients requiring complex *in situ* thermal controls. As these controls entail significant non-trivial effort (if even possible), others have even suggested the observation of features of the biological system itself as a reporter for the microenvironment thermal conditions (Wang et al., 2002).

2.10 Temperature Differentials' Potential Biological Relevance

Biological systems are typically allowed to work under very narrow functional temperature ranges (Sen and Murray, 2014), and optimization can be such that the organisms death may be only a few degrees away from its optimal physiological set point (Dill et al., 2011). In principle, such sensitivity stems from the fact that chemical reactions are particularly sensitive to small temperature changes as they are governed by forms of the Arrhenius equation, where the temperature dependency is exponential (van't Hoff, 1884; Arrhenius, 1889; Jorjani and Ozturk, 1999; Dewey, 2009). As a result, the metabolism and cell cycle rate from microbes (Baumgärtner and Tolić-Nørrelykke, 2009; Goranov et al., 2009) to unorganized cells in culture from complex biological systems (Ratkowsky et al., 1982, 1983; Zwietering et al., 1991; Membré et al., 2005) to complex vertebrates is exponentially dependent on environmental temperature affecting most of the biochemical processes directly or indirectly (Gillooly et al., 2001). In the case of microbes, both their growth and biosynthesis can have very narrow optimal ranges (<1–2°C), which typically appear as a "J-shape" function of temperature (Ratkowsky et al., 1982, 1983; Zwietering et al., 1991; Honglin et al., 2001; Membré et al., 2005). In addition, unlike other cellular systems, cell division in microorganisms like bacteria can be completed in the order of minutes, which can severely amplify the effect of small and transient environmental differentials, like temperature, potentially introducing well-known significant deviations from prediction models that are routinely found in practice (Zwietering et al., 1994). In the same way, unorganized cell cultures from complex biological systems may also be susceptible to thermal challenges in the absence of control loops originally found in the complex biological system. In the case of cultures of animal origin, the effects of mild hyperthermic and hypothermic exposures (~≥±3°C) have been long documented to induce both temporary and lasting effects on several aspects of cellular structure and physiology (phenotype and biosynthetic activity) (Fairchild et al., 2000; Kalamida et al., 2015; Malyshkina et al., 2016) depending on both cell type and initial conditions (Pickering and Johnson, 1987). For example, human umbilical vein endothelial cells showed a much more favorable mitogenic response (and required much lower seeding densities) at 33°C when compared with 37°C in contrast to the findings for pigment epithelial cells and fibroblasts, for which 37°C was more favorable (Yang and Berghe, 1995). In the sense of multiple exposures, such exposures have also shown potential to cause detrimental effects or enhancing thermotolerance for both transient (Ju et al., 2005; Xiao et al., 2011) and continual (Ito et al., 2015) exposures.

In the case of smaller differences (~<±3°C), early observations of significant biological effects were also reported. For example, differences on the order of up to 25%/48 h per °C (around 37°C) were observed for human lymphocyte proliferation in a differentiable manner for interdonor differences (Purrott et al., 1981). Neutrophils have also been shown to have differentiable velocities (resolutions ~1°C) around 37°C (Aly et al., 2008). Temperature differences of the same order as those found *in vivo* (as small as 0.5°C) (Luck et al., 2001; Einer-Jensen and Hunter, 2005; Hunter and Einer-Jensen, 2005; Hunter et al., 2006; Ye et al., 2007) appear to significantly affect gamete competency, embryo development, and offspring survival in culture. For example, several aspects of the oocyte-to-embryo transition have earlier been shown to be closely tied to temperature down to at least 1.5°C for several mammal species (Lenz et al., 1983; Eng et al., 1986; Lavy et al., 1988; Shi et al., 1998). Some studies have also uncovered some of the dependence of such cells to their respective *in vivo* temperature gradients during stages of maturation (McKiernan and Bavister, 1990) and more recently showed the resulting discernible delays in maturation due to cell cycle–temperature mismatches of the order of 12 h/°C (Ye et al., 2007). In this regard, adverse effects of temperature fluctuations have been identified since the inception of *in vitro* fertilization in embryo development (Ulberg and Burfening, 1967), and embryo development success has shown strong indications to be inversely proportional to the frequency of incubator door openings (Abramczuk and Lopata, 1986), and both the early-stage good embryo formation rate and the good blastocyst formation rate were significantly higher in the mini-incubator (40% and 15%) versus a conventionally sized incubator (28% and 8%) (Fujiwara et al., 2007).

Additionally, several reports have provided evidence of neural sensitivity to the rate of change of temperature, while experiments to study such effects require extraordinary effort on the design and monitoring of exposure systems. For example, changes in the firing rate of pacemaker cells from the ganglion of *Aplysia californica* have been induced by total temperature changes of as little as 0.1°C when the rates of change are ~1°C/s, corresponding to an injection of ~1 nA into the cell (Chalker, 1982; Barnes and Greenebaum, 2006). Similar observations have also been made for the large parietal ganglion of the central nervous system of *Limnaea stagnalis*, where a slow increase in temperature of 1°C/min or less increases the firing rate of the pacemaker cell, and a rapid increase in temperature of 0.1°C/s or faster decreases or stops the firing (Bol'shakov and Alekseyev, 1986). In this regard, contemporary investigations are yielding physics-based insights on the nature of neural pulse transmission according to the soliton model theory, for instance, in which transmission is described as inherently thermal phenomena that can be measured electrically (Heimburg, 2007; Andersen et al., 2009) predicting sensitivity to spatiotemporal thermal stimuli (Hasani et al., 2015). Similarly, classical theory also describes neural extreme temperature sensitivity. For example, the Nernst equation is used in classical electrophysiology as a simplified starting point to model the passive equilibrium of membrane potentials (especially those excitable, e.g., neural) as a function of ion concentrations and temperature as it describes the relationship between chemical and electrical gradients across a semipermeable membrane (Kandel et al., 2000). Interestingly, little work is needed to show the fundamental temperature rate-of-change ($\Delta T/\Delta t$) dependence of current flow through cell membranes predicted by the same equation (Barnes, 1984) in the order of a few nA per °C/s depending on the specific cell in question.

Finally, several reports have shown that interaction with unorganized cellular cultures is possible by introducing thermal oscillations of the same order as those generated *in vivo* (*e.g., as low as* ±0.3°C/0.4°C for humans) (Hammel and Pierce, 1968; Keatinge et al., 1986; Rubin, 1987; Shiraki et al., 1988; Webb, 1992; Kelly, 2006, 2007; Kräuchi et al., 2014). This responsiveness may come from the absence of synchronizing biochemical signals found *in vivo*, making the cells more susceptible to those of thermal origin. For example, human fibroblasts' autonomous oscillators have been shown to be able to be synchronized *in vitro* by small temperature fluctuations (~±0.5°C) mimicking daily body temperature fluctuations as effectively as with other chemical signals (Saini et al., 2012). Similar results were observed for rat SCN cells (~±0.75°C) and fibroblasts (~±1.25°C), mice fibroblasts (~±1.5°C), and even cyanobacteria (~±7.5°C) (Herzog and Huckfeldt, 2003; Yoshida et al., 2009; Sládek and Sumová, 2013). Furthermore, in view of the effects reported, it is reasonable to speculate that thermal oscillations may be able to induce analogous effects on the biochemistry of other existing cellular oscillatory biochemical systems (Adler, 1964; Paciorek, 1965; Razavi, 2004). In this regard, apart from the circadian-related biochemical oscillatory systems, there are many other known sustained autonomous oscillatory biochemical processes (Rapp, 1979; Ehrengruber et al., 1996; Petty, 2006; Stark et al., 2007; Novák and Tyson, 2008; Brasen et al., 2010) with periods ranging from fractions of a second to days (Maroto and Monk, 2008). Oscillations in intracellular and extracellular calcium (Falcke and Malchow, 2003), insulin production (Bergsten, 2002), and electrical potentials (Wachtel, 1985; Koshiya and Smith, 1999) are widely documented and studied examples. Some authors have highlighted the importance of the study of the effects of stimulatory oscillations since growing evidence indicates that biochemical signaling encodes information in oscillatory signals rather than in constant ones (Tostevin et al., 2012). However, the emergence of exposure systems with the capability to accurately control thermal fluctuations (Portelli et al., 2010a, 2011; 2012a; Saini et al., 2012; Kausik, 2014) promises better understanding of thresholds described here.

2.11　Measures for Electric, Magnetic, and Temperature Experimental Environment Decoupling and Transient Mitigation

Although most of these observations and general guidelines for *in vitro* bioelectromagnetics experiments have been previously delineated and revised several times for more than 40 years (Bassett et al., 1974; Bassen et al., 1992; Valberg, 1995; Kuster and Schönborn 2000; Negovetic et al., 2005), this section provides some concrete advices to help the cautious researcher. As discussed in the previous sections, the contribution to the total electric and magnetic fields to which cultures are ultimately exposed could be relevant in practice (especially for bioelectromagnetics experiments) if the culture space is not properly decoupled by active or passive shielding as required (Portelli et al., 2013). Electric and magnetic shielding is an active field of research (Umurkan et al., 2010). In general, there are three ways in which MFs (both SMFs and TVMFs) can be attenuated in a designated volume, namely, compensation, shielding, or their combination. Compensation refers to the use of a feedback system to generate MFs in inverse direction to the ones existing in the volume of interest, causing the resultant MF to be attenuated. As more spatial uniformity is needed in the CDV, an active system requires increasing numbers of coils, precision current sources to drive those coils, and sensors distributed in the space (Reta-Hernández and Karady, 1998). However, special attention must be taken in that the noise in the injected currents be below acceptable levels (<60 dB) in order not to generate unwanted harmonics or coupling to the power distribution lines (Portelli, 2012). The combination of passive shielding and active compensation has been utilized successfully in cases where the space available for sensors and actuating coils is reasonably smaller than the volume of interest and for reducing MFs generated in WMFS (Malmivuo et al., 1987; Bryś et al., 2005). Unlike the active compensation approach, shielding is passive, and so feedback systems are not necessary when the maximum magnitudes of the MF sources are known. Soft magnetic materials with high magnetic permeability can concentrate high magnetic flux densities at static and low-frequency MF environments. Therefore, an assembly consisting of a geometrically closed surface made of such material will concentrate the impinging MF into the alloy, bypassing or "shielding" its interior. Additionally, induced eddy current reflection makes these materials capable of shielding incident MFs more efficiently as their frequency increases (Seeber, 1998). Typically resourced architectures amount to a 5-sided box (body) with one face acting as a lid or door to provide culture access. Such architecture is unavoidably supplemented with permanent connections necessary for gas exchange, instrumentation connectivity. In order to diminish the negative effect in the structure shielding ability, such modifications are usually supplemented with flanges (lowering the reluctance introduced by the interrupted structure) (Blackman et al., 2001; Schuderer et al., 2004). Additionally, the lids or doors must be supplemented with flanges (sometimes several cm long) that overlap on the body of the shield. In case any such decoupling measures are taken, proper numerical modeling and validation of the system is imperative as both types of decoupling can generate steep gradients (Lekner, 2013). In the case of shields based on partially saturated, high-permeability materials, these can also modify the imposed signal's spectrum (Buschow, 2003). Therefore, upon numerical modeling and validation of the shielding system, it is imperative to characterize such perturbations by a combination of wideband measurements and simulations when possible. In this regard, care must be taken in choosing the appropriate instrumentation as both artifact types may be undetectable by commonly utilized measurement systems. Such signals may contain frequencies which are different from those at which these devices are typically optimized for (e.g., 50 Hz/60 Hz). Additionally, the introduced gradients can be too steep to be detected or properly surveyed by direct measurement with commonly utilized sensors. In such cases, replacement of some of these parts by plastic ones may be desired (Publicover et al., 1999). The PEF can be easily mitigated in generating coils via conductive shielding that can be put around the CDV or around the exposure system coils as long as this shielding is interrupted to stop the flow of Eddy currents. Such measures have shown to be able to reduce the PEF original field by a factor of 300 (Caputa and Stuchly, 1996; Craviso et al., 2003; Schuderer et al., 2004). Additionally, PEF can be minimized by a crisscross winding pattern (in the case of a solenoid = top-to-bottom and then back to top), with the additional benefit of adjacent coil connectors (Lyle et al., 1997). For the case of temperature, adequate control and monitoring of the thermal history of culture containers in real time could not only potentially lead to certain unprecedented experimental precision, replication, but also minimize the amount of experimental resources needed to achieve statistical significance (Weaver et al., 1999).

In the case of conventional incubators, some simple measures can be taken to minimize the thermal perturbation effects on culture container thermal history. For example, some laboratories separate incubator use depending on the sensitivity of the biological experiment. In general, a "working incubator" (shared, with high-frequency perturbation) and a "long term incubator" (dedicated, with low frequency perturbations) can be set up (Lane et al., 2007). Moreover, these incubators can be set up with a solid metal plate insert (or any other material with adequate heat capacity) that acts as a direct conduction and radiation heat transfer of retained heat absorbed during thermal equilibrium minimizing perturbation effects (Cooke et al., 2002). Short stacks, or not stacking at all, are recommended to also minimize such variation (Portelli et al., 2013). However, ideally, a new paradigm in thermal control should be implemented for thermal control in experimental setups. Such an approach should allow for localized and real time monitoring and control of temperature in such a way that the introduction of unwanted environmental artifacts is minimized. For example, some common architectures and strategies utilized in *ad hoc* systems utilize conduction (from water convection) as a primary way of controlling temperature by using single-dish water jackets (Walleczek and Liburdy, 1990; Liburdy, 1992) or baths (Blackman et al., 1993; Carson and Prato, 1996; Blackman et al., 2001; Capstick et al., 2013) or via forced air convection (Schuderer et al., 2004; Portelli et al., 2010b), with some of these configurations reaching control of up to within 0.1°C–0.2°C. Special care must be taken for any of these architectures as details tend to be important. For example, in the case of air heating, it might be important to acknowledge that the air pump utilized might introduce variable amounts of heat that depend on external variables such as the room temperature and the time of use, in case this pump is utilized to push air into the culture container vicinity. This heat might escape the researcher's control and therefore make control difficult or even impossible. On the other hand, it might be more adequate to utilize its pumping action in a way in which it sucks air from the culture container vicinity, eliminating the self-generated heat away from culture containers.

2.12 Conclusions

In order for scientific observations to mature into established effects through the requisite of independent experimental replication, experiments and their associated reports must comply with a thorough exposure system, protocol, and method description that uncertainty budgets will allow for satisfactory replication of the conditions claimed. The facts presented in this chapter put electric, magnetic, and temperature uncertainties under a quantitative perspective, which can aid the careful researcher to implement contingency plans if necessary in their own research. Although it is clear that individual experiments and protocols differ on an endpoint to endpoint basis, and therefore exposure systems and adequate exposure conditions (including relevant biophysical factors, controls, etc.) cannot be standardized, the general aim in all exposures in bioelectromagnetics research is to increase the likelihood of evoking small effects, should there be any. In view of this, although the contribution of confounders that are not under investigation may not be substantial, experiments need to be made as a function of those confounders so as to rule out their significance in the same study. In the not-preferred case that such an evaluation is undertaken in subsequent studies, adequate retrospective dosimetry is an important factor that should at least be facilitated by appropriate documentation on numerical design, validation, monitoring, and reporting, which has shown to be seldom possible from contemporary literature. In the case of general biological experimentation, attention to the details presented in this chapter could not only further minimize the amount of experimental resources needed to achieve statistical significance but also potentially lead to unprecedented experimental precision and successful replication rates.

Acknowledgments

I would like to acknowledge my family, Prof. Frank Barnes, and all other coauthors and friends cited in previous manuscripts and the ones who are not recognized in writing but will always be remembered. These teachers gave me the necessary tools to explore past the established truths and showed me that uphill battles are always the best ones.

References

Abramczuk J.W., Lopata A. (1986) Incubator performance in the clinical *in vitro* fertilization program: Importance of temperature conditions for the fertilization and cleavage of human oocytes. *Fertility and Sterility.* 46(1):132–4.

Adair R.K. (1998) A physical analysis of the ion parametric resonance model. *Bioelectromagnetics.* 19:181–91.

Adair R.K. (2003) Biophysical limits on athermal effects of RF and microwave radiation. *Bioelectromagnetics.* 24:39–48.

Adair R.K., Astumian R.D., Weaver J.C. (1998) Detection of weak electric fields by sharks, rays, and skates. *Chaos: An Interdisciplinary Journal of Nonlinear Science.* 8(3):576–87.

Adler R. (1964) A study of locking phenomena in oscillators. *Proceedings of the IRE.* 34(6):351–7.

Ahlbom A., Day N., Feychting M., Roman E., Skinner J., Dockerty J., Linet M., McBride M., Michaelis J., Olsen J.H., Tynes T, Verkasalo P.K. (2000) A pooled analysis of magnetic fields and childhood leukaemia. *British Journal of Cancer.* 83(5):692–8.

Aly A.A., Cheema M.I., Tambawala M., Laterza R., Zhou E., Rathnabharathi K., Barnes F.S. (2008) Effects of 900-MHz radio frequencies on the chemotaxis of human neutrophils *in vitro. IEEE Transactions on Biomedical Engineering.* 55(2):795–7.

Andersen S.S., Jackson A.D., Heimburg T. (2009) Towards a thermodynamic theory of nerve pulse propagation. *Progress in Neurobiology.* 88(2):104–13.

Arrhenius S. (1889) Über die Reaktionsgeschwindigkeit bei der Inversion von Rohrzucker durch Säuren. *Zeitschrift für physikalische Chemie.* 4:226–48.

ASHRAE Handbook. (1997) *Fundamentals.* SI Edition. The American Society of Heating. Ventilation and Air-Conditioning Inc., Atlanta, GA.

Barnes F.S. (1984) Cell membrane temperature rate sensitivity predicted from the Nernst equation. *Bioelectromagnetics.* 5(1):113–5.

Barnes F.S., Greenebaum B., editors. (2006) *Handbook of Biological Effects of Electromagnetic Fields.* CRC Press, Boca Raton, FL.

Barnes F.S., Greenebaum B. (2015) The effects of weak magnetic fields on radical pairs. *Bioelectromagnetics.* 36(1):45–54.

Basile A., Zeppa R., Pasquino N., Arra C., Ammirante M., Festa M., Barbieri A., Giudice A., Pascale M., Turco M.C., Rosati A. (2011) Exposure to 50 Hz electromagnetic field raises the levels of the anti-apoptotic protein BAG3 in melanoma cells. *Journal of Cellular Physiology.* 226:2901–7.

Bassen H., Litovitz T., Penafiel M., Meister R. (1992) ELF *in vitro* exposure systems for inducing uniform electric and magnetic fields in cell culture media. *Bioelectromagnetics.* 13(3):183–98.

Bassett C.A., Pawluk R.J., Pilla A.A. (1974) Acceleration of fracture repair by electromagnetic fields. A surgically noninvasive method. *Annals of the New York Academy of Sciences.* 238(1):242–62.

Baumgärtner S., Tolić-Nørrelykke I.M. (2009) Growth pattern of single fission yeast cells is bilinear and depends on temperature and DNA synthesis. *Biophysical Journal.* 96(10):4336–47.

Belova N., Panchelyuga V. (2010) Lednev's model: Theory and experiment. *Biophysics.* 55:661–74.

Belyaev I.Y., Alipov Y.D., Harms-Ringdahl M. (1997) Effects of zero magnetic field on the conformation of chromatin in human cells. *Biochimica et Biophysica Acta.* 1336:465–73.

Bender M.A., Brewen J.G. (1969) Factors influencing chromosome aberration yields in the human peripheral leukocyte system. *Mutation Research/Fundamental and Molecular Mechanisms of Mutagenesis.* 8(2):383–99.

Bergsten P. (2002) Role of oscillations in membrane potential, cytoplasmic Ca^{2+}, and metabolism for plasma insulin oscillations. *Diabetes.* 51(suppl 1):S171–6.

Binhi V.N. (2000) Amplitude and frequency dissociation spectra of ion-protein complexes rotating in magnetic fields. *Bioelectromagnetics.* 21:34–45.

Binhi V.N., Alipov Y.D., Belyaev I.Y. (2001) Effect of static magnetic field on *E. coli* cells and individual rotations of ion-protein complexes. *Bioelectromagnetics.* 22(2):79–86.

Blackman C.F., Benane S.G., House D.E. (1991) The influence of temperature during electric and magnetic-field-induced alteration of calcium-ion release from *in vitro* brain tissue. *Bioelectromagnetics.* 12(3):173–82.

Blackman C.F., Benane S.G., House D.E., Pollock M.M. (1993) Action of 50 Hz magnetic fields on neurite outgrowth in pheochromocytoma cells. *Bioelectromagnetics.* 14(3):273–86.

Blackman C.F., Benane S.G., House D.E. (2001) The influence of 1.2 μT, 60 Hz magnetic fields on melatonin- and tamoxifen-induced inhibition of MCF-7 cell growth. *Bioelectromagnetics.* 22:122–8.

Blanchard J.P., Blackman C.F. (1994) Clarification and application of an ion parametric resonance model for magnetic field interactions with biological systems. *Bioelectromagnetics.* 15:217–38.

Blank M., Goodman R. (2011) DNA is a fractal antenna in electromagnetic fields. *International Journal of Radiation Biology.* 87:409–15.

Bol'shakov M.A., Alekseyev S.I. (1986) Change in the electrical activity of the pacemaker neurons of L. stagnalis with the rate of their heating. *Biophysics.* 31:569–71.

Bounds P.L., Kuster N. (2015) Is Cryptochrome a primary sensor of extremely low frequency magnetic fields in childhood leukemia? *Biophysical Journal.* 108(2):562a.

Borys P. (2012) On the biophysics of cathodal galvanotaxis in rat prostate cancer cells: Poisson–Nernst–Planck equation approach. *European Biophysics Journal.* 41(6):527–34.

Brasen J.C., Barington T., Olsen L.F. (2010) On the mechanism of oscillations in neutrophils. *Biophysical Chemistry.* 148(1):82–92.

Buchachenko A. (2016) Why magnetic and electromagnetic effects in biology are irreproducible and contradictory? *Bioelectromagnetics.* 37(1):1–3.

Buchachenko A.L., Kuznetsov D.A. (2014) Magnetic control of enzymatic phosphorylation. *Journal of Physical Chemistry & Biophysics.* 4:142.

Burda H., Begall S., Červený J., Neef J., Němec P. (2009) Extremely low-frequency electromagnetic fields disrupt magnetic alignment of ruminants. *Proceedings of the National Academy of Sciences.* 106(14):5708–13.

Buschow K.H., editor. (2003) *Handbook of Magnetic Materials.* Elsevier, New York.

Butler J.M., Johnson J.E., Boone W.R. (2013) The heat is on: room temperature affects laboratory equipment-an observational study. *Journal of Assisted Reproduction and Genetics.* 30(10):1389–93.

Bryś T., Czekaj S., Daum M., Fierlinger P., George D., Henneck R., ... Pichlmaier A. (2005) Magnetic field stabilization for magnetically shielded volumes by external field coils. *Nuclear Instruments and Methods in Physics Research Section A: Accelerators, Spectrometers, Detectors and Associated Equipment.* 554(1):527–39.

Capstick M., Schär P., Schuermann D., Romann A., Kuster N. (2013) ELF exposure system for live cell imaging. *Bioelectromagnetics.* 34(3):231–9.

Caputa K., Stuchly M.A. (1996) Computer controlled system for producing uniform magnetic fields and its application in biomedical research. *IEEE Transactions on Instrumentation and Measurement.* 45:701–9.

Carson J.J., Prato F.S. (1996) Fluorescence spectrophotometer for the real time detection of cytosolic free calcium from cell suspensions during exposure to extremely low frequency magnetic fields. *Review of Scientific Instruments.* 67(12):4336–46.

Casadevall A., Fang F.C. (2010) Reproducible science. *Infection and Immunity.* 78(12):4972–5.

Cellini L., Grande R., Di Campli E., Di Bartolomeo S., Di Giulio M., Robuffo I., Trubiani O., Mariggio M.A. (2008) Bacterial response to the exposure of 50 Hz electromagnetic fields. *Bioelectromagnetics.* 29:302–11.

Chalker R.B. (1982) The effect of microwave absorption and associated temperature dynamics on nerve cell activity in Aplysia, M.S. thesis. University of Colorado Boulder.

Chatterjee I., Hassan N., Craviso G.L., Publicover N.G. (2001) Numerical computation of distortions in magnetic fields and induced currents in physiological solutions produced by microscope objectives. *Bioelectromagnetics.* 22(7):463–9.

Chen G., Upham B.L., Sun W., Chang C.C., Rothwell E.J., Chen K.M., Yamasaki H., Trosko J.E. (2000) Effect of electromagnetic field exposure on chemically induced differentiation of friend erythroleukemia cells. *Environmental Health Perspectives.* 108:967–72.

Cooke S., Tyler J.P., Driscoll G. (2002) Objective assessments of temperature maintenance using *in vitro* culture techniques. *Journal of Assisted Reproduction and
Genetics.* 19(8):368–75.

Craviso G.L., Chatterjee I., Publicover N.G. (2003) Catecholamine release from cultured
bovine adrenal medullary chromaffin cells in the presence of 60-Hz magnetic
fields. *Bioelectrochemistry.* 59(1–2):57–64.

Creanga D.E., Poiata A., Morariu V.V., Tupu P. (2004) Zero-magnetic field effect in patho-
gen bacteria. *Journal of Magnetism and Magnetic Materials.* 272–276:2442–4.

Davanipour Z., Sobel E. (2009) Long-term exposure to magnetic fields and the risks of
Alzheimer's disease and breast cancer: Further biological research. *Pathophysiology.*
16(2–3):149–56.

Del Re B., Garoia F., Mesirca P., Agostini C., Bersani F., Giorgi G. (2003) Extremely
low frequency magnetic fields affect transposition activity in *Escherichia coli.*
Radiation and Environmental Biophysics. 42(2):113–8.

Dewey W.C. (2009) Arrhenius relationships from the molecule and cell to the clinic.
International Journal of Hyperthermia. 25(1):3–20.

Dill K.A., Ghosh K., Schmit J.D. (2011) Physical limits of cells and proteomes. *Proceedings
of the National Academy of Sciences.* 108(44):17876–82.

Ehrengruber M.U., Deranleau D.A., Coates T.D. (1996) Shape oscillations of human neutrophil leukocytes: Characterization and relationship to cell motility. *The Journal
of Experimental Biology.* 199(4):741–7.

Einer-Jensen N., Hunter R.H.F. (2005) Counter-current transfer in reproductive biology.
Reproduction. 129(1):9–18.

Eng L.A., Kornegay E.T., Huntington J., Wellman T. (1986) Effects of incubation
temperature and bicarbonate on maturation of pig oocytes *in vitro. Journal of
Reproduction and Fertility.* 76(2):657–62.

Engström S. (2007) Magnetic field effects on free radical reactions in biology. In: Barnes
F., Greenebaum B., editors, *Handbook of Biological Effects of Electromagnetic
Fields: Biological and Medical Aspects of Electromagnetic Fields*, 3rd edition. CRC
Press, Boca Raton, FL.

EPA/DoE. (2008) *Laboratories for the 21st Century: An Introduction to Low-Energy
Design.* U.S. Environmental Protection Agency in partnership with the U.S.
Department of Energy, Washington, DC, 1–12.

Fairchild K.D., Viscardi R.M., Hester L., Singh I.S., Hasday J.D. (2000) Effects of
hypothermia and hyperthermia on cytokine production by cultured human mononuclear phagocytes from adults and newborns. *Journal of Interferon & Cytokine
Research.* 20(12):1049–55.

Falcke M., Malchow D., editors. (2003) *Understanding Calcium Dynamics: Experiments
and Theory.* Springer, Berlin and Heildelberg.

Finlay C.C., Maus S., Beggan C.D., Bondar T.N., Chambodut A., Chernova T.A.,
Chulliat A., Golovkov V.P., Hamilton B., Hamoudi M., Holme R., Hulot G., Kuang
W., Langlais B., Lesur V., Lowes F.J., Lühr H., Macmillan S., Mandea M., McLean
S., Manoj C., Menvielle M., Michaelis I., Olsen N., Rauberg J., Rother M., Sabaka

T.J., Tangborn A., Tøffner Clausen L., Thébault E., Thomson A.W.P., Wardinski I., Wei Z., Zvereva T.I. (2010) International geomagnetic reference field: The eleventh generation. *Geophysical Journal International.* 183:1216–30.

Fitzsimmons R.J., Ryaby J.T., Magee F.P., Baylink D.J. (1994) Combined magnetic fields increased net calcium flux in bone cells. *Calcified Tissue International.* 55(5):376–80.

Foster K.R. (2000) Thermal and nonthermal mechanisms of interaction of radio-frequency energy with biological systems. *IEEE Transactions on Plasma Science.* 28(1):15–23.

Freshney R.I. (2015) *Culture of Animal Cells: A Manual of Basic Technique and Specialized Applications,* 7th edition. Wiley-Blackwell, New York.

Frix W.M., Karady G.G., Venetz B.A. (1994) Comparison of calibration systems for magnetic field measurement equipment. *IEEE Transactions on Power Delivery.* 9(1):100–8.

Fu J.P., Mo W.C., Liu Y., He R.Q. (2016) Decline of cell viability and mitochondrial activity in mouse skeletal muscle cell in a hypomagnetic field. *Bioelectromagnetics.* 37(4):212–22.

Fujiwara M., Takahashi K., Izuno M., Duan Y.R., Kazono M., Kimura F., Noda Y. (2007) Effect of micro-environment maintenance on embryo culture after *in-vitro* fertilization: Comparison of top-load mini incubator and conventional front-load incubator. *Journal of Assisted Reproduction and Genetics.* 24(1):5–9.

Gaetani R., Ledda M., Barile L., Chimenti I., De Carlo F., Forte E., Ionta V., Giuliani L., D'Emilia E., Frati G., Miraldi F., Pozzi D., Messina E., Grimaldi S., Giacomello A., Lisi A., (2009) Differentiation of human adult cardiac stem cells exposed to extremely low-frequency electromagnetic fields. *Cardiovascular Research.* 82:411–20.

Gardner D.K., Weissman A., Howles C.M., Shoham Z., editors. (2012) *Textbook of Assisted Reproductive Techniques, Volume 2: Clinical Perspectives,* 4th edition. CRC Press, Boca Raton, FL.

Gegear R.J., Casselman A., Waddell S., Reppert S.M. (2008) Cryptochrome mediates light-dependent magnetosensitivity in *Drosophila. Nature.* 454(7207):1014–8.

Gillooly J.F., Brown J.H., West G.B., Savage V.M., Charnov E.L. (2001) Effects of size and temperature on metabolic rate. *Science.* 293(5538):2248–51.

Girgert R., Emons G., Hanf V., Gründker C. (2009) Exposure of MCF-7 breast cancer cells to electromagnetic fields up-regulates the plasminogen activator system. *International Journal of Gynecological Cancer.* 19:334–8.

Girgert R., Gründker C., Emons G., Hanf V. (2008) Electromagnetic fields alter the expression of estrogen receptor cofactors in breast cancer cells. *Bioelectromagnetics.* 29:169–76.

Girgert R., Hanf V., Emons G., Gründker C. (2010) Signal transduction of the melatonin receptor MT1 is disrupted in breast cancer cells by electromagnetic fields. *Bioelectromagnetics.* 31:237–45.

Girgert R., Schimming H., Körner W., Gründker C., Hanf V. (2005) Induction of tamoxifen resistance in breast cancer cells by ELF electromagnetic fields. *Biochemical and Biophysical Research Communications.* 336:1144–9.

Golbach L.A., Portelli L.A., Savelkoul H.F., Terwel S.R., Kuster N., de Vries R.B., Verburg-van Kemenade B.L. (2015) Calcium homeostasis and low-frequency magnetic and electric field exposure: A systematic review and meta-analysis of *in vitro* studies. *Environment International.* 92–93:695–706.

Goranov A.I., Cook M., Ricicova M., Ben-Ari G., Gonzalez C., Hansen C., Tyers M., Amon A. (2009) The rate of cell growth is governed by cell cycle stage. *Genes & Development.* 23(12):1408–22.

Greenhill G. (1984) *A Treatise on Hydrostatics.* MacMillan, London, 385–422.

Greenland S., Sheppard A., Kaune W., Poole C., Kelsh M. (2000) A pooled analysis of magnetic fields, wire codes, and childhood leukemia. *Epidemiology.* 11:624–34.

Grellier J., Ravazzani P., Cardis E. (2014) Potential health impacts of residential exposures to extremely low frequency magnetic fields in Europe. *Environment International.* 62:55–63.

Gresits I., Necz P.P., Jánossy G., Thuróczy G. (2015) Extremely low frequency (ELF) stray magnetic fields of laboratory equipment: A possible co-exposure conducting experiments on cell cultures. *Electromagnetic Biology and Medicine.* 34(3):244–50.

Grosberg A.Y. (2003) A few remarks evoked by Binhi and Savin's review on magnetobiology. *Physics-Uspekhi* 46(10):1113–6.

Grzelak A., Rychlik B., Bartosz G. (2001) Light-dependent generation of reactive oxygen species in cell culture media. *Free Radical Biology and Medicine.* 30(12):1418–25.

Hammel H.T., Pierce J.B. (1968) Regulation of internal body temperature. *Annual Review of Physiology.* 30(1):641–710.

Hansson Mild K., Greenebaum B. (2006) Environmental and occupationally encountered electromagnetic fields. In: Barnes F., Greenebaum B., editors, *Handbook of Biological Effects of Electromagnetic Fields: Biological and Medical Aspects of Electromagnetic Fields*, 3rd edition. CRC Press, Boca Raton, FL.

Hansson Mild K., Mattsson M.O., Hardell L. (2003) Magnetic fields in incubators a risk factor in IVF/ICSI fertilization? *Electromagnetic Biology and Medicine.* 22(1):51–3.

Hansson Mild K., Wilén J., Mattsson M.O., Simko M. (2009) Background ELF magnetic fields in incubators: A factor of importance in cell culture work. *Cell Biology International.* 33(7):755–7.

Hart F.X. (1996) Cell culture dosimetry for low-frequency magnetic fields. *Bioelectromagnetics.* 17(1):48–57.

Hart F.X., Laird M., Riding A., Pullar C.E. (2013) Keratinocyte galvanotaxis in combined DC and AC electric fields supports an electromechanical transduction sensing mechanism. *Bioelectromagnetics.* 34(2):85–94.

Hasani M.H., Gharibzadeh S., Farjami Y., Tavakkoli J. (2015) Investigating the effect of thermal stress on nerve action potential using the soliton model. *Ultrasound in Medicine & Biology.* 41(6):1668–80.

Hassan N., Chatterjee I., Publicover N.G., Craviso G.L. (2003) Numerical study of induced current perturbations in the vicinity of excitable cells exposed to extremely low frequency magnetic fields. *Physics in Medicine and Biology.* 48(20):3277.

Heimburg T. (2007) *Thermal Biophysics of Membranes.* Wiley-VCH Verlag, Weinheim. 99–122.

Herzog E.D., Huckfeldt R.M. (2003) Circadian entrainment to temperature, but not light, in the isolated suprachiasmatic nucleus. *Journal of Neurophysiology.* 90(2):763–70.

Higdon H.L., Blackhurst D.W., Boone W.R. (2008) Incubator management in an assisted reproductive technology laboratory. *Fertility and Sterility.* 89(3):703–10.

Honglin Z., Xiufang Y., Li Y., Fenghua L., Zhaodong N., Haitao S. (2001) Study of a biological oscillation system by a microcalorimetric method. *Journal of Thermal Analysis and Calorimetry.* 65(3):755–60.

Hunter R.H.F., Einer-Jensen N. (2005) Pre-ovulatory temperature gradients within mammalian ovaries: a review. *Journal of Animal Physiology and Animal Nutrition.* 89(7-8):240–3.

Hunter R.H.F., Einer-Jensen N., Greve T. (2006) Presence and significance of temperature gradients among different ovarian tissues. *Microscopy Research and Technique.* 69(6):501–7.

Huss A., Spoerri A., Egger M., Kromhout H., Vermeulen R.; For the Swiss National Cohort (2014) Occupational exposure to magnetic fields and electric shocks and risk of ALS: The Swiss National Cohort. *Amyotroph Lateral Scler Frontotemporal Degeneration.* 17:1–6.

Huveneers C., Rogers P.J., Semmens J.M., Beckmann C., Kock A.A., Page B., Goldsworthy S.D. (2013) Effects of an electric field on white sharks: in situ testing of an electric deterrent. *PLoS One.* 8(5):e62730.

IARC Working Group on the Evaluation of Carcinogenic Risks to Humans, World Health Organization, and International Agency for Research on Cancer. (2002) *Non-ionizing radiation, Part I: Static and extremely low-frequency (ELF) electric and magnetic fields*, Volume 80. IARC Press, Lyon, France.

International Commission on Non-Ionizing Radiation Protection (ICNIRP). (2010) Guidelines for limiting exposure to time-varying electric and magnetic fields (1 Hz to 100 kHz). *Health Physics.* 99(6):818–36.

International Electrotechnical Commission (IEC). (1996) Assessment of emission limits for distorting loads in MV and HV power systems. IEC Technical Report IEC 61000-3-6.

Ito A., Aoyama T., Iijima H., Tajino J., Nagai M., Yamaguchi S., Zhang X., Kuroki H. (2015) Culture temperature affects redifferentiation and cartilaginous extracellular matrix formation in dedifferentiated human chondrocytes. *Journal of Orthopaedic Research.* 33(5):633–9.

Ito Y., Kimura T., Ago Y., Nam K., Hiraku K., Miyazaki K., Masuzawa T., Kishida A. (2011) Nano-vibration effect on cell adhesion and its shape. *Bio-medical Materials and Engineering.* 21(3):149–58.

Jayaraman A., Hahn J. (2009) *Methods in Bioengineering: Systems Analysis of Biological Networks.* Artech House, Boston, MA.

Jorjani P., Ozturk S.S. (1999) Effects of cell density and temperature on oxygen consumption rate for different mammalian cell lines. *Biotechnology and Bioengineering.* 64(3):349–56.

Ju J.C., Jiang S., Tseng J.K., Parks J.E., Yang X. (2005) Heat shock reduces developmental competence and alters spindle configuration of bovine oocytes. *Theriogenology.* 64(8):1677–89.

Kajiura S.M., Holland K.N. (2002) Electroreception in juvenile scalloped hammerhead and sandbar sharks. *Journal of Experimental Biology.* 205(23):3609–21.

Kalamida D., Karagounis I.V., Mitrakas A., Kalamida S., Giatromanolaki A., Koukourakis M.I. (2015) Fever-range hyperthermia vs. hypothermia effect on cancer cell viability, proliferation and HSP90 expression. *PLoS One.* 10(1):e0116021.

Kandel E.R., Schwartz J.H., Jessell T.M., editors. (2000) *Principles of Neural Science.* McGraw-Hill, New York.

Kausik A. (2014) Investigation of the biological effects of low-level temperature oscillations and magnetic-field pulses. PhD dissertation, University of Colorado.

Keatinge W.R., Mason A.C., Millard C.E., Newstead C.G. (1986) Effects of fluctuating skin temperature on thermoregulatory responses in man. *The Journal of Physiology.* 378:241.

Kelly G.S. (2006) Body temperature variability (Part 1): A review of the history of body temperature and its variability due to site selection, biological rhythms, fitness, and aging. *Alternative Medicine Review.* 11(4):278–293.

Kelly G.S. (2007) Body temperature variability (Part 2): Masking influences of body temperature variability and a review of body temperature variability in disease. *Alternative Medicine Review.* 12(1):49–62.

König H.L., Krueger A.P., Lang S., Sönning W. (2012) *Biologic Effects of Environmental Electromagnetism.* Springer Science & Business Media, New York.

Koshiya N., Smith J.C. (1999) Neuronal pacemaker for breathing visualized *in vitro. Nature.* 400(6742):360–3.

Kräuchi K., Konieczka K., Roescheisen-Weich C., Gompper B., Hauenstein D., Schoetzau A., Fraenkl S., Flammer J. (2014) Diurnal and menstrual cycles in body temperature are regulated differently: A 28-day ambulatory study in healthy women with thermal discomfort of cold extremities and controls. *Chronobiology International.* 31(1):102–13.

Krider E.P., Roble R.W. (1986) *The Earth's Electrical Environment.* National Academy Press, Washington, DC.

Kuster N., Schönborn F. (2000) Recommended minimal requirements and development guidelines for exposure setups of bio-experiments addressing the health risk concern of wireless communications. *Bioelectromagnetics.* 21(7):508–14.

Lane M., Mitchell M., Cashman K.S., Feil D., Wakefield S., Zander-Fox D.L. (2007) To QC or not to QC: The key to a consistent laboratory? *Reproduction, Fertility and Development.* 20(1):23–32.

Lavy G., Diamond M.P., Pellicer A., Vaughn W.K., Decherney A.H. (1988) The effect of the incubation temperature on the cleavage rate of mouse embryos in vitro. *Journal of Assisted Reproduction and Genetics.* 5(3):167–70.

Lekner J. (2013) Conducting cylinders in an external electric field: Polarizability and field enhancement. *Journal of Electrostatics.* 71(6):1104–10.

Lenz R.W., Ball G.D., Leibfried M.L., Ax R.L., First N.L. (1983) *In vitro* maturation and fertilization of bovine oocytes are temperature-dependent processes. *Biology of Reproduction.* 29(1):173–9.

Liburdy R.P. (1992) Calcium signaling in lymphocytes and ELF fields evidence for an electric field metric and a site of interaction involving the calcium ion channel. *FEBS Letters.* 301(1):53–9.

Liedvogel M., Mouritsen H. (2010) Cryptochromes—a potential magnetoreceptor: what do we know and what do we want to know?. *Journal of the Royal Society Interface.* 7(Suppl 2):S147–S62.

Lin J.C. (2014) Reassessing laboratory results of low-frequency electromagnetic field exposure of cells in culture [telecommunications health and safety]. *IEEE Antennas and Propagation Magazine.* 56(1):227–9.

Luck M.R., Griffiths S., Gregson K., Watson E., Nutley M., Cooper A. (2001) Follicular fluid responds endothermically to aqueous dilution. *Human Reproduction.* 16(12):2508–14.

Lyle D.B., Fuchs T.A., Casamento J.P., Davis C.C., Swicord M.L. (1997) Intracellular calcium signaling by jurkat T-lymphocytes exposed to a 60 Hz magnetic field. *Bioelectromagnetics.* 18(6):439–45.

Malmivuo J., Lekkala J., Kontro P., Suomaa L., Vihinen H. (1987) Improvement of the properties of an eddy current magnetic shield with active compensation. *Journal of Physics E: Scientific Instruments.* 20(2):151.

Malyshkina S., Diedukh N., Vishnyakova I., Poshelok D., Samoilova K. (2016) Structural-metabolic characteristics of bones and cartilage under the influence of hypothermia (the review of literature). *Ortopaedics. Traumatology and Prosthetics.* 1:124–33.

Mangiacasale R., Tritarelli A., Sciamanna I., Cannone M., Lavia P., Barberis M.C., Lorenzini R., Cundari E. (2001) Normal and cancer-prone human cells respond differently to extremely low frequency magnetic fields. *FEBS Letters.* 487:397–403.

Mannerling A., Simko M., Hansson Mild K., Mattsson M. (2010) Effects of 50-Hz magnetic field exposure on superoxide radical anion formation and HSP70 induction in human K562 cells. *Radiation and Environmental Biophysics.* 49:731–41.

Maroto M., Monk N., editors. (2008) *Cellular Oscillatory Mechanisms.* Springer, New York.

Martino C.F. (2011) Static magnetic field sensitivity of endothelial cells. *Bioelectromagnetics.* 32(6):506–8.

Martino C.F., Castello P.R. (2011) Modulation of hydrogen peroxide production in cellular systems by low level magnetic fields. *PLoS One.* 6:e22753.

Martino C.F., Perea H., Hopfner U., Ferguson V. L., Wintermantel E. (2010a). Effects of weak static magnetic fields on endothelial cells. *Bioelectromagnetics.* 31(4):296–301.

Martino C.F., Portelli L., McCabe K., Hernandez M., Barnes F. (2010b) Reduction of the earth's magnetic field inhibits growth rates of model cancer cell lines. *Bioelectromagnetics.* 31(8):649–55.

Mather J., Barnes D., editors. (1998) *Methods in Cell Biology.* Academic Press, San Diego, CA, 1–368.

Mattsson M.O., Simko M. (2014) Grouping of experimental conditions as an approach to evaluate effects of extremely low-frequency magnetic fields on oxidative response in *in vitro* studies. *Front Public Health.* 2:132.

McKiernan S.H., Bavister B.D. (1990) Environmental variables influencing *in vitro* development of hamster 2-cell embryos to the blastocyst stage. *Biology of Reproduction.* 43(3):404–13.

McLeod B.R., Pilla A.A., Sampsel M.W. (1983) Electromagnetic fields induced by Helmholtz aiding coils inside saline-filled boundaries. *Bioelectromagnetics.* 4(4):357–70.

Membré J.M., Leporq B., Vialette M., Mettler E., Perrier L., Thuault D., Zwietering M. (2005) Temperature effect on bacterial growth rate: quantitative microbiology approach including cardinal values and variability estimates to perform growth simulations on/in food. *International Journal of Food Microbiology*. 100(1):179–86.

Mo W., Liu Y., Bartlett P.F., He R. (2014) Transcriptome profile of human neuroblastoma cells in the hypomagnetic field. *Science China Life Sciences*. 57(4):448–61.

Moriyama K., Sato H., Tanaka K., Nakashima Y., Yoshitomi K. (2005) Extremely low frequency magnetic fields originating from equipment used for assisted reproduction, umbilical cord and peripheral blood stem cell transplantation, transfusion and hemodialysis. *Bioelectromagnetics*. 26(1):69–73.

Muehsam D., Ventura C. (2014) Life rhythm as a symphony of oscillatory patterns: Electromagnetic energy and sound vibration modulates gene expression for biological signaling and healing. *Global Advances in Health and Medicine*. 3(2):40–55.

Muehsam D.J., Pilla A.A. (2009a) A Lorentz model for weak magnetic field bioeffects: Part I—Thermal noise is an essential component of AC/DC effects on bound ion trajectory. *Bioelectromagnetics*. 30(6):462–75.

Muehsam D.J., Pilla A.A. (2009b) A Lorentz model for weak magnetic field bioeffects: Part II—Secondary transduction mechanisms and measures of reactivity. *Bioelectromagnetics*. 30(6):476–88.

Müller P., Ahmad M. (2011) Light-activated cryptochrome reacts with molecular oxygen to form a flavin–superoxide radical pair consistent with magnetoreception. *Journal of Biological Chemistry*. 286(24):21033–40.

Müller P., Bouly J.P., Hitomi K., Balland V., Getzoff E.D., Ritz T., Brettel K. (2014) ATP binding turns plant cryptochrome into an efficient natural photoswitch. *Scientific Reports*. 4:5175.

Negovetic S., Samaras T., Kuster N. (2005) EMF health risk research: Lessons learned and recommendations for the future. Workshop in Monte Verita.

Novák B., Tyson J.J. (2008). Design principles of biochemical oscillators. *Nature Reviews Molecular Cell Biology*. 9(12):981–91.

Okano H. (2008) Effects of static magnetic fields in biology: Role of free radicals. *Front Bioscience*. 13(1):610–25.

Paciorek L.J. (1965) Injection locking of oscillators. *Proceedings of the IEEE*. 53(11):1723–7.

Patching J.W., Rose A.H. (1970) The effects and control of temperature. *Methods in Microbiology*. 2:23–38.

Pedersen C., Bräuner E.V., Rod N.H., Albieri V., Andersen C.E., Ulbak K., Hertel O., Johansen C., Schüz J., Raaschou-Nielsen O. (2014) Distance to high-voltage power lines and risk of childhood leukemia: An analysis of confounding by and interaction with other potential risk factors. *PLoS One*. 9(9):e107096.

Petty H.R. (2006) Spatiotemporal chemical dynamics in living cells: from information trafficking to cell physiology. *Biosystems*. 83(2):217–24.

Pickering S.J., Johnson M.H. (1987) The influence of cooling on the oranization of meiotic spindle of the mouse oocyte. *Human Reproduction*. 2(3):207–16.

Poiata A., Creanga D.E., Morariu V.V. (2003) Life in zero magnetic field. V. *E. coli* resistance to antibiotics. *Electromagnetic Biology and Medicine*. 22(2–3):171–82.

Portelli L., Kausik A., Barnes F. (2011) Study of the effects of pulsed temperature stimulus on fibrosarcoma HT1080 cells. The Bioelectromagnetics Society (BEMS) Annual Meeting, Halifax, Canada.

Portelli L., Kausik A., Barnes F. (2012a) Pulsed temperature stimulus limits growth of fibrosarcoma HT1080 cells. The Bioelectromagnetics Society (BEMS) Annual Meeting, Brisbane, Australia.

Portelli L., Kausik A., Semwal H., Barnes F. (2013a) Temperature variations in biological samples is a potential confounder for experimental variability and reproducibility. The Bioelectromagnetics Society (BEMS) Annual Meeting, Thessaloniki, Greece.

Portelli L., Martino C., Kausik A., Barnes F. (2010a) Study of the effects of pulsed temperature stimulus on fibrosarcoma HT1080 cells. The Bioelectromagnetics Society (BEMS) Annual meeting, Seoul, Korea.

Portelli L., Martino C., Whitesell K., Barnes F. (2010b) Study of the effects of static magnetic fields on chemotaxis of fibrosarcoma HT1080 cells. The Bioelectromagnetics Society (BEMS) Annual Meeting, Seoul, Korea and Bioelectrochemistry Gordon Research Conference, Biddeford, ME.

Portelli L.A. (2012) Device for controlling the electric, magnetic and electromagnetic fields in biological incubators. PhD dissertation, University of Colorado.

Portelli L.A., Madapatha D.R., Martino C., Hernandez M., Barnes F.S. (2012b) Reduction of the background magnetic field inhibits ability *Drosophila melanogaster* to survive ionizing radiation. *Bioelectromagnetics.* 33(8):706–9.

Portelli L.A., Schomay T.E., Barnes F.S. (2013b) Inhomogeneous background magnetic field in biological incubators is a potential confounder for experimental variability and reproducibility. *Bioelectromagnetics.* 34(5):337–48.

Prato F.S. (2015) Non-thermal extremely low frequency magnetic field effects on opioid related behaviors: Snails to humans, mechanisms to therapy. *Bioelectromagnetics.* 36(5):333–48. doi:10.1002/bem.21918.

Publicover N.G., Marsh C.G., Vincze C.A., Craviso G.L., Chatterjee I. (1999) Effects of microscope objectives on magnetic field exposures. *Bioelectromagnetics.* 20(6):387–95.

Purrott R.J., Vulpis N., Lloyd D.C. (1981) The influence of incubation temperature on the rate of human lymphocyte proliferation *in vitro. Experientia.* 37(4):407–8.

Qin S., Yin H., Yang C., Dou Y., Liu Z., Zhang P., Yu H., Huang Y., Feng J., Hao J., Hao J., Deng L., Yan X., Dong X., Zhao Z., Jiang T., Wang H.W., Luo S.J., Xie C. (2016) A magnetic protein biocompass. *Nature Materials.* 15(2):217–26.

Rapp P.E. (1979) An atlas of cellular oscillators. *Journal of Experimental Biology.* 81(1):281–306.

Ratkowsky D.A., Lowry R.K., McMeekin T.A., Stokes A.N., Chandler R.E. (1983) Model for bacterial culture growth rate throughout the entire biokinetic temperature range. *Journal of Bacteriology.* 154(3):1222–6.

Ratkowsky D.A., Olley J., McMeekin T.A., Ball A. (1982) Relationship between temperature and growth rate of bacterial cultures. *Journal of Bacteriology.* 149(1):1–5.

Razavi B. (2004) A study of injection locking and pulling in oscillators. *IEEE Journal of Solid-State Circuits.* 39(9):1415–24.

Redmayne M., Johansson O. (2014) Could myelin damage from radiofrequency electromagnetic field exposure help explain the functional impairment electrohypersensitivity? A review of the evidence. *Journal of Toxicology and Environmental Health, Part B.* 17(5):247–58.

Reta-Hernández M., Karady G.G. (1998) Attenuation of low frequency magnetic fields using active shielding. *Electric Power Systems Research.* 45(1):57–63.

Riederer P. (2002) Thermal room modelling adapted to the test of HVAC control systems. Doctoral dissertation, École Nationale Supérieure des Mines de Paris.

Riederer P., Marchio D., Visier J.C., Husaunndee A., Lahrech R. (2002) Room thermal modelling adapted to the test of HVAC control systems. *Building and Environment.* 37(8):777–90.

Robison J.G., Pendleton A.R., Monson K.O., Murray B.K., O'Neill K.L. (2002) Decreased DNA repair rates and protection from heat induced apoptosis mediated by electromagnetic field exposure. *Bioelectromagnetics.* 23:106–12.

Rodgers C.T., Hore P.J. (2009) Chemical magnetoreception in birds: The radical pair mechanism. *Proceedings of the National Academy of Sciences.* 106(2):353–60.

Rodgers C.T., Norman S.A., Henbest K.B., Timmel C.R., Hore P.J. (2007) Determination of radical re-encounter probability distributions from magnetic field effects on reaction yields. *Journal of American Chemical Society.* 129:6746–55.

Rosenspire A.J., Kindzelskii A.L., Petty H.R. (2001) Pulsed DC electric fields couple to natural NAD(P)H oscillations in HT-1080 fibrosarcoma cells. *Journal of Cell Science.* 114(8):1515–20.

Rosenspire A.J., Kindzelskii A.L., Simon B.J., Petty H.R. (2005) Real-time control of neutrophil metabolism by very weak ultra-low frequency pulsed magnetic fields. *Biophysical Journal.* 88(5):3334–47.

Rubin S.A. (1987) Core temperature regulation of heart rate during exercise in humans. *Journal of Applied Physiology.* 62(5):1997–2002.

Saini C., Morf J., Stratmann M., Gos P., Schibler U. (2012) Simulated body temperature rhythms reveal the phase-shifting behavior and plasticity of mammalian circadian oscillators. *Genes & Development.* 26(6):567–80.

Savitz D.A., Wachtel H., Barnes F.A., John E.M., Tvrdik J.G. (1988) Case-control study of childhood cancer and exposure to 60-Hz magnetic fields. *American Journal of Epidemiology.* 128(1):21–38.

Schomay T.E., Portelli L.A., Barnes F.S. (2011) Study of the static and low frequency magnetic fields in incubators. In: 33rd Annual Meeting of the Bioelectromagnetics Society, Halifax, Nova Scotia, Canada.

Schreier N., Huss A., Röösli M. (2006) The prevalence of symptoms attributed to electromagnetic field exposure: A cross-sectional representative survey in Switzerland. *Sozial-Und Preventivmedizin.* 51(4):202–9.

Schuderer J., Kuster N. (2003) Effect of the meniscus at the solid/liquid interface on the SAR distribution in Petri dishes and flasks. *Bioelectromagnetics.* 24(2):103–8.

Schuderer J., Oesch W., Felber N., Spät D., Kuster N. (2004) *In vitro* exposure apparatus for ELF magnetic fields. *Bioelectromagnetics.* 25(8):582–91.

Schüz J., Dasenbrock C., Ravazzani P., Röösli M., Schär P., Bounds P.L., Erdmann F., Borkhardt A., Cobaleda C., Fedrowitz M., Hamnerius Y., Sanchez-Garcia I., Seger R., Schmiegelow K., Ziegelberger G., Capstick M., Manser M., Müller M., Schmid C.D., Schürmann D., Struchen B., Kuster N. (2016) Extremely low-frequency magnetic fields and risk of childhood leukemia: A risk assessment by the ARIMMORA consortium. *Bioelectromagnetics*. 37(3):183–9.

Seeber B., editor. (1998) *Handbook of Applied Superconductivity*. CRC press, Boca Raton, FL.

Sen S., Murray R.M. (2014) Negative feedback facilitates temperature robustness in biomolecular circuit dynamics.

Sheppard A.R., Swicord M.L., Balzano Q. (2008) Quantitative evaluations of mechanisms of radiofrequency interactions with biological molecules and processes. *Health Physics*. 95(4):365–96.

Shi D.S., Avery B., Greve T. (1998) Effects of temperature gradients on *in vitro* maturation of bovine oocytes. *Theriogenology*. 50(4):667–74.

Shikata T., Shiraishi T., Tanaka K., Morishita S., Takeuchi R. (2007) Effects of acceleration amplitude and frequency of mechanical vibration on osteoblast-like cells. In: *ASME 2007 International Mechanical Engineering Congress and Exposition* (pp. 229–36). American Society of Mechanical Engineers, Seattle, WA.

Shiraki K., Sagawa S., Tajima F., Yokota A., Hashimoto M., Brengelmann G.L. (1988) Independence of brain and tympanic temperatures in an unanesthetized human. *Journal of Applied Physiology*. 65(1):482–6.

Simko M. (2007) Cell type specific redox status is responsible for diverse electromagnetic field effects. *Current Medicinal Chemistry*. 14(10):1141–52.

Sládek M., Sumová A. (2013) Entrainment of spontaneously hypertensive rat fibroblasts by temperature cycles. *PLoS One*. 8(10):e77010.

Slusky D.A., Does M., Metayer C., Mezei G., Selvin S., Buffler P.A. (2014) Potential role of selection bias in the association between childhood leukemia and residential magnetic fields exposure: A population-based assessment. *Cancer Epidemiology*. 38(3):307–13.

Solov'yov I.A., Schulten K. (2009) Magnetoreception through cryptochrome may involve superoxide. *Biophysical Journal*. 96(12):4804–13.

Stark J., Chan C., George A.J. (2007). Oscillations in the immune system. *Immunological Reviews*. 216(1):213–31.

Tanaka S.M., Li J., Duncan R.L., Yokota H., Burr D.B., Turner C.H. (2003) Effects of broad frequency vibration on cultured osteoblasts. *Journal of Biomechanics*. 36(1):73–80.

Timmel C.R., Henbest K.B. (2004) A study of spin chemistry in weak magnetic fields. *Philosophical Transactions of the Royal Society of London A: Mathematical. Physical and Engineering Sciences*. 362(1825):2573–89.

Tokalov S.V., Gutzeit H.O. (2004) Weak electromagnetic fields (50 Hz) elicit a stress response in human cells. *Environmental Research*. 94(2):145–51.

Tostevin F., de Ronde W., Ten Wolde P.R. (2012) Reliability of frequency and amplitude decoding in gene regulation. *Physical Review Letters*. 108(10):108104.

Úbeda A., Martínez M.A., Cid M.A., Chacón L., Trillo M.A., Leal J. (2011) Assessment of occupational exposure to extremely low frequency magnetic fields in hospital personnel. *Bioelectromagnetics.* 32(5):378-87.

Ulberg L.C., Burfening P.J. (1967) Embyro death resulting from adverse environment on spermatozoa or ova. *Journal of Animal Science.* 26(3):571-577.

Umurkan N., Koroglu S., Kilic O., Adam A.A. (2010) A neural network based estimation method for magnetic shielding at extremely low frequencies. *Expert Systems with Applications.* 37(4):3195-3201.

Usselman R.J., Chavarriaga C., Castello P.R., Procopio M., Ritz T., Dratz E.A., Singel D.J., Martino C.F. (2016) The quantum biology of reactive oxygen species partitioning impacts cellular bioenergetics. *Scientific Reports.* 6.

Valberg P.A. (1995) Designing EMF experiments: What is required to characterize "exposure"? *Bioelectromagnetics.* 16(6):396-401.

van't Hoff J.H. (1884) *Etudes de Dynamique Chemique.* Muller & Co., Amsterdam.

Vaughan T.E., Weaver J.C. (2005) Molecular change signal-to-noise criteria for interpreting experiments involving exposure of biological systems to weakly interacting electromagnetic fields. *Bioelectromagnetics.* 26(4):305-22.

Vera J., Bachmann J., Pfeifer A.C., Becker V., Hormiga J.A., Darias N.V., Timmer J., Klingmüller U., Wolkenhauer O. (2008) A systems biology approach to analyse amplification in the JAK2-STAT5 signalling pathway. *BMC Systems Biology.* 2(1):38.

Wachtel H. (1985) Synchronization of neural firing patterns by relatively weak ELF fields. In: *Biological Effects and Dosimetry of Static and ELF Electromagnetic Fields.* Springer, New York.

Walker M.W., Butler J.M., Higdon III H.L., Boone W.R. (2013) Temperature variations within and between incubators—a prospective, observational study. *Journal of Assisted Reproduction and Genetics.* 30(12):1583-5.

Walleczek J., Liburdy R.P. (1990) Nonthermal 60 Hz sinusoidal magnetic-field exposure enhances $^{45}Ca^{2+}$ uptake in rat thymocytes: Dependence on mitogen activation. *FEBS Letters.* 271(1-2): 157-60.

Wang W., Litovitz T.A., Penafiel L.M., Meister R. (1993) Determination of the induced ELF electric field distribution in a two layer in vitro system simulating biological cells in nutrient solution. *Bioelectromagnetics.* 14(1):29-39.

Wang W.H., Meng L., Hackett R.J., Oldenbourg R., Keefe D.L. (2002) Rigorous thermal control during intracytoplasmic sperm injection stabilizes the meiotic spindle and improves fertilization and pregnancy rates. *Fertility and Sterility.* 77(6):1274-7.

Wang D.L., Wang X.S., Xiao R., Liu Y., He R.Q. (2008) Tubulin assembly is disordered in a hypogeomagnetic field. *Biochemical and Biophysical Research Communications.* 376(2):363-8.

Weaver J.C., Vaughan T.E., Martin G.T. (1999) Biological effects due to weak electric and magnetic fields: The temperature variation threshold. *Biophysical Journal.* 76(6):3026-30.

Webb P. (1992) Temperatures of skin, subcutaneous tissue, muscle and core in resting men in cold, comfortable and hot conditions. *European Journal of Applied Physiology and Occupational Physiology.* 64(5):471-6.

Wei M., Guizzetti M., Yost M., Costa L.G. (2000) Exposure to 60-Hz magnetic fields and proliferation of human astrocytoma cells *in vitro*. *Toxicology and Applied Pharmacology*. 162(3):166–76.

Wu L.Q., Dickman J.D. (2012) Neural correlates of a magnetic sense. *Science*. 336(6084):1054–7.

Xiao B., Coste B., Mathur J., Patapoutian A. (2011) Temperature-dependent STIM1 activation induces Ca^{2+} influx and modulates gene expression. *Nature Chemical Biology*. 7(6):351–8.

Yang Q.R., Berghe D.V. (1995) Effect of temperature on *in vitro* proliferative activity of human umbilical vein endothelial cells. *Experientia*. 51(2):126–32.

Ye J., Coleman J., Hunter M.G., Craigon J., Campbell K.H., Luck M.R. (2007) Physiological temperature variants and culture media modify meiotic progression and developmental potential of pig oocytes *in vitro*. *Reproduction*. 133(5):877–86.

Yoshida T., Murayama Y., Ito H., Kageyama H., Kondo T. (2009) Nonparametric entrainment of the *in vitro* circadian phosphorylation rhythm of cyanobacterial KaiC by temperature cycle. *Proceedings of the National Academy of Sciences*. 106(5):1648–53.

Zhang J., Richardson I.G., Webb D.F., Gopalswamy N., Huttunen E., Kasper J.C., Nitta N.V., Poomvises W., Thompson B.J., Wu C.-C., Yashiro S., Zhukov A.N. (2007) Solar and interplanetary sources of major geomagnetic storms ($Dst \leq -100\,nT$) during 1996–2005. *Solar and Heliospheric Physics*. 112(A10). doi:10.1029/2007JA012321.

Zmyslony M., Rajkowska E., Mamrot P., Politanski P., Jajte J. (2004) The effect of weak 50 Hz magnetic fields on the number of free oxygen radicals in rat lymphocytes *in vitro*. *Bioelectromagnetics*. 25(8):607–12.

Zwietering M.H., De Koos J.T., Hasenack B.E., De Witt J.C., Van't Riet K. (1991) Modeling of bacterial growth as a function of temperature. *Applied and Environmental Microbiology*. 57(4):1094–101.

Zwietering M.H., De Wit J.C., Cuppers H.G.A.M., Van't Riet K. (1994) Modeling of bacterial growth with shifts in temperature. *Applied and Environmental Microbiology*. 60(1):204–13.

3

Potential Causes for Nonreplication of EMF BioEffect Results, and What to Do about It

Carl F. Blackman
Wake Forest Baptist
Medical Center

3.1 Introduction

Replication of experimental reports is especially important for any new and unexpected findings in the electromagnetic field (EMF) bioeffects research because such study frequently requires the collaborative contributions from scientists from different disciplines, who may be from different groups and locations. In this situation, miscommunication can occur.

I was hired as a radiation biophysicist by the Bureau of Radiological Health to perform laboratory tests to determine whether the radiofrequency (RF) radiation leakage that was allowed to occur from microwave ovens could cause mutations. At that time, it was widely accepted that microwave radiation could heat water, including water in biological materials. Our initial experimental results supported the prevalent opinion regarding biological changes resulting from RF-induced temperature change [1]. However, we also attempted to replicate the millimeter-wave radiation-induced biological effects reported by Webb and Booth [2], but lacking details of their experimental design, our similar but not exact replication failed [1].

Then a report appeared, by Bawin et al. [3], that an RF field, at intensities too low to cause heating in the biological sample, and only at certain amplitude modulation (AM) frequencies, could cause changes in calcium ion release from avian brain tissue exposed in a test tube. Further, they reported that there was a narrow range of radiation intensities that was effective.

This surprising report certainly met the criteria for independent replication to determine whether there were issues, such as specialized equipment or environmental features, which might be unintentionally involved. Three of us visited the Adey laboratory for a day to observe their experimental setups and procedures and to solicit help from them for an attempt to replicate their experiments, which they gave. Upon returning home, a colleague attempted over a 6- to 8-month period to replicate the exact experiment, without success, while I unsuccessfully tried to extend the experiment using mammalian cell culture.

It was then decided that the original experiment needed to be examined in more detail and performed as faithfully as possible. I had a few phone conversations with Dr. Bawin that helped us establish more detailed procedures and conditions, but we were still unable to replicate the findings. In desperation, we examined exposure conditions further, monitoring pH and temperature during treatment, made adjustments in procedures, and just before a scheduled return trip to Adey's lab, we obtained results similar to the reported results. By this time, we had worked for about nine months, using 64 chicks per week for each experiment in this replication attempt. When we arrived at Adey's lab, we had a very productive time. We subsequently published our independent replication of their report [4], and a number of extension studies (some discussed later, and in Blackman [5]). Independent replication experiments are essential to resolve controversial results, but care must be taken to ensure cooperation on both sides even if the results cannot be replicated.

Replication experiments are not easy to perform successfully, especially if all the critical factors necessary for success are not known. In general, other investigators need to contact successful groups for advice. For example, I was approached by a scientist at a meeting where I revealed that the earth's magnetic field was involved in setting the effective frequency parameter in our extremely low frequency (ELF) exposure tests with the chick-brain calcium-efflux system [6,7]. The scientist said he had been using exposures similar to mine, but was testing an immune system endpoint in cell culture, and was getting biological results with respect to frequency and intensity patterns similar to what I was reporting. Then, due to renovations he had to move his incubator to another side of the room, and in the new location the effects had disappeared. On the basis of my presentation, he asked if the earth's magnetic field would be different in another part of his laboratory; I indicated that it was likely in multistory buildings with steel supports because I found that situation in my laboratory. Unfortunately, that knowledge was too late for him as the contract period for his study had ended. The lesson is that he should have called me for advice; most scientists are willing to help if asked.

The following text describes additional issues that can compromise replication attempts that I have become aware of, and highlights a new issue through a recent literature search.

3.2 EMF Exposure Conditions for Receptivity

3.2.1 Intentionally Imposed EMF Exposures

3.2.1.1 Frequency and Intensity

Intentionally imposed EMF, from ELF through the RF region into the gigahertz range, is generally segregated into categories described by primary frequency and intensity ranges.

Since the advent of radar development in the 1940s, corollary studies have been performed to determine whether the various frequencies of EMF can interfere with biological systems, and also if such exposures can provide new biological information particularly in medical imaging. One concern in such studies is to avoid high-exposure intensities that can have known hazardous effects.

Unanticipated biological responses have been reported in the ELF and RF frequency ranges, where regions (termed "windows") of intensity and/or frequency produce biological changes, while in the immediate vicinity above and below those windows no biological changes are observed. In some cases, multiple effect "windows" have been observed over a range of intensities and frequencies [5].

In the RF range, amplitude-modulated at selected ELF, it is possible to observe a series of power-density windows, which shift to different power densities at higher or lower carrier frequencies. A clear example of this phenomenon is the power-density windows for calcium ion release from chick brain tissue, *in vitro*. When sample exposures occurred to 50, 147, or 450 MHz carrier waves, all amplitude-modulated at 16 Hz, the power-density effect-windows were not at the same value. This misalignment was resolved when the intensity metric was converted from free-field power density to electric field value at the tissue surface [8,9].

In the ELF range, a similar biological response has been observed as a function of frequency, producing one or more frequency "windows" separated by frequencies where no biological changes occur. In some cases, these frequency windows are accompanied by intensity windows [10,11].

3.2.1.2 Modulation

Modulation of primary frequencies, such as AM, frequency modulation (FM), and pulsed modulation (PM), using sine or pulsed waves of various shapes to contain information, can also produce biological changes. Further, modulations are used in medical imaging [5]. See the next sections for further details.

3.2.1.3 Orientation between EMF and DC Magnetic Fields

Additional features of the interaction of EMF, from ELF through GHz, with biological systems included the presence and orientation of a static magnetic field with respect to the oscillating EMF fields that can influence biological outcomes. For example, the biological responses could change from enhancement to inhibition in some cases [12].

3.2.1.4 Orientation of EMF Radiation to Biological Samples

In laboratory tests, especially with cell and tissue culture preparations, it is very important to design experiments to optimize the desired exposure conditions, because unintentionally high-intensity induced fields can lead to incorrect exposure interpretations [13].

The underlying lesson in this section, for attempts at replication or extension of published results, is to follow the exposure conditions as closely as possible, and even to contact the original authors to get suggestions, perhaps even unpublished precautionary advice, so as to enhance the probability of success.

3.2.2 Unintentional Exposures

3.2.2.1 Spacing of Multiple Samples

In our early studies using 147-MHz radiation, AM at ELF, to replicate the results of the Bawin et al. [3] study, we noticed but did not study the influence of sample spacing on the width of the power-density response-window [4]. When we studied the influence of 50-MHz radiation, using the same AM ELF frequencies, and also refined our system to examine the spacing of dielectrically equivalent samples around a four brain–sample configuration, we discovered experimentally [14] and theoretically demonstrated [15] that there can be unintentional interactions between closely spaced samples that can produce broadening of the widths of power-density windows as sample spacing narrowed.

3.2.2.2 Ambient Fields

3.2.2.2.1 Natural Fields

Unobserved ambient electric and magnetic fields have been identified as a source of disruption of bird migration [16]. Another source of variation is irregular sunspot activity that can affect the natural geomagnetic field, influencing migration of birds and other animals [17]. Such variations can presumably also influence laboratory experiments.

3.2.2.2.2 Industrial Sources

One source of unintentional exposure can arise from "turnkey" exposure systems delivered by engineering firms to biologists, without adequate quality control instructions to prevent unwanted exposure conditions. In one such case, I was asked to examine such an exposure system because reported biological effects could not be reproduced. The exposure system was set up as per instructions, but no quality control information was included. When an oscilloscope was attached, I discovered that the sine wave was clipped near the peaks, thereby producing a spread of additional frequencies twice each cycle; this unintended "feature" may have contributed to the irreproducibility.

Another source of unintentional exposures can be the ground return current in laboratories that can result in pseudo-DC fields or oscillating fields with harmonics that will not be noticed unless routine monitoring is performed. Further, the incident angle of the ambient DC magnetic field and its flux density in a laboratory are not always constant; the fields need to be measured before and after each exposure. For example, in one 4-week experiment, I measured the static magnetic field at the beginning and the end and found a 10% change. I was very surprised, but thought I must have incorrectly measured the earlier reading because the wire to the Hall effect probe was notorious for altering the DC offset when it was handled. The uncertain DC magnetic field value meant that the 4-week experiment was worthless, but it sensitized me to the need to purchase a fluxgate magnetometer and to always measure the DC magnetic field conditions at the beginning and the end of each day of an experiment.

3.2.2.2.3 Static Magnetic Fields

Several years later, after faithfully measuring the DC magnetic field conditions at the start and the completion of the daily exposure periods, I observed in an incubator a change in the static magnetic field from 360 mG (36.0 µT) to 403 mG (40.3 µT) in a 3-h period, from 11:30 am to 2:30 pm just before the start of an experiment. Postdoctoral fellows, Drs Alejandro Ubeda and Angeles Trillo, had started their 23-h experiments, each in their own incubators, by 11:30 am. At the conclusion of their experiments the next day, Dr. Ubeda's incubator, which was closest to mine, showed a 7-mG (0.7 µT) increase in the ambient static magnetic field, and Dr. Trillo's incubator showed a zero to 1 mG (0.1 µT) increase. The locations of the incubators allowed us to triangulate the source to the electrical circuit box on the wall that was next to my incubator, on the side away from Dr. Ubeda's incubator. We found no changes in the three-story building, such as movement of large steel object or construction that could account for the change, other than the auxiliary air conditioners in the building that were turned on for the season that day. We surmise that the altered static magnetic field was caused by a pseudo-DC ground current return from air conditioners that were operating out of phase so that the oscillating return currents would be randomized—this hypothesis was not tested, but the static field intensities were constant until the air conditioners were turned off.

I subsequently talked to Dr. Robert Liburdy at Lawrence Berkeley Laboratories about this observation. He confirmed the general circumstance. When he was housed in a High Bay physics building, he found it impossible to conduct experiments in the summer time when equipment was energized for use by college faculty, who arrived to run experiments during that period. He reported that this event was correlated with elevation of ambient magnetic fields up to 8 G (0.8 mT) appearing in his laboratory area.

3.2.2.2.4 Incubator-Generated Fields

Other sources of unintentional electric and magnetic fields are associated with the cell culture CO_2 incubators. The first source we discovered was the fan motor that keeps the air circulating. Our original water-jacketed incubator, manufactured by Forma, was a "hand-me-down" that generated up to 3-mG (0.3 μT) AC magnetic fields in locations where cell cultures were placed. When Drs. Ubeda and Trillo decided to come to the laboratory for postdoctoral training, I bought two new Forma incubators for them to use once they had achieved the biological results we had in our original Forma. When the new incubators arrived, I measured approximately 200 mG (20 μT) in the cell culture area. The cause of this high value was the 2-pole motors the manufacturer had installed in the fans, when their supply of 4-pole motors started throwing oil. This high-ambient, 60-Hz field strength was unsuitable for our work, so we replaced the motors with 4-pole motors, thereby resolving that problem.

When Drs. Ubeda and Trillo arrived in 1991, they tried to reproduce some of our EMF effects, e.g., nerve growth factor (NGF)–stimulated neurite outgrowth from PC-12 cells, in the older incubator. After several attempts, they were successful, and so they moved to the new incubators and again were immediately successful. I subsequently began testing the influence of AC/DC magnetic field combinations on neurite outgrowth from NGF-stimulated PC-12 cells, and found interesting periodic responses as a function of intensity. When they tried to reproduce the PC-12 cell response in their newer incubators, the cell response was muted and required higher AC intensities. This response led us to the discovery of another potential source of undesirable EMF exposure. The older incubator had a primitive on/off heater circuit that caused periodic temperature excursions to maintain the set point temperature. The newer incubators had a more sophisticated temperature controller that reduced the current going to the heating coils by chopping the 60-Hz sine wave, apparently designed to diminish temperature ripple inside the incubator. The immediate solution was to use the one older incubator for EMF studies that did not distort the results. To ultimately resolve this undesirable exposure problem, we followed Dr. Liburdy's solution, by ordering the fabrication of Mu-metal boxes to be placed inside all the incubators to shield from the ambient electromagnetic noise [18,19].

I discussed this general issue with Dr. Frank Barnes and Dr. Portelli, who was then his student. They did an extensive review of incubator environment and stability [20] that should be included in resources identifying potential anomalies in cell culture replication experiments.

3.3 Biological Conditions for Receptivity

3.3.1 Genetics

Genetically based sensitivities to various environmental factors and to some medications are well known to medical professionals. The same genetic sensitivities can be found in biological surrogates used to test response to EMF. For example, we used an established cell culture model to test the influence of EMF on the action of NGF to induce neurite outgrowth in PC-12 cells and in a genetic variant, PC-12D cells. When those two cell types were treated with NGF and also to an intensity series of 50-Hz magnetic fields, the PC-12 cells displayed inhibited neurite outgrowth [21], whereas the PC-12D cells responded by showing enhanced neurite outgrowth [22].

Scarfi and colleagues have shown that when lymphocytes from humans with and without a chromosome abnormality (Turner's syndrome) were exposed to 50-Hz pulsed magnetic fields, only those cells from the patients with Turner's syndrome displayed increased micronucleus formation [23].

Animal study results have also produced an "inability-to-replicate" puzzle that was resolved by identifying genetic dissimilarities. A highly visible dispute occurred at the IARC meeting in 2001 convened to decide whether ELF fields in the environment could be implicated as a causal factor in cancer incidence. The dispute resolution was critical because IARC procedures use human data and animal data as the two fundamental pillars upon which to rate the probability that an agent under investigation could cause cancer. The particular dispute paired two research groups who studied the growth of induced mammary tumors in rats exposed to 50/60 Hz EMF; one group in Germany claimed that 50-Hz exposures were implicated in enhanced tumor growth [24,25], whereas the research group in the United States, whose experimental design was to replicate the German finding, reported they could not replicate the results using similar 60-Hz fields [26].

The lack of agreement in what was assumed to be the same animal model and similar exposure conditions was certainly a cause for reviewers at the IARC meeting to put less stress on the animal data when they decided that low-frequency EMF was a possible class 2B carcinogen, rather than at a higher ranking of concern. In a model of appropriate interaction to try to examine the nonreplication effort, the two principal investigators and colleagues published a joint paper identifying the potential differences in the two studies [27]. Subsequent resolution of this disagreement was experimentally provided by Fedrowitz and coworkers to be due to slight genetic differences in the rat model used in Germany compared to that used in the United States [28,29]. Although the rats used in both countries were of the Sprague-Dawley strain, the lack of attention to specific substrain led to the lack of replication. Thus, the slight differences in genetic model, and the strong disagreement it generated at the meeting, likely had ramifications for the conclusions reached by the IARC panel (I was present as a voting member).

Finally, pluripotent embryonic mouse stem cells, with and without an active tumor suppressor (p53) gene, were found to respond differently to a GSM signal when the 217-Hz square-wave AM was present. Only the cell model lacking p53 tumor suppressor activity responded to that EMF exposure by displaying permanent upregulation of mRNA level of the heat-shock protein, hsp-70, and transient upregulation of c-jun, c-myc, and p21 mRNA levels [30].

3.3.2 Exposure Time

Another variable that can lead to nonreplication of results is attention to the duration of exposure of the samples before an effect is observed. At the Department of Energy (DOE) meeting in San Antonio in 1996, another investigator reported that contrary to our earlier reports, he could not observe any changes in the gap junction communication (GJC) in cells different from what we used, but nevertheless were GJC competent, during exposure to ELF EMF fields at normal growth temperature (37°C). Moreover, he had employed a sophisticated, real-time detection system that allowed him to monitor changes continuously, but only for up to 10 min of total exposure time. Immediately following that presentation, we reported (Blackman, 1996, DOE Contractors Review, San Antonio, TX, "Dynamics of gap junction changes in rat liver cells as revealed by magnetic fields") time-dependent changes in GJC measured after ELF EMF exposure at 37°C. Our procedure was to expose a monolayer of rat liver epithelial cells (Clone-9 from ATCC) for different time periods at 37°C, and after each exposure period, we performed a standard "scrape-load" assay at room temperature in a solution that contained Lucifer Yellow (LY) dye; the dye would enter the partially damaged cells along the scrape and move through operational gap junctions into neighboring cells and on into the monolayer. We reported no changes in GJC after 10 min of ELF EMF exposure (in agreement with the previous speaker), but we reported a partial reduction in GJC after 20 min of exposure, and complete, reduced response after 30 min of exposure that remained unchanged through 45 min of exposure. It was apparent that allowing time for a biological response to be manifested is important.

To expand on this result, another postdoctoral fellow in my lab, Dr. Xin Wang, performed a similar gap-junction functional assay using mouse primary hepatocytes. He injected LY dye into a single primary cell in a cluster of cells and counted the percentage of nearest neighbors, surrounding cells receiving the dye after a defined time period. He found no change in dye transfer until EMF exposure times were at least 45 min and longer, with a maximum reduction in dye transfer at 2.5 h of ELF EMF exposure (Wang, 2000, Bioelectromagnetics Society Meeting, Munich, Germany, abstract book page 178, "Time-dependent changes in gap junction function in primary hepatocytes"). Thus, the biological response time must be considered to be a function of the timescales of biological processes in the study population, and any study that employs a different biological system (cells and/or media) in a replication test must take different timescales into consideration.

3.3.3 Recovery Time

The biological system recovery time for GJC following the termination of EMF exposure can also provide interesting data (Blackman, 1996, DOE Contractors Review, San Antonio, TX, "Dynamics of gap junction changes in rat liver cells as revealed by magnetic fields"). Using the same Clone-9 rat liver epithelial cells system (ATCC) described earlier in Section 3.3.2, we also exposed the cell monolayers to ELF EMF for 30 min at 37°C to produce maximal reduction in GJC, and then waited different periods of time before assaying for functional GJC, via the spread of LY dye, to define the kinetics of any cell recovery of the LY dye-transfer response. There were two components in the recovery process: after 5 min of fully suppressed GJC, partial recovery occurred within 5 min, which was unchanged for 20 more min before recovering to original level of dye transfer (full GJC activity) within 5 min. This result was ripe for more investigation of the biochemical mechanism(s) responsible, but we were not able to pursue this interesting observation.

3.4 Blinded Conditions

One frequent concern that is raised when independent replications fail to obtain the original result is to blame the procedures in the original report, and to raise the banner of the need to have blinded exposure conditions in experiments, that is, the person doing the assays is not aware of the treatment condition. This position has validity when subjective judgment of outcomes needs to be made by the personnel and when there are just two results to choose from. However, the blind-assay approach frequently requires more personnel and can also introduce errors in the outcome. In our case, we exposed PC-12 cells to a series of flux densities where the neurite outgrowth responses could change in either direction because of the EMF intensity conditions, which were unknown to the judging personnel; thus, the exposures were essentially analyzed blind. However, Dr. Joseph Kirschvink strongly encouraged us to perform an explicitly blinded experiment in our study of the effects of AC/DC magnetic field conditions on neurite outgrowth from PC-12 cells that we had reported. We decided to accommodate his request by performing the test under double-blinded conditions: First, we had two identical exposure chambers, labeled A and B, and someone outside the lab personnel secretly selected the exposure chamber to energize via a hidden selection box. Second, we had six identical dishes of cells with labels (one through six) written on their removable tops.

At the conclusion of the 23-h exposure, the dishes were removed from the incubators and aligned in numerical order. A second person, not a laboratory worker, made a record of the original dish numbers that were in each incubator, and then, in secret, exchanged some of the dish tops from dishes between incubators, except for controls, according to new codes based on a random number generator, and rearranged those dishes to display the original numerical order before the cells in each dish were evaluated. A third incubator housed four control dishes, two with NGF and two without, to act as positive and negative controls known to the scoring person. Scoring was performed and the results kept secret. This experiment was repeated four times, before any decoding was done. The results of the experiment gave the confirmatory response and were published [18]. This paper has been cited 17 times, but it has had little if any impact on those scientists who doubt that the effect can occur without any discernable impact on doubters. My conclusion is the most scientifically useful way to resolve nonconfirmatory results is similar to the actions we took in our attempt to replicate the publication of Bawin et al. [3], described previously, combined with the suggestion made in Section 3.6.

3.5 Physical and Chemical Conditions for Receptivity

3.5.1 Temperature Conditions

Early in the attempt to replicate the calcium ion efflux results of Bawin et al. [3], there was an understanding that the brain temperature had to be maintained near the *in vivo* temperature. The temperature control we had during the EMF exposure was explicitly shown in Blackman et al. [31]. Subsequently, it was found that a narrow range of temperatures, around 36°C and 37°C, was necessary for EMF to induce calcium efflux change, because the effect did not occur at 35°C or 38°C [5,32].

3.5.2 Ionic Concentrations in Artificial Cerebral Spinal Fluid

We tested the concentrations of calcium and potassium in the synthetic cerebral spinal fluid that Bawin used, but could not observe an enhanced effect, so we decided to use the published ionic concentrations [5].

3.5.3 pH Change/Buffers

We had observed that the pH of the artificial cerebral spinal fluid changed from approximately 7.6 to about 6.6 during the course of the 20-min exposures, under which EMF-induced effects were observed. In an effort to stabilize the pH, 10-mM HEPES or 10-mM Tris was added to the artificial cerebral spinal fluid, but then the EMF-induced calcium efflux could not be detected [9,33]. Bawin, in a personal communication cited in Blackman [33], observed no EMF-induced change in calcium efflux when bicarbonate was replaced with imidazole or MOPS buffers.

3.5.4 Greater Calyx on the Brain Surface

Adey had proposed that the EMF-induced calcium ion release from chick brain tissue was associated with the interaction of calcium with the greater calyx (extracellular matrix) on the surface of the tissue [34]. We decided to test that hypothesis by replacing radioactive calcium ion tracers with one of the three radioactive tracers, tritiated mannitol, tritiated sucrose, or tritiated inulin, because these compounds had been traditionally used to measure the volume of extracellular matrix associated with the cell/tissue surface. Mannitol and sucrose are small molecules that would become entangled in the extracellular matrix, whereas inulin was large enough to be excluded. Our experiments showed that EMF fields could alter the efflux of mannitol and of sucrose, in a manner similar to the efflux of calcium ions, but inulin efflux was not altered by those fields (Blackman, 1987, Bioelectromagnetics Society Meeting, Portland, OR, abstract book page 62, "ELF electromagnetic fields cause enhanced efflux of neutral sugars from brain tissue in vitro"). This result supported some of Adey's speculations about the involvement of the extracellular matrix.

3.5.5 Growth Media and Serum Source

Our PC-12 cell response to EMF was not fully replicated, as reported in a scientific meeting (Bergquist, 1995, Bioelectromagnetics Society Meeting, Boston, MA, abstract book pages 21–22, "ELF magnetic field effects on neurite outgrowth and extracted peptide fragments in PC-12 cells"). The scientists described a partial replication of our AC/DC magnetic field–induced changes in NGF-stimulated PC-12 neurite outgrowth. Although not mentioned in the abstract, my meeting notes indicated that they grew and tested their line of PC-12 cells in a different medium, Iscoves, which contains HEPES buffer.

We reported at a subsequent meeting that when HEPES buffer was added to the medium we used, RPMI-1640, the EMF results were substantially attenuated (Blackman, 1999, Bioelectromagnetics Society, Abstract book page 100, "Magnetic fields as an agent in biological studies: complications and complexities") see Figure 3.1. The authors did not claim this was a replication, but it is apparent that the conditions were significantly different.

In another situation involving research funded by the U.S. DOE, one group led by Liburdy reported that very-low-intensity 60-Hz magnetic fields [12 mG (1.2 µT)] could

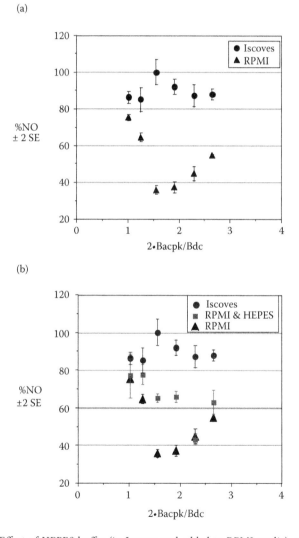

FIGURE 3.1 Effect of HEPES buffer (in Iscoves and added to RPMI media) on 45-Hz EMF-induced changes in percent neurite outgrowth (NO) in PC-12 cells as a function of intensity. (a) Effect of Iscoves versus RPMI medium on NO from PC-12 cells. (b) Direct comparison of media components on NO from PC-12 cells.

reduce the inhibition of growth caused by melatonin and by Tamoxifen on MCF-7 breast cancer cells in culture when compared to control conditions 2 mG (0.2 µT) [35,36]. Several DOE-funded groups were asked to replicate the results. We initially had diffi-culty until we obtained the exact cell type from Liburdy, and then used the same supplier of medium and serum to grow and test the cells. Under these conditions, we replicated their results [19]. Subsequently, Ishido, also using Liburdy's cells, replicated Liburdy's findings [37], and Girgert, using locally available cells, replicated and extended those findings [38–40].

3.5.6 Handling of NGF

We published several reports on EMF alteration of neurite outgrowth (NO) from NGF-stimulated PC-12 cells [41–43], and one group tried to replicate the findings, but could not. We tried to help, but even we did not delve deeply enough into details of handling NGF, when we visited their lab. They finally decided to come to our lab, and in the discussions that followed, as we went through the step-by-step procedures, equipment, and supplies we used; then, a new technician from their lab realized that they were using the Eppendorf microfuge tubes to store their NGF solutions (NGF is known to bind to the walls of those tubes). Apparently, they had unknowingly been using noisy concentration/neurite outgrowth profiles when they tested their primed PC-12 cells for response; this quality control test is a fundamental measurement upon which all the results depend. To confirm this evaluation, they performed in our lab a 23-h, NGF-stimulated PC-12 NO assay using our cells, stock of NGF, supplies, etc. and obtained a perfect relationship between NGF concentration and percentage NO—something they had never seen in their laboratory. They became convinced that this oversight was very likely the problem in their attempted replication. This was an unfortunate outcome, but at least a resolution was reached.

3.5.7 Melatonin Modulation of 50-Hz Action on Neurite Outgrowth in PC-12 Cells

PC-12 cells treated with NGF and assayed for NO after 23 h showed no change under melatonin treatment (Mel) over 10^{-5} to 10^{-12} M. When this procedure included 50-Hz 230 mG (23 µT) magnetic field treatment, which would normally reduce NO, there was an additional reduction of NO at Mel concentrations of 10^{-5} and 10^{-6} M (phar-macological levels), but enhancements at 10^{-7}, 10^{-8}, and 10^{-9} M (physiological levels). See Figure 3.2, redrawn from Figure 2 in [44]. This combination of hormone activity and 50-Hz magnetic field exposure should be studied further for potential health implications.

3.5.8 Primary Hepatocyte Time Delay for Melatonin Enhancement

The melatonin-induced enhancement of intercellular communication in mouse primary hepatocytes under normal incubator conditions was shown to start at 10 h of incubation

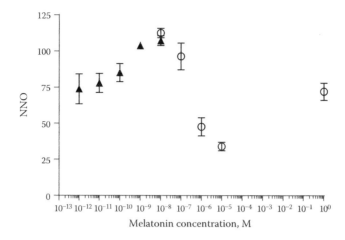

FIGURE 3.2 Normalized neurite outgrowth (NNO) as a function of the molarity of melatonin present. There was no change in neurite outgrowth (NO) when PC-12 cells were stimulated with NGF while simultaneously being treated with a range of melatonin from 10^{-5} to 10^{-12} M; it was a level line (data not shown). Results shown are from that addition of a magnetic field treatment of 50 Hz and 230 mG (23 μT) for the standard 23-h treatment. It reveals a relative reduction in this normalized NO, statistically significant at 10^{-5} and 10^{-6} M, and statistically significant enhancements at 10^{-8} and 10^{-9} M melatonin compared to no melatonin treatment.

and reach maximum response by 12 h [45]. This cell response time to melatonin treatment provides guidance for replication efforts using primary cells in culture.

3.5.9 Polyethylene versus Glass Containers

The composition of the media and serum containers appeared to have unexpected effects on some of our EMF-induced results. This transition from glass to polyethylene (PETE) took us by surprise, compared to other research groups, because we had an abundant supply of medium and serum in glass containers, and thus we did not notice the change to PETE when it occurred in the marketplace. The first clue I had that something might be wrong resulted from an interaction with a scientist who wanted to replicate our EMF/GJC results. I mailed the scientist two T-25 flasks of confluent Clone 9 cells filled with media. A few weeks later I received a call telling me the cells had not responded, and requesting another shipment. I mailed that as well. A month later, I received a call from the scientist indicating that when the package arrived and the cells had settled over night, the scientist withdrew and saved much of the media that had been in the flasks, then propagated the cells, and performed a successful test with that medium. When the scientist continued the cell passaging with local media (same supplier, medium type, catalogue number), the cells again did not respond. Neither of us had a clue about the cause.

About 6–12 months later, I was informed by one of our technical support staff that our Clone 9 cells had entered into a "not responding" phase. I ordered new cells from ATCC, but had to wait because they had received many requests for that cell type (perhaps because we were not the only lab with this problem). I had not paid much attention to the media and serum, relying on my support staff to purchase more supplies when needed. In searching for answers, I wanted to confirm that we had received the correct catalogue numbers. We had received the correct supplies, but they were no longer in 500-mL cylindrical glass bottles; they were in 500-mL rectangular plastic bottles. Thus, it is likely that other investigators, without the large supply of media and serum we initially had, would likely have used media supplied in the plastic containers, and therefore had experienced the same difficulties we were then having replicating our results.

We were never able to restore the original Clone 9 cell response. Subsequent literature searching revealed that plasticizers, with estrogenic properties, are released in small quantities from PETE containers. This situation had an unfortunate outcome: Three years of work testing the action of a series of water disinfection by-products on reduction of GJC, along with the ameliorating role of melatonin and the action of EMF on that melatonin action, was near completion but could not be published.

3.6 Failure of Independent Replication, and Potential Solutions

During the preparation of this manuscript, I located two published papers I had not seen that demonstrate how issues beyond our control can influence events. We had congenial and collaborative interactions with scientists from Oak Ridge National Laboratory. They published two excellent attempts to replicate our work on the influence of EMF exposure on GJC [46,47]. Essentially all of the recommendations for such "replication" interactions were followed: we worked with them as openly as we could by hosting visits to our laboratory to show them our techniques, exposure, and shielding equipment and providing our operating procedure notes, as attested to in their publications. This was all before we became aware of the impact of the replacement of glass with plastic (PETE) bottles for media and serum.

Having just read the two Griffin papers, I now suspect that the failed replication effort by Griffin and coworkers fell victim to the introduction of plastic containers to replace the glass containers for media and serum. This "improvement" by the industry to reduce potential harm from broken glass and to economize on occupied refrigerator space, and perhaps to save money, is the strongest clue to the lack of replication of our EMF reports. The most direct way this minor change, in an otherwise complex experiment, could have been detected is by having Dr. Griffin actually perform the experiment in our laboratory and get the results we reported. Assuming his experiment in our laboratory replicated our results, only that experience would have been enough to have the investigator question everything that might be different in their laboratory compared to ours. Alternatively, if the replication experiment failed, we would have been watching closely and would have possibly identified different procedural actions that might have contributed to the failure. We have confirmed that major manufacturers of the medium and serum for the growth of Clone-9 cells can be obtained on special order in glass bottles rather than in PETE bottles. Thus, it is theoretically possible to test the PETE hypothesis.

3.7 Conclusions

For any attempt to replicate an unusual experimental outcome, it is important to identify the sources of all supplies used, including catalogue numbers, and to describe the exposure and measurement (both electrical and biological) instruments/equipment. Examples given earlier highlight some of the operational and environmental details that need to be controlled/monitored.

Caution must also be exercised when nonreplication occurs. To reiterate, the lessons describing the interactions between Blackman and Bawin in the beginning of this chapter, and the Anderson/Loscher interactions with their joint paper, are prime examples that can be the beginning of the resolution of the conflict. Our interaction with Griffin and colleagues highlights the need to exchange workers.

It is in the best interests of science for each research group that will or has attempted replication of presented or published results to contact and work with those original scientists to identify possible critical variables to be controlled, or to offer possible explanation(s) for any differences in results. If the initial finding is important enough, there should be funding allocated to allow for the performance of the experiment by each group in the other group's laboratory.

Acknowledgments

The revised Figure 3.2. was created with the assistance of a PhD candidate, Hugo Jimenez, whom I interact with in the Cancer Biology Department of Wake Forest Baptist Medical Center.

References

1. C.F. Blackman, S.G. Benane, C.M. Weil, J.S. Ali (1975) Effects of nonionizing electromagnetic radiation on single-cell biologic systems. *Annals of the New York Academy of Sciences* 247: 352–366.
2. S.J. Webb, A.D. Booth (1969) Absorption of microwaves by microorganisms. *Nature* 222(5199): 1199–1200.
3. S.M. Bawin, L.K. Kaczmarek, W.R. Adey (1975) Effects of modulated VHF fields on the central nervous system. *Annals of the New York Academy of Sciences* 247: 74–81.
4. C.F. Blackman, J.A. Elder, C.M. Weil, S.G. Benane, D.C. Eichinger, D.E. House (1979) Induction of calcium ion efflux from brain tissue by radio-frequency radiation: Effects of modulation-frequency and field strength. *Radio Science* 14(6S): 93–98.
5. C.F. Blackman (2015) Replication and extension of Adey group's calcium efflux results. In: M.S. Markov (ed.): *Electromagnetic Fields in Biology and Medicine*. CRC Press, Boca Raton, FL, pp. 7–14.
6. C.F.Blackman, S.G. Benane, D.E. House, J.R. Rabinowitz, W.T. Joines (1984) A role for the magnetic component in the field-induced efflux of calcium ions from brain tissue. *Bioelectromagnetics Society Sixth Annual Meeting*, Atlanta, GA, USA, Abstract Book, pp. 29.

7. Slesin, L. (1984) ELF bioeffects studies at BEMS. *Microwave News* 4(7): 2.
8. W.T. Joines, C.F. Blackman (1980) Power density, field intensity, and carrier frequency determinants of RF-energy-induced calcium-ion efflux from brain tissue. *Bioelectromagnetics* 1(3): 271–275.
9. C.F. Blackman, W.T. Joines, J.A. Elder (1981) Calcium-ion efflux in brain tissue by radiofrequency radiation. In: K.H. Illinger (ed.): *Biological Effects of Nonionizing Radiation*. Washington, DC, American Chemical Society, 157, pp 299–314.
10. C.F. Blackman, S.G. Benane, D.E. House, W.T. Joines (1985) Effects of ELF (1–120 Hz) and modulated (50 Hz) RF fields on the efflux of calcium ions from brain tissue *in vitro*. *Bioelectromagnetics* 6(1): 1–11.
11. C.F. Blackman, S.G. Benane, D.J. Elliott, D.E. House, M.M. Pollock (1988) Influence of electromagnetic fields on the efflux of calcium ions from brain tissue *in vitro*: A three-model analysis consistent with the frequency response up to 510 Hz. *Bioelectromagnetics* 9(3): 215–227.
12. C.F. Blackman, J.P. Blanchard, S.G. Benane, D.E. House (1996) Effect of AC and DC magnetic field orientation on nerve cells. *Biochemical and Biophysical Research Communications* 220(3): 807–811.
13. M. Misakian, A.R. Sheppard, D. Krause, M.E. Frazier, D.L. Miller (1993) Biological, physical, and electrical parameters for in vitro studies with ELF magnetic and electric fields: A primer. *Bioelectromagnetics* Suppl 2: 1–73.
14. C.F. Blackman, S.G. Benane, J.A. Elder, D.E. House, J.A. Lampe, J.M. Faulk (1980) Induction of calcium-ion efflux from brain tissue by radiofrequency radiation: Effect of sample number and modulation frequency on the power-density window. *Bioelectromagnetics* 1(1): 35–43.
15. W.T. Joines, C.F. Blackman, M.A. Hollis (1981) Broadening of the RF power-density window for calcium-ion efflux from brain tissue. *IEEE Transactions on Bio-medical Engineering* 28(8): 568–573.
16. S. Engels, N.L. Schneider, N. Lefeldt, C.M. Hein, M. Zapka, A. Michalik, D. Elbers, A. Kittel, P.J. Hore, H. Mouritsen (2014) Anthropogenic electromagnetic noise disrupts magnetic compass orientation in a migratory bird. *Nature* 509(7500): 353–356.
17. R. Wiltschko, W. Wiltschko (1995) *Magnetic Orientation in Animals*. Springer-Verlag GmbH, Berlin.
18. C.F. Blackman, J.P. Blanchard, S.G. Benane, D.E. House, J.A. Elder (1998) Double blind test of magnetic field effects on neurite outgrowth. *Bioelectromagnetics* 19(4): 204–209.
19. C.F. Blackman, S.G. Benane, D.E. House (2001) The influence of 1.2 microt, 60 Hz magnetic fields on melatonin- and tamoxifen-induced inhibition of MCF-7 cell growth. *Bioelectromagnetics* 22(2): 122–128.
20. L.A. Portelli, T.E. Schomay, F.S. Barnes (2013) Inhomogeneous background magnetic field in biological incubators is a potential confounder for experimental variability and reproducibility. *Bioelectromagnetics* 34(5): 337–348.
21. C.F. Blackman, S.G. Benane, D.E. House (1993) Evidence for direct effect of magnetic fields on neurite outgrowth. *FASEB J* 7(9): 801–806.

22. C.F. Blackman, S.G. Benane, D.E. House, M.M. Pollock (1993) Action of 50 Hz magnetic fields on neurite outgrowth in pheochromocytoma cells. *Bioelectromagnetics* 14(3): 273–286.

23. M.R. Scarfi, F. Prisco, M.B. Lioi, O. Zeni, M. Della Noce, R. Di Pietro, C. Fanceschi, D. Iafusco, M. Motta, B. F. (1997) Cytogenetic effects induced by extremely low frequency pulsed magnetic fields in lymphocytes from turner's syndrome subjects. *Bioelectrochemistry & Bioenergetics* 43: 221–226.

24. S. Thun-Battersby, M. Mevissen, W. Loscher (1999) Exposure of Sprague-Dawley rats to a 50-Hertz, 100-microtesla magnetic field for 27 weeks facilitates mammary tumorigenesis in the 7,12-dimethylbenz[a]-anthracene model of breast cancer. *Cancer Research* 59(15): 3627–3633.

25. W. Loscher, M. Mevissen, W. Lehmacher, A. Stamm (1993) Tumor promotion in a breast cancer model by exposure to a weak alternating magnetic field. *Cancer Letters* 71(1–3): 75–81.

26. L.E. Anderson, G.A. Boorman, J.E. Morris, L.B. Sasser, P.C. Mann, S.L. Grumbein, J.R. Hailey, A. McNally, R.C. Sills, J.K. Haseman (1999) Effect of 13 week magnetic field exposures on DMBA-initiated mammary gland carcinomas in female Sprague-Dawley rats. *Carcinogenesis* 20(8): 1615–1620.

27. L.E. Anderson, J.E. Morris, L.B. Sasser, W. Loscher (2000) Effects of 50- or 60-Hertz, 100 microt magnetic field exposure in the dmba mammary cancer model in Sprague-Dawley rats: Possible explanations for different results from two laboratories. *Environmental Health Perspectives* 108(9): 797–802.

28. M. Fedrowitz, K. Kamino, W. Loscher (2004) Significant differences in the effects of magnetic field exposure on 7,12-dimethylbenz(a)anthracene-induced mammary carcinogenesis in two substrains of Sprague-Dawley rats. *Cancer Research* 64(1): 243–251.

29. M. Fedrowitz, W. Loscher (2005) Power frequency magnetic fields increase cell proliferation in the mammary gland of female fischer 344 rats but not various other rat strains or substrains. *Oncology* 69(6): 486–498.

30. J. Czyz, K. Guan, Q. Zeng, T. Nikolova, A. Meister, F. Schonborn, J. Schuderer, N. Kuster, A.M. Wobus (2004) High frequency electromagnetic fields (GSM signals) affect gene expression levels in tumor suppressor p53-deficient embryonic stem cells. *Bioelectromagnetics* 25(4): 296–307.

31. C.F. Blackman, S.G. Benane, L.S. Kinney, W.T. Joines, D.E. House (1982) Effects of ELF fields on calcium-ion efflux from brain tissue in vitro. *Radiation Research* 92(3): 510–520.

32. C.F. Blackman, S.G. Benane, D.E. House (1991) The influence of temperature during electric- and magnetic-field-induced alteration of calcium-ion release from in vitro brain tissue. *Bioelectromagnetics* 12(3): 173–182.

33. C.F. Blackman (1985) The biological influences of low-frequency sinusoidal electromagnetic signals alone and superimposed on RF carrier waves. In: A. Chiabrera, C. Nicolini, H.P. Schwan (eds.): *Interactions Between Electromagnetic Fields and Cells*. Plenum Publishing Corp, London, pp. 521–535.

34. W.R. Adey (1993) Biological effects of electromagnetic fields. *J Cell Biochem* 51(4): 410–416.

35. R.P. Liburdy, T.R. Sloma, R. Sokolic, P. Yaswen (1993) Elf magnetic fields, breast cancer, and melatonin: 60 Hz fields block melatonin's oncostatic action on ER+ breast cancer cell proliferation. *J Pineal Res* 14(2): 89–97.

36. J.D. Harland, R.P. Liburdy (1997) Environmental magnetic fields inhibit the antiproliferative action of tamoxifen and melatonin in a human breast cancer cell line. *Bioelectromagnetics* 18(8): 555–562.

37. M. Ishido, H. Nitta, M. Kabuto (2001) Magnetic fields (MF) of 50 Hz at 1.2 microt as well as 100 microt cause uncoupling of inhibitory pathways of adenylyl cyclase mediated by melatonin 1a receptor in MF-sensitive MCF-7 cells. *Carcinogenesis* 22(7): 1043–1048.

38. R. Girgert, H. Schimming, W. Körner, C. Gründker, V. Hanf (2005) Induction of tamoxifen resistance in breast cancer cells by ELF electromagnetic fields. *Biochemical and Biophysical Research Communications* 336(4): 1144–1149.

39. R. Girgert, C. Grundker, G. Emons, V. Hanf (2008) Electromagnetic fields alter the expression of estrogen receptor cofactors in breast cancer cells. *Bioelectromagnetics* 29(3): 169–176.

40. R. Girgert, V. Hanf, G. Emons, C. Grundker (2010) Signal transduction of the melatonin receptor MT1 is disrupted in breast cancer cells by electromagnetic fields. *Bioelectromagnetics* 31(3): 237–245.

41. C.F. Blackman, J.P. Blanchard, S.G. Benane, D.E. House (1994) Empirical test of an ion parametric resonance model for magnetic field interactions with PC-12 cells. *Bioelectromagnetics* 15(3): 239–260.

42. C.F. Blackman, J.P. Blanchard, S.G. Benane, D.E. House (1995) The ion parametric resonance model predicts magnetic field parameters that affect nerve cells. *Faseb J* 9(7): 547–551.

43. M.A. Trillo, A. Ubeda, J.P. Blanchard, D.E. House, C.F. Blackman (1996) Magnetic fields at resonant conditions for the hydrogen ion affect neurite outgrowth in PC-12 cells: A test of the ion parametric resonance model. *Bioelectromagnetics* 17(1): 10–20.

44. C.F. Blackman, S.G. Benane, D.E. House (1997) Action of melatonin on magnetic field inhibition of nerve-growth-factor-induced neurite outgrowth in PC-12 cells. In: R.G. Stevens, B.W. Wilson, L.E. Anderson (eds.): *The Melatonin Hypothesis: Breast Cancer and the Use of Electric Power.* Battelle Press, Columbus, OH, pp. 173–185.

45. C.F. Blackman, P.W. Andrews, A. Ubeda, X. Wang, D.E. House, M.A. Trillo, M.E. Pimentel (2001) Physiological levels of melatonin enhance gap junction communication in primary cultures of mouse hepatocytes. *Cell Biology and Toxicology* 17(1): 1–9.

46. G.D. Griffin, M.W. Williams, P.C. Gailey (2000) Cellular communication in clone 9 cells exposed to magnetic fields. *Radiation Research* 153(5 Pt 2): 690–698.

47. G.D. Griffin, W. Khalaf, K.E. Hayden, E.J. Miller, V.R. Dowray, A.L. Creekmore, C.W. Carruthers, Jr., M.W. Williams, P.C. Gailey (2000) Power frequency magnetic field exposure and gap junctional communication in clone 9 cells. *Bioelectrochemistry* 51(2): 117–123.

4

Physical Dynamics: The Base for the Development of Biophysical Treatments

Ruggero Cadossi
and Stefania Setti
IGEA

Matteo Cadossi
University of Bologna

Leo Massari
University of Ferrara

4.1 Introduction

Dosimetry in bioelectromagnetics often refers to the means and procedures of calculating the amount of energy transferred to cell, tissue, and body. Thus, biological effects are considered to depend exclusively, or at least largely, on this parameter. Increasing energy levels may lead to tissue heating, cell death, denaturation of tissue macromolecules, and ultimately, death of the exposed subject. In pharmacological terms, increasing doses may be described as toxic and poisoning. High-energy effects are nonspecific; they completely hide those effects that can trace back to the specific interactions with cell structures and depend on the physical parameters of the biophysical stimuli applied. Actually, only for physical agents with low-energy content is it possible to establish a specific relationship among the values of the parameters of the physical agent and the resulting biological response. As for pharmacological research for bioelectromagnetics too, dose–response effects are of paramount importance for the development of clinical applications.

The therapeutic options that are available for the physician include the use of both chemical and physical agents. The use of chemicals (drugs) and high-energy physical agents for disease treatment has been well defined and structured in the different branches of medicine and surgery. Whereas the application and clinical indications for physical agents with low-energy content are not well defined.

The property of biological systems to absorb energy has drawn attention to the dosage (amount of energy absorbed) as a fundamental parameter on which "all" the effects depend, although the different biological targets may evince specific susceptibilities and sensitivities. Widely alien to this approach was the principle of specificity of the signal, i.e., the possibility that the biological effects could be dependent, not solely, and not so much, on the total energy introduced into the system as by other characteristics of the physical agent.

4.2 Biophysical Stimulation

Biophysics is an interdisciplinary scientific area that:

a. Employs methods and theories belonging to physics to study biological systems;
b. Investigates how nonionizing physical stimuli (electrical, magnetic, mechanical) interact with biological systems, modifying their behaviour: "biophysical stimulation." The biological effects of biophysical stimulation are related to the specific physical parameters employed: frequency, amplitude, waveform of the signal, duration of exposure (Figure 4.1).

The capability of modifying the activity of a biological target through exposure to a physical agent, independently of its energy content, is a recent acquisition. Several

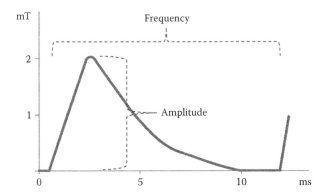

FIGURE 4.1 Waveform of the electric field induced in a coil probe by PEMFs used for bone growth stimulation.

reasons underlie the delay in recognition of "signal-specific effects." The complexity of studying the interaction among physical agents and biological systems has rendered the work of researchers especially hard. Above all, when very low energies are used, it may be difficult not only to control the physical amounts in action, but also to verify the biological effects; the latter often appearing elusive and, in any case, not comparable with the effects observed with chemical agents. Only now sufficient knowledge has been acquired so that physics can assist significantly in the development of biology and, ultimately, open new windows for the clinical employment of physical energy.

"Biophysical stimulation" techniques can be used alone in clinical practice to increase and enhance reparative and anabolic activities of tissue, or they can be used in association with drugs, growth factors, to potentiate their activity and reduce side effects. These characteristics make biophysical stimulation an effective "new pharmacology."

4.3 Clinical Biophysics

Here, we focus on the area of medicine dealing with physical agents with low-energy content, which we define as "clinical biophysics." Clinical biophysics is that branch of medical science that studies the action process and the effects of nonionizing physical agents utilized for therapeutic purposes. Only during the last century was it clear that it is possible to selectively modulate biological processes.

Clinical biophysics applies the fundamental principles of pharmacological research and adopts its methods. In particular, clinical biophysics studies the specificity of the physical agents, investigates their mechanisms of action and the metabolic pathways involved, and assesses their employment in relevant pathological conditions.

Clinical biophysics is based on a new pharmacology that employs physical stimuli for treatment of various human pathologies. Likewise, for drugs, the effects will be specific to function rather than being cell or tissue specific. This will foresee its use in all conditions that are positively influenced by the activation or modulation of a specific cell function. According to the new pharmacology, the effects of physical agents must be described in terms of modulation of a cell function upon which its clinical employment will be based. The cell membrane has been identified as the target and site of interaction of biophysical stimulation. Through the cell membrane, the physical signal triggers a series of intracellular events that result in the biological response; the mechanisms of transduction through the cell membrane differ based on the energy involved—whether it is mechanical, electrical, or electromagnetic. A recent review has evidenced how different biophysical stimulation methods are able to activate membrane receptors and transmembrane channels, suggesting that a transitory increase in cytoplasmic calcium and activation of calmodulin are common to all these physical forces (Brighton et al., 2001) (Figure 4.2).

No direct intracellular effects have been observed. Unlike for drugs, the effects of biophysical stimulation are local and limited to the site of application, and no systemic effects have been observed following exposure to low-energy physical signals.

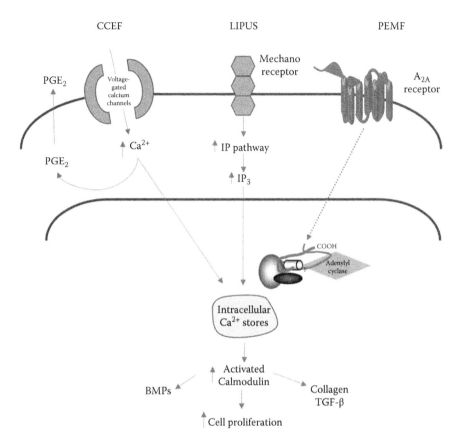

FIGURE 4.2 Cell membrane targets of biophysical stimuli. Abbreviation: coupled capacitive electrical field, CCEF; low-intensity pulsed ultrasound, LIPUS; pulsed electromagnetic fields, PEMF.

The principles on which clinical biophysics is based are *recognition* and *specificity* of the energy applied:

As to *recognition*, we mean the capacity of the biological target to sense the presence of the physical agent: this aspect becomes more and more important with the lowering of the energy applied;

As to *specificity*, we mean the capacity of the physical signal applied to the biological target to trigger a specific biological response that usually depends on its characteristics: waveform, frequency, duty cycle, amplitude.

Pharmacodynamics is fundamental for drug research and therapy development; similarly, clinical use of low-energy signals should be based on "physical dynamics": investigation of dose–response curves for the physical parameters characterizing the signals. Physical dynamics studies will identify the "effective dose" to be adopted in clinical practice by selection of the most effective physical signal and length of exposure (Varani et al., 2002) (Figure 4.3).

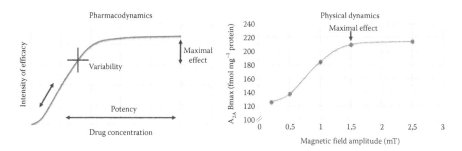

FIGURE 4.3 Dose–response. Pharmacodynamics (left): Biological response depends on drug concentration. Physical dynamics (right): Biological response depends on pulsed electromagnetic field amplitude.

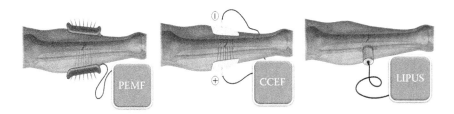

FIGURE 4.4 Biophysical stimulation treatments for bone healing. Left: pulsed electromagnetic fields (PEMF); center: capacitive coupled electric fields (CCEF); right: low-intensity pulsed ultra-sounds (LIPUS).

During the last century, physical energy has been introduced, particularly in the form of electric currents, electromagnetic fields, capacitive coupled electric fields, and low-intensity pulsed ultrasound in orthopedic practice. During the past two decades, the focus has been on the employment of physical energy to promote or enhance healing of nonunion and recent fractures, and its clinical use is now based on sound scientific evidence. The methods of biophysical stimulation used in clinical practice are: inductive (pulsed electromagnetic fields, PEMF), capacitive (capacitive coupled electrical field, CCEF), or mechanical (low-intensity pulsed ultrasound, LIPUS) (Figure 4.4).

Electromagnetic energy can be applied to the human body by direct contact through electrodes or by irradiation. Otherwise, the biological target can be exposed to the electromagnetic energy through electric or magnetic field generators placed in proximity of the body. Mechanical energy is applied directly, subjecting the tissue to forces of traction, compression, and rotation. The tissue can undergo mechanical vibration by irradiation, e.g., by means of ultrasound or infrasound.

4.4 Clinical Biophysics in Orthopedics

The community of orthopedic surgeons has certainly played a central role in the development and understanding of the importance of the physical stimuli to control biological activities and their clinical relevance. Today, several clinicians are interested in utilizing nonchemical means to treat different pathologies (Di Lazzaro et al., 2013). The development of new clinical treatments with physical agents having low-energy represents an important opportunity.

4.5 Physical Dynamics and Pulsed Electromagnetic Fields

The most commonly used method for biophysical stimulation is the inductive one, affording a great number of clinical experiences based on numerous solid *in vitro* and *in vivo* experimental researches. PEMFs have been widely studied and have led to a series of significant observations on their biological effect, especially on bone and cartilage cells.

Studies of transduction of the signal focus on two considerations: activation of transmembrane Ca^{++} channels and interactions with membrane receptors (Brighton et al., 2001; Varani et al., 2002).

Receptors have exhibited contradictory responses to the interaction of the electromagnetic fields with various transmembrane channels (Lacy-Hulbert et al., 1998). Dosage studies on static and low-intensity fields have not demonstrated any effect on ATP-sensitive potassium transmembrane channels (Wang and Hladky, 1994). Although the existence of transmembrane calcium flows has been verified, it remains uncertain whether the transmembrane channels are, or are not, the transduction mechanisms of the signal.

We reported that exposure to PEMFs favors the response of human lymphocytes to the proliferative stimulus induced by PHA (phytohemoagglutinin) and the release of IL-2 growth factor. The response of PHA-stimulated lymphocytes to electromagnetic fields was shown to be dose dependent both on peak value of the magnetic field and on pulse frequency (Figure 4.5) (Conti et al., 1985).

It has been hypothesized that the effects of electrical and electromagnetic fields may be mediated by interaction with the parathyroid hormone (PTH) receptors, suggesting that the effects of pulsed electromagnetic fields occur through conformational changes in the transmembrane portion of the receptor (Luben et al., 1982; Cain et al., 1987).

Recently, detailed research in physical dynamics has demonstrated the existence of interactions between PEMFs and adenosine receptors, more specifically the A_{2A} and the A_3 (Vincenzi et al., 2013). Notably, A_1 and A_{2B} receptors are not influenced by the same exposure conditions. Experiments in saturation kinetics have revealed a significant augment in the density of adenosine A_{2A} receptors in human neutrophils treated with PEMFs of increasing amplitude.

In vitro, Chang et al. have shown that different intensities of pulsed electromagnetic fields, 4.8 and 13.3 µV/cm, regulate osteoclastogenesis, bone reabsorption, osteoprotegerin (OPG), the receptor activator of the nuclear kB factor ligand (RANKL), and the concentration of the macrophage colony stimulating factor (M-CSF) differently (Chang et al., 2005). Sakai et al. have demonstrated the different effects on extracellular matrix synthesis using three different wave forms of PEMFs; waveforms substantially differ in the time–amplitude domain, which are reflected in the distribution of power in the

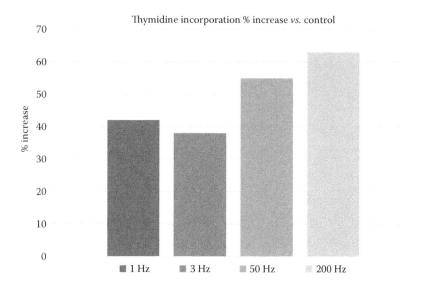

FIGURE 4.5 Signal frequency-dependent thymidine incorporation in PHA-stimulated lymphocytes.

frequency–amplitude domain (Sakai et al., 2006). Borsje et al. demonstrated different effects of PEMFs on the expression of OPG and RANKL in cultures of human osteoblasts when evaluated at different exposure times, 10 min or 3 h (Borsje et al., 2010).

Cartilage tissue proved to be highly sensitive to PEMFs, and it has been shown how specific parameters of physical signals are able to modify the cartilage metabolism. In a dose–response study of explanted bovine joint cartilage, De Mattei et al. showed that the signal frequency does not significantly affect proteoglycan synthesis, whereas response depends on exposure time (at least 4 h) and amplitude of impulse (1.5 mT) (De Mattei et al., 2007).

In vitro, the effect of PEMFs has therefore been well characterized in cartilage explants. Furthermore, explants whose culture medium included proinflammatory cytokines faced inhibition of proteoglycan synthesis, which was effectively countered by exposure to PEMFs. In the presence of IL-1b and TGFb3 a recovery of proteoglycan (PG) synthesis, PG content and aggrecan and type II collagen mRNA expression in the PEMF-exposed compared to unexposed pellets was observed (Ongaro et al. 2012).

Preclinical research has shown that the *in vitro* effects translate *in vivo*; significant increase in the velocity of bone deposition occurs such that, in a model of reparative osteogenesis in horses, the growth of bone trabeculae accelerates from 1.75 to 3.05 μ/day when the lesion undergoes exposure to PEMFs (Canè et al., 1993). The effect observed was dependent on the duration of daily exposure (Figure 4.6).

In vivo, Pienkowski et al., in an experimental model of peroneal osteotomy of rabbits, showed that the intensity and a portion of the sharp impulse of an asymmetric form of the PEMF are the main components of the signal for obtaining osteogenetic activity (Pienkowski et al., 1992). Midura exposed fibular osteotomies in rats to two PEMFs

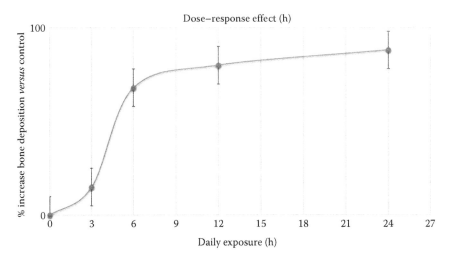

FIGURE 4.6 Daily length of exposure to PEMF on bone deposition in horses.

employing different pulse parameters but the same exposure time, 3 h/day (Midura et al., 2005). The beneficial effect of the stimulation was observed for one of the two PEMFs only. The authors concluded that there is a specific relationship between waveform characteristics and biological outcomes.

In vivo, the first clear demonstration of the ability of PEMFs to limit the progress of arthritic degeneration of cartilage was provided by Ciombor et al., in Dunkin Hartley guinea pigs. The histologic score showed that the joint cartilage of the medial tibial plateau was well preserved in the animals exposed to PEMFs, and the immunohistochemical study showed that the chondrocytes in the stimulated animals were positive for TGF-β, whereas the positivity in the control animals was greatest for IL-1β (Ciombor et al., 2003). The chondroprotective effect was dependent on the exposure length (Fini et al., 2005) and on frequencies (Veronesi et al., 2014) (Figure 4.7).

The effects observed on joint cartilage, subchondral bone, and synovia indicate a role for physical stimuli, and in particular for electromagnetic fields, on various tissues that make up the functional unit (the joint), and hence the treatment can be collocated within the framework of articular biophysics.

In clinical experience, PEMFs enhance fracture healing, limit complications, and reduce pain, ultimately leading to a better quality of life; PEMF stimulation represents an important adjunct to surgery, resulting, in the long term, in substantial cost-savings attributed to earlier return to work of the treated patients. Faldini et al., 2010 conducted a randomized double blind study in patients suffering from femoral neck fracture. Patients were instructed to use PEMF 6 h/day for 90 days. Patients' compliance was monitored. In the group of patients with active devices, they observed clinical and radiographic higher healing rates among the compliant patients as against the noncompliant (8.3 ± 2.0 vs. 2.6 ± 1.7 average h/day). No difference was found between compliant and noncompliant patients of the placebo group (6.9 ± 1.5 vs. 2.5 ± 1.6 average h/day) (Figure 4.8).

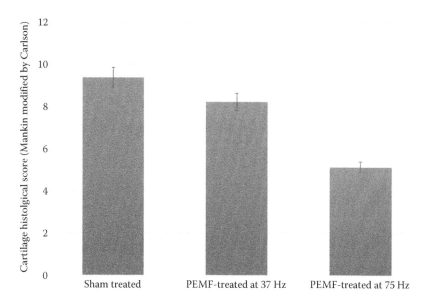

FIGURE 4.7 Histological score of cartilage knee of guinea pigs exposed to PEMFs of different frequencies. Lower scores are related to healthy cartilage.

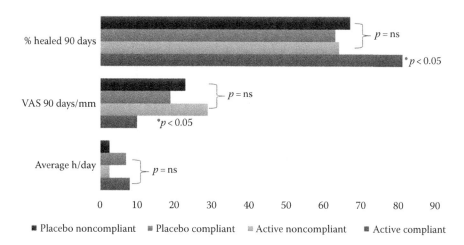

FIGURE 4.8 Dose–response effects are observed in the active group and not in the placebo group.

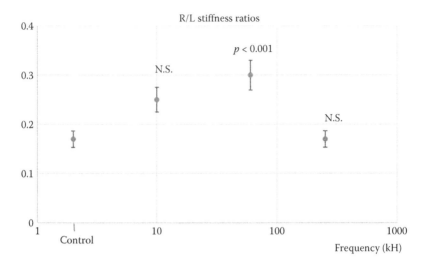

FIGURE 4.9   Frequency-dependent osteogenetic effect of CCEF on fracture healing in rabbits.

4.6 Physical Dynamics and Capacitive Coupled Electric Fields

Exposure to an electrical fields produces an increase in the proliferation of osteogenetic cells and an increase in the extracellular matrix. The interaction of the electrical field occurs at the level of the cell membrane: the increase in intracellular calcium result-ing from opening the voltage-dependent calcium channels activates calmodulin that, in turn, through gene transcription and cell proliferation, determines production of BMPs, TGF-β, and various matrix proteins, among which are collagen, BMP-2, BMP-4, BMP-6, BMP-7, TGF-β1, FGF-2, and VEGF (Aaron et al., 2004). Dose–response curves show that waveform, amplitude, frequency, and treatment time can induce different biological effects on bone tissue (Wiesmann et al., 2001).

In vivo, stimulation with capacitive coupling electric fields has shown effectiveness in accelerating repair of bone defects in animal models. Brighton et al. demonstrated the ability of this method to stimulate and shorten the consolidation time of experi-mental fractures of rabbit fibula, evidencing the most effective combination of physical amounts of the signal (60 kHz–0.33 V/cm–33 μA/cm) (Brighton et al., 1985) (Figure 4.9).

4.7 Physical Dynamics, Mechanical Stimulation, and Low-Intensity Pulsed Ultrasound

In vitro, Leung et al. studied the effect of LIPUS on cell cultures of human periostal cells, along with the effects in relation to time and dosage. Cell cultures were treated with ultrasound for 5, 10, and 20 min. Total number of live cells, cell proliferation, alka-line phosphatase activity, osteocalcin secretion, and expression of vascular endothelial growth factor were evaluated. The authors demonstrated a clear dose-dependent effect, the greatest efficacy being recorded at 20 min exposure (Leung et al., 2004).

In vivo, Chan et al., in an experimental model of rabbit distraction osteogenesis demonstrated that 20-min treatment with LIPUS accelerates the formation of the bone callus in a dose-dependent manner (Chan et al., 2006); repeated treatments (two sessions of 20 min/day) turned out less effective.

The formation of endochondral bone and, in particular, the healing of a fracture, can be modified by exposure to mechanical stresses directly applied to the fractured bone. The speed and quality of the formation of the extracellular matrix can be regulated by application of mechanical stresses of various amplitudes and frequencies. The extent of the deformation, the frequency, and the size of the fracture gap are of importance for determining the healing process. An interfragmentary cyclical movement of short duration, applied at high frequency, produces a greater quantity of periostal callus than that produced by application at low frequency.

Bone response to mechanical stimulation varies with both the extent of deformation and the number of cycles. Typically, as the load applied or the intensity of deformation diminishes, the number of cycles required for activation and formation of the bone tissue increases. For example, Claes et al., in a model of sheep, have shown that the amplitudes of the deformations regulate the synthesis of molecules of the extracellular matrix in the bone, the cartilage, and the fibrous tissue: this indicates the regulatory capacity of the mechanical forces with reference to the dosage, and also the effects of the mechanical factors on the clinical outcomes of the fracture healing (Claes et al., 1998).

4.8 Conclusions

The last century witnessed the introduction of physical energy to favor bone healing.

Orthopedic surgeons with their researches played a central role in the development and understanding of the importance of physical stimulation to control repair processes in bone and cartilage. They deserve recognition for having been the first to evidence the clinical importance of clinical biophysics Today, the encounter between physics and biology and between physics and clinical practice is a topic of paramount importance, and it may play a significant role well beyond orthopedics.

Currently, several other areas are addressing, or at any rate evincing interest in, the possibilities of employing nonchemical means for intervention on various pathologies. Some applications are still at the initial stages or involved in preliminary *in vitro* investigation; however, it can safely be predicted that these treatment possibilities will find increasing use—for example, in neurology and angiology (Capone et al., 2009; Clausen, 2012).

As compared with pharmacological treatment, biophysical therapy enjoys the undoubted advantage of being relatively easy to administer locally, and hence to achieve maximum "concentration" and effectiveness while precluding systemic negative side effects. Biophysical therapy would seem to be indicated for prolonged treatment in the presence of chronic degenerative phenomena, but it would not be suitable for use in systemic diseases. One important feature of treatment with physical agents is transferability of the effects observed in preclinical research to clinical practice. This high efficiency can be appreciated by considering that biophysical therapy is confined to local treatments, and the effective doses can be easily achieved through appropriate focus. Physical signals are not altered by the surrounding tissue, nor, as it happens in the case of drugs, are modified and metabolized by the body. Unlike for drugs whose effectiveness, based on stoichiometric reactions, lessens as concentration diminishes, the activity of the physical stimuli can be compared to that of an endogenous modulator in which the effects remain constant over time in the absence of negative side effects. Therefore, PEMFs represent a physiological alternative to adenosine agonists; they can mediate tissue-specific effects without desensitization and downregulation. These considerations prove that PEMF may represent an important example of "soft pharmacology" (Borea et al., 2015).

As the clinical employment of physical agents develops, the ability to recognize and define the area of clinical biophysics marks a moment of fundamental and necessary synthesis for creation of a common meeting place where researchers and clinicians in different areas can recognize each other.

The biophysical treatment can be an important integration to pharmacology to potentiate locally the effects of drugs; electroporation techniques for electrochemotherapy and electrogenetherapy are part of the clinical routine in skin cancer treatment.

Today, several new areas of medicine are interested in the possibility of utilizing nonchemical means of treating different pathologies. Some applications are simply at their beginning or still at the preliminary stage of research; however, every preliminary result provides the basis for new therapeutic possibilities (ClinicalTrials. gov NCT02767778).

As for drugs, for biophysical stimulation, too, the knowledge of the mechanism of action should always underpin the rationale for clinical use, with the aim to design clinical trials and relative endpoints in a coherent way. Indeed, it may be affirmed that if the study of physical stimuli for therapeutic purposes had adopted the research protocols of clinical pharmacology and borrowed from its methodology, this would undoubtedly have benefited the entire sector and would rapidly have led to an adequate scientific structuring. Physical dynamics is the fundamental step to be adopted.

References

Aaron RK, Boyan BD, Ciombor DM, Schwartz Z, Simon BJ. Stimulation of growth factor synthesis by electric and electromagnetic fields. *Clin Orthop Relat Res.* 2004;419:30–7.

Borea PA, Varani K, Vincenzi F, Baraldi PG, Tabrizi MA, Merighi S, Gessi S. The A3 adenosine receptor: History and perspectives. *Pharmacol Rev.* 2015;67(1):74–102.

Borsje MA, Ren Y, de Haan-Visser HW, Kuijer R. Comparison of low-intensity pulsed ultrasound and pulsed electromagnetic field treatments on OPG and RANKL expression in human osteoblast-like cells. *Angle Orthod.* 2010;80:498–503.

Brighton CT, Hozack WJ, Brager MD, Windsor RE, Pollack SR, Vreslovic EJ, Kotwick JE. Fracture healing in the rabbit fibula when subjected to various capacitively coupled electrical fields. *J Orthop Res.* 1985;3(3):331–40.

Brighton CT, Wang W, Seldes R, Zhang G, Pollack SR. Signal transduction in electrically stimulated bone cells. *J Bone Joint Surg Am.* 2001;83(10):1514–23.

Cain CD, Adey WR, Luben RA. Evidence that pulsed electromagnetic fields inhibit coupling of adenylate cyclase by parathyroid hormone in bone cells. *J Bone Miner Res.* 1987;2(5):437–41.

Canè V, Botti P, Soana S. Pulsed magnetic fields improve osteoblast activity during the repair of an experimental osseous defect. *J Orthop Res.* 1993;11:664–70.

Capone F, Dileone M, Profice P, Pilato F, Musumeci G, Minicuci G, Ranieri F, Cadossi R, Setti S, Tonali PA, Di Lazzaro V. Does exposure to extremely low frequency magnetic fields produce functional changes in human brain? *J Neural Transm.* 2009;116(3):257–65.

Chan CW, Qin L, Lee KM, Zhang M, Cheng JC, Leung KS. Low intensity pulsed ultrasound accelerated bone remodeling during consolidation stage of distraction osteogenesis. *J Orthop Res.* 2006;24(2):263–70.

Chang K, Chang WH, Huang S, Huang S, Shih C. Pulsed electromagnetic fields stimulation affects osteoclast formation by modulation of osteoprotegerin, RANK ligand and macrophage colony-stimulating factor. *J Orthop Res.* 2005;23:1308–14.

Ciombor DM, Aaron RK, Wang S, Simon B. Modification of osteoarthritis by pulsed electromagnetic field: A morphological study. *Osteoarthritis Cartilage.* 2003;11:455–62.

Claes LE, Heigele CA, Neidlinger-Wilke C, Kaspar D, Seidl W, Margevicius KJ, Augat P. Effects of mechanical factors on the fracture healing process. *Clin Orthop Relat Res.* 1998;(355 Suppl.):S132–47.

Clausen F. Exploring a new approach to treating brain injury: Anti-inflammatory effect of pulsed electromagnetic fields. *Neurosci Lett.* 2012;519(1):1–3.

Conti P, Gigante GE, Cifone MG, Alesse E, Reale M, Angeletti PU. Effects induced "in vitro" by extremely low frequency electromagnetic fields (E.L.F.) on blastogenesis of human lymphocytes and on thromboxane B-2-release by ionophore-stimulated neutrophils. In: Grandolfo M, Michaelson SM, Rindi A, editors. *Biological Effects and Dosimetry of Static and ELF Electromagnetic Fields.* 1985; Plenum Press, New York, pp. 335–8.

De Mattei M, Fini M, Setti S, Ongaro A, Gemmati D, Stabellini G, Pellati A, Caruso A. Proteoglycan synthesis in bovine articular cartilage explants exposed to different low-frequency low-energy pulsed electromagnetic fields. *Osteoarthritis Cartilage.* 2007;15(2):163–8.

Di Lazzaro V, Capone F, Apollonio F, Borea PA, Cadossi R, Fassina L, Grassi C, Liberti M, Paffi A, Parazzini M, Varani K, Ravazzani P. A consensus panel review of central nervous system effects of the exposure to low-intensity extremely low-frequency magnetic fields. *Brain Stimul.* 2013;6(4):469–76.

Faldini C, Cadossi M, Luciani D, Betti E, Chiarello E, Giannini S. Electromagnetic bone growth stimulation in patients with femoral neck fractures treated with screws: Prospective randomized double-blind study. *Curr Orthop Pract.* 2010;21(3):282–7.

Fini M, Giavaresi G, Torricelli P, Cavani F, Setti S, Cane V, Giardino R. Pulsed electromagnetic fields reduce knee osteoarthritic lesion progression in the aged Dunkin Hartley guinea pig. *J Orthop Res.* 2005;23(4):899–908.

Lacy-Hulbert A, Metcalfe JC, Hesketh R. Biological responses to electromagnetic fields. *FASEB J.* 1998;12(6):395–420.

Leung KS, Cheung WH, Zhang C, Lee KM, Lo HK. Low intensity pulsed ultrasound stimulates osteogenic activity of human periosteal cells. *Clin Orthop Relat Res.* 2004;(418):253–9.

Luben RA, Cain CD, Chen MC, Rosen DM, Adey WR. Effects of electromagnetic stimuli on bone and bone cells *in vitro* inhibition of responses to parathyroid hormone by low-energy, low-frequency fields. *Proc Natl Acad Sci USA.* 1982;79(13):4180–4.

Midura RJ, Ibiwoye MO, Powell KA, Sakai Y, Doehring T, Grabiner MD, Patterson TE, Zborowski M, Wolfman A. Pulsed electromagnetic field treatments enhance the healing of fibular osteotomies. *J Orthop Res.* 2005;23(5):1035–46.

Ongaro A, Varani K, Masieri FF, Pellati A, Massari L, Cadossi R, Vincenzi F, Borea PA, Fini M, Caruso A, De Mattei M. Electromagnetic fields (EMFs) and adenosine receptors modulate prostaglandin E(2) and cytokine release in human osteoarthritic synovial fibroblasts. *J Cell Physiol.* 2012;227(6):2461–9.

Pienkowski D, Pollack SR, Brighton CT, Griffith NJ. Comparison of asymmetrical and symmetrical pulse waveforms in electromagnetic stimulation. *J Orthop Res.* 1992;10(2):247–55.

Sakai Y, Patterson TE, Ibiwoye MO, Midura RJ, Zborowski M, Grabiner MD, Wolfman A. Exposure of mouse preosteoblasts to pulsed electromagnetic fields reduces the amount of mature, type I collagen in the extracellular matrix. *J Orthop Res.* 2006;24:242–53.

Varani K, Gessi S, Merighi S, Iannotta V, Cattabriga E, Spisani S, Cadossi R, Borea PA. Effect of low frequency electromagnetic fields on A2A adenosine receptors in human neutrophils. *Br J Pharmacol.* 2002;136(1):57–66.

Veronesi F, Torricelli P, Giavaresi G, Sartori M, Cavani F, Setti S, Cadossi M, Ongaro A, Fini M. In vivo effect of two different pulsed electromagnetic field frequencies on osteoarthritis. *J Orthop Res.* 2014;32(5):677–85.

Vincenzi F, Targa M, Corciulo C, Gessi S, Merighi S, Setti S, Cadossi R, Goldring MB, Borea PA, Varani K. Pulsed electromagnetic fields increased the anti-inflammatory effect of A_2A and A_3 adenosine receptors in human T/C-28a2 chondrocytes and hFOB 1.19 osteoblasts. *PLoS One.* 2013;8(5):e65561.

Wang KW, Hladky SB. An upper limit for the effect of low frequency magnetic fields on ATP-sensitive potassium channels. *Biochim Biophys Acta.* 1994;1195:218–222.

Wiesmann H, Hartig M, Stratmann U, Meyer U, Joos U. Electrical stimulation influences mineral formation of osteoblast-like cells in vitro. *Biochim Biophys Acta.* 2001;1538(1):28–37.

5

Dose and Exposure in Bioelectromagnetics

Kjell Hansson Mild
Umeå University

Mats-Olof Mattsson
*Austrian Institute of
Technology GmbH*

5.1　Introduction

A great difficulty in determining possible health risks associated with exposure to electromagnetic fields (EMF) is our lack of knowledge of what constitutes "dose" in this context. Present exposure guidelines (ICNIRP, 1998, 2010) are focused on eliminating acute thermal effects from high-frequency (radiofrequency) fields and neuroexcitation from low-frequency fields. The radiofrequency exposure is referred to as "specific absorption rate," SAR (W/kg), which denotes how much energy per time and mass unit is absorbed by the tissue, where SAR is proportional to the square of the electric field (E^2). For low-frequency fields, induced current density (A/m^2) is used, which is directly proportional to the E-field in the tissue. Both these variables are exposure measures that give a real-time image of how the electric field is distributed within the tissue. However, it may be too simplistic to equal the dose taken up by a biological structure (cell, tissue, organ, organism) with the strength of the electric field. Our argument is that other factors may need to be considered as well to end up with a measure of dose that is biologically relevant. So far, this discussion has not really been pursued, and there is no agreed upon unit describing dose for exposure to EMF.

In contrast, the respective dose concepts are well established in other areas. Thus, dose in chemical toxicology and pharmacology refers to the amount of a chemical substance that is given to an animal [including humans; e.g., see Klaasen (2001) for a detailed discussion]. It is typically given as mass units per animal (or per bodyweight, or per surface area), such as x mg/g bodyweight.

Research in the field of bioelectromagnetics, especially for radiofrequency electromagnetic fields, has adopted some words from ionizing radiation and also from chemical toxicology: dose, dosage, and dosimetry. In ionizing radiation, dose in its original sense is the measurement of the absorbed dose of ionizing radiation, measured in Grays, which also can be expressed as J/kg. Dosage is the rate at which the dose is absorbed, given as Gray/s. If we convert the Joule to its equivalent Ws, we are left with the unit W/kg for dose rate. This unit is what has been called SAR (specific absorbtion rate) in nonionizing radiation, where, by convention, it refers to the anatomical distribution of the SAR, or for low-frequency fields the distribution of the current density.

That the concepts *exposure* and *dose* are not clearly defined in EMF research is due to several factors, primarily because the interaction mechanism(s) are not well understood, especially concerning weak fields and nonthermal effects. Presumably, we also need to look at different meanings of the dose concept with regard to different symptoms and diseases and also to different organ/tissue sensitivities. Probably, for acute thermal effects we will need one way of expressing the *dose*, in cancer studies another meaning might be given to *dose*, and for nervous system effects still another. For mobile phone studies, we need to be more detailed with the use of SARs and not just use the highest value found anywhere near the phone without paying attention to the anatomical localization. We need more information from the medical/biological side as to what sites are of interest for which symptom and/or disease. To some extent, such considerations are made regarding doses for ionizing radiation. In that area, different types of doses are used that involve weighting of the effects, for example, by tissue or organ sensitivity (Wrixon, 2008).

For extremely low frequency (ELF) fields, several possible exposure characteristics have been discussed as relevant for dose description (Morgan and Nair, 1992; Valberg et al., 1995). These include, for instance, intensity (strength) given as *rms* or peak value; duration of exposure; single versus repetitive exposures or intermittent exposure; transients; frequency content of the signal; field orientation; combination of ELF and static field. However, in many of the epidemiological studies, particularly regarding EMF in the low-frequency range, due to the lack of a generally accepted interaction mechanism, the exposure assessment has been based on time-averaged field strengths and similar parameters.

In an overview on 21st-century magnetotherapy, Markov (2015) discusses the problem of "dose" when applying different forms of magnetic fields for therapy. He states that the question of "dosage" here is much more complex than the dosage in the case of prescribing pharmaceuticals. This is mainly due to the large variety of electromagnetic signals available. He lists the following parameters needed for characterization of device and study/clinical trial: type of field, frequency, pulse shape, intensity or induction, gradient (dB/dt), vector (dB/dx), component (E or H), time of exposure (duration of session).

Most epidemiological studies of health effects have, as a measure of dose, used various forms of mean values of exposure, such as the time average of the magnetic field during a working day or a 24-h day. Others have used the highest value found in a day as a measure. Only a few studies have attempted to create a dose in the form of an integration of exposure over time, such as in the form of microtesla-hours. We do not know, however, if this is the correct dimension; is an exposure to $20\,\mu T$ for $0.5\,h$ equivalent to $0.5\,\mu T$ for $20\,h$? Is the effect linearly dependent on the "dose" or is it as in many biological contexts, a nonlinear relationship? Are various frequencies—basic tones and harmonics—to be included when assessing the weighted average of a dose? Before these fundamental questions of what constitutes "dose" in these contexts are better understood, our risk assessment will be more or less arbitrary when it comes to long-term effects.

Depending on the type of disease that is studied, the exposure assessment in the epidemiological studies needs to be very different. For instance, for effects depending on only short-term exposures, effects of more acute character, the field strengths or the power density might be the useful measures to obtain. This can be exemplified by skin rashes and video display terminal (VDT) work, subjective symptoms and mobile phone use, and early miscarriages (Sandström et al., 1995, 2001; Wilén et al., 2003), which will all be discussed later. When it comes to diseases with long latencies like cancer and Alzheimer's, it becomes much more difficult, because then it is the exposure a number of years prior to the onset of disease that is of interest. This may not be easy or at all possible to measure today with reasonable accuracy due to the many unknown factors. In particular, questions about how the exposure is accumulated over years—i.e., is μT-years a reasonable effect measure?—need to be answered before the ultimate exposure assessment can be made. When calculating accumulated exposure over the years, another important question is if there is a threshold under which no effect occurs, i.e., how low should values be counted? For comparison, in ionizing radiation one is distinguishing between stochastic (linear nonthreshold) and deterministic effects (i.e., with thresholds; Wrixon, 2008).

The question of intermittent exposure also leaves many unanswered aspects, like the spacing of the repeated exposure, and if there is an effect of that exposure, what is the biological reset time, i.e., when is the system fully recuperated? There are indications that repeated exposure in the minute-to-hour scale can be much more efficient than continuous exposure (e.g., Nordenson et al., 1994; Rannug et al., 1994), but much remains to be investigated before this is properly taken into account in epidemiological studies.

The aim of this chapter is to address the question of whether we have a basis for defining a dose concept for exposure to electromagnetic fields. We will give some examples of studies (epidemiological, *in vivo*, *in vitro*) where attempts have been made to measure exposure or estimate a dose. Furthermore, we assess if these studies provide support for one or more dose concepts, or if specific research can close crucial knowledge gaps. Our starting point is that we currently have more questions about what constitutes *dose* than we have answers, and it is imperative to define what a biologically relevant dose is.

Critical questions that we have identified and that we search answers for in the considered studies include:

- For how long did the exposure last, and how often was it repeated?
- Is there a connection between high and low exposure levels, respectively, and biological and/or health-relevant effects? Is there any monotonous increase in response to increasing exposure levels, such as a classic dose–response relationship?

- What kind of data are there regarding short-term exposure and biological effects? What is the connection between long-lasting exposure and effects? Is there any data on intermittent exposure?
- Is it reasonable to expect a specific biological response after a certain exposure time?
- Is there any biological "reset time" among reported effects; how much time must pass before the investigated end point is back to initial levels?

Importantly, we have focused on studies that are dealing with exposure levels that are below the ones that are known to cause tissue heating or tissue excitation.

5.2 ELF Epidemiological Studies

In the first study of childhood leukemia and residential magnetic fields, Wertheimer and Leeper (1979) introduced the concept "wiring code" as a proxy for the magnetic field exposure. The homes of the cases and controls were classified according to the wiring configuration outside their homes, ranging from the so-called "high wire configuration" to "low wire configuration," and later modifications of the scheme even introduced a lower class in the form of buried wires. The argument was that the higher the classification, the higher the magnetic field that was encountered in the dwelling. Many studies have been devoted to the "wire code" and how well it mimics the magnetic field, and we refer the reader to Kaune et al. (2001) and IARC (2002) for further details on this.

When Feychting and Ahlbom (1993) performed their large epidemiological study on Swedish children living near high-voltage power lines, they introduced calculated values for the magnetic field. By using the distance between the house and the power line and the knowledge of the historical annual average load on the line, the annual average flux density was calculated. Then the average was taken over the years in that residency. These time-weighted-average (TWA) values were then divided into different categories, for instance above or below $0.2\,\mu T$. This "magic" number was introduced by Tomenius (1986) in one of the first epidemiological studies to follow after the Wertheimer and Leeper study. He measured the field strengths and came up with a dividing point set at $0.2\,\mu T$. Ever since, that particular number has been used as a cutoff point in many epidemiological studies, and in the minds of many lay people this now denotes the upper limit for "allowed" magnetic field exposure.

To date, many studies have been published on residential exposure and cancer outcome, both for children and adults. Measurements have been made in the form of both spot measurements as well as long-term measurements from 24 to 48 h. Personal measurements using logging devices have also been tried, and the discussion is still ongoing on how to deal with the exposure from different electrical appliances. However, there is still no general consensus about what measure to use that best describes the exposure situation and thus the possible health hazard associated with the magnetic field exposure. For an overview of the epidemiological studies, we refer the reader to Ahlbom et al. (2000) for a pooled analysis of some of the studies and also to the study by the IARC working group (2002), which evaluated the cancer risk and also discussed exposure assessment for a low-frequency magnetic field.

5.2.1 Occupational Exposure

In many occupations, quite strong magnetic fields can be encountered, for instance, in electrical welding, train operation, induction heating, and in other electrosteel work settings. Wertheimer and Leeper (1979) also looked at some "electrical occupations" in their childhood leukemia study and found an indication that there was an increased risk of cancer for those in these occupations. This first "occupational" study has been followed by many more from all over the world. The exposure assessment in these studies has varied from the very crude estimate of just the job title to sophisticated measurements with logging instruments capable of following the exposure over one or more working days.

On the basis of these measurements, a job-exposure matrix (JEM) was often created. Floderus et al. (1996) measured the ELF exposure of 1000 Swedish men with an EMDEX recorder over one working day. Each job was then classified according to the TWA over the working day. They had up to four people in each job category. This JEM has been used in several studies and has been modified as more data have been entered (see for instance Håkansson et al., 2002). The method is usually combined with an expert panel of occupational hygienists who have good knowledge of the specific working situations and can therefore give a good estimate of the exposure value from the JEM to be used in the study at hand. See for instance Flynn et al. (1991) for a further discussion on this.

Several studies have been undertaken on workers in the electric utility industry. To describe the exposure, an expert group has classified the exposure in the various occupational tasks and given a value to each task. Then, the number of years of work in each task was added up to give a measure of the total exposure in μT-years, and this measure has in turn been used in the risk evaluation; see Kheifets et al. (2008).

In a study of Swedish engine drivers, the exposure to magnetic field was measured with EMDEX logging instruments adapted to the low frequency used in trains (Nordenson et al., 2001). The peak values during a working shift could be over a 100 μT, and the daily average was in the order of 2–15 μT depending on the type of engine they were driving. This once again illustrates the problem we are facing in these type of studies when we are lacking the interaction mechanism(s). Should we go for the maximum encountered flux density, or is the daily TWA value of interest? Or should we rather use instruments that record the time derivative of the flux density and thereby more closely give an estimate of induced current? Until we have a better knowledge of the way low-intensity magnetic field interacts with the biological organism, we have to make do with these uncertainties.

5.2.2 Combined Residential and Occupational Exposure

Most studies of residential exposure to magnetic fields and the risk of cancer have only assessed the residential exposure. This may be adequate for small children who spend most of their time at home, but as soon as they enter day care or school, the exposure there must also be included in the total assessment. In studies of adults, both Feychting et al. (1997) and Tynes and Haldorsen (2003) have tried to include both the residential and the occupational exposure. However, the exposure assessment is still rather coarse in that the occupational exposure is based on the JEM and a panel of experts' judgment of the level and duration of the exposure to magnetic field. However, this is still our best estimate in lack of a better "dose" understanding.

The residential exposure in the Tynes and Haldorsen (2003) study was assessed by calculation since the cohort in the study were people living near high-voltage power lines. By using the exact position of the house (GIS), and the historical annual average load on the line, the annual average flux density was calculated. Then this average was multiplied with the number of years in that residency. The same procedure was then applied to the occupational exposure, i.e., first an estimate of the yearly exposure and then an average over the number of years in the occupation. The authors cumulated exposure categories and multiplied by number of years in that occupation, and then the TWA was calculated.

5.3 ELF Experimental Studies

The reported effects due to ELF MF exposure from both *in vitro* and *in vivo* studies include virtually all kinds of end points. Studies carried out at the cellular level have found effects on DNA structure, gene expression, cell growth and survival, cellular metabolism, motility, protein functions, etc.. Studies on intact animals have included *inter alia* end points related to reproduction and development, endocrinology and metabolism, immune system functions, nervous system functions, and tumour formation. It has to be noted, however, that despite the considerable number of studies with observed effects, studies not finding any effects are also very common. Unlike the situation in, e.g., chemical toxicology, in the case of observed effects, many studies have not addressed the issue of dose–response relationships. There is also a clear lack of systematic approaches to other important characteristics such as exposure duration, intermittency, etc. that are mentioned in other sections of this chapter.

Mattsson and Simkó (2012, 2014) investigated the experimental support for an ELF effect on radical homeostasis relevant for neurodegenerative diseases (NDD) (Mattsson and Simkó, 2012; both *in vivo* and *in vitro* studies) and a grouping effort to find out which exposure characteristic is most important for effects on ROS formation *in vitro* (Mattsson and Simkó, 2014). The earlier of the studies focused on studies employing ELF-MF exposures to see if there is mechanistic support for any causal connection between NDD and MF exposure. The hypothesis was that ELF–MF exposure can promote inflammation processes and thus influence the progression of NDD. A firm conclusion regarding this hypothesis was difficult to draw based on available data since there is a lack of specific experimental studies. Furthermore, the heterogeneity of the performed studies regarding, e.g., exposure duration, flux density, biological end point, and the cell type and time point of the investigation, is substantial and makes conclusions difficult to draw. The second of the studies (Mattsson and Simkó, 2014) was a systematic grouping approach where available *in vitro* studies were grouped according to cell type, flux density, exposure duration, modulation, and study quality. Some of the results are relevant for the dose discussion. Thus, it was found that exposures at or above 1 mT in most cases generated significant increases in ROS-formation-related end points, whereas pulse form or modulations as well as exposure duration had no significant influence on the outcome.

One of the groups that contributed to solving the problem of "dose" was Ted Litovitz's Bioelectromagnetics group at Catholic University of America (CUA), Washington, D.C. Their work has been devoted to investigating the physical parameters that are needed to obtain an effect of EMF exposure on biological systems and also on how to inhibit that effect. Their work on bioeffects caused by low-level EMF show a dependence on coherence time, constancy, and spatial averaging, and also how the effects can be modified by an applied ELF noise magnetic field. The group has been using early chick embryos as well as L929 and Daudi cells as their main experimental systems. For further details, see a review by Hansson Mild and Mattsson (2010).

In conclusion, there are very few systematic approaches to address what characterizes and constitutes dose in ELF experimental studies. The occupational studies may have included different flux density levels and/or exposure durations, but the current foundations for making general conclusions regarding dose are weak.

5.4 RF Epidemiological Studies

5.4.1 RF Occupational Workers

In a study of RF sealer workers, Kolmodin-Hedman et al. (1988) measured the E and H fields at various workplaces. Since no near-field dosimetry was available at the time, and the electromagnetic fields varied widely both in strengths and in combination of E and H fields, it was impossible to combine the measured values into a single dose-value for each individual. The best classification of exposure that could be obtained was a categorization of the workers according to the various RF sealers they worked with: tarpaulin machines, machines for the ready-made clothing industry, and automatic machines. In the first category, the workers are prone to whole-body exposure with the highest field on the trunk of the body, and since the workers usually operate the machine in a standing position with little or no separation from the factory floor, which can be considered to be RF ground, the induced body current can be rather substantial [see study by Williams and Hansson Mild (1991)]. Working with ready-made machines, the highest exposure usually occurs on the hands, but both the head and the trunk of the body can also get substantial exposure. However, the operator is usually sitting down when operating the machine and the body current is therefore much lower than that in the previous case. The third category, the automatic machines, often gives rise to very low field strengths at the operator's position, which is normally located 1–3 m away from the electrodes. Both sitting and standing positions are used when working with these machines. A significantly impaired two-point discrimination on the fingertip of the index finger and in the palm of the hand was found in the exposed operators of ready-made machines as compared to a control group consisting of sewing machine operators. The total group of welders reported a higher frequency (19%) of neurasthenic symptoms in comparison with the control group (9%). The consumption of headache pills was also much higher in the exposed group, on average 1 pill per week versus 0.2 in the control group.

In a study of RF sealer by Wilén et al. (2004), measurements of E and H fields, contact currents, and induced currents were taken. It was not possible to combine all of these measurements into a single value for each worker, and the old classification used above, i.e. by type of machine used, seemed to be the best categorization variable to use. A large part of the RF sealers—almost 25%—exceeded the ICNIRP guideline reference values (ICNIRP, 1998), and thus this occupational group can be considered as highly exposed, in principle, regardless of which of the earlier measured variables were used.

Tynes et al. (1996) studied breast cancer among female telegraph operators on ships. They measured the RF fields in the radio room on some of the vessels. An excess risk was seen for breast cancer (SIR = 1.5) but not for all cancer (SIR = 1.2). However, they never used the measured values in the evaluation of the outcome; the measurements were merely used as a confirmation that there was RF exposure in the radio room, and the values found were not particularly high and were within present standards.

Bortkiewicz et al. (1995, 1997) studied workers occupationally exposed to RF in broadcast stations and in radio services of mobile radio communication networks. The exposure was assessed by measurements of the electric field and the dose was given as maximum E field encountered during the day or as a daily dose given by Vh/m. In the highest exposed group—ARE station workers—the abnormalities in ECG recordings (resting and/or 24-h) were significantly higher than among both Radio Services and the control group Radio Link Stations. The authors concluded by saying that at the present stage it is impossible to determine which of the parameters (frequency, max E field, daily dose, etc.) may be responsible for the observed cardiovascular impairment.

5.4.2 Radio Communication Systems

Hocking et al. (1996) studied cancer incidence and mortality and the proximity to TV towers. On the basis of calculations, they estimated the power density in the exposed areas to be $8\,\mu W/cm^2$ near towers, $0.2\,\mu W/cm^2$ at a radius of 4 km, and $0.02\,\mu W/cm^2$ at 12 km. Leukemia incidence and mortality were significantly increased in the inner area (near the tower). Again no *dose* measure was created, but merely an *exposure* measure.

Some recent studies have addressed the question of health complaints and living near mobile phone base stations. To assess the exposure, various measures have been used. Santini et al. (2003) obtained information from 530 people responding to a questionnaire about nonspecific health symptoms and about distance to the base station. Respondents living at various distances (<300 m) from base stations were compared with a reference group living more than 300 m from the base stations. The authors also asked the subjects to estimate the distance to the nearest base station. However, it has been shown by Schüz and Mann (2000), among others, that distance to a base station cannot be regarded as a relevant surrogate for RF exposure.

In Austria, Hutter et al. (2006) investigated symptoms, sleep quality, and cognitive performance of 336 people living in the proximity of base stations. The subjects were told that the investigation was about environment and health problems. The exposure was classified by RF emission measurements in the subject's bedrooms.

Regarding measurements, not only should the spatial averaging be taken into account but the temporal variation is also of importance. There will be variation in the output power of the base station on a daily as well as weekly basis, all depending on the number of ongoing calls at the station.

5.4.3 Mobile Phones

Different phones have different designs of the antenna position and physical dimensions; for instance, dipole antenna or helical antenna. Kuster (1997) measured 16 different European digital phones and found a very wide variation in the SAR values. The phone with the lowest value, when averaged over 10 g tissue, had a SAR of 0.28 W/kg, while the one with the highest value had 1.33 W/kg (all normalized to an antenna input power of the maximum 0.25 W). If the averaging was done over 1 g tissue, the span was from a low of 0.42 W/kg to a high of 2.0 W/kg. The SAR measurements were taken under normal user conditions. However, when the phone was slightly tilted towards the head of the user, the value ranged from 0.2 to 3.5 W/kg. Thus, for different phones under maximal output, SAR values can differ with a factor of about 5 between the extremes, which can be further affected by the personal handling of the phone, so a factor >10 can be found.

Anger (2003) reported on measurement of SAR on 21 different phones and his results are similar to the ones obtained by Kuster (1997). The SAR ranged from 0.3 to 1.7 W/kg over 10 g of tissue. He also reported the so-called TCP (telephone communication power) value, i.e., how much of the output power that can be used for communication, and he found values ranging from only about 5% in the phone with the lowest value to just <50% in the "best" phone. Thus, more than half of the output power from the phone was lost, some due to mismatch in the phone and the antenna and some being deposited as SAR in the user.

It should be noted that all given values are maximum SAR values found regardless of the anatomical localization. An equal weight is given to the values regardless of whether they are obtained on the external, middle, or inner ear, or behind the ear. In the future, it will be necessary to make the comparison at the same anatomical localization. Presumably, then the values as given by Kuster (1997) might differ even more. Taken together, we have a rather large uncertainty in the actual SAR determination for a specific situation depending on, for instance, nearness to the base station and make and model and personal style of use.

The currents from the battery also give rise to magnetic fields near the phone. For GSM phones, magnetic flux densities of a few μT near the phone have been measured (Bach Andersen et al., 1995; Linde and Hansson Mild, 1997). The fields are pulsed DC fields with a main frequency of 217 Hz. Jokela et al. (2004) measured seven different GSM phones and looked at the frequency content of the magnetic pulse, and he found that a considerable amount is found in the low kHz range. It was even found that some phones exceeded the ICNIRP guideline reference values when the multiple frequency formula was applied, but calculations showed that the basic restrictions were not exceeded.

5.4.4 Short-Term Effects of Mobile Phones

In an epidemiological study of subjective symptoms among mobile phone (MP) users, Hansson Mild et al. (1998) used several factors to assess the exposure and to get an estimate of dose. The study was set up to see if there were differences between users of the analogue (NMT) and the digital (GSM) systems. GSM users reported warmth sensation on the ear and behind or around the ear less frequently than did NMT users. There was also a statistical association between both *Calling time* and *Number of calls per day* and the occurrence of warmth sensation as well as headache and fatigue both among NMT users and GSM users. When calling time per day was used as the exposure parameter, it was found that people using phones for 15–60 min/day were 1.6 times more likely to complain of fatigue and 2.7 times more likely to complain of headache than people who used their phones for 2 min or less per day. Users of phones who talked more than 60 min/day were 4.1 times more likely to complain of fatigue and 6.3 times more likely to have a headache than those who talked for <2 min/day.

In a follow-up study, Wilén et al. (2003) looked at ~2500 of the users in the study, distributed over four different phone models. On the basis of the distribution of the SAR values over the area (Figure 5.1), exposure was assessed in three different areas/volumes: above the ear, on the ear, and below the ear. At each site, the *dose* was expressed as specific absorption (SA), expressed in J/kg instead of SAR. By integrating the estimated "use time" with the SAR value for each device, the mean SA for all devices was obtained. Two other different dosimetric quantities were also used, specific absorption per day (SAD)

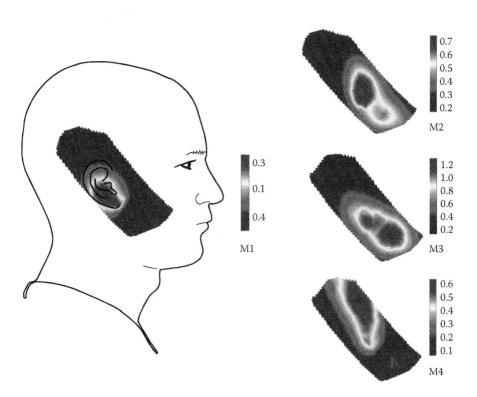

FIGURE 5.1   SAR distributions (W/kg) for four different GSM (digital) mobile phone models (M1 through M4), measured with maximum output power on the right hand side of the phantom. The picture is not to exact scale. Modified from Wilén, J., et al., *Bioelectromagnetics*, 24, 152–159, 2003.)

and specific absorption per call (SAC). Both SAD and SAC are time-integrated quantities expressed in J/kg, but with different time scales; SAD and SAC were calculated by the use of the total calling time per day and the average calling time per call, respectively. The results indicated that SAR values >0.5 W/kg may be an important factor for the prevalence of some of the symptoms, especially in combination with long calling times per day.

5.4.5 Mobile Phone Use and Health

In the studies of brain cancer and mobile phone use, all of the problems discussed earlier apply, but since for this type of disease it is the exposure 5–10 years or more ago that is of interest, exposure assessment becomes an even greater problem than for the acute effects. Most users of mobile phones have not been using just one single telephone, and it is even more likely that if they have been mobile phone users for more than a few years, they have also been changing their phone a few times. Many early adopters will also have been using different phone systems, such as analog and digital, and probably many of them have also been using a cordless phone at home or at work. The problem we are facing is then: how to integrate the—probably different—SAR distributions from the different devices.

The epidemiological studies on mobile phone use and brain tumours have not taken this into account, and the main reason is that at the moment it is not clear how to combine the use of different phones with different power output, different operating systems, different frequencies, and different anatomical SAR distribution, into one *exposure* and *dose* measure. The difficulties arise from not knowing the interaction mechanism(s) between the electromagnetic fields emitted from the phone and from the biological organism.

However, in spite of this we need to start a discussion on how to take into account the use of several phones, and Hansson Mild et al. (2004) have made such an attempt toward a combined exposure measure starting from the data from Hardell et al. (2002). This study included over 1600 brain tumor cases and an equal number of controls and assessed the relationship between mobile phone use and brain tumor incidence. The conditional (matched) logistic regression analysis showed that users of NMT phones with 1-year latency had an odds ratio (OR) for brain tumors of 1.3 (95% CI 1.02–1.6), and with 5 and 10 years latency the OR increased to 1.4 (95% CI 1.04–1.8) and 1.8 (95% CI 1.1–2.9), respectively. For users of GSM and cordless phones, no clearly increased risks were found. Further results from the study using unconditional (unmatched) logistic regression analysis have been presented (Hardell et al., 2003). Many of the subjects in the study had been using all three types of phones: NMT, GSM, and cordless. One obvious combination of the use of different phones is to add the total time on each phone without putting different weights to each of them.

Different types of phones have different output power. The NMT phone operates with a maximum power of 1 W and very seldom downregulates this; the GSM 900 phone operates with a maximum of 0.25 W but can downregulate the power to a few mW depending on the distance to the base station, and a typical value would be 0.1 W; cordless phones operate at 10 mW. One selection of weighting factors according to mean output power of the phones could then be NMT = 1, GSM = 0.1, and cordless = 0.01. Using these factors and the time spent on each of the phone types multiplied with these factors before adding them into one score, the results differ depending on how the combination is done, but not so much. The main trend of an increased risk with increased hour of use also remains. This could be due to the large weight put on the NMT phone due to their high output power.

The authors of the Interphone study on brain tumor and mobile phone use did attempt to go further than just considering hours of use or years of use for the exposure assessment. Cardis et al. (2011) developed an algorithm to evaluate the total absorbed RF dose at specific locations in the brain. They estimated the dose as cumulative specific energy in joules per kilogram (J/kg) absorbed at a given location. They had information about the SAR distribution from each phone and tumour localization from radiological images. An increased risk for glioma was seen in individuals at the highest quintile RF dose, estimated at more than 3123 J/kg.

IARC (2012) evaluated carcinogenicity of radiofrequency EMF and the Working group concluded that there is "limited evidence in humans" for carcinogenicity of RF-EMF based on positive associations between glioma and acoustic neuroma and the exposure for RF-EMF from wireless phones.

Roser et al. (2015), in the HERMES study, estimated whether the RF exposure from mobile phones and other wireless communication devices affects cognitive functions and nonspecific health disturbances in adolescents. From a questionnaire, data on usage of phone and other exposure predictors was obtained. From that, they calculated the doses in the brain and whole body combining the different sources and the dose given as the sum of SAR × time. They found mean cumulative brain and whole-body doses of 1560 and 340 mJ/kg a day, respectively.

5.5 RF Experimental Studies

Many experimental studies on both animals and eukaryotic cells have studied whether RF exposures at "nonthermal" levels can cause biological effects. A large number of experimental systems as well as experimental end points have been employed. Some of the studies have also expressed the aim to focus on "dose-" related questions. Veskovic et al. (2002) published a review where they analysed 12 *in vitro* studies and 8 *in vivo* studies that fulfilled certain quality criteria and provided data relevant for the discussion of dose. The review concluded that, from the *in vitro* studies, it is difficult to see a pattern in terms of common exposure parameters that cause a specific effect. Cited studies reported exposure-related effects at low as well as high SAR values, and also effects due to single exposures as well as due to repeated exposures. The importance of modulation and also a transient effect of exposure were reported. However, none of the *in vitro* studies systematically looked for thresholds where effects start to occur, and there have been no systematic replication studies of dose-relevant findings.

The same review (Veskovic et al., 2002) also analysed a number of animal studies. Similar to the studies conducted at the cellular level, these studies also documented a number of responses attributable to exposure. However, the authors concluded that classical dose–response studies were either lacking or did not find support for SAR-dependent effects, and furthermore that exposure duration was not systematically studied.

5.6 Discussion

In epidemiological studies, regardless of the frequency range of interest, the important consideration of which time scale to use must be based on hypotheses about induction and progression of the studied end point variable. One needs to set up a clear hypothesis about e.g. how the adsorption to RF from mobile phones could affect the end point variable in terms of anatomical localization of the absorption, the duration of the exposure, and the induction and progression of the end point variable before choosing an appropriate dosimetric quantity.

For mobile phone studies, more detailed SAR measurements are needed, not just using the highest value found anywhere near the phone without paying attention to the anatomical localization. We need more information from the medical/biological side as to what sites are of interest for specific symptoms and/or diseases.

It is difficult to see how the "billing record" can be used as a proxy for actual exposure (if measured as SAR) a procedure that in Europe would only give the total time for outgoing calls, not showing incoming calls or the power settings of the phone. An alternative approach to exposure assessment has been to determine which transmitter system has been used and combine this with the user's own estimate of number of minutes on the phone per day and the number of calls per day. This is also an indirect approach that does not provide the physical characterization of the exposure. However, such an approach can possibly be used in future studies of MP users exposures. Thus, it would be useful to initially perform a study of the base station regulation for a number of users. This cannot be done retrospectively, but it could be done prospectively for a selected number of users. When done prospectively, it would be possible to obtain detailed records of their phone use, both number of calls, length of each call, and the actual power settings of the phone as ordered by the base station. These records could then be compared with the subjective estimates of number and time of calls from the users themselves, so as to validate the usefulness of the latter procedure.

As shown with examples from the RF epidemiological studies, none of the earlier studies have assessed the relationship between the effect of the SAR and its distribution in the body. However, this issue is of great importance for continued bioelectromagnetics research. One of the questions we need to address is, for instance, how time comes into the connection between exposure and dose, and here we need to distinguish between different aspects of time: very short times—order of minutes, daily averages, and total time in the actual occupation—and number of years with exposure.

Let us assume that we only have to worry about thermal effects and that we know all we need about the SAR-distribution in an RF-sealer operator. In the present guidelines, there is a 6-min time-averaging based on thermal factors and for preventing heating effects. How would we best construct a dose from this? Would the thermal effects have an accumulating effect and the dose simply be an integration of the SAR over time, or do we here deal with a threshold effect, i.e., if the 6-min averaging gives whole body values below 0.4 W/kg, then it does not contribute to the total daily dose?

The major concern with RF exposure is tissue heating and its thermal effects. It has been a common practice to relate laboratory findings to measured or estimated SAR values. Most current Western standards and guidelines are based on levels of SAR <0.4 W/kg for whole-body average. The SAR and its distribution in the body depends on several factors such as intensity of the electric and magnetic fields; frequency, polarization, and shape and orientation of the body; and possible contact with RF ground. The situation of estimating SAR in an occupational study is thus difficult since each situation needs individual attention. For instance, near RF sealers and glue dryers, measurements of field strengths in the air need to be combined with measurements of induced currents. In the use of mobile phones, local SARs need to be addressed. For low-level RF field effects, SAR might not be generally applicable, but at present it is the best starting point for the dose concept.

References

Ahlbom A, Day N, Feychting M, Roman E, Skinner J, Dockerty J, Linet M, McBride M, Michaelis J, Olsen JH, Tynes T, Verkasalo PK. (2000). A pooled analysis of magnetic fields and childhood leukemia. *Br J Cancer.* 83:692–698.

Anger G. (2003). SAR and transmitted power from 21 mobile phones. The Bioelectromagnetics Society 25th Annual Meeting, Wailea, Maui, Hawaii, p. 98.

Bach Andersen J, Johansen C, Frölund Pedersen G, Raskmark P. (1995). *On the possible health effects related to GSM and DECT transmissions.* Report to the European Commission, Aalborg University, Aalborg, Denmark.

Bortkiewicz A, Zmyslony M, Gadzicka E, Palczynski C, Szmigielski S. (1997). Ambulatory ECG monitoring in workers exposed to electromagnetic fields. *J Med Eng Technol.* 21:41–46.

Bortkiewicz A, Zmyslony M, Palczynski C, Gadzicka E, Szmigielski S. (1995). Dysregulation of autonomic control of cardiac function in workers at AM broadcast stations (0.738–1.503 MHz). *Electro Magnetobiol.* 14:177–191.

Cardis E, Armstrong BK, Bowman JD, Giles GG, Hours M, Krewski D, McBride M, Parent ME, Sadetzki S, Woodward A, Brown J, Chetrit A, Figuerola J, Hoffmann C, Jarus-Hakak A, Montestruq L, Nadon L, Richardson L, Villegas R, Vrijheid M. (2011). Risk of brain tumours in relation to estimated RF dose from mobile phones: Results from five Interphone countries. *Occup Environ Med.* 68(9):631–640.

Feychting M, Ahlbom A. (1993). Magnetic fields and cancer in children residing near Swedish high-voltage power lines. *Am J Epidemiol.* 138(7):467–481.

Feychting M, Forssen U, Floderus B. (1997). Occupational and residential magnetic field exposure and leukemia and central nervous system tumors. *Epidemiology.* 8:384–389.

Floderus B, Persson T, Stenlund C. (1996). Magnetic-field exposures in the workplace: Reference distribution and exposures in occupational groups. *Int J Occup Environ Health.* 2(3):226–238.

Flynn MR, West S, Kaune WT, Savitz DA, Chen CC, Loomis DP. (1991). Validation of experts judgement in assessing occupational exposure magnetic fields in the utility industry. *Appl Occup Environ Hyg.* 6:141–145.

Håkansson N, Floderus B, Gustavsson P, Johansen C, Olsen JH. (2002). Cancer incidence and magnetic field exposure in industries using resistance welding in Sweden. *Occup Environ Med.* 59:481–486.

Hansson Mild K, Mattsson MO. (2010). ELF noise fields: A review. *Electromag Biol Med.* 29:72–97.

Hansson Mild K, Oftedal G, Sandström M, Wilén J. (1998). Comparison of analogue and digital mobile phone users and symptoms. A Swedish-Norwegian epidemiological study. Nordic Radio Symposium, Arbetslivsrapport, 23, Arbetslivsinstitutet, Umeå, Sweden.

Hansson Mild K, Wilén J, Carlberg M, Hardell L. (2004). "Exposure" and "dose" in mobile phone health studies. In: *Mobile Communication and Health: Medical, Biological and Social Problems. International WHO Conference*, Moscow, Russia, September 22–24, pp. 70–79.

Hardell L, Hallquist A, Hansson Mild K, Påhlson A, Lilja A. (2002). Cellular and cordless telephones and the risk for brain tumours. *Eur J Cancer Prev.* 11:377–388.

Hardell L, Hansson Mild K, Carlberg M. (2003). Further aspects on cellular and cordless telephones and brain tumours. *Internat. J. Oncol.* 22:399–407.

Hocking B, Gordon, IR, Grain HL, Hatfield GE. (1996). Cancer incidence and mortality and proximity to TV towers. *Med J Aust.* 165:601–605.

Hutter HP, Moshammer H, Wallner P, Kundi M. (2006). Subjective symptoms, sleeping problems, and cognitive performance in subjects living near mobile phone base stations. *Occup Environ Med.* 63(5):307–313.

International Agency for Research on Cancer. (2002). *IARC Monographs on the Evaluation of Carcinogenic Risks to Humans, Vol. 80, Non-ionizing Radiation, Part 1: Static and Extremely Low frequency Electric and Magnetic Fields.* IARC, Lyon, France.

International Agency for Research on Cancer. (2012). *IARC Monographs on the Evaluation of Carcinogenic Risks to Humans, Vol. 102, Non-ionizing Radiation, Part II: Radiofrequency electromagnetic field.* IARC, Lyon, France.

International Commission for Non-ionising Radiation Protection (ICNIRP). (1998). ICNIRP guidelines for limiting exposure to time-varying electric, magnetic and electromagnetic fields (up to 300 GHz). *Health Phys.* 74(4):494–522.

International Commission for Non-ionising Radiation Protection (ICNIRP). (2010). ICNIRP guidelines for limiting exposure to time-varying electric and magnetic fields (1 Hz–100 kHz). *Health Phys.* 99(6):818–836.

Jokela K, Purananen L, Sihvonen A-P. (2004). Assessment of the magnetic field exposure due to the battery current of digital mobile phones. *Health Phys.* 86:56–66.

Kaune WT, Dovan T, Kavet RI, Savitz DA, Neutra RR. (2001). Studies of high- and low-current-configuration homes from the 1988 Denver childhood cancer study. *Bioelectromagnetics.* 23:177–188.

Kheifets L, Monroe J, Vergara X, Mezei G, Afifi AA. (2008). Occupational electromagnetic fields and leukemia and brain cancer: An update to two meta-analyses. *J Occup Environ Med.* 50(6):677–688.

Klaasen CD. (2001). *Casarett & Doulls's Toxicology: The Basic Science of Poisons.* 6th ed. McGraw-Hill, Chicago, IL.

Kolmodin-Hedman B, Hansson Mild K, Hagberg M, Jönsson E, Andersson M-C, Eriksson E. (1988). Health problems among operators of plastic welding machines and exposure to radio frequency electromagnetic fields. *Int Arch Occup Environ Health.* 60:243–247.

Kuster N. (1997). Swiss tests show wide variation in radiation exposure from cell phones. *Microw News.* 1:10–11.

Linde T, Hansson Mild K. (1997). Measurement of low frequency magnetic fields from digital cellular telephones. *Bioelectromagnetics.* 18:184–186.

Markov M. (2015). XXIst century magnetotherapy. *Electromagn Biol Med.* 34(3):190–196.

Mattsson MO, Simkó M. (2012). Is there a relation between extremely low frequency magnetic field exposure, inflammation and neurodegenerative diseases? A review of in vivo and in vitro experimental evidence. *Toxicology.* 301(1–3):1–12.

Mattsson MO, Simkó M. (2014). Grouping of experimental conditions as an approach to evaluate effects of extremely low-frequency magnetic fields on oxidative response in *in vitro* studies. *Front Public Health.* 2:132.

Morgan G, Nair I. (1992). Alternative functional relationships between ELF field exposure and possible health effects: Report on an expert workshop. *Biolelectromagnetics.* 13(5):335–350.

Nordenson I, Hansson Mild K, Andersson G, Sandström M. (1994). Chromosomal aberrations in human amniotic cells after intermittent exposure to fifty hertz magnetic fields. *Bioelectromagnetics.* 15(4):293–301.

Nordenson I, Hansson Mild K, Järventaus H, Hirvonen A, Sandström M, Wilén J, Blix N, Norppa H. (2001). Chromosomal aberrations in peripheral lymphocytes of train engine drivers. *Bioelectromagnetics.* 22:306–315.

Rannug A, Holmberg B, Ekström T, Hansson Mild K, Gimenez-Conti I, Slaga TJ. (1994). Intermittent 50 Hz magnetic field and skin tumor promotion in SENCAR mice. *Carcinogenesis.* 15(2):153–157.

Roser K, Schoeni A, Bürgi A, Röösli M. (2015). Development of an RF-EMF exposure surrogate for epidemiologic research. *Int J Environ Res Public Health.* 12(5):5634–5656. doi:10.3390/ijerph120505634.

Sandström M, Hansson Mild K, Stenberg B, Wall S. (1995). Skin symptoms among VDT workers related to electromagnetic fields: A case referent study. *Indoor Air.* 5:29–37.

Sandström M, Wilén J, Oftedal G, Hansson Mild K. (2001). Mobile phone use and subjective symptoms. Comparison of symptoms experienced by users of analogue and digital phones. *Occup Med (Lond).* 51:25–35.

Santini R, Santini P, Le Ruz P, Danze JM, Seigne M. (2003). Survey of people living in the vicinity of cellular phone base stations. *Electromagn Biol Med.* 22:41–49.

Schüz J, Mann S. (2000). A discussion of potential exposure metrics for use in epidemiological studies on human exposure to radiowaves from mobile phone base stations. *J Expo Anal Environ Epidemiol.* 10(6 Pt 1):600–605.

Tomenius L. (1986). 50-Hz electromagnetic environment and the incidence of childhood tumors in Stockholm County. *Bioelectromagnetics.* 7:191–207.

Tynes T, Haldorsen T. (2003). Residential and occupational exposure to 50 Hz magnetic fields and hematological cancers in Norway. *Cancer Causes Control.* 14:715–720.

Tynes T, Hannevik M, Andersen A, Vistnes AI, Haldorsen T. (1996). Incidence of breast cancer in Norwegian female radio and telegraph operators. *Cancer Causes Control.* 7:197–204.

Valberg PA, Kaune WT, Wilson BW. (1995). Designing EMF experiments: What is required to characterize "exposure"? *Bioelectromagnetics.* 16:396–406.

Veskovic M, Mattsson M-O, Hansson Mild K. (2002). Dose and exposure in mobile phone research. National Institute for Working Life, Arbetslivsrapport 2002:04. Available at http://libris.kb.se/bib/8411627, accessed January 27 2017.

Wertheimer N, Leeper E. (1979). Electrical wiring configurations and childhood cancer. *Am J Epidemiol.* 109(3):273–284.

Wilén J, Hörnsten R, Sandström M, Bjerle P, Wiklund U, Stensson O, Lyskov E, Hansson Mild K. (2004). Electromagnetic field exposure and health among RF plastic sealer operators. *Bioelectromagnetics.* 25:5–15.

Wilén J, Sandström M, Hansson Mild K. (2003). Subjective symptoms among mobile phone users: A consequence of absorption of radiofrequency fields. *Bioelectromagnetics.* 24:152–159.

Williams P, Hansson Mild K. (1991). *Guidelines for the measurement of RF welders.* Undersökningsrapport, Arbetsmiljöinstitutet, 8.

Wrixon AD. (2008). New ICRP recommendations. *J Radiol Prot.* 28:161–168.

6

Physical Aspects of Radiofrequency Radiation Dosimetry

Marko S.
Andjelković and
Goran S. Ristić
University of Niš

6.1 Introduction

During the past two decades, environmental radiofrequency (RF) exposure has increased tremendously due to the rapid development and widespread use of wireless communication technologies such as mobile telephony, wireless networking, radar systems, TV and radio broadcasting, microwave ovens, and so on. As a consequence of this increased RF radiation exposure, concerns regarding possible adverse health effects associated with exposure to RF radiation have become a serious issue and a motivation for extensive scientific research [1–3].

However, RF radiation has many useful applications in medicine. It has been shown through numerous research studies that specific RF radiation exposure can be used under controlled conditions for diagnosis and treatment of various diseases [4–7]. Use of RF radiation in medicine is noninvasive and safe, and is therefore often regarded as an alternative to conventional medicine. For this reason, substantial research is being devoted to the exploration of medical uses of nonionizing radiation sources.

To ensure maximum protection from the adverse health effects of electromagnetic radiation exposure, the national and international regulatory bodies have introduced reference levels for general public exposure and occupational exposure [8]. It is therefore necessary to conduct systematic monitoring of RF radiation exposure not only to check compliance with the prescribed limits but also to investigate the effects of RF exposure on human health.

In this chapter, the fundamental physical aspects of RF radiation, which are essential for conducting practical RF dosimetric measurements and understanding the effects of RF radiation exposure, are presented.

6.2 Applications of RF Radiation Sources

Among the sources of nonionizing radiation, RF radiation sources have attracted significant interest in the past few decades because of the rapid proliferation of electronic devices operating in the RF range. In the electromagnetic spectrum, RF radiation covers the frequency range up to 300 GHz and is divided into a number of subranges that are utilized for various commercial and special purpose applications [8–14].

RF radiation is employed in a wide range of applications. Typical applications can be divided into four major groups [13]:

1. General applications
2. Safety applications
3. Industrial applications
4. Medical applications

General applications are related to the use of RF radiation in everyday life, and typical examples include:

1. Mobile communication services (mobile telephony)
2. Wireless communication services (Bluetooth, WLAN)
3. Household devices (microwave ovens)
4. Radio and TV broadcasting services
5. Satellite communications

Safety applications are related to the use of RF radiation for safety purposes, and typical examples are:

1. RFID (radiofrequency identification) systems used in the identification and tracking of objects.

2. Electronic article surveillance for prevention of theft in establishments such as shops.
3. Military radar for controlling/tracking airborne objects.
4. Radar used in air traffic control for guidance and surveillance of civil planes.
5. Weather radar used in weather forecasting to identify precipitation data.

Industrial applications are related to the use of RF radiation sources in manufacturing processes such as:

1. Induction heating: RF induction heaters are used extensively in industries for various applications such as surface hardening, zone hardening, and brazing.
2. Dielectric heating: Machines for dielectric heating are widely used in industries for welding.
3. Plasma discharge equipment: Plasma etchers are used in various stages of the semiconductor manufacturing process to break down polymer etch-resistant coatings.

Medical applications refer to the use of RF radiation for diagnostic and therapeutic purposes in medicine. Some typical examples are:

1. Magnetic resonance imaging (MRI): A technique used in radiology to image the anatomy and the physiological processes of the body in both health and disease. MRI scanners use strong magnetic fields, radio waves, and field gradients to form images of the body.
2. RF ablation: A technique that uses contact electrodes to deliver low-frequency voltages in a wide variety of medical therapies. For over half a century, an electro-surgical knife (electrosurgery) has been used by surgeons as a replacement for the scalpel to cut and cauterize tissues.
3. RF telemetry: RF telemetry transmitters encapsulated in a small pill have been used to monitor internal body temperature and other physiological parameters.
4. Hyperthermia: A type of cancer treatment involving exposure of the human body to high temperatures, i.e., the human tissue is exposed to intense electromagnetic radiation to damage and kill the cancer cells.

6.3 Physical Characterization of RF Fields

For accurate measurement of RF radiation exposure, the physical aspects of the RF fields have to be understood and their impact on RF exposure has to be taken into account. Thus, this section discusses the most important physical parameters of the RF fields [8,15].

6.3.1 Electric and Magnetic Field Vectors

RF radiation consists of waves of electric and magnetic energy moving together (radiating) through space. Therefore, the RF field is characterized by a pair of vector fields of electric field **E** and magnetic field **H** (or magnetic induction **B**). Each field vector has three components that can vary throughout space and over time in terms of their magnitude and direction. Therefore, it is necessary to determine six functions of time at each point in space to characterize the field completely.

6.3.2 Emitted and Received Power

The RF signal emitter by an RF source is characterized by the power, which decreases as the distance from the source increases in accordance with the inverse square law, $P = k/r^2$. Thus, it is important to differentiate between the power emitted by the source (radiated power) and the power absorbed by an object (received power).

6.3.3 Modulation

One of the fundamental aspects of any RF transmission system is modulation, i.e., the way in which the information is superimposed on the radio carrier. For a steady radio signal or "radio carrier" to carry information, it must be changed or modulated so that the information can be conveyed from one place to another.

6.3.4 Polarization

The orientation of an electric/magnetic field vector in the plane orthogonal to the direction of propagation is called polarization. If the electric/magnetic field vector is oriented in a given direction, the wave is linearly polarized. If the electric field vector rotates around the direction of propagation, maintaining a constant magnitude, the wave is circularly polarized. If the extremity of the electric field vector traces an ellipse, the wave is elliptically polarized. The rotation of the electric field vector occurs in one of two directions, clockwise or counterclockwise.

6.3.5 Radiation Pattern

Electromagnetic waves are radiated into space by means of antennas. The radiation pattern of the antenna determines the spatial distribution of the radiated energy. A pattern taken in a plane containing the electric field vector is referred to as an E-plane pattern, and a pattern taken in a plane perpendicular to an E-plane is called an H-plane pattern. The directional pattern of an antenna describes how much energy is concentrated in one direction in preference to radiation in other directions.

6.3.6 Fading

Obstacles such as buildings, trees, etc. may cause an RF signal to be reflected during its propagation, thereby resulting in the change of its amplitude. Fading of the RF field signal is an important aspect to be considered in the estimation of RF exposure.

TABLE 6.1 RF Exposure Parameters

	Parameter Name (symbol)	Unit of Measurement
Densitometric	Received RF power (P)	W or dBm
Parameters	RF power density (S or P_D)	W/m^2
	Electric field strength (E)	V/m
	Magnetic field strength (H)	A/m
Dosimetric Parameters	Specific absorption rate (SAR)	W/kg
	Specific absorption (SA)	J/kg
	Current density (I_d)	A/m^2

6.4 RF Exposure Parameters

Besides the fundamental parameters (frequency, wavelength, and amplitude), RF radiation is also characterized by a set of physical parameters that determine the level of RF exposure. The exposure parameters can be divided into two groups:

• Densitometric parameters, and
• Dosimetric parameters.

Densitometric parameters describe the RF exposure in free space, whereas the dosimetric parameters provide information on the absorbed RF radiation in tissue. The most important densitometric and dosimetric RF radiation parameters, and the corresponding measurement units, are listed in Table 6.1 [8].

By measuring the densitometric RF field parameters, it is possible to only estimate the intensity of RF radiation in free space, thereby enabling the determination of the possible risks from RF radiation exposure. However, for estimation of the dosimetric aspects of RF radiation, i.e., the influence of RF radiation on the human body in terms of thermal and nonthermal excitation, the energy deposited in tissue has to be evaluated. Specific absorption rate (SAR) is used for quantifying the thermic effects of RF radiation, which are dominant at frequencies above 100 kHz, and is determined by measuring either the electric field in the body or the induced change of temperature. The current density is essential for characterization of the stimulation (nonthermal) effects for frequencies below 100 kHz, and is determined by measuring the current induced in certain parts of the body such as hands and wrists.

6.5 RF Field Zones

For practical assessment of RF radiation exposure, an understanding of the RF field variations and relations between the RF field parameters is essential. The intensity of the RF field decreases with the distance from the source in accordance with the inverse square law ($\sim 1/r^2$) [16].

Besides the intensity of the field, the RF radiation pattern changes depending on the distance between the source of the RF field and the receiver of the RF field. The RF field zones can thus be divided into two categories [17,18] (Figure 6.1):

- Near-field zone
- Far-field zone.

In the near zone, i.e., close to the source, the RF field is complex and the relationship between the electric and magnetic field components is not constant. Therefore, it is necessary to measure electric and magnetic fields separately to determine the contribution of each component to the total field. The near zone can be further divided into two zones: reactive zone and radiative zone. The reactive zone is closest to the source (transmission antenna), and in this zone the electric and magnetic fields are independent of each other. On the other hand, the radiative zone is further away from the source. The electric and magnetic fields are well correlated in the radiative zone, but there are changes in the angular distribution of electric and magnetic fields with increasing distance.

In the distant (far) zone, the magnetic and electric fields are orthogonal to one another, and therefore it is much easier to determine their interdependence. The relationship between the electric and magnetic fields is constant in the distant (far) zone, as a result of the plane wave character of the field, which means that it is sufficient to measure only one component of the field and based on this the other component can be calculated.

Of course, in practical applications, it is important to know the boundaries between the RF field zones. In principle, there is no clearly defined border, but it depends on the characteristics of the field, i.e., the wavelength of the signal to be transmitted, as well as the dimensions of the transmitting antenna. According to the international conventions, the criteria for determining the boundaries between RF field zones were established, and these are listed in Table 6.2.

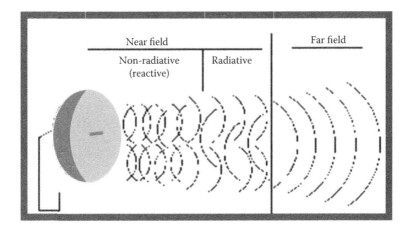

FIGURE 6.1 RF fields in near and far zones.

TABLE 6.2 Criteria for Boundaries of RF Field Zones

	Reactive Near-Field	Radiating Near-Field	Far-Field
Distance between transmitting and reception antenna (r)	$r < \lambda$	$\lambda < r < 2D^2/\lambda$	$r > 2D^2/\lambda$
Relation between E and H	$E/H \neq Z_0$	$E/H \approx Z_0$	$E/H = Z_0$
Parameter to be measured	E and H	E or H	E or H

In Table 6.2, λ is the wavelength of the emitted electromagnetic signal, while Z_0 is the characteristic impendace of air and its value is $377\,\Omega$. D is the largest linear dimension of the antenna.

The data given in Table 6.2, with the parameter D, are related to larger antennas such as transmitting GSM antennas. For smaller antennas, i.e., antennas whose dimensions are smaller than the wavelength of the emitted electromagnetic signal, the following relations are used,

$$\lambda < r < 10\lambda, \quad \text{for near field}$$

$$r > 10\lambda, \quad \text{for far field}$$

According to the criteria defined in Table 6.2, the relation between the electric and magnetic field components is constant in the far field and can be expressed as,

$$\frac{E}{H} = Z_0,$$

where Z_0 denotes the characteristic impedance of space, its value is $377\,\Omega$.

With known E or H parameters for the far-field condition, the power density can easily be determined from the relation [9],

$$P_D = \frac{E^2}{377} = H^2 \times 377.$$

Measurement instruments or other objects in the reactive near field of a source can alter the field strengths and phases of E and H. For example, the presence of measurement personnel or an instrument at an arbitrary location in the reactive near field of a source may change the E- and H-fields at any other nearby locations. Therefore, sensors that are used to measure fields in this region must be very small compared not only to the wavelength but also to the field gradients.

6.6 RF Exposure Measurement

The instruments for the measurement of electromagnetic radiation are classified as instruments for measuring the field levels in free space, i.e., densitometric parameters, and instruments for measuring the radiation absorbed dose, i.e., dosimetric parameters.

From the aspect of the operating frequency band, the instruments for measuring RF radiation parameters can be classified into two distinct groups: narrowband (frequency-selective) instruments and broadband instruments [8]. The frequency selective instruments measure both the field level and frequency of the received signal, and thus can determine the contributions of all frequencies to the total field level. On the other hand, the broadband instruments measure the total field levels within a designated frequency range, but cannot distinguish between different frequencies. Frequency-selective instruments, also known as spectrometers, are suitable for use when the frequency of the source is unknown, while the broadband instruments are useful when the frequency of the RF radiation source is known. Exposimeters are small instruments worn on the body that are used for measuring the RF exposure parameters in free space and are intended for use by professional personnel exposed to RF.

Basically, every RF radiation measurement system is composed of three main components: an RF field sensor, an RF processing unit, and a data acquisition unit [11]. Most commercially available general-purpose RF field meters are designated for measuring the densitometric RF field parameters in free space, while special instruments are used for measuring the dosimetric parameters (Figure 6.2).

The sensor interacts with the field propagating through space, extracts the useful information, and converts it into an electrical signal (current or voltage). The most common field sensors are conventional antennas, but other types of field sensors are also available, such as the Hall sensor for detecting magnetic field and Pockels sensor for detecting electric fields. The antennas are categorized as electric field antennas or magnetic fields antennas. Monopole and dipole antennas are typically used for sensing electric fields, while loop antennas are used for detecting magnetic fields. Antennas are usually directional, i.e., they detect only the field propagating through one axis, but the use of three orthogonal antennas can detect RF field in all three directions—such a configuration is known as an isotropic antenna.

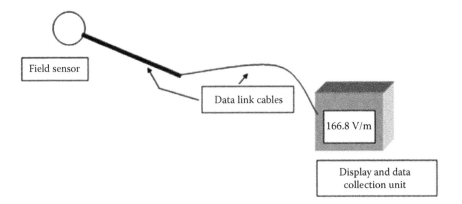

FIGURE 6.2 General architecture of an RF radiation meter. (From ICNIRP. Exposure to high-frequency electromagnetic fields, biological effects and health consequences (100 kHz–300 GHz), International Commission on Non-Ionizing Radiation Protection. Available at http://www.emf. ethz.ch/archive/var/ICNIRP_effekte_RFReview.pdf., 2009.)

The output of the RF field sensor requires further processing such as amplification or conversion into appropriate electric form. When an antenna is used as a field sensor, the conditioning circuit is basically an RF detector that converts the RF signal from the antenna into a DC voltage suitable for further processing, where the DC voltage is proportional to a certain parameter of the input signal (voltage or power) [19]. The two basic types of RF detectors are: thermal based (thermistor and thermocouple) and diode based. Besides, currently, there is a wide range of advanced RF detectors in the form of integrated circuits based on complex amplification chains. These detectors have far better performance in comparison to conventional diode and thermal detectors. The two common types of integrated RF detectors are logarithmic and rms RF detectors [19].

The data acquisition unit measures the DC voltage generated at the output of the RF processing unit, performs appropriate processing of the measured data to obtain the corresponding RF field parameters, and displays results on a suitable display. The core of the data acquisition unit is a processor-based element such as a microcontroller. Depending on the type of measurement instrument, the data acquisition unit can provide only the conversion of electrical parameters into RF field parameters, or it may be equipped with frequency selective logic to determine the frequency of the received signal. Besides, the data acquisition unit can store the measured values as well as act as an interface for communication with a personal computer.

Practically, RF measurement instruments can be either compact or modular. A compact instrument includes all the elements embedded into a single unit. On the other hand, in the modular design, the sensor and the RF processing unit are considered as one unit, commonly known as the probe, and the processing unit is separate, while the connection between these two units is established via cable or through direct connection.

The conventional instruments used for measuring the EMF parameters can be classified as:

- Electric and magnetic field meters
- Spectrum analyzers
- Power and power density meters
- Induced current meters
- Contact current meters

6.6.1 Electric and Magnetic Field Strength Meters

Electric and magnetic field strength meters are narrowband devices. They consist of an antenna, cable(s) to carry the signal from the antenna, and a signal conditioning/readout instrument. Field strength meters may use linear antennas, such as monopoles, dipoles, loops, biconical or conical log spiral antennas, horns, or parabolic reflectors. The appropriate field parameters can be determined from a measurement of voltage or power at the selected frequency and at the antenna terminal. The electric (or magnetic) field strength can be derived from information on the antenna gain or antenna factor and the loss in the connecting cable.

6.6.2 Spectrum Analyzers

Spectrum analyzers are essentially broadband tunable receivers whose reception bandwidth may be set over a wide range of frequencies. They are used to measure the power at the antenna terminal at the selected frequency(ies). If used in combination with a narrowband selective antenna, the overall device becomes conceptually similar to a field strength meter. However, spectrum analyzers can also be connected to relatively short antennas to produce a broad response over a given frequency range. In this case, the analyzer will display the spectrum of ambient signals and permit ascertaining the frequencies involved and their relative contribution to the overall power density.

6.6.3 Power and Power Density Meters

Power and power density meters are generally isotropic and broadband devices. However, there are conceptual differences among these devices in the way the fields are detected and processed. The instruments described in the following sections have essentially the same basic components (i.e., a probe, a connecting cable, and a conditioning display unit). They are limited to those types that are currently available and can provide reasonable accuracy in both near-field and far-field situations. Measurements conducted with a power density meter may produce erroneous readings when the connecting cables are inadvertently aligned with the electric field. This is due to the fact that high-resistance leads carrying the signal act as a more efficient antenna.

6.6.4 Induced Current Meters

Induced current meters display the current induced through the body to the ground when an individual is exposed to an electric field created by a high-power transmitter. These currents can provide an indication of energy absorbed by the body. Induced current meters are generally stand-on devices that measure the induced current flowing through the subject's feet to the ground. The stand-on baseplate is made of two stainless steel plates and is in fact a capacitor/resistor network. The meter reads the current flowing through the resistor connected between the capacitor plates. The size of the baseplate is kept small to minimize any pickup of electric field from the sides of the baseplate. There are also induced current meters that can measure the induced current in arms and legs directly using clamp-on sensors. The typical frequency range of commercial meters is from 10 kHz to 100 MHz.

6.6.5 Contact Current Meters

Contact current meters measure the amount of current through the body caused by contact with a "hot" metallic object located in the vicinity of a high-power transmitter. Contact current meters generally feature an insulated contact probe for contact with the "hot" object. Together with a stainless steel baseplate and internal circuitries, the measured current simulates the equivalent induced current by a barefoot individual gripping the "hot" metallic object. The typical frequency range of this type of meter is from 3 kHz to 30 MHz.

6.7 Uncertainty in RF Measurement

Despite the fact that a wide range of RF measurement instrumentation is available, measuring RF radiation exposure is not an easy task. The measured results are subject to a certain degree of uncertainty, which must be taken into account to obtain precise information on the RF exposure levels [20–23].

The accuracy of the measured field strength levels can be affected by the uncertainties of the actual measurement and the uncertainties of the instrument used to perform the measurement. Actual measurement uncertainties can be minimized by following proper measurement procedures, and instrumentation uncertainties can be reduced by correct calibration and careful selection of the instrument.

In carrying out calibrations in standard laboratory environments, the effect of scattering objects and the conducting parts of the RF instrumentation being calibrated will disturb the incident field. In general, it would be expected that the uncertainty should not exceed 2 dB and in some circumstances may be less. Uncertainty for TEM cell calibrations may be as little as 5%, but 10% is more typical. For Gigahertz Transverse Electromagnetic cell (GTEM), where the field strength cannot simply be calculated from the power and cell geometry, it is likely that a transfer standard field sensor will provide the lowest uncertainty for calibration.

In addition to the uncertainty in the calibration procedures, there are other measurement factors that will affect the overall uncertainty when using RF field instrumentation in particular situations. These will include temperature and drift effects, resolution of the display, issues related to the relative location of the RF source and the measurement probe, positioning of the sensor, nature of polarization, perturbation of measurement by people, and the degree of repeatability. All of these will contribute to the derivation of the expanded uncertainty budget, which may be much larger than the calibration uncertainty but may be reduced by adopting approaches to minimize the uncertainty on some of the abovementioned factors.

The measurement given by the instrument is only an estimate of the measured (the subject to measurement), and thus it is complete only if associated with a statement of the uncertainty parameter that characterizes the dispersion of the values that could be reasonably attributed to the measured value. All the components contributing to uncertainty should then be identified with reference to both the measuring instruments used and the measurement procedures and conditions. The evaluation of uncertainty becomes crucial when comparing a result of measurement with a field limit value fixed by a standard. Besides the uncertainty associated with the use of a field meter, other contributions also have to be considered when evaluating uncertainty of a field measurement. These contributions depend both on the measurement procedures and conditions and on the characteristics of the field source.

The following are some of the most common uncertainty contributions in RF measurements [20,21]:

- Probe calibration, which should be carried out in an accredited laboratory.
- Frequency interpolation, due to the fact that the probe calibration curve is determined for discrete frequencies of the reference EMF.

- The measuring procedure followed to estimate the measured quantity, and differences arising due to different staff carrying out the same type of measurement.
- The effects of environmental conditions (i.e., temperature, humidity) in the measurement.
- The anisotropy of the RF antenna.

6.8 RF Dosimetry

RF dosimetry establishes the relationship between an EMF distribution in free space and the induced fields inside biological tissues, generally the human body. The dosimetric properties of electromagnetic exposure are evaluated using two standard dosimetric quantities:

- SAR
- Current density

6.8.1 SAR Measurement

SAR is used as a primary indicator of RF energy absorbed in the body when quantifying the biological effects and thus defining the basic exposure limits. It is defined as the absorbed power per unit mass and is expressed in the unit of watt per kilogram (W/kg). In other words, SAR can be expressed as the time derivative of the incremental energy (dW) absorbed by an incremental mass (dm) contained in a volume element (dV) of given mass density (ρ),

$$\text{SAR} = \frac{d}{dt}\left(\frac{dW}{dm}\right) = \frac{d}{dt}\left(\frac{dW}{\rho dV}\right).$$

For practical assessment of SAR, it is possible to use two alternative equations. One is,

$$\text{SAR} = \frac{\sigma_{\text{eff}}}{\rho}(E_{\text{local}})^2,$$

where E_{local} denotes the electric field strength (expressed in V/m) in the tissue, σ_{eff} is the effective conductivity of the tissue in S/m, and ρ is the density of tissue in kg/m³. The other equation is,

$$\text{SAR} = c_{\text{p}}\frac{\Delta T}{\Delta t},$$

where c_{p} is the heat capacity of the tissue expressed in J/kg K, and ΔT is the temperature change within the time interval Δt.

Calculations of SAR from temperature rise can be done only if the temperature rise is linear with time. This method is appropriate for local SAR measurement when the exposure levels (irradiating fields) are intense enough so that heat transfer within and out of the body does not influence temperature rise. On the other hand, determining the SAR from the electric field is appropriate for local SAR measurement when the exposure levels (irradiating fields) are intense enough that heat transfer within and out of the body does not cause significant fluctuations of the body temperature.

From these equations, it is clear that for SAR measurement it is necessary to measure the strength of an electric field or temperature changes in the organism being tested. There are special instruments for this purpose, but such a procedure is invasive and is not easy, especially when it comes to scientific research. Therefore, an alternative method is based on using a special artificial simulation model human tissue as well as appropriate mathematical models that allow SAR estimation analytically, i.e., without measuring (Figure 6.3).

SAR can be classified as average and local. Average SAR represents the ratio of the total power absorbed throughout the body and the weight of the body. On the other hand, the local SAR refers to a specific part of the body, i.e., to a particular mass. Values of SAR depend on the following parameters:

- Parameters of the field and the distance between the source and body.
- Physical dimensions of the body.
- The effects of the grounding and reflection for other bodies in the vicinity of the body exposed to radiation.

The average SAR, i.e., SAR for the whole body, depends on the position of the body in relation to the electric and magnetic field components. The highest SAR value is achieved when the body is parallel to the propagation direction of the electric field in the far zone.

Usually, SAR is measured only in research laboratories because they are relatively difficult and require specialized equipment and conditions. Three basic techniques are used for measuring SARs. One is to measure the E-field inside the body, using implantable E-field probes, and then to calculate the SAR; this requires knowing the conductivity of the material. This technique is suitable for measuring the SAR only at specific

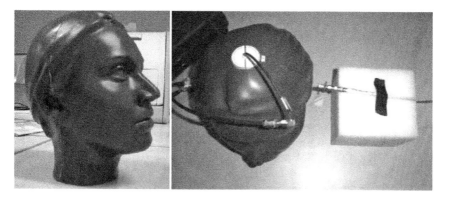

FIGURE 6.3 Human head phantom (left), and measurement probe inserted in the phantom (right). (From Psenakova, Z., and Benova, M., *Adv. Electr. Electron. Eng.*, 7, 1–2, 350, 2008.)

points in an experimental animal. Even in models using tissue-equivalent synthetic material, measuring the internal E-field at more than a few points is often not practical.

A second basic technique for measuring SAR is to measure the temperature change due to heat produced by the radiation, and then to calculate the SAR from that. Probes inserted into experimental animals or models can measure local temperatures, and then the SAR at a given point can be calculated from the temperature rise. Such a calculation is easy if the temperature rise is linear with time; that is, the irradiating fields are intense enough so that heat transfer within and out of the body has but negligible influence on the temperature rise. Generating fields intense enough is sometimes difficult. If the temperature rise is not linear with time, calculation of the SAR from temperature rise must include heat transfer, which is much more difficult. Another problem is that the temperature probe sometimes perturbs the internal E-field patterns, thus producing artifacts in the measurements. This problem has led to the development of temperature probes using optical fibers or high-resistance leads instead of ordinary wire leads.

A third technique for measuring SAR is to calculate the absorbed power as the difference between incident power and scattered power in a radiation chamber. This is called the differential power method.

Whole-body (average) SAR in small animals and small models can be calculated from the total heat absorbed, as measured with whole-body calorimeters. Whole-body SARs have also been determined in saline-filled models by shaking them after irradiation to distribute the heat and then measuring the average temperature rise of the saline.

6.8.2 Current Density Measurement

Current density is the electrical current per unit area and is expressed in A/m^2. This parameter determines stimulatory effects of RF radiation. Tests have shown that the largest current density is induced in the hands and feet. This is explained by the fact that the hands and feet have the smallest cross section. Accordingly, special instruments are developed to measure the current density in these parts of the body.

The current measurement is undertaken in one of three ways:

1. By placing a human subject on a conducting, standard-size plate and measuring the current between the plate and the surface (of the earth) using a thermocouple,
2. By measuring the voltage drop on a resistance between the plate and the ground, or
3. By using a current transformer.

These principles of current measurement are illustrated in Figure 6.4, and a practical current measurement procedure is illustrated in Figure 6.5.

It should be noted that the measurement of the induced currents has two important advantages. First, there is the possibility of comparing the current induced by the RF field with the currents that occur naturally in the body, i.e., in the nerves. Second, comparing the current that is induced in the body with currents that occur in the surrounding conductive bodies can be determined by the marginal value of electric shocks that may occur in contact between the human body and any other conductive object.

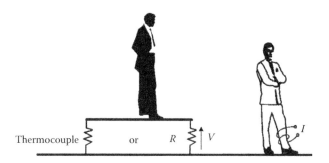

FIGURE 6.4 Principles of measuring induced current in the body.

FIGURE 6.5 Instrument for measuring induced current.

6.8.3 Experimental Dosimetry

Experimental dosimetry uses instrumentation and measurements to directly measure the dosimetric quantities in exposed subjects or in artificial models called phantoms.

In vivo measurements of induced current densities and SAR in humans exposed to EMFs are highly invasive and are thus almost impossible for ethical reasons. Measurements on animals pose fewer ethical problems, but their results cannot be easily extended to humans. Thus, researchers have to resort to:

- Measurement of basic quantities in phantoms
- *In vivo* measurement of some derived quantities such as current.

6.8.4 Analytical Dosimetry

Analytical dosimetry is aimed at finding a solution to the set of Maxwell equations that describe the coupling of the EMF with the exposed body, taking source characteristics and environmental properties into account, with reference to some particularly simplified geometries, like the sphere, the cylinder, the spheroid, and the ellipsoid, in free space or over an infinite, perfectly conducting ground plane.

6.9 Biological Effects of RF Radiation

When an RF field interacts with a biological body, it is reflected, transmitted, refracted, or scattered by the biological body. The refracted and scattered fields may proceed in directions different from that of the incident RF field. These phenomena are described and governed by the well-known Maxwell's equations of electromagnetic theory. The transmitted and refracted fields from the RF exposure induce electric and magnetic fields in the biological systems that interact with cells and tissues in a variety of ways depending on the frequency, waveform, and strength of the induced fields, and the energy deposited or absorbed in the biological systems. Thus, to achieve a biological impact, the electric, magnetic or electromagnetic field must exert its influence on the biological system (tissue) in such a way that the deposited energy produces a detectable change in the biological system.

An important consideration in RF exposure is the coupling of RF fields and their distribution inside the body. This association is also valuable in human epidemiological investigations on the health effects of RF field usage. The coupling of RF electromagnetic energy into biological systems may be quantified by the induced electric and magnetic fields, power deposition, energy absorption, and their distribution and penetration into biological tissues. These quantities are all functions of the source and its frequency or wavelength, and their relationship to the physical configuration and dimension of the biological body. Furthermore, the coupling is more complicated in that the same exposure or incident field does not necessarily provide the same field inside biological systems of different species, size, or constitution. Additionally, the interaction of RF energy with biological systems depends on electric field polarization, especially for elongated bodies with a large height-to-width ratio.

It is emphasized that the quantity of induced field is the primary driving force underlying the interaction of electromagnetic energy with biological systems. The induced field in biological tissue is a function of body geometry, tissue property, and exposure conditions. Moreover, determination of the induced field is important because: (1) it relates the field to specific responses of the body, (2) it facilitates understanding of biological phenomena, and (3) it applies to all mechanisms of interaction. Once the induced field is known, quantities such as current density (J) and specific energy absorption rate (SAR) are related to it by simple conversion formulas.

It is important to assess the health and safety risk of RF energy to determine not only the fields induced in biological tissues but also the mechanisms underlying its biological interactions with cells, tissues, and the human body. RF radiation causes health effects that can be divided into two main groups [24–28]:

1. Thermal effects and
2. Stimulating (nonthermal) effects.

Thermal effect is pronounced at frequencies >100 kHz (typically defined in the range from 100 kHz to 10 GHz). Basically, thermal (or heat) effect is the change in temperature of the body that is exposed to radiation. In other words, the tissue is heated. The thermal EMFs can be shown by measurement with thermovision camera, measurement with special phantoms of biological tissue, or measurement with water phantoms. The organs most sensitive to temperature are the eye lens, brain, and seminal vesicles.

A very high percentage of the human body is made up of water, and water molecules are polar molecules liable to be influenced by impinging EMFs. Hence, those tissues having a significant water content are most liable to be influenced by the RF fields. The effect of RF on such body tissues is to cause polar molecules to attempt to follow the reversals of the cycles of RF energy. Due to the frequency and inability of the polar molecules to follow these alternations, the vibrations lag on them, resulting in a gain of energy from the field in the form of heat, which causes an increase in the temperature of the tissue concerned [2].

This stimulatory effect is pronounced at frequencies lower than 100 kHz and is reflected in the appearance of irritation of nerve and muscle cells, which occurs as a result of current flow that is induced in the tissue under the influence of radiation. Nonthermal effects include nonspecific symptoms that are caused by mobile radiation. It is believed that nonthermal effects can lead to headache, dizziness, and insomnia. These types of symptoms might result from a unilateral influence on the vestibular system in the middle ear, which arise from absorption of the telephone's EMF [29].

The existence of thermal effects is scientifically proven and is the subject of numerous studies whose primary objective is to determine the exact consequences of thermal effects of RF radiation and measures for the protection of the same.

The depth of penetration of electromagnetic waves into the tissue is not sufficient to explore the biological effects caused by EMF penetration, because the distribution of energy in the tissues is very complex. Different local distribution of electromagnetic energy in the tissue leads to thermal and non-thermal excitation which may be associated with a variety of effects that have not been yet fully explored. Therefore, the most common parameter for characterization of biological effects of electromagnetic radiation is the absorbed energy or SAR.

6.10 Exposure Standards

To define guidelines for RF radiation safety, a number of reference exposure limits have been specified both for environmental and occupational conditions. The RF radiation limits are defined by national and international regulatory bodies, and the most important regulatory organizations are:

1. ICNIRP (International Commission on Non-Ionizing Radiation Protections)
2. ANSI (American National Standards Institution)
3. CENELEC (European Committee for Electrotechnical Standardization)
4. FCC (U.S. Federal Communications Commission)

Electromagnetic radiation standards, such as the ICNIRP Guidelines, usually indicate the maximum allowable values (called exposure limit values) for the basic quantities. These values are set with reference to the thresholds of biological effects, applying proper safety margins.

While exposure guidelines such as ICNIRP do not apply to exposures to patients for medical purposes, they do apply to occupational exposures to medical staff; and compliance with the guidelines and associated possible health risks need to be examined.

6.11 Conclusion

As a result of the rapid expansion of wireless communication technologies and services, exposure to RF radiation has increased tremendously, and this has in turn raised the concerns of both public and general specialists regarding the possible adverse health effects of RF radiation. On the other hand, it has been confirmed through extensive research that controlled RF radiation exposure can be very helpful in medical treatment. Thus, the investigation of the physical aspects and biological effects of RF radiation is very important for better understanding of both the undesired and the beneficial effects of RF radiation exposure.

This chapter presents a review of the fundamental physical aspects of electromagnetic RF radiation dosimetry. The basic physical characteristics, measurement procedures, and dosimetric aspects of RF radiation have been analyzed, providing fundamental knowledge for conducting general RF exposure measurements.

References

1. Röösli M, Frei P, Bolte J, Neubauer G, Cardis E, Feychting M, Gajsek P, Heinrich S, Joseph W, Mann S, Martens L, Mohler E, Parslow RC, Poulsen AH, Radon K, Schüz J, Thuroczy G, Viel JF, Vrijheid M (2010), Conduct of a personal radio-frequency electromagnetic field measurement study: Proposed study protocol, *Environmental Health*, Volume 9, p. 23.
2. Kitchen R (2001), *RF and Microwave Radiation Safety Handbook*, Newnes, Oxford, UK.
3. Cooper TG, Allen SG, Blackwell RP, Litchfield I, Mann SM, Pope JM, van Tongeren MJA (2004), Assessment of occupation exposure to radiofrequency fields and radiation, *Radiation Protection Dosimetry*, Volume 11, Issue 2, pp. 191–203.
4. Markov MS (2015), 21st century magnetotherapy, *Electromagnetic Biology and Medicine*, Volume 340, pp. 190–196.
5. Markov MS (2015), *Electromagnetic Fields in Biology and Medicine*, CRC Press, Taylor & Francis Group, Boca Raton, FL.
6. Rosch P, Markov M (2004), *Bioelectromagnetic Medicine*, Marcel Dekker, New York.
7. Shupak NM (2003), Therapeutic uses of pulsed magnetic-field exposure: A review, *Radio Science Bulletin*, Volume 307, pp. 9–32.
8. International Commission on Non-ionizing Radiation Protection (ICNIRP) (2009), *Exposure to High Frequency Electromagnetic Fields, Biological Effects and Health Consequences (100 kHz–300 GHz)*, International Commission on Ionizing Radiation Protection. Available at http://www.emf.ethz.ch/archive/var/ICNIRP_effekte_RFReview.pdf.
9. Frei P (2010), Personal exposure to radio frequency electromagnetic fields and implications for health, PhD dissertation, University of Basel.
10. WHO (2007), *Extremely Low Frequency Fields*. Environmental Health Criteria, Monograph 238. World Health Organization, Geneva.

11. Sanchez-Hernandez DA (2009), *High Frequency Electromagnetic Dosimetry*, Artech House, Norwood, OH.

12. Mantiply ED, Pohl KR, Poppell SW, Murphy JA (1997), Summary of measured radiofrequency electric and magnetic fields (10 kHz to 30 GHz) in the general and work environment, *Bioelectromagnetics*, Volume 18, Issue 8, pp. 563–577.

13. Health Protection Agency (2012), *Health Effects from Radiofrequency Electromagnetic Fields*, Report of the Independent Advisory Group on Non-Ionizing Radiation, Health Protection Agency.

14. Markov MS (2007), Expanding use of pulsed electromagnetic field therapies, *Electromagnetic Biology and Medicine*, Volume 26, Issue 3, pp. 257–274.

15. Taki M, Watanabe S, Wake K (2001), Characteristics, dosimetry and measurement of EMF, WHO meeting on EMF biological effects, Seoul, Korea.

16. Goiceanu C, Danulescu R (2006), Principles and methods of measuring environmental levels of high frequency electromagnetic fields, *Journal of Preventive Medicine*, Volume 14, Issue 3–4, pp. 79–86.

17. Dlugosz T, Trzaska H (2010), How to measure in the near field and in the far field, *Communication and Network*, Volume 2, Issue 1, pp. 65–68.

18. Bienkowski P, Trzaska H (2010), *Electromagnetic Measurements in the Near Field*, Scitech Publishing, Raleigh, NC.

19. Frenzel LE (2007), RF detectors for wireless devices, frequently asked questions, analog devices. Available at http://www.analog.com/media/en/technical-documentation/frequently-asked-questions/201551981Detector_FAQ.pdf.

20. Vulevic B, Osmokrovic P (2010), Evaluation of uncertainty in the measurement of environmental electromagnetic fields, *Radiation Protection Dosimetry*, Volume 141, Issue 2, pp. 173–177.

21. Stratakis D, Miaoudakis A, Katsidis C, Zacharopoulos V, Xenos T (2009), On the uncertainty estimation of electromagnetic field measurements using field sensors: A general approach, *Radiation Protection Dosimetry*, Volume 133, Issue 4, pp. 240–247.

22. Borsero M, Crotti G, Anglesio L, d'Amore G (2001), Calibration and evaluation of uncertainty in the measurement of environmental electromagnetic fields, *Radiation Protection Dosimetry*, Volume 97, Issue 4, pp. 363–368.

23. International Commission on Non-Ionizing Radiation Protection (ICNIRP) (1998), ICNIRP guidelines for limiting exposure to time-varying electric, magnetic and electromagnetic fields (up to 300 GHz), *Health Physics*, Volume 74, Issue 4, pp. 494–522.

24. Repacholi MH (1998), Low level exposure to radiofrequency electromagnetic fields: Health effects and research needs, *Bioelectromagnetics*, Volume 19, Issue 1, pp. 1–19.

25. Chou CK, Bassen H, Osepchuk J, Balzano Q, Petersen R, Meltz M, Cleveland R, Lin JC, Heynick L (1996), Radio frequency electromagnetic exposure: Tutorial review on experimental dosimetry, *Bioelectromagnetics*, Volume 17, Issue 3, pp. 195–208.

26. Sowa P, Rutkowska-Talipska J, Sulkowska U, Rutkowski K, Rutkowski R (2012), Electromagnetic radiation in modern medicine: Physical and biophysical properties, *Polish Annals of Medicine*, Volume 19, Issue 2, pp. 139–142.

27. Challis LJ (2005), Mechanisms for interaction between RF fields and biological tissue, *Bioelectromagnetics*, Volume 26, Issue Suppl. 7, pp. S98–S106.
28. Litvak E, Foster KR, Repacholi MH (2002), Health and safety implications of exposure to electromagnetic fields in the frequency range 300 Hz to 10 MHz, *Bioelectromagnetics*, Volume 23, Issue 1, pp. 68–82.
29. Psenakova Z, Benova M (2008), Measurement evaluation of EMF effect by mobile phone on human head phantom, *Advances in Electrical and Electronics Engineering*, Volume 7, Issue 1–2, p. 350.

7

Necessary Characteristics of Quality Bioelectromagnetic Experimental Research

Ben Greenebaum
*University of
Wisconsin Parkside*

7.1 Introduction

Bioelectromagnetics is a highly interdisciplinary field since it studies the interactions of all types of electromagnetic fields with various types of biological systems. Since it is concerned with both the fundamental mechanisms involved in these interactions, their implications for health and their possibilities for use as medical and research tools, it involves disciplines across physical, engineering, chemical and biochemical, biological, and medical sciences. On the physical side, engineering and physics show how to generate, apply, and measure the intensity, frequency content, pulse characteristics, etc., of the fields, while chemistry analyzes the many reactions that occur within a living organism. In addition, physics, chemistry, and engineering supply theoretical insights and experimental tools to probe biology in new ways. At the same time, each biological subspecialty that concentrates on a particular organism, suborganismal system, or basic biological process, has a large number of measurable parameters available that might show field effects. As in any interdisciplinary area of study, when the insights and the intellectual and experimental tools of one field are brought to bear on the unsolved problems of another, they can give great insight.

A lack of reproducibility of many results in bioelectromagnetics research has been a problem, especially for small effects presumably caused by exposure to low-intensity fields. Outcomes of nominally equivalent experiments have sometimes been consistent, but it is also common for different researchers to see increases, decreases, and no changes in a parameter under similar exposures. In addition, the effects reported can be small, often of the same order of magnitude as the experiment's "noise," the fluctuations of a series of measurements among similar samples. This lack of reproducibility and reported effects close to background "noise" has delayed progress in bioelectromagnetics and led many, both scientists and nonscientists, to doubt the soundness of reports of effects and of bioelectromagnetics itself.

The goal of this chapter is to point out some features common to a good bioelectromagnetics experiment that have often have been overlooked (Greenebaum, 2003). These omissions often occur because, in common with any interdisciplinary area of study, each of the participating specialties and subspecialties has its own terminology, techniques, ways of experimentation, and pitfalls that are not well-understood outside that discipline. Over the years, many authors in many journals have reviewed what should be considered in planning and executing bioelectromagnetics experiments, including Misakian (1984, 1991, 1997), Valberg (1995), Misakian and Kaune (1990), Misakian and Fenimore (1997), and Kuster and Schönborn (2000). *Bioelectromagnetics* has published supplementary issues on dosimetry and related matters (Rafferty et al., 1992; Misakian et al., 1993; Gandhi, 1999). This chapter depends upon these and other publications and on this author's experience as the editor of *Bioelectromagnetics* and as the reviewer for various publications and granting agencies. Often, older rather than more current references are cited to indicate how persistent this issue has been.

In bioelectromagnetics, "dosimetry" generally means characterizing the fields and other conditions of an experiment. The goal of this chapter, however, is to collect ways one experiment may inadvertently differ from another and/or suggest ways to reduce or eliminate the differences.

7.2 Planning and Experimental Design

Before any experimentation begins, planning is important. And unless the investigator is fortunate enough to have the all-important funding in hand from the start, any proposal will have to show that the project has considered, as thoroughly as possible, the factors necessary for its likely success.

7.2.1 Involve a Multidisciplinary Team Throughout

Because bioelectromagnetics involves so many scientific disciplines, it is crucial that a multidisciplinary team be involved from the start of the planning process through to publication of the results. It is rare that an individual scientist has expertise in all of the relevant specialties and subspecialties. All too often, when physical scientists perform a bioelectromagnetic experiment, the field exposures may be well executed, but the field-generating apparatus may inadvertently affect growing conditions or an important aspect of the biological assays or some important aspect of data analysis or of the growth environment may be overlooked. Similarly a team of biologists may execute the biological part of an experiment with great skill and insight, but overlook some important physical aspects of the electromagnetic or nonelectromagnetic environment. The inherent variability of biological organisms means that statistical analysis of data is almost always necessary, and bioelectromagnetic experiments very often involve several variable factors that require complicated statistical techniques. Therefore, someone well versed in biostatistics is also needed, either as an additional person or as part of the core team.

7.2.2 Review the Literature

This obvious prerequisite includes looking for previous publications relevant to the planned experiment in both the physical and biological science literature. A properly diverse multidisciplinary team will be of advantage in performing an adequately broad search.

7.2.3 Blinding

Especially when expected effects are small, subtle differences in handling cultures, samples, or live animal or human subjects during all parts of the experiment may affect the outcome (Valberg, 1995). Therefore, it is important that those scientists handling the biological aspects of the experiment, whether feeding cell cultures, handling animals, taking samples, or running tests, are not aware of the exposure or control status of what they are handling. One implication is that by using an exposure system and a dummy but nonidentical control position for the biological samples, blinding is not possible. All positions should be capable of generating fields, and their roles should be interchanged randomly. Everything handled by the biological team should be recorded according to which system it came from, but the exposure status should not be revealed until the data analysis stage.

In addition to blind handling of the samples and performing the assays, the data analysis should also be handled blindly. To accomplish this, the team members responsible for the fields must only partially break the code indicating which apparatus was providing exposure and which was providing control for each experimental run when the data was given to the statistical team. Results from samples subjected to the same conditions of exposure and other variables should be placed in coded groups, but this code should not be broken for the statisticians until the analysis is complete.

7.2.4 Data Analysis and Outcomes

Statistical data analysis techniques are almost always called for. In the planning stage, it is important to decide not only what is to be measured, but which analysis techniques will be appropriate, and to choose a sample size that will likely lead to useful results. While Student's *t*-test may be satisfactory for some situations, many experiments involve multiple factors or have field effects small enough, compared to other sources of variation, that more sophisticated techniques may be necessary. In addition, the often-used criterion of $p < 0.05$ for statistical significance is not always appropriate, at least not without further interpretation. This may be especially true where the difference between exposed and unexposed samples may be significant by the criterion of $p < 0.05$ but may not be particularly large or biologically significant.

As noted earlier, the research team should include a member who can help ensure that the experimental design and data analysis methods are appropriate. Many references exist that could be helpful in planning appropriate data analysis, deciding what criterion for significance is appropriate, and estimating the sample size likely needed to detect an estimated small effect, as discussed in a later section. Two such book-length ones are Whitlock and Schluter (2015), an undergraduate-level text, and Sokal and Rohlf (2012), which provides more detail. A much shorter book, Hector (2015), also introduces R, a recently developed, free statistical software package that is becoming widely used in biology.

Logically, an experiment giving a null result can only show that any field effect is smaller than the level of the noise, though many published reports simply say, "no effect." Here, "noise" refers to all variations in repeated measurements of the same parameter not directly related to the fields' influence on it. The noise may include random fluctuations in the apparatus, biological variation among the organisms or systems being tested, uncontrolled environmental changes, and other confounders.

Performing an experiment that cannot reasonably be expected to detect an effect is clearly a waste of time and resources. Good experimental design therefore requires an analysis of whether the intended number of samples and measurements of a given parameter will be sufficient to yield a significant result, given the estimated size of the effect being considered and the estimated "noise" in the system. The "power" or "sensitivity" of a statistical test refers to its ability to detect an experimental effect of some particular magnitude. The planning stage of an experiment is when the research team must ensure that it will have the desired sensitivity, given the expected magnitude of the effect(s) to be detected. Whitlock and Schluter (2015) and Sokal and Rohlf (2012) discuss the determination of the sample size needed to detect an estimated small effect, as do many other authors, including Lenth (2001), Cohen (1992), and Nakagawa and Cuthill (2007). Blume (2002) also provides an introductory discussion and references to the analogous approach of minimizing the chances of an experiment leading to weak or misleading evidence.

7.3 Experimental Issues

Several other chapters in this volume discuss generating and measuring electric and magnetic fields for various types of experiments in considerable detail, and readers are referred to them to identify many of the important ideas that pertain to a specific type of experiment. This section will discuss a few more general issues that are often overlooked in bioelectromagnetic work. Field-related issues specific to certain types of biological subjects are discussed in the following sections.

As noted earlier, blinding requires that instead of an exposure and a dummy control apparatus, all sets should be identical and each should be capable of generating fields or serving as a control. This also prevents minor differences in construction that might inadvertently cause differences in results not caused by the fields. In any case, experiments without fields in any apparatus should be run to ensure that results from all sets of apparatus are the same. To ensure blinding, there should be a method of covertly changing the activation status of the systems between replicate experiments. Those responsible for the fields should use random numbers to decide which system(s) to activate before each replicate experiment and make a separate record that is consulted at the data analysis stage.

Although the magnetic and electric fields at all frequencies are linked, as described by Maxwell's equations (Lorrain and Corson, 1970), the linkage is weak at relatively low frequencies, where they can be considered and generated separately. At high frequencies, the linkage is stronger, and electric and magnetic fields must be considered generated together, using separate techniques.

7.3.1 Low-Frequency Exposure Environment

7.3.1.1 Low-Frequency Magnetic Field Generation

The most common way of generating low-frequency magnetic fields is with current-carrying coils, though permanent magnets are also available for steady (DC) fields. For very strong DC fields, an iron core or superconducting windings can be needed. Most experiments seek to create fields that are uniform over the region occupied by the sample. Since the field from a single coil changes in intensity and direction as one moves away from its center, multiple coils in various configurations are necessary for uniformity. The intensity and direction of permanent magnets also vary with position. The best-known coil configuration, suitable for small regions of relative uniformity, is the Helmholtz pair (DeTroye and Chase, 1994), but the Merritt et al. (1983) grouping of several square coils and other configurations can often provide better uniformity, a larger region of relatively uniform field while occupying less space overall, or other advantages. The field near a current-carrying plane, whether a conducting sheet (Gundersen et al., 1986) or closely spaced parallel wires (Rogers et al., 1995), is also uniform; but high currents are needed for relatively weak fields. Kirschvink (1992) recommends that winding each magnet coil with two identical current-carrying wires, allowing generating a field if the two currents are in the same direction, but essentially no field if the currents are in opposition. This system reduces differences in sound and heat generation between exposed and control systems, though any vibration due to the magnetic attraction between coils in the exposure system is still present.

A calculation of the field from a system of currents at any point is easily possible using the Biot–Savart law (Lorrain and Corson, 1970), but one should always measure the field of the particular apparatus being used to check the calculation and to find any distortions. Distortions may be due to uneven coil windings or other construction issues, as well as from the coils' surroundings, including metal parts of the exposure system or steel in nearby furniture or in the building. During the full period of the experiment, the field intensity, direction, and waveform should be monitored periodically.

The waveform of both steady and time-varying magnetic fields should be examined. With the exception of steady fields from permanent magnets or battery-driven or superconducting coils, a current power supply is needed. Even the best supplies of DC current cannot filter all of the AC power grid's 50/60 Hz; what leaks through is often called "ripple." Most laboratory-grade supplies specify ripple at less than 0.04–0.06%, which for a DC current generating 1 mT means that there can be an AC component of 0.4–0.6 µT, most commonly at 50/60 or 100/120 Hz, a level at which biological effects have been reported in some instances. Instructional grade power supplies often have 1–100 higher levels of ripple (Greenebaum, 2008). The field-generating coil may provide some additional filtering to reduce the ripple. Similarly, the coil generating a time-varying field may distort the waveform produced by the power supply, especially if the waveform involves high frequencies or very fast step-like changes. A field-detecting device connected to an oscilloscope is probably the best way to detect and monitor the actual field waveform.

7.3.1.2 Low-Frequency Magnetic Field Measurement

Several types of instruments can detect magnetic fields; the general capabilities of each are discussed in more detail in other chapters in this volume. The discussion here is limited to understanding the characteristics of the instruments to be considered when choosing them, considering the characteristics of the field. Consulting the manufacturer's published specifications can help in any purchase decision.

The range of frequencies and intensities to which the instrument is sensitive must obviously include the experimental field conditions. For DC or single-frequency sinusoidal AC fields, comparing the frequency to the manufacturer's specifications is easy. For fields combining multiple frequencies of pulsed fields, all frequencies—or at least all contributing most of the power—must lie within the specifications. The full Fourier spectrum of pulsed fields should be considered; the Fourier spectrum of any waveform is the combination of single sinusoidal waves that can be summed to produce it (Feynman et al., 1963). This is either an infinite series or an integral over a range of frequencies, but as noted, the lowest several frequencies usually contribute most of the power and higher ones can ignored. For a train of pulses, important frequencies are both those related to the rate at which the pulses are repeated and those related to the shape of the pulse itself. For instance, a cell phone signal might consist of pulses of about 900 MHz that last about 0.6 ms and are repeated about every 4.6 ms. The repetition rate of the pulses corresponds to about 217 Hz, while the pulse width corresponds to about 1.7 kHz. Both those frequencies, as well as some integer multiples of them, and 900 MHz are important, though two different measuring instruments are likely to be necessary.

When measuring a time-varying field, instruments, unless they display the actual waveform, generally show the intensity as a root mean square (RMS) average, about 0.7 of the peak voltage of a sinusoidal waveform. For continuous sinusoidal fields of a single frequency, either the RMS or peak value is sufficient, though one should state which. However, the RMS value for any more complicated fields, whether a combination of sinusoidal fields or a train of pulses, is not sufficient information; it is necessary to give the intensity of each sinusoidal component or a pulsed field's repetition rate, pulse width, height and shape, and how fast it rises and falls. These can be summarized by presenting its Fourier frequency content.

Any measurement must consider the entire system, including all electronics that receive and display or read out the measurement. For instance, it may be necessary to use a detector whose output is displayed on both a voltmeter, which could be interpreted as intensity, and an oscilloscope, to determine details of the waveform. The frequency range of the field should not be near the upper or lower end of the specification limits of any of the three devices, whereas the field intensity should be within the detector's range and its output should be in the range of the other two instruments. For fields containing some combination of AC, DC, high and low frequency, or electric and magnetic fields, it is also important to understand how the instrument's reading of one aspect, for instance, a low frequency, may be affected by the others, including aspects outside the instrument's specifications, for instance, a radio frequency.

Other, perhaps less obvious specifications include being sure that the instrument's accuracy is good enough, which it usually is, and that its drift with temperature or other environmental conditions is small enough, which it also usually is. Also, consider whether the physical size of the detector will allow taking measurements at closely enough spaced points, which it may not be if, for instance, a voltage induced by a time-varying field in a coil is used. Independent of the manufacturer's specifications, it is wise to check the instrument upon receipt and periodically thereafter by measuring a known field, such as from a coil/current combination that can be calculated, using a current meter that has also been checked.

It is also extremely important to measure and periodically monitor the background AC and DC fields at the position of the biological samples with all apparatus present and operating. The earth's own DC field, about $50\,\mu T$ at an angle of about $70°$ below the horizontal in our laboratory, will be affected by the building and nearby furniture, and nearby items are sometimes magnetized themselves. In addition, there is evidence that the DC field can produce bioelectromagnetic effects, either by itself or as a cofactor with AC fields, as discussed later. Natural time-varying fields are quite weak and generally can be neglected. But considerable background AC magnetic fields with varying frequency content can arise from the building's power system and from apparatus that is nearby or part of the experiment, especially cell culture incubators.

7.3.1.3 Low-Frequency Electric Field Generation and Measurement

While experiments applying low-frequency electric fields are uncommon at present, many were performed a few decades ago, though with fewer positive results except when magnetic fields were also present. Homogeneous low-frequency electric fields are generally generated in a space by applying the appropriate voltages to a pair of electrodes at opposite sides of the region. Parallel, planar electrodes produce uniform fields between them, though less uniform near or beyond their ends. Fields between parallel wires are relatively uniform except within a few diameters of one of them. Other configurations produce less uniformity. Field lines spread from the surface of an electrode of any shape, filling the interelectrode space but remaining strongest directly between them and relatively uniform when the electrodes are not too near. Field intensity may be calculated relatively easily for parallel plate electrodes in air or vacuum. But electrodes that are not parallel plates or have the presence of anything else between them, including the biological samples being exposed, their containers, or their growth medium, present complications. Computer calculations for these situations are quite possible, however.

The electric field at the surface of a biological sample cannot be assumed to be the same as that set up by electrodes in an empty space. Whenever an object, whether electrically conducting or not, is placed in an electric field, the field will exert a pull on its molecular charges. Depending on the material, molecules or structures may orient or become distorted, becoming polarized; and if they are able, charged molecules or structures may move in the direction of the field, forming a concentration at the surface. Both these actions serve to reduce the intensity of the field within the material, sometimes only a very little, sometimes to zero in the case of a good conductor (Lorrain and Corson, 1970). Furthermore, these changes in the material's charge distributions will themselves create electric fields that combine with the initially imposed fields to affect any other nearby object. The field inside a culture dish between external electrodes is attenuated first by the walls of the dish and then by the conducting medium inside it. Adding the sample material itself introduces local distortions. These distortions can be significant; and direct measurement of an electric field, even if isolated in air, is difficult (Rafferty et al., 1992). Even the presence of a measurement probe produces at least some distortion (Misakian, 1984). In conducting media, the electric field can often be measured with a small dipole pair of electrodes (Bassen et al., 1992).

If the electrodes generating a field are placed in contact with the medium inside the dish, which generally is conducting, the picture is more complicated, though the approximate field in the medium can be calculated as above. In addition, the presence of the conducting medium means that an electric current is established and the electric field in the immediate vicinity of the electrodes is distorted by the clustering of charges at the interface with the electrode (Gundersen and Greenebaum, 1985). Depending on the voltages and materials involved, electrolysis products may be produced from the electrodes or medium that over time can diffuse to the position of the biological sample, not only distorting the field but also interacting chemically with the sample. Possible non-field interactions may be overcome by placing the electrodes at one end of a long tube, with its other end in the culture medium and filled with agar made from the medium (Kaune et al., 1984), increasing the time before electrolysis products diffuse to the sample and also removing from the region of the sample any electrochemical distortion of the field at the electrodes. Since an electric charge in a time-changing magnetic field or a moving charge, even in a static field, is subject to at least a weak induced electric field according to Maxwell's equations (Lorrain and Corson, 1970), at least an upper limit to this influence should be estimated and reported.

7.3.2 High-Frequency Exposure Environment

The linkage between electric and magnetic fields becomes increasingly important as the frequency increases. There is no set dividing line between low- and high-frequency fields, but this linkage has to be considered when generating and measuring above hundreds of kilohertz or a few megahertz and is certainly relevant in the radiofrequency (RF) and microwave frequency ranges of mobile phones and a growing number of other devices.

Fields are generally no longer generated with simple coils or electrodes, and the electromagnetic absorption properties of biological and other materials change drastically from lower frequencies, as do the ability of the fields' surroundings to absorb, reflect, or otherwise affect intensity at a given location, either outside or inside a biological sample. Energy deposition per unit mass at various points inside the sample (SAR-Specific Absorption Ratio) is the generally accepted specification of field intensity, even if some interaction other than heating is suspected. It is most often calculated from a model, though measurements based on heating are also possible (Chou et al., 1996).

In general, samples are exposed to fields by either placing them near an antenna or inside a metal chamber that acts as an expansion of a transmission waveguide. Unless the sample is close to an antenna, fields are usually in the mode where electric and magnetic fields are perpendicular to each other and to the direction of travel of the RF waves (TEM mode). Chou et al. (1996) and Guy et al. (1999) review a number of exposure systems and calculate the pattern of energy deposition by the fields in these configurations.

Paffi et al. (2010) reviewed various methods of exposing culture dishes, primarily by placing them in a TEM cell or other cavity or next to an antenna or stripline. Animals have often been restrained in small chambers clustered around an antenna (Moros et al., 1998; Capstick et al., 2016), especially when head exposure similar to that from a mobile phone is desired. In this instance, the restraint may act as a cofactor. But cages allowing motion have also been placed near a larger antenna (Wilson et al., 2002) or inside a waveguide (Wasoontarajaroen et al., 2012). As with lower frequency fields, care must be taken between exposed and control samples to maintain similar environmental fields, ventilation, temperature, and other nonexposure conditions and to avoid inadvertent electrical currents, especially in the case of animals exposed for long periods—long enough to require drinking water (Capstick et al., 2016).

Interesting results and applications have resulted when fast, strong electric field pulses are applied to cell cultures, resulting in increases in permeability (commonly called electroporation) of the external membrane and sometimes also of internal structures. The pulses are generated through closely spaced electrode pairs in a cuvette (Kolb et al., 2006; Weaver and Chizmadzhev, 2007).

Various instruments can make direct measurement of ambient fields at these frequencies (Gandhi, 1999). The same cautions about proper use of instruments to measure low-frequency electric and magnetic fields apply to these instruments as well. However, because the specifics of what biological sample and what surroundings are involved, the high-frequency fields actually affecting a biological sample can differ substantially from the fields in air near its surface.

Therefore, a great deal of effort has gone into determining the RF fields and their level of absorption inside a mobile phone user or test organism, much of it devoted to numerical modeling of the phone's fields and the anatomy of the subject (Kuhn and Kuster, 2007). Much of the effort has focused on temperature rise in the subject from deposition of RF energy, but other biological endpoints have been shown to be affected by fields at lower intensities (Barnes and Greenebaum, 2007).

With the rise of mobile telephone communication and concerns about potential effects of the fields from a handset held close to the user's head, many experiments have used the phone itself to generate RF fields for experiments on various biological organisms, sometimes carefully but often in a naïve manner. In general, the phones are not well-suited for well-controlled experiments for a number of reasons: The fields emitted change rapidly with distance from the phone and with orientation about it. As noted earlier, measurement in air next to the sample does not give an indication of what the interior field actually is. In addition, unless the phone has been modified from its intended use in communication, the fields can be varying in frequency, may include a pulsing pattern, and can vary over seconds in intensity (Pandya, 2000). The specific characteristics of these changes depend upon the communication network for which the phone was built. Finally, some authors have not measured the fields at all, but assume that it is emitting continuously at the maximum level for which it is built or that others have measured. Experiments like these do not add to the understanding.

7.3.3 Additional Environmental Conditions

Clearly, except for the fields themselves, the environments of exposed and control samples should ideally be identical in all respects. Environmental conditions should be measured at the actual locations of the exposed subjects with the full exposure system in operation and compared with the same measurement at the actual location of the controls. In addition, the biological endpoints should be measured for subjects with no fields in the exposure or control systems to ensure that there are no inadvertent differences between the two sets of apparatus.

One obviously necessary environmental condition is maintaining equal temperatures in exposed and control samples, especially since chemical reaction rates and other processes important in biology are temperature-dependent. Also, any significant gradients or cycling of temperature should be similar, not only during exposure or sham exposure, but also during performing assays or general handling of samples.

Other possible confounders are not always considered. When animals are exposed, they have sometimes been confined in spaces unlike those in which they often live, subjecting both exposed and control (no exposure) subjects to conditions that can be possible confounders. For this reason, both exposed and control animals must be compared to another set of "cage controls" that are not moved from their home cages into exposure apparatus, with or without fields.

Sound or vibration may occur in an exposure system, both due to forces between time-varying fields or currents, but also from mechanical systems in an incubator or the building. Many species of animals are sensitive to sound or vibration at levels well below the range of human perception, and vibration may also affect cultures (Spiegel et al., 1986). Air flow differences have caused complications in experiments that examine when animals sense AC electric fields through movement of hair or whiskers (Stell et al., 1993).

7.3.4 Biological Sample Maintenance and Assays

The biological subjects of bioelectromagnetics experiments range from subcellular systems and cells through tissues to a broad array of whole organisms, including humans. Careful planning with an interdisciplinary team fully involved from the start, proper consideration of sample size to help ensure detection of an effect if it is present, and controlling the fields and environmental factors discussed earlier, such as temperature, noise, and vibration, are essential for any type of subjects. However, each subject will have its own requirements for being maintained with as little biological change as possible, and each set of endpoints will have requirements for being measured successfully. Any experiment, independent of whether fields are included, must follow the full protocol expected when using a particular biological subject and measuring a particular set of endpoints. This section can make only a few general comments about things that should not be overlooked for essentially all types of bioelectromagnetics experiments.

7.3.4.1 Cultures

A great many bioelectromagnetics experiments expose *in vitro* cultures of cells, subcellular systems, or tissues. There are well-developed protocols for culture of each type of biological sample, which should be followed carefully. However, if an experiment is trying to replicate a previous one, it is important to follow exactly the protocol as well as the materials of the earlier work, including any nonstandard procedures. Minor departures, including changes in culture medium, variations in timing of feeding the culture or taking data, or following the samples for different lengths of time, have been known to affect outcomes.

In particular, most culture experiments use incubators to control temperature. Unfortunately, most incubators affect external fields and can create internal fields themselves. In particular, many incubators' temperature control systems generate often considerable fields that vary strongly throughout their interior; see, for example, Chapter 2. Shutting off the incubator heating/cooling system and regulating culture temperature with water circulating from an external temperature-controlled bath is one way to overcome this problem (Jones and Sheppard, 1992). Incubators' steel shells also will cause interior fields to differ from ambient AC and DC fields in the room. Motors from shakers that aerate cultures or fans that circulate air in large incubators can also generate significant AC fields. Since the fields from motors decrease when farther away from them, changing motorized apparatus to increase the distance is often possible by placing culture shelves on stands, increasing the length of the motor shaft, or using other methods. In our laboratory, replacing an incubator fan with one using a hysteresis motor reduced the stray field significantly (unpublished). Vibration may also affect cultures through the stirring mechanism (Spiegel et al., 1986).

As noted earlier, a time-varying magnetic field will induce an electric field and, in conducting media or objects, a current that depends strongly on the geometry of the culture dish and the position within it. The induced field pattern has been calculated for many situations (Bassen et al., 1992); in general, it is zero at the center and strongest at the outer edge. Cell cultures in a shaken or rolled aeration system will also experience the added complication of a modulation of applied fields at the frequency of the mechanical agitation of the liquid medium.

7.3.4.2 Animals and Humans

In both animal and human experiments, blinding throughout all aspects of the protocol is extremely important. Animals can pick up clues as to differences in status from laboratory personnel, including animal handlers, who are unconscious that they are doing something differently (Villescas et al., 1977; Quinlan et al., 1985; Gottlieb et al., 2013). The same holds true for human subjects. Attention to minor differences in environment, including temperature, noise, and vibration, is also important. As noted earlier, many species of animals are sensitive to noise and vibrations at levels well below the range of human perception. Air flow differences between exposed and control cages have caused complications in experiments that examine when animals sense AC electric fields through movement of hair or whiskers (Stell et al., 1993). Many of the comments concerning reducing background fields for cultures also apply to animal experiments.

Most materials that are not electrically conducting are suitable for animal cages or other apparatus for animal or human experiments, but there are exceptions. As discussed, metal or other electrically conducting or polarizable material will affect an electric field. Magnetic fields pass through most materials without noticeable perturbation, the main exceptions being ferromagnetic materials such as iron, steel, and nickel. Most stainless steels, however, are not ferromagnetic, but a few are. Metallic parts of animal cages may pick up electric charges from the fields that could result in animals receiving unintended shocks (Kaune et al., 1978; Capstick et al., 2016).

If animals or humans are placed in an electric field, the conducting nature of their tissues creates a charge layer at the skin, which drastically reduces the electric field inside the body (Kaune and Gillis, 1981). Internal fields and currents will vary with the anatomy and will also depend on whether the subjects are on a grounded surface or not (Kaune and Forsythe, 1985, 1988). Furthermore, the presence of one animal or human near another also distorts the field (Kaune, 1981). In both static and time-varying magnetic fields, an electric field will be induced in the body of animals moving within their cages, but it is likely to be small. A greater problem with animals or humans in an electric field is that a subject touching a metallic object in the field may receive a noticeable electric shock that could confound any experiment looking for a field effect. This was first noticed when exposed animals were seen avoiding metal drinking apparatus (Kaune et al., 1978).

7.3.5 Epidemiology

Epidemiological experiments have their own very important and separate set of dosimetry issues, quite different from those of laboratory experiments, but these will not be discussed here. These problems arise from needing to discern field exposures that occur over time, usually in the past, for very large numbers of subjects and then to aggregate and draw conclusions from the exposure and biological information using sophisticated statistical techniques. Epidemiological dosimetry has been discussed in many other references, such as in the volume edited by Roosli (2014).

7.3.6 Cofactors and Confounders

Many bioelectromagnetic experiments investigate whether electromagnetic fields, either electric or magnetic, affect the influence of a drug or some other cofactor in the biological system. It is important in designing an experiment, with or without an identified cofactor, to consider as fully as possible whether any other influence on the system could confound the results by acting as an additional cofactor. For example, although many early papers reporting results using AC magnetic fields did not report the DC magnetic field, it is now considered to be important in cell growth, birds' migration compass, and in other processes. If radicals or other molecules with short lifetimes are involved in a reaction, the total instantaneous magnetic field, AC and DC, is important. Several experiments have shown that reducing the earth's field significantly has a significant biological influence, both with and without an applied time-varying magnetic field (Prato et al., 2009). Theoretical explanations have been proposed for how fields may affect radicals by Brocklehurst and McLauchlan (1996) and Barnes and Greenebaum (2015); the latter theory also accommodates the effects of very weak fields.

Prato et al. (2009) have also shown that light can be a cofactor in magnetic shielding experiments. Light and time of day of exposure are known to be important with animals. Since light is often not measured, it could be a confounder in some cases where there are conflicting results. Some of the many other possible confounders include the way field frequencies or exposure cycles mesh with various biological and biochemical time cycles, ranging from very short chemical reaction time constants through cell cycle times ranging from minutes to hours to animals' diurnal or longer cycles. Animal gender may or may not be important in various cases. The reader can undoubtedly think of other potential confounders.

7.4 Conclusion

This chapter has presented a number of considerations that have often been overlooked by scientists performing bioelectromagnetics experiments. In many cases, the absence of attention to one or more of these factors may have led to conflicting results between nominally identical experiments or null results due to uncertainties in the data being too large to detect a small, but real effect. The factors are presented with the hope that plans for future work will consider these and similar factors in order that the overall picture on bioelectromagnetics effects may be clarified instead of generating further confusion. In addition, it is hoped that, to the same ends, these factors will be considered by those examining funding proposals or manuscripts submitted for publication.

Acknowledgments

During the writing of this chapter, Professors Frank Barnes, Greg Mayer, and Betty Sisken provided very helpful comments and observations.

References

Barnes, F. S., and B. Greenebaum, eds. 2007. *Biological and Medical Aspects of Electromagnetic Fields*. Boca Raton, FL: CRC Press.

Barnes, F. S., and B. Greenebaum. 2015. The effects of weak magnetic fields on radical pairs. *Bioelectromagnetics* 36: 45–54.

Bassen, H., T. Litovitz, M. Penafiel, and R. Meister. 1992. ELF in vitro exposure systems for inducing uniform electric and magnetic fields in cell culture media. *Bioelectromagnetics* 13: 183–198. DOI: 10.1002/bem.2250130303.

Blume, J. D. 2002. Likelihood methods for measuring statistical evidence. *Stat. Med.* 21: 2563–2599. DOI: 10.002/sim.1216.

Brocklehurst, B., and K. McLauchlan. 1996. Free radical mechanism for the effects of environmental electromagnetic fields on biological systems. *Int. J. Radiat. Biol.* 69: 3–24.

Capstick, M., Y. Gong, B. Pasche, and N. Kuster. 2016. An HF exposure system for mice with improved efficiency. *Bioelectromagnetics* 37: 223–233. DOI: 10.1002/bem.

Chou, C. K., H. Bassen, J. Osepchuk, Q. Balzano, R. Petersen, M. Meltz, R. Cleveland, J. C. Lin, and L. Heynick. 1996. Radio frequency electromagnetic exposure: Tutorial review on experimental dosimetry. *Bioelectromagnetics* 17: 195–208.

Cohen, J. 1992. A power primer. *Psychol. Bull.* 112: 155–159.

DeTroye, D. J., and R. J. Chase. 1994. *The calculation and measurement of Helmholtz coil fields*. Adelphi, MD: Army Research Laboratory AD-A286–081, 22 pp. http://dtic. mil/dtic/tr/fulltext/u2/a286081.pdf. Accessed online April 30, 2016.

Feynman, R. P., R. B. Leighton, and M. Sands. 1963. *The Feynman Lectures on Physics*, vol. 1. Reading, MA: Addison-Wesley.

Gandhi, O. P., ed. 1999. A collection of some of the papers presented at the symposium RF dosimetry: 25 Years of progress. *Bioelectromagnetics* 20(Suppl. S4): 1–2.

Gottlieb D. H., K. Coleman, and B. McCowan. 2013. The effects of predictability in daily husbandry routines on captive rhesus macaques (*Macaca mulatta*). *Appl. Anim. Behav. Sci.* 143(2–4): 117–127. DOI: 10.1016/j.applanim.2012.10.010.

Greenebaum, B. 2003. Editorial: "It's déjà vu all over again." *Bioelectromagnetics* 24: 529–530. DOI: 10.1002/bem.10174.

Greenebaum, B. 2008. Comment: Did undetected AC components affect a DC magnetic field experiment? (Vashisth A, Nagarajan S. 2008. *Bioelectromagnetics* 29: 571–578). *Bioelectromagnetics* 30: 249. DOI: 10.1002/bem.20467.

Gundersen, R., and B. Greenebaum. 1985. Low-voltage ELF electric field measurements in ionic media. *Bioelectromagnetics* 6: 157–168.

Gundersen, R., B. Greenebaum, and M. Schaller. 1986. Intracellular recording during magnetic field application to monitor neurotransmitter release events: Methods and preliminary results. *Bioelectromagnetics* 7: 271–281.

Guy, A. W., C. K. Chou, and J. A. McDougall. 1999. A quarter century of in vitro research: A new look at exposure methods. *Bioelectromagnetics* 20(Suppl. S4): 21–39.

Hector, A. 2015. *The New Statistics with R: An Introduction for Biologists*. Oxford: Oxford University Press.

Jones, R. A., and A. R. Sheppard. 1992. An integrated ELF magnetic-field generator and incubator for long-term in vitro studies. *Bioelectromagnetics* 13: 199–207. DOI: 10.1002/bem.2250130304.

Kaune, W. T. 1981. Interactive effects in 60-Hz electric-field exposure systems. *Bioelectromagnetics* 2: 33–50.

Kaune, W. T., and W. C. Forsythe. 1985. Current densities measured in human models exposed to 60-Hz electric fields. *Bioelectromagnetics* 6: 13–32. DOI: 10.1002/bem.2250060103.

Kaune, W. T., and W. C. Forsythe. 1988. Current densities induced in swine and rat models by power-frequency electric fields. *Bioelectromagnetics* 9: 1–24. DOI: 10.1002/bem.2250060103.

Kaune, W. T., and M. F. Gillis. 1981. General properties of the interaction between animals and ELF electric fields. *Bioelectromagnetics* 2: 1–11.

Kaune, W. T., R. D. Phillips, D. L. Hjeresen, and R. L. Richardson. 1978. A method for the exposure of miniature swine to vertical 60 Hz electric fields. *IEEE Trans. Biomed. Eng.* BME-25: 276–283. DOI:10.1109/TBME.1978.326333.

Kaune, W. T., M. E. Frazier, A. J. King, J. E. Samuel, F. P. Hungate, and S. C. Causey. 1984. System for the exposure of cell suspensions to power-frequency electric fields. *Bioelectromagnetics* 5: 117–129. DOI: 10.1002/bem.2250050202.

Kirschvink, J. L. 1992. Uniform magnetic fields and double-wrapped coil systems: Improved techniques for the design of bioelectromagnetics experiments. *Bioelectromagnetics* 13: 401–411.

Kolb, J. F., S. Kono, and K. H. Schoenbach. 2006. Nanosecond pulsed electric field generators for the study of subcellular effects. *Bioelectromagnetics* 27: 172–187.

Kuhn, S., and N. Kuster. 2007. Experimental EMF exposure assessment. In F. S. Barnes and B. Greenebaum, eds. *Biological and Medical Aspects of Electromagnetic Fields.* Boca Raton, FL: CRC Press.

Kuster, N., and F. Schönborn. 2000. Recommended minimal requirements and development guidelines for exposure setups of bio-experiments addressing the health risk concern of wireless communications. *Bioelectromagnetics* 21. 508–514.

Lenth, R. V. 2001. Some practical guidelines for effective sample size determination. *Am. Stat.* 55: 187–193. http://homepage.divms.uiowa.edu/~rlenth/Power/. Accessed online January 26, 2017.

Lorrain, P., and D. Corson. 1970. *Electromagnetic Fields and Waves.* New York: W. H. Freeman and Company.

Merritt, R., C. Purcell, and G. Stroink. 1983. Uniform magnetic fields produced by three, four, and five square coils. *Rev. Sci. Instrum.* 54: 879–882.

Misakian, M. 1984. Calibration of flat 60-Hz electric field probes. *Bioelectromagnetics* 5: 447–450. DOI: 10.1002/bem.2250050410.

Misakian, M. 1991. In vitro exposure parameters with linearly and circularly polarized elf magnetic fields. *Bioelectromagnetics* 12: 377–381. DOI: 10.1002/bem.2250120606.

Misakian, M. 1997. Vertical circularly polarized ELF magnetic fields and induced electric fields in culture media. *Bioelectromagnetics* 18: 524–526.

Misakian, M., and C. Fenimore. 1997. Distributions of measurement errors for single-axis magnetic field meters during measurements near appliances. *Bioelectromagnetics* 18: 273–276.

Misakian, M., and W. T. Kaune. 1990. Optimal experimental design for in vitro studies with ELF magnetic fields. *Bioelectromagnetics* 11: 251–255. DOI: 10.1002/bem.2250110306.

Misakian, M., A. R. Sheppard, D. Krause, M. E. Frazier, and D. L. Miller. 1993. Biological, physical, and electrical parameters for in vitro studies with ELF magnetic and electric fields: A primer. *Bioelectromagnetics* 14(Suppl. 2): 1–73. DOI: 10.1002/bem.2250140703.

Moros, E. G., W. L. Straube, and W. F. Pickard. 1998. A compact shielded exposure system for the simultaneous long-term UHF irradiation of forty small mammals: I. Electromagnetic and environmental design. *Bioelectromagnetics* 19: 459–468.

Nakagawa, S., and I. C. Cuthill. 2007. Effect size, confidence interval and statistical significance: A practical guide for biologists. *Biol. Rev.* 82: 591–605. DOI:10.1111/j.1469–185X.2007.00027.x.

Paffi, A., F. Apollonio, G. A. Lovisolo, C. Marino, R. Pinto, M. Repacholi, and M. Liberti. 2010. Considerations for developing an RF exposure system: A review for in vitro biological experiments. *IEEE Trans. Microwave Theory* 58: 2702–2714. DOI: 10.1109/TMTT.2010.2065351.

Pandya, R. 2000. *Mobile and Personal Communication Services and Systems.* New York: IEEE Press.

Prato, F. S., D. Desjardins-Holmes, L. D. Keenliside, J. C. McKay, J. A. Robertson, and A. W. Thomas. 2009. Light alters nociceptive effects of magnetic field shielding in mice: Intensity and wavelength considerations. *J. R. Soc. Interface* 6: 17–28. doi:10.1098/rsif.2008.0156.

Quinlan, W. J., D. Petrondas, N. Lebda, S. Pettit, and S. M. Michaelson. 1985. Neuroendocrine parameters in the rat exposed to 60-Hz electric fields. *Bioelectromagnetics* 6: 381–389.

Rafferty, C. N., R. D. Phillips, and A. W. Guy. 1992. Dosimetry workshop: Extremely-low-frequency electric and magnetic fields. *Bioelectromagnetics* 13(Suppl. S1): 1–10.

Rogers, W. R., J. H. Lucas, W. E. Cory, J. L. Orr, and D. Smith. 1995. A 60 Hz electric and magnetic field exposure facility for nonhuman primates: Design and operational data during experiments. *Bioelectromagnetics* 16(Suppl. S3): 2–22. DOI: 10.1002/bem.2250160703.

Roosli, M., ed. 2014. *Epidemiology of Electromagnetic Fields.* Boca Raton, FL: CRC Press.

Sokal, R. R., and F. J. Rohlf. 2012. *Biometry,* 4th ed. New York: W. H. Freeman and Company.

Spiegel, R. J., J. S. Ali, J. F. Peoples, and W. T. Joines. 1986. Measurement of small mechanical vibrations of brain tissue exposed to extremely-low-frequency electric fields. *Bioelectromagnetics* 7: 295–306.

Stell, M., A. R. Sheppard, and W. R. Adey. 1993. The effect of moving air on detection of a 60-Hz electric field. *Bioelectromagnetics* 14: 67–78. DOI: 10.1002/bem.2250140109.

Valberg, P. 1995. How to plan EMF experiments. *Bioelectromagnetics* 16: 396–401.

Villescas, R., R. W. Bell, L. Wright, and M. Kufner. 1977. Effect of handling on maternal behavior following return of pups to the nest. *Dev. Psychobiol.* 10: 323–329. DOI: 10.1002/dev.420100406.

Wasoontarajaroen, S., A. Thansandote, G. B. Gajda, E. P. Lemay, J. P. McNamee, and P. V. Bellier. 2012. Dosimetry evaluation of a cylindrical waveguide chamber for unrestrained small rodents at 1.9 GHz. *Bioelectromagnetics* 33: 575–584. DOI: 10.1002/bem.217.

Weaver, J. C., and Y. Chizmadzhev. 2007. Electroporation. In F. S. Barnes and B. Greenebaum, eds. *Biological and Medical Aspects of Electromagnetic Fields.* Boca Raton, FL: CRC Press.

Whitlock, M. C., and D. Schluter. 2015. *The Analysis of Biological Data*, 2nd ed. Greenwood Village, CO: Roberts and Company.

Wilson, B. W., A. Faraone, D. Sheen, M. Swicord, W. Park, J. Morrissey, J. Creim, and L. E. Anderson. 2002. Space efficient system for small animal, whole body microwave exposure at 1.6 GHz. *Bioelectromagnetics* 23: 127–131. DOI: 10.1002/bem.105.

8

External Electric and Magnetic Fields as a Signaling Mechanism for Biological Systems

Frank S. Barnes
University of
Colorado Boulder

8.1 Introduction

Dosimetry is usually thought of as the measurement of the rate or the integral of the exposure to high-energy radiation or a chemical. In most instances, the measurement is of the amount of whatever is being assessed. In the case of electric and magnetic fields at microwave or lower frequencies, it is not clear how to define "amount" in a biologically significant, measurable way. At present, the specific absorption rate (SAR), the integral of the exposure to the fields, or the power deposited, are three commonly used ways to describe electric or magnetic fields that may be affecting a biological system. These parameters are useful in predicting the temperature rise of the biological material at radio and microwave frequencies. Changes in temperature are important in calculating chemical reaction rates, which are exponential in temperature, and many biological functions. The timing of exposures leading to these biological changes with respect to the state of the biological system is important, as exposures at different times may lead to different results.

However, the concept of dose may need to be modified to include a more complete description of the electric and magnetic fields that are being applied to the biological system. This is particularly true for fields with frequency components near a resonance or a biological cycle time or for fields with magnitudes and time characteristics similar to those occurring naturally within a system. Biological systems are electrochemical systems, and electrical fields are one of the ways information is transmitted from one part of the system to other parts. Under normal conditions, the direction and the amplitude of the internally generated electric fields change with time and location, both within cells and between cells. Thus, it may be better to think of externally applied fields as a signaling mechanism that may be superimposed on the naturally occurring fields. For example, both the natural and externally applied fields can control the direction of growth or the migration of cells. It should not be a surprise that if electric fields of approximately the same size as the natural fields are superimposed on top of the natural fields, biological processes such as cell growth are modified. In this case, the direction of the applied field, amplitude, timing, and duration of the field with respect to the ongoing natural fields are likely to be important. In general, it is to be expected that when external signals are applied it will take some time for the biological system to respond to these signals. Delayed responses can be expected, for example, in exciting the immune system so that the biological system can manufacture enough neutrophils and other defense mechanisms to fight the infection.

Additionally, it has been shown that weak magnetic fields can modify radical pair lifetimes and radical concentrations of molecules such as reactive oxygen species (ROS). These molecules and nitrous oxide species (NOS) are both signaling molecules that are important parts of many biological processes such as metabolism and the activation of muscles; they also can do damage that leads to aging, cancer, and Alzheimer's disease (Droge, 2002). Again, the length of exposure, times, amplitude, pulse repetition rate, and angle between the alternating and static components of the magnetic fields are important. For example, changing the value of the static magnetic field can change the resonant frequency for coupling of electronic spin states in radicals and thus the recombination rate for radical pairs and radical concentrations (Barnes and Greenebaum, 2015).

There is a substantial history showing that externally applied electric and magnetic fields can affect biological processes such as repair of nonunion bone fractures and the healing of wounds (Pila, 2007). More recently, it has been shown that both static and radiofrequency (RF) magnetic fields can accelerate and inhibit the growth rates of cancer cells in culture (Castello et al., 2014; Usselman et al., 2014). However, systematic explorations of the effects of changing amplitude, frequency, and exposure times have rarely been done.

Biological processes are not in thermal equilibrium. They are time dependent and include many parallel paths, feedback, and feed-forward loops. Thus, a signal applied at one time may not lead to the same effects as the signal applied at another time. For example, the action potential spikes traveling along a nerve fiber are approximately the same size and shape. Externally injected currents during the peak of the spike do little to change its size or the repetition rate. However, if the current is injected during the exponential recovery phase in a depolarizing direction (reducing the negative potential between the outside and inside of the cell), it shortens the time between spikes, and if it is in a hyperpolarizing direction (making the interior of the cell more negative with respect to the outside), it increases the time between pulses. Information is carried in the pattern of the nerve pulses, and so the timing of an injected signal makes a difference in its effect. Pacemaker cells can be locked to an external signal by injecting the signal at frequencies that are near the natural oscillating frequency, at harmonics or subharmonics. For nonharmonically related signals at higher frequencies than the natural oscillation frequency, the nonlinearities in the amplification process rectify the external signal and shift the oscillating frequency. Biological oscillators seem to behave just as they would be expected to behave if they were electronic oscillators (Barnes, 2007). Similarly, growing cells are affected differently by external signals at different stages of the growth cycle.

Thus, we suggest that the effects of externally applied electric and magnetic fields should be systematically varied in amplitude, timing, and other parameters so as to be investigated as a way of introducing signals into biological systems. In this chapter, we will assume that the electric and magnetic fields at the site of interest can either be measured or calculated with sufficient accuracy to allow them to be taken as given. The remainder of the chapter will be devoted to reviewing some experimental results on modifying the behavior of biological systems with externally applied electric or magnetic fields with the purpose of suggesting that in specifying exposures the definition of dose should be expanded to include a more complete description of the fields and their timing with respect to the biological system parameters that are being impacted.

8.2 Electric Field Effects

Endogenous biological electric fields vary in both time and space over a very wide range of values. Typical fields across membranes are often in the range of megavolts per meter, while the electric fields parallel to the membrane surfaces are in the range of volts per meter. Externally applied electric fields as low as 10^{-3} V/m have been shown to affect growth patterns for exposures of *Xenopus* neurons over a period of 6 h (Patel and Poo, 1982). Additionally, the voltages in the interior of a cell are not uniform and vary in time and position. Normally growing cells and cancer cells have smaller negative potential differences between the interior and the exterior of the cell than resting or quiescent cells. See Figure 8.1.

Shifts in membrane potentials resulting from externally applied fields are expected to modify the growth rate of cells and be different both for different cells and for cells in different stages of their life span. Voltage differences from a few tenths of a millivolt to several millivolts are sufficient to change firing rates of nerve cells. The corresponding electric fields across the membranes are large or on the order of 10^5 V/m. In contrast, externally applied fields of 100 V/m can lead to the rearrangement of proteins on the surface of a cell when applied for a period of 10 min (Poo, 1982). Natural fields of 10–20 V/m are seen along the axis in growing chick embryos. We have shown that electric fields of 2 kV/m change the direction of the migration of neutrophils 78% of the time when they are migrating up a concentration gradient of c-AMP. Reversing the direction of the electric field will change the direction of migration again about 55% of the time with a delayed response time of about 2 min (Rathnabharathi et al., 2005).

Electrical fields of 40 to 200 V/m are measured near fresh wounds fields, which decrease as the wound heals. In all these cases, it is not surprising that the introduction of external fields of approximately the same magnitude for a reasonable fraction of the normal biological process time will modify the biology. Another parameter of interest

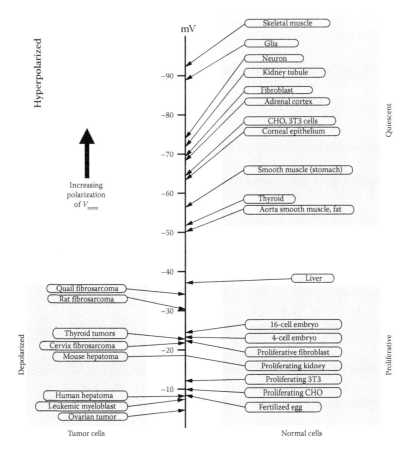

FIGURE 8.1 Membrane potential (V_m) scale, showing that proliferating cancer cells possess depolarized V_m, while the V_m of quiescent cells is generally more negative (hyperpolarized). (From Levin, M., *The Physiology of Bioelectricity in Development, Tissue Regeneration and Cancer*, CRC Press, Boca Raton, FL, 2010. With permission.)

is the currents that flow both within and between cells. Transcellular current densities are in the range of 1–10 μA/cm², and epithelial current densities are in the range of 10–100 μA/cm² (Nuccitelli, 1988, 1990). It is to be noted that nerve cells can be used to activate a large number of biological processes, including preventing overstimulation of the immune system that leads to inflammation, pain, swelling, and tissue damage, which are a consequence of many diseases. In this case, small electrical pulses applied to the vagus nerve can block the generation of tumor necrosis factor (TNF), which in turn can turn off the inflammatory process (Tracey, 2015).

From the point of view of a dose response, we need to know what biological processes we are modifying, the variations over time of the magnitude, and the direction of the endogenous fields. Given this, it is reasonable to expect that if externally induced fields or currents are a reasonable fraction of the natural fields over similar periods of time, they will modify the biological process. A number of parameters of interest for bacterial *Escherichia coli* cells, *Saccharomyces cerevisiae* (Yeast), and human fibroblasts are given in Table 8.1 (Alon, 2007). This table shows that we have times from the order of 1 μs for transitions between protein states to over 10 h for the lifetime of mRNA. The diffusion time required for a small molecule to cross a cell may be on the order of 0.2 s, while for a protein it may be more nearly 100 s. Thus, we might expect electric fields of 1 V/m to move

TABLE 8.1   Typical Parameter Values for the Bacterial *E. coli* Cell, the Single-Celled Eukaryote *Saccharomyces cerevisiae* (Yeast), and a Mammalian Cell (Human Fibroblast) (Alon, 2007)

Property	Bacteria (*E. coli*)	Yeast (*S. cerevisiae*)	Mammalian Cell (Human Fibroblast)
Cell volume (μm³)	~1	~1000	~10,000
Proteins/cell	~4×10^6	~4×10^9	~4×10^{10}
Mean size of protein	5 nm		
Size of genome	4.6×10^6 bp	1.3×10^7 bp	3×10^9 bp
	4500 genes	6600 genes	~30,000 genes
Size of: regulator binding site (bp)	~10	~10	~10
Promoter (bp)	~100	~1000	~10^4–10^6
Gene (bp)	~1000	~1000	~10^4–10^6 (with introns)
Concentration of one protein/cell	~1 nM	~1 pM	~0.1 pM
Diffusion time of protein across cell	~0.1 s $D = 10 \,\mu m^2/s$	~10 s	~100 s
Diffusion time of small molecule across cell	~1 ms $D = 1000 \,\mu m^2/s$	~10 ms	~0.1 s
Time to transcribe a gene	~1 min 80 bp/s	~1 min	~30 min (including mRNA processing)
Time to translate a protein	~2 min 40 aa/s	~2 min	~30 min (including mRNA nuclear exp)
Typical mRNA lifetime	2–5 min	~10 min to over 1 h	~10 min to over 10 h
Cell generation time	~30 min (rich medium) to several hours	~2 h (rich medium) to several hours	20 h—nondividing
Ribosomes/cell	~10^4	~10^7	~10^8
Transitions between protein states (active/inactive)	1–100 μs	1–100 μs	1–100 μs
Timescale for equilibrium binding of small molecule to protein (diffusion limited)	~1 ms (1 μM affinity)	~1 s (1 nM affinity)	~1 s (1 nM affinity)
Timescale of transcription factor binding to DNA site	~1 s		
Mutation rate	~10^{-9}/bp/generation	~10^{-10}/bp/generation	~10^{-8}/bp/year

bp, base-pair (DNA letter).

a potassium ion with a mobility of $7.6 \times 10^{-8}\,m^2/V\text{-}s$ to cross a $10\,\mu m$ cell in about $130\,s$. However, if the fields are on the order of those associated with internal membranes and the distances are on the order of $5\,nm$, this time could decrease by a factor of a million. Thus, the value of $0.2\,s$ is in the middle of this range. The time to transcribe a gene is on the order of $30\,min$, and if we are going to need that gene to generate many proteins it may need to be activated for hours. On the other hand, recombination times for radical pairs can take from 10^{-10} to $10^{-6}\,s$.

Therefore, the exposure times that are effective in modifying a biological process can be expected to vary over a wide range, depending on the process that is being modified. The typical reaction time to trigger the movement of your hand or foot is likely to be on the order of tenths of a second. However, the time to heal a cut it is likely to be days or weeks. For a shock, we might expect a jump response in tenths of a second. If we only jump, the recovery period may only be seconds or minutes. But recovery from the burn could take weeks. To help recover from a burn or a cut, it is likely that the electrical stimulation may need to be applied for at least periods of minutes to initiate the attraction of cells like neutrophils to fight the infection and last long enough so that phase of the recovery has occurred. The overall process is expected to take hours or days, and the applied current to accelerate wound healing is likely to be most effective if it varies in time.

8.3 Magnetic Field Effects

There are several mechanisms by which magnetic fields can affect biological systems. The first is by applying forces to ions or molecules that have magnetic dipole moments. The second is by modifying the energy levels and the angular momentum of the ions or molecules in the system. The third mechanism for time varying magnetic fields is via the induced electric field and the corresponding changes in the local electrical currents.

Typically, the forces applied by magnetic fields to biological systems are weak compared to other forces such as thermal vibrations, except for large magnetic field flux densities with large magnetic field flux density gradients. Forces on magnetic dipoles are proportional to the gradient of the magnetic field flux densities, and for many biological materials the dipoles are induced and proportional to the magnetic field flux density so that the forces are proportional to the magnetic flux density times its gradient. With the exception of exposures for extended periods of time such as from an NMR machine where peak fields are in the range of $1.5-4\,T$ and the field gradients may be large, the displacement of molecules is likely to be small. Most of this section will concentrate on the effects of weak magnetic fields in modifying the concentration of radicals by shifting the energy levels and changing the angular momentum of the fragments of radical pairs. The third mechanism results from the induced electric fields that in turn lead to induced currents which can modify ion transport and the direction of cell growth, as discussed in Section 8.2. In all these cases the timing axes the length of the exposure is important.

That magnetic fields can affect biological systems is now well established (Kaptein, 1968, 1972; Steiner and Ulrich, 1989; Grissom, 1995; Hayashi, 2004; Ritz et al., 2004; Zmyslony et al., 2004; Okano, 2008; Usselman et al., 2014; Yakymenko et al., 2016 and many more). For example, it has been shown that birds, salmon, and other animals use the earth's magnetic field for navigation, and in some cases they have been shown to sense fields as weak as 15 nT (Hoben et al., 2009). It has also been shown that weak magnetic fields in the order of microtesla can change the lifetime for the recombination of radical pairs, and thus radical concentrations (Batchelor et al., 1993; Brocklehurst and McLauchlan, 1996). Concentrations of radicals such as O^{-2*}, NO_x, and molecules such as H_2O_2 may be modified by weak magnetic fields (Henbest et al., 2004, 2006; Simko, 2007; Mannerling et al., 2010; Martino et al., 2010; Usselman et al., 2014). Theoretical work shows a quantum limit for sensitivity to be in the range around 10 nT (Cai et al., 2012). Theoretical work leading to changes in radical concentration has been extended for time-varying fields by Wang and Ritz (2006), Woodward et al. (2001), Barnes and Greenebaum (2015), Rodgers et al (2007), and Liu et al (2005).

Radicals are widely present throughout biochemistry (Cheeseman and Slater, 1993). Various radicals have key roles in reactions involving enzymes and electron transfer. ROS and molecules such as H_2O_2 can both act as signaling molecules and can do damage to lipids and DNA (Droge, 2002; Kipriyanov et al., 2015; Lee et al., 2015). These molecules are generated as a part of the normal metabolic processes, and so their concentrations are normally tightly controlled. Weak magnetic fields have been shown to modify radical concentration of O_2^- and H_2O_2, which is not a radical but which is also a signaling molecule that is readily converted to the radical OH^- (Castello et al., 2014; Usselman et al., 2014). It has also been shown that changes in the static and low-frequency magnetic fields can lead to both inhibition and acceleration of the growth rate of some cancer cells. See Figures 8.2 and 8.3 (Bingham, 1996).

A mechanism by which radicals and H_2O_2 can modify the production of proteins is by activating transcription protein AP1, which in turn can bind to a cis-activating TRE site in promoter genes. Binding at promoter sites can function as AND, OR, or NOT gates in controlling the rate of production of messenger RNA by DNA and thus the required proteins for cell growth, etc. Increases in H_2O_2 increase AP1 activity (Rhee et al., 2003). H_2O_2 and AP1 are a part of many biological activities. For example, they can modify

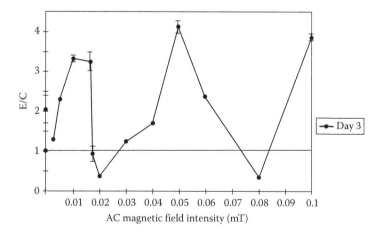

FIGURE 8.2 Growth of P815 mastocytoma cells, Bdc = 38 μT, f = 60 Hz. (From Bingham, C., The effects of DC and ELF AC magnetic fields on the division rate of mastocytoma cells. PhD Thesis, University of Colorado Boulder, 1996.)

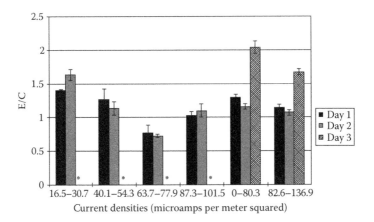

FIGURE 8.3 Effects of current density on cell division rate with Bac = 0.16 G at 60 Hz and Bdc = 0.38 G.

Fos and Jun proteins, and the exposure of neutrophils to H_2O_2 can increase the activity of MAPK/ERK kinases. H_2O_2 can also inactivate protein tyrosine-phosphatages (PTPs) and phosphoinositides (PTEN) growth factors (Rhee et al., 2003). Additionally, it has been shown that H_2O_2 in the presence of 60-Hz magnetic fields of 5 mT for 4 h can increase the mutation rate of supF in *E. coli* by approximately 2.5 times over treatment with H_2O_2 alone at a concentration of 1 μM alone (Koyama et al., 2004).

Li et al. (2014) have shown that exposures to 50-Hz magnetic flux densities at 30, 100, 150, and 200 μT for 30 days can inhibit the growth of tilapia. The body length was significantly decreased after being exposed to 30, 100, 150, and 200 μT on the 20th, 30th, and 50th days in comparison with the sham group. The activities of pepsin and intestinal protease of tilapia were significantly decreased after exposure to certain intensities of extremely low-frequency magnetic fields (ELFMF), but recovered when the ELFMF exposure was removed 20 days later; however, the fish size remained smaller. The growth rates of *Planaria* have been decreased by applying a static magnetic flux density of 200 μT, and the concentration of ROS was also decreased (Mortonetal, 2016). Similar results were obtained with concentrations of antioxidants. At this same field of 200 μT, it has been shown that there is an increase in Ca^{2+}–CaM–dependent myosin phosphorylation and neurite length from embryonic chick ganglia explants. This data coupled with the other data clearly show that exposures to magnetic fields can increase cell growth rates and bone growth (Pilla, 2007). Other data show that growth rates of some cancer cells can be inhibited by reducing the static magnetic field from the range of the earth's magnetic field of from 23 to 65 μT to <18 μT (Martino et al., 2010). The exposure time for these studies range from several hours to 4 days. The exposure times were set by the requirements of several cell cycle times needed to get statically significant difference in the cell counts between the controls and the exposed.

At high concentrations over extended periods of time, ROS and other radicals have been shown to lead to cancers, aging, and Alzheimer's disease (Droge, 2002). A review by Vijayalaxmi et al. (2014) reports data that show human blood lymphocytes *in vitro* exposed to RF fields and then treated with a genotoxic mutagen or subjected to ionizing radiation showed significantly decreased genetic damage. Similar studies in tumor cells showed significantly increased viability, decreased apoptosis, increased mitochondrial membrane potential, decreased intracellular free Ca^{2+}, and increased Ca^{2+}–Mg^{2+} ATPase activity. Thus, the net result on the cancer growth rates depended on the relative timing of the introduction of the magnetic fields and the mutagen. *In vivo* studies on the exposure of rodents to RF fields and then to lethal/sublethal doses of γ-radiation showed that RF produced a survival advantage with significantly decreased damage in hematopoietic tissues, decreased genetic damage in blood leukocytes and bone marrow cells, increased numbers of colony-forming units in bone marrow, increased levels of colony-stimulating factor and interleukin-3 in the serum, and increased expression of genes related to cell cycle. These observations suggested the ability of RF fields to induce adaptive responses and also indicated some potential mechanisms for the induction of such responses (Vijayalaxmi et al., 2014).

Related responses have been observed at 50 Hz by Korneva et al. (1999) where the exposure of mice for an hour a day for 5 days at 21 µT decreased the recovery time after a heavy dose of x-rays in comparison to the unexposed mice. Cells have different sensitivities at different parts of the cell cycle. NO_x has been shown to act as an important regulator role in many biological processes. Variations of NO_x have been shown to regulate ROS, including H_2O_2 (Arnold and Lambeth, 2003). Magnetic fields and the radical NO_x have both been shown to change the vascular tone of capillaries.

8.4 Discussion

But what do these diverse data on the biological effects of electric and magnetic fields mean with respect to dosimetry? Both theoretical and experimental data show that biological sensor systems can detect static magnetic fields as weak as 10–15 nT. As shown in Figure 8.2, changes in the magnitude of the magnetic field of microtesla at 60 Hz can inhibit or accelerate the growth of mastocytoma cells. Given precise enough knowledge of the biology, this indicates that we may be able to signal a biological system to either accelerate or inhibit growth rates as desired. It is likely that other biological processes can be modified by sending electric or magnetic signals that are then translated into chemical signals to preform desired functions. The ranges of frequencies that are important go from static through RF to optical or x-ray. The mechanisms by which these fields affect the biological systems differ with the type of field, the amplitude, frequency, wave shape, polarization, pulse repetition rates, the biological process, the biological system, and the time of the exposure with respect to the life cycle of the biological system.

At RFs, the SAR and corresponding temperature changes are clearly important, as many chemical reaction rates increase exponentially with temperature. Additionally, there are now data showing that fields that are too small to raise the temperature by 1° may also modify biological processes by changing radical concentrations. The energy for changing the radical concentration may well come from the metabolic processes and is controlled by the magnetic field polarization of the electron plus nuclear spins. In these cases, the angle between the alternating fields and static magnetic field is important, as well as the amplitudes and the length of the exposure.

The timing of the externally applied signal with respect to oscillating processes such as nicotinamide adenine dinucleotide phosphate (NADPH) waves in a biological system can lead either to an increase or a decrease in the biological response (Slaby and Lebiedz, 2009). This can be related to the fact that the timing for pushing a swing can either accelerate it or slow it down.

For static or very slowly varying electric fields, the direction and amplitude of the electric field will be important. The current flow can transport nutrients and ions needed for growth and control the direction of growth. Both the strength of the electric field and gradient will affect the orientation and motion of proteins and other large molecules.

All these external fields must be applied long enough for the biological system to sense the fields and process the data. The timing of the signal makes a difference. Thus, a single measure of dose such as the signal amplitude, its integral, or SAR is usually not adequate to tell us what we want to know about what these fields are likely to do to humans.

There is clearly a lot of work that needs to be done for us to be able to use both electric and magnetic fields as a way of communicating with biological systems. At the same time the use of these fields has the potential for providing a very precise way of controlling important biological processes such as cell growth, the immune system, and wound repair.

Acknowledgments

The author appreciates the many helpful discussions and comments by Ben Greenebaum and his many students who asked leading questions and found papers he would never have discovered. Special thanks go to Lucas Portelli, Carlos Martino, and Julius Cyrus who carried out a lot of measurements that improved his understanding of this subject. The author also greatly appreciates the financial support from the University of Colorado and the Milheim Foundation.

References

Alon U. 2007. *An Introduction to Systems Biology: Design Principles of Biological Circuits.* Chapman and Hall/CRC Press, Boca Raton, FL.

Arnold R and Lambeth J. 2003. The NOX enzymes and the regulated generation of reactive oxygen species, Chapter 6. In *Signal tansduction by reactive oxygen and nitrogen species*, Eds. Forman J, Fukuto J, Torres M, Kluwer Academic Publishers, pp. 102–118.

Barnes F. 2007. Interaction of direct current and extremely low-frequency electric fields with biological materials and systems, Chapter 5. In *Bioengineering and Biophysical Aspects of Electromagnetic Fields*, Eds. Barnes FS and Greenebaum B, CRC Press, Boca Raton, FL, pp. 136–137.

Barnes F and Greenebaum B. 2015. The effects of weak magnetic fields on radical pairs. *Bioelectromagnetics* 36(1):45–54.

Batchelor S, Kay C, McLauchlan K, and Shkrob I. 1993. Time-resolved and modulation methods in the study of the effects of magnetic fields on the yields of free-radical reactions. *J Phys Chem* 97:13250–13258.

Bingham C. 1996. The effects of DC and ELF AC magnetic fields on the division rate of mastocytoma cells. PhD Thesis, University of Colorado Boulder.

Brocklehurst B and McLauchlan K. 1996. Free radical mechanism for the effects of environmental electromagnetic fields on biological systems. *Int J Radiat Biol* 69:3–24.

Cai J, Caruso F, and Plenio MB. 2012. Quantum limits for the magnetic sensitivity of a chemical compass. *Phys Rev A* 85:040304(R).

Castello P, Hill I, Portelli L, Barnes F, Usselman R, and Martino C. 2014. Inhibition of cellular proliferation and enhancement of hydrogen peroxide production in fibrosarcoma cell line by weak radio frequency magnetic fields. *Bioelectromagnetics* 35(8):598–602.

Cheeseman KH and Slater TF. 1993. An introduction to free radical biochemistry. *Br Med Bull* 49:481–493.

Droge W. 2002. Free radicals in the physiological control of cell function. *Physiol Rev* 82:47–95.

Grissom CB. 1995. Magnetic field effects in biology: A survey of possible mechanisms with emphasis on radical-pair recombination. *Chem Rev* 95:3–24.

Hayashi H. 2004. *Introduction to Dynamic Spin Chemistry: Magnetic Field Effects upon Chemical and Biochemical Reactions*. World Scientific Publishing Co., Singapore. p. 268.

Henbest KB, Kukura P, Rodgers CT, Hore PJ, and Timmel CR. 2004. Radio frequency magnetic field effects on a radical recombination reaction: A diagnostic test for the radical pair mechanism. *J Am Chem Soc* 126:8102–8103.

Henbest KB, Maeda K, Athanassiades E, Hore PJ, and Timmel CR. 2006. Measurement of magnetic field effects on radical recombination reactions using triplet–triplet energy transfer. *Chem Phys Lett* 421:571–576.

Hoben HJ, Efimova O, Wagner-Rundell N, Timmel C, Hore PJ. 2009. Possible involvement of superoxide and dioxygen with cryptochome in avian magnetoreception: Origin of Zeeman resonances observed by in vivo EPR spectroscopy. *Chem Physics Letters* 480:118–122, Singapore: World Scientific Publishing Co. p. 268.

Kaptein R. 1968. Chemically induced dynamic nuclear polarization in five alkyl radicals. *Chem Phys Lett* 2:261–267.

Kaptein R. 1972. Chemically induced dynamic nuclear polarization. X. On the magnetic field dependence. *J Am Chem Soc* 94:6269–6280.

Kipriyanov A Jr, Doktorov A, and Purtov P. 2015. Magnetic field effects on bistability and bifurcation phenomena in lipid peroxidation. *Bioelectromagnetics* 36:485–493.

Korneva H, Grigoriev V, Isaeva E, Kaloshina S, and Barnes F. 1999. Effects of low-level 50 Hz magnetic fields on the level of host defense and on spleen colony formation. *Bioelectromagnetics* 20:57–62.

Koyamaa S, Nakaharaa T, Hirosec H, Dinga G, Takashimaa Y, Isozumib Y, Miyakoshia J. 2004. ELF electromagnetic fields increase hydrogen peroxide (H_2O_2)-induced mutations in pTN89 plasmids. *Mutation Research/Genetic Toxicology and Environmental Mutagenesis* 560(1): 27–32.

Lee HC, Hong MN, Jung SH, Kim BC, Suh YJ, Ko YG, Lee YunSil, Lee B-Y, Cho Y-G, Myung S-H, Lee J-S. 2015. Effect of extremely low frequency magnetic fields on cell proliferation and gene expression. *Bioelectromagnetics* 36:506–516.

Levin M. 2010. Endogenous bioelectric signals as morphogenetic controls of development, regeneration, and neoplasm, Chapter 3. In *The Physiology of Bioelectricity in Development, Tissue Regeneration and Cancer*, Ed. Pullar C, CRC Press, Boca Raton, FL.

Li Y, Ru B, Liu X, Miao W, Zhang K, Han L, Ni H, and Wu H. 2014. Effects of extremely low frequency alternating-current magnetic fields on the growth performance and digestive enzyme activity of tilapia *Oreochromis niloticus*. *Environ Biol Fish* 98(1):337–343.

Liu Y, Edge R, Henbest K, Timmel C, Hore P, and Gasta P. 2005. Magnetic field effect on singlet oxygen production in a biochemical system. *Chem Commun* 41:174–176.

Mannerling AC, Simkó M, Mild KH, and Mattsson MO. 2010. Effects of 50-Hz magnetic field exposure on superoxide radical anion formation and HSP70 induction in human K562 cells. *Radiat Environ Biophys* 49:731–741.

Martino C, McCabe K, Portelli L, Hernandez M, and Barnes F. 2010. Reduction of the earth's magnetic field inhibits growth rates of model cancer cell lines. *Bioelectromagnetics* 8:649–655.

Morton J, Czajka J, Cyrus J, Barnes F, Beane, W. 2016. "Weak Magnetic Field Manipulation Disrupts Regenerative Outgrowth in Planaria" Annual Meeting of Bioelectromagnetics Society.

Nuccitelli R. 1988. Ionic currents in morphogenesis. *Experientia* 44:657–666.

Nuccitelli R. 1990. The vibrating probe technique for studies of ion transport. In *Noninvasive Techniques in Cell Biology*, Eds. Foskett JK and Grinstein S, Wiley-Liss, New York, pp. 273–310.

Okano H. 2008. Effects of static magnetic fields in biology: Role of free radicals. *Frontiers in Bioscience* 13(13):6106-6125, February 2008.

Patel N and Poo M. 1982. Orientation of neurite growth by extracellular electric fields. *J Neurosci* 2:483–496.

Pilla A. 2007. Mechanisms and therapeutic applications of time-varying and static magnetic fields, Chapter 11. In *Handbook of Biological Effects of Electromagnetic Fields Biological and Medical Aspects of Electromagnetic Fields*, Eds. Barnes FS and Greenebaum B, CRC press, Boca Raton, FL, pp. 351–413.

Poo M. 1982. Rapid lateral diffusion of functional ACh receptors in embryonic muscle cell membrane. *Nature* 295:332–334.

Rathnabharathi K, Zhou R, Aly A, Cheema MI, Anchala D, Laterza R, and Barnes F. 2005. Some effects of electric and magnetic fields on the movement of white blood cells. Bioelectromagnetics Annual Meeting, University of Colorado Boulder, Boulder.

Rhee S, Lee S, Yang K, Kwon J, and Kuang S. 2003. Hydrogen peroxide as intercellular messenger: Idenification of protein phosophatases and PTEN as H_2O_2, Chapter 9. In *Signal Transduction by Reactive Oxygen and Nitrogen Species: Pathways and Chemical Properties*, Eds., Forman J, Fukuto J, and Torres M, Kluwer Academic Publishers, Dordrecht, pp. 167–179.

Ritz T, Thalau P, Phillips JB, Wiltschko R, Wiltschko W. 2004. Resonance effects indicate a radical-pair mechanism for avian magnetic compass. *Nature* 429:177–180.

Rodgers C, Norman S, Henbest K, Timmel C, and Hore P. 2007. Determination of radical re-encounter probability distributions from magnetic field effects on reaction yields. *J Am Chem Soc* 129:6746–6755.

Simkó M. 2007. Cell type specific redox status is responsible for diverse electromagnetic field effects. *Gen Med Chem* 14:1141–1152.

Slaby O and Lebiedz D. 2009. Oscillatory NAD(P)H waves and calcium oscillations in neutrophils? A modeling study of feasibility. *Biophys J* 96(2):417–428.

Steiner U and Ulrich T. 1989. Magnetic field effects in chemical kinetics and related phenomena. *Chem Rev* 89:147–151.

Tracey KJ. 2015. Shock medicine. *Sci Am* 312:28–35, |doi:10.1038/scientificamerican0315-28.

Usselman RJ, Hill I, Singel DJ, and Martino CF. 2014. Spin biochemistry modulates reactive oxygen species (ROS) production by radio frequency magnetic fields. *PLoS One* 9:e101328.

Vijayalaxmi, Cao Y, and Scarfi MR. 2014. Adaptive response in mammalian cells exposed to non-ionizing radiofrequency fields: A review and gaps in knowledge. *Mutat Res Rev Mutat Res* 760:36–45.

Wang K and Ritz T. 2006. Zeeman resonances for radical-pair reactions in weak static magnetic fields. *Mol Phys* 104:1649–1658.

Woodward JR, Timmel CR, McLauchlan KA, and Hore PJ. 2001. Radio frequency magnetic field effects on electron-hole recombination. *Phys Rev Lett* 87:077602-1–077602-4.

Yakymenko I, Tsybulin O, Sidorik E, Henshel D, Kyrylenko O, Sergiy Kyrylenko S. 2016. Oxidative mechanisms of biological activity of low intensity radio frequency radiation. *Electromagn Biol* Med 1–17.

Zmyslony M, Rajkowska E, Mamrot P, Politanski J, and Jajte J. 2004. The effect of weak 50 Hz magnetic fields on the number of free oxygen radicals in rat lymphocytes *in vitro*. *Bioelectromagnetics* 25:607–6612.

9

Duration of Exposure and Dose in Assessing Nonthermal Biological Effects of Microwaves

Igor Belyaev
Slovak Academy of Sciences and Russian Academy of Sciences

9.1 Introduction

Radiofrequency (RF) radiation (3 kHz–300 GHz) or microwaves (MW, 300 MHz–300 GHz) induce a variety of biological and health effects that are usually classified into thermal and nonthermal (NT) effects. This classification is not based on physics of interaction between MW and biological tissues but rather reflects experimental observation of heating induced by MW exposure. The thermal effects are defined as those induced by elevation of temperature in the MW-exposed tissue. The thermal effects of acute exposure to MW are well described by the dose rate, which is defined as specific absorption rate (SAR) in bioelectromagnetics. SAR is often called "dose," which is physically incorrect. In physics, the dose (J/kg) is defined as specific absorbed energy and is calculated as the dose rate (W/kg) multiplied by exposure time. SAR is a main determinant for thermal MW effects, which usually correlate with SAR value. The SAR-based safety limits protect from thermal MW effects only (ICNIRP, 1998). Along with thermal MW effects, diverse biological NT responses to MW, which are observed at SAR values well below any measurable elevation of temperature, have been described by many research groups all over the world (Binhi, 2011; IARC, 2013; Belyaev, 2015a,b; Morgan et al., 2015). The main issue in assessing the risks from NT MW is finding out feasible parameters that would correlate with the exposure risks. It has been known for a long time that SAR value is not a relevant parameter and cannot be applied for safety limits dealing with NT MW; e.g., from mobile communication systems (Belyaev, 2010; Binhi, 2011). In contrast to the International Commission on Non-Ionizing Radiation Protection (ICNIRP), Russian safety standards, which are based on NT effects, do not use SAR values but instead limit the duration of exposure and power flux density (PD, W/cm^2) (SanPiN, 1996).

For chemical substances in a toxicology model, a dose–response relationship is usually observed: the greater the dose, the greater the effect. In analogy with toxicology, MW experiments tended to be designed with high doses and with little regard for other parameters such as modulation, frequency, and duration of exposure, which were shown to be important for the NT MW bioeffects. On the basis of mechanistic consideration of the NT MW effects, Frey has suggested that the toxicology model used by many investigators was not an appropriate model on which the experiments with MW should be designed (Frey, 1993). This may be one reason why many MW studies yielded so little useful information (Frey, 1993).

In analogy with ionizing radiation, where dose is a definitive factor for safety limits, the dose was also considered for the NT MW effects. The role of exposure duration and dose in assessing the NT MW effects has been analyzed by many research groups using various endpoints. Laboratory *in vitro* studies, animal *in vivo* studies, and human studies were performed. A survey of studies, which evaluate the dependence of the NT MW effects on exposure duration and dose, is provided in this chapter. Only studies in which duration and dose of MW exposure was varied and a positive effect was found at least under one exposure condition to be comparable with other exposure conditions were considered.

9.2 Laboratory *In Vitro* Studies

Bozhanova et al. reported that the cell synchronization induced by MW depended on duration of exposure and power density (PD) (Bozhanova et al., 1987). The dependence on duration of exposure fit to exponential function. The important observation was that to achieve the same MW effect, decrease in PD could be compensated for by the increase in the duration of exposure.

Exposure of *Escherichia coli* cells and rat thymocytes to MW at the PD of 10^{-5}–10^{-3} W/cm^2 resulted in significant changes in chromatin conformation when exposure was performed at specific "resonance" frequencies during 5–10 min (Belyaev et al., 1992a,b; Belyaev and Kravchenko, 1994). Decrease in the MW effects due to lowering the PD by orders of magnitude down to 10^{-14}–10^{-17} W/cm^2 could be compensated for by a severalfold increase of exposure duration to 20–40 min (Belyaev et al., 1994). At a relatively longer duration of exposure, more than 1 h, and the lowest PD of 10^{-19} W/cm^2, the same effect was induced as at highest PD and shorter durations (Belyaev et al., 1994). These data have shown that duration of exposure may be even more important for biological effects of NT MW than PD.

Kwee and Raskmark studied the effects of MW at 960 MHz and NT SAR of 0.021, 0.21, and 2.1 mW/kg on proliferation of human epithelial amnion cells (Kwee and Raskmark, 1998). Linear correlations between duration of exposure to MW and MW-induced changes in cell proliferation were observed at 0.021 and 2.1 mW/kg, while no such correlation was seen at 0.21 mW/kg.

Peinnequin et al. studied the effects of 24- or 48-h MW exposure at 2.45 GHz and NT level 5 mW/cm², on apoptosis induced by Fas in human Jurkat cells (Peinnequin et al., 2000). MW affected Fas-induced apoptosis. This effect was observed only on 48-h exposure and depended on exposure duration.

Gerner et al. exposed human fibroblasts to modulated MW from Global System for Mobile Communications (GSM) 1800 MHz at a SAR value of 2 W/kg (Gerner et al., 2010). While short-term exposure within 2 h did not significantly alter the proteome, an 8-h exposure caused a significant and reproducible increase in protein synthesis. Most of the proteins found to be induced were chaperones, which are mediators of protein folding. Heat-induced proteome alterations detectable with the used proteome methodology would require heating >1°C. Because GSM-induced heating was <0.15°C, a heat-related response was excluded. These data further supported the notion that exposure time can be a critical factor.

Differentiated astroglial cell cultures were exposed for 5, 10, or 20 min to either 900 MHz continuous waves (CW) or amplitude-modulated at 50 Hz MW (Campisi et al., 2010). The strength of the electric field was 10 V/m (rms). The exposure conditions excluded any thermal effect. A significant increase in levels of reactive oxygen species (ROS) and DNA fragmentation was found only after exposure of the astrocytes to modulated MW for 20 min. No effects were detected at shorter exposure durations.

Schrader et al. analyzed mitotic spindle disturbances in FC2 cells, a human hamster hybrid A(L) cell line, induced by MW at a frequency of 835 MHz with a field strength of 90 V/m (Schrader et al., 2008). Sigmoid dependence on duration of exposure was observed a with linear increase up to 30 min of exposure and saturation at longer exposures up to 2 h.

Markova et al. exposed human mesenchymal stem cells (MSC) and fibroblasts to MW from GSM 915 MHz and Universal Mobile Telecommunication System (UMTS) 1947 MHz at a SAR of 37 and 39 mW/kg, respectively (Markova et al., 2010). The authors found that MW inhibited formation of DNA repair foci and that the inhibitory effect of MW levelled off at 1-h exposure. No further increase in MW effect was observed in MSC and fibroblasts by prolongation of exposure to 3 h. This data was in agreement with results obtained by the same research group with human peripheral blood lymphocytes showing that MW inhibitory effects were the same at 1- and 2-h exposures (Belyaev et al., 2005; Markova et al., 2005).

Campisi et al. exposed primary rat neocortical astroglial cells to MW at either 900 MHz CW or 900 MHz waves modulated in amplitude at 50 Hz (Campisi et al., 2010). Duration of exposure was 5, 10, or 20 min. A significant increase in ROS levels and DNA fragmentation was found after exposure of the astrocytes to modulated MW for 20 min only. No effects were detected with shorter exposure durations. The irradiation conditions excluded thermal effect.

Esmekaya et al. exposed human peripheral blood lymphocyte to GSM-modulated MW radiation at 1.8 GHz and SAR of 0.21 W/kg for 6, 8, 24, and 48 h (Esmekaya et al., 2011). The authors reported morphological changes in exposed lymphocytes. Longer exposure periods resulted in destruction of organelle and nucleus structures. Chromatin change and the loss of mitochondrial crista, which occurred in cells exposed to MW for 8 and 24 h, were more pronounced in cells exposed for 48 h. This effect was not thermal as exposure did not increase the temperature. The authors concluded that greater damage occurred after a longer duration of MW exposure.

Lu et al. demonstrated that ROS plays an important role in the process of apoptosis in human peripheral blood mononuclear cells, which can be induced by exposure to MW at a frequency of 900 MHz and a SAR of 0.4 W/kg (Lu et al., 2012). An almost linear response was observed in ROS induction depending on duration of MW exposure within 1–6 h, while ROS slightly decreased after 8-h exposure. In line with this data, induction of caspase 3 and apoptotic rate strongly depended on the exposure time.

Naziroglu et al. measured cytosolic free Ca^{2+} in human leukemia cells during 1–24 h exposure to 2.45 GHz electromagnetic radiation at the average SAR of 1.63 W/kg (Naziroglu et al., 2012). Radiation-induced increase of cytosolic free Ca^{2+} concentration was time-dependent and reached its maximum at 24-h exposure.

Trivino Pardo et al. exposed acute T-lymphoblastoid leukemia cells to MW (900 MHz, 3 V/m, 2- and 48-h exposure times) generated by a transverse electromagnetic (TEM) cell (Trivino Pardo et al., 2012). The results showed a statistically significant decrease in the total cell number after 48 h of exposure. No significant effect on cellular viability was observed at the shorter exposure time.

Valbonesi et al. studied stress response in rat PC12 cells exposed for 4, 16, or 24 h to MW GSM 1.8 GHz modulated at 217 Hz, 2 W/kg (Valbonesi et al., 2014). After exposure to the GSM 217 Hz signal for 16 or 24 h, transcription of HSP70 stress protein significantly increased, whereas no effect was observed in cells exposed for 4 h.

Manta et al. measured ROS in *Drosophila melanogaster* following exposure for 0.5, 1, 6, 24, and 96 h to a wireless Digital Enhanced Cordless Telephone (DECT) base radiation (1.88–1.90 GHz, 2.7 V/m, and 0.009 W/Kg) (Manta et al., 2014). ROS levels increased twofold in male and female bodies after 6 h of exposure, slightly increasing thereafter at 24 and 96 h. Ovaries of exposed females had a 1.5-fold increase in ROS after 0.5 h, reaching 2.5-fold after 1 h with no elevation at 6, 24, and 96 h. ROS levels returned to normal in the male and the female bodies 24 h after 6 h of exposure of the flies and in the ovaries 4 h after 1 h exposure of the females.

Sefidbakht et al. exposed a stable cell line (HEK293T) harboring the firefly luciferase gene to MW at 940 MHz, 0.09 W/kg (Sefidbakht et al., 2014). Cell cultures were lysed along with their matched nonexposed control cells after exposure for 15, 30, 45, 60, and 90 min. Luciferase activity and oxidative response elements were investigated. Endogenous luciferase activity was reduced after 30 and 45 min of continuous exposure, while after 60 min, the exposed cell lysate showed higher luciferase activity compared with the nonexposed control. MW exposure induced ROS reaching maximum at 30 min of continuous exposure. MW effect on catalase (CAT) activity, glutathione (GSH), caspase 3/7 activity, and malondialdehyde (MDA) depended on duration of exposure with maximum observed at 45 min. The data suggested that an increase in the activity of luciferase after 60 min of continuous exposure could be associated with a decrease in ROS level caused by activation of the oxidative response.

Liu et al. exposed mouse spermatocyte-derived cells to GSM MW in GSM-Talk mode at 1800 MHz and SAR values of 1, 2 or 4 W/kg for 12, 24, and 24 h (Liu et al., 2014). Expression of autophagic marker LC3-II increased in a SAR- and time-dependent manner with MW exposure, showing significant change at 4 W/kg. To avoid thermal effects, the exposure system was kept a constant temperature and the temperature difference between real- and sham-exposed samples did not exceed 0.1°C. Slightly increased ROS production was detected after 12-h exposure, while significant increase was seen after 24-h exposure, and ROS levels did not increase after 3- and 6-h exposure. The authors concluded that intracellular ROS levels significantly increased in a SAR- and time-dependent manner upon exposure to MW.

Sarapultseva et al. analyzed the effect of low-level MW exposure for 0.05–10 h at 1 and 10 GHz with a PD of 0.05, 0.1, and 0.5 W/m^2 on ciliates *Spirostomum ambiguum* (Sarapultseva et al., 2014). Almost identical dose-dependent inhibition of the motility of ciliates was observed at both frequencies. At the highest PD of 0.1 and 0.5 W/m^2, the maximum effect of about 50% was observed at a dose of 300 J/m^2. The value of the effect was the same at a wide range of higher doses. Lower PD was less efficient, and the maximum reduction of motility was observed at a dose of 1400 J/m^2. This study indicated that the effect can be described by dose within the limited PD values.

Hou et al. investigated ROS, DNA damage, and apoptosis in mouse embryonic fibroblasts (NIH/3T3) after intermittent exposure (5 min on/10 min off, for various durations from 0.5 to 8 h) to MW at 1800-MHz GSM talk mode, 2 W/kg (Hou et al., 2015). Significant increase in intracellular ROS levels was observed, reaching the highest level at 1-h exposure time and followed by a slight decrease when the exposure continued for 8 h. The number of late-apoptotic cells increased significantly after exposure for 1, 4, or 8 h, while no effects on DNA damage measured with γH2AX foci was observed. Altogether, this study indicated that exposure should be performed for at least 1 h to observe the effects under study in the NIH/3T3 cells.

In some studies, prolonged MW exposures were associated with less prominent effects than shorter exposure (Nikolova et al., 2005; Tkalec et al., 2007; Markova et al., 2010). This type of dependence on exposure duration was explained by a specific adaptation of the exposed biosystems to MW exposure (Markova et al., 2010).

9.3 Animal *In Vivo* Studies

Koveshnikova et al. exposed rats to pulsed MW (carrier frequency 3 GHz, pulse repetition 400 Hz, rectangular pulses of 2 μs, PD of 100, 500, and 2500 μW/cm^2), during 60 days, 12 h/day (Koveshnikova and Antipenko, 1991). Chromosomal aberrations, the "gold standard" for estimation of mutagenicity and biological dosimetry, were analyzed in hepatocytes. Exposure was performed in three arrays of pulses so that 16, 29, or 48 arrays of pulses per min were generated. The ratio of the obtained doses per animal was 1: 1.8: 3, correspondingly. Increased level of CA was generally observed at PD >100 μW/cm^2. Importantly, the differences between PD disappeared when the dose per animal increased. In particular, even the PD of 100 μW/cm^2 induced CA at higher absorbed doses. The data indicated that the dose may be an important parameter for estimation of mutagenicity and cancer risks.

Persson et al. studied the effects of CW and pulse-modulated MW at 915 MHz and at various exposure durations on permeability of the blood–brain barrier (BBB) in Fischer 344 rats (Persson et al., 1997). In these experiments, the CW pulse power varied from 0.001 to 10 W and the exposure duration ranged from 2 to 960 min. Albumin and fibrinogen were demonstrated immunochemically and classified as normal versus pathological BBB leakage. The frequency of pathological BBB leakage significantly increased in all exposed rats. The authors grouped the exposed animals according to the level of specific absorbed energy/dose (J/kg) and found significant differences in all doses above 1.5 J/kg. The effect was not dependent on dose; however, different doses might be obtained with different durations of exposure.

Gapeyev et al. evaluated the anti-inflammatory effect of low-intensity MW exposure at 0.1 mW/cm^2 on mice with acute footpad edema induced by zymosan (Gapeyev et al., 2008). Whole-body MW exposure of mice at a frequency of 42.2 GHz reduced both the footpad edema and local hyperthermia induced by zymosan injection. This effect had a sigmoid dependence on exposure duration with a maximum at 20–80 min. A linear dependence on the exposure duration with significantly lower increment was observed at the tenfold reduced intensity of 0.01 mW/cm^2. This decrease in the effect was compensated by an increase in duration of exposure from 80 to 120 min. In line with studies by Belyaev et al. (1994), this data has shown that duration of exposure may be even more important for biological effects of NT MW than PD.

Adang et al. exposed Wistar albino rats to low-intensity MW at two different frequencies and exposure modes, 2 h a day, 7 days a week, during 21 months (Adang et al., 2009). Exposure affected the immune system, which reacted to this stressor by increasing the percentage of monocytes in the peripheral blood. After 14 and 18 months of exposure, a significant increase in white blood cells and neutrophils of about 15% and 25%, respectively, was found. Lymphocyte levels reduced after 18 months of exposure with about 15% compared to the sham-exposed group. No effects were observed at shorter exposure intervals.

Panagopoulos and Margaritis studied the effects of MW exposure from GSM 900 MHz and DCS 1800 MHz (Digital Cellular System—referred to also as GSM 1800 MHz) mobile phones at an intensity of about 10 μW/cm^2 on the reproductive capacity of *D. melanogaster* (Panagopoulos and Margaritis, 2010). For both radiation types, GSM 900 and DCS 1800, the reproductive capacity of insects decreased almost linearly with increasing duration of exposure from 1 to 21 min. The data suggested that short-term exposure to mobile phone radiation had a cumulative effect.

Jorge-Mora et al. studied expression of c-Fos in paraventricular nucleus (PVN) of the hypothalamus extracted from brains of rats exposed once or repeatedly (10 times in 2 weeks) to MW at 2.45 GHz and NT SAR of 0.776 and 0.301 W/kg (Jorge-Mora et al., 2011). High SAR triggered an increase of the c-Fos from 90 min to 24 h after exposure, while low SAR resulted in c-Fos increase after 24 h. Repeated MW exposure at 0.776 W/kg increased cellular activation of the PVN by more than 100% compared with acute exposure and repeated sham exposures. The data are in line with other studies, which suggest that the time of exposure to single or repeated MW exposure is one of the determining factors in establishing power levels that can produce a response.

Kim et al. exposed rats to MW at 915 MHz from radiofrequency identification (RFID) (whole-body SAR of 4 W/kg) for 1 or 8 h/day, 5 days/week, for 4 weeks during the nighttime) and investigated melatonin biosynthesis and the activity of rat pineal arylalkylamine *N*-acetyltransferase (AANAT) (Kim et al., 2015). This exposure had no significant effect on body temperature. Urinary concentration of melatonin and 6-OHMS was decreased in the 8-h nocturnal RFID-exposed group but not in the 1-h RFID-exposed group. The authors concluded that nocturnal exposure duration of 8 h is required to have an impact on melatonin, while 1 h does not suffice.

9.4 Human Studies with Volunteers and Epidemiological Studies

Croft et al. tested 24 human volunteers for resting EEG and phase-locked neural responses to auditory stimuli during real and sham exposure to MW from a mobile phone (Croft et al., 2002). MW exposure altered resting EEG, decreasing 1–4 Hz activity, and increasing 8–12 Hz activity as a function of exposure duration. The mobile phone exposure also altered early phase-locked neural responses, attenuating the normal response decrement over time in the 4–8 Hz band, decreasing the response in the 12–30 Hz band globally and as a function of time, and increasing midline frontal and lateral posterior responses in the 30–45 Hz band. The data indicated that the mobile phone exposure affected neural function in humans depending on the exposure duration.

Using alkaline comet assay, Cam and Seyhan analyzed DNA damage in hair root cells of volunteers before and after they used 900 GHz GSM mobile phone for 15–30 min. The 900 GHz GSM exposure significantly increased single-strand DNA breaks in cells of hair roots close to the position of the phone at the heads of volunteers. Exposure during 30 min induced more DNA damage than 15 min exposure (Cam and Seyhan, 2012).

Agarwal et al. reported that the negative impact of mobile phone usage on semen quality in human males correlates with the duration of exposure (Agarwal et al., 2008; Agarwal et al., 2009).

In a recent German study, the level of stress hormones was studied in urine of persons exposed to MW from a base station at a power density of $<60 \mu W/m^2$ (24 out of 60 participants), $60–100 \mu W/m^2$ (20 participants), and to more than $100 \mu W/m^2$ (16 participants) (Buchner and Eger, 2011). The values of adrenaline and noradrenaline increased significantly during the first 6 months after exposure to the GSM base station; the values of the precursor substance dopamine substantially decreased in this time period. The decrease in dopamine was stronger for higher exposure levels. This decrease was not restored to the initial level even after 1.5 years. The phenylethylamine levels dropped in the group exposed above $100 \mu W/m^2$. These effects showed a dose–response relationship.

Kucer and Pamukcu analyzed the association of different self-reported symptoms with exposure to electromagnetic fields from mobile phones using a questionnaire on 350 people aged +9 years (Kucer and Pamukcu, 2014). Users of mobile phones more often complained of headache, joint and bone pain, hearing loss, vertigo/dizziness, and tension–anxiety symptoms. A significant increase in headache, hearing loss, and joint and bone pain was reported for the users of a mobile phone daily for >16 min in comparison to those using a mobile phone daily for <16 min.

Epidemiologic studies have consistently shown an increase in brain cancer risks at prolonged (usually ≥10 years) duration of exposure to mobile phones while the risks are significantly lower at shorter durations (Hardell et al., 2012). Recently reported indication of a dose–response relationship between chronic exposure to cellular phone MW and parotid gland malignancy indicated the necessity of the dose approach at the epidemiological level (Duan et al., 2011). For the first time in the epidemiology of MW-induced tumors, Cardis et al. used estimates of RF energy depositions at the center of tumors in the brain as a measure of MW dose (Cardis et al., 2011). An increased risk of glioma was seen in individuals at the highest quintile of RF dose, though reduced risks were seen in the four lower quintiles. When risk was examined as a function of dose received in different time windows before diagnosis, increasing trend was observed with increasing MW dose (for exposures 7 years or more in the past).

Li et al. performed a population-based case–control study in Taiwan considering cancer incidence cases aged 15 years or less and admitted in 2003–2007 for all neoplasms ($n = 2606$), including 939 leukemia and 394 brain neoplasm cases (Li et al., 2012). The cancer risks were estimated versus dose, which was defined as the annual summarized power (ASP, W-year) and was calculated for each of the 71,185 mobile phone base stations (MPBS) in service between 1998 and 2007. Then, the annual power density (APD, W-year/km^2) of each township ($n = 367$) was computed as a ratio of the total ASP of all MPBS in a township to the area of that particular township. Exposure of each study subject to MW was indicated by the averaged APD within 5 years prior to the neoplasm diagnosis (cases). A higher-than-median averaged APD (approximately 168 W-year/km^2) was significantly associated with an increased OR for all neoplasms, but not for leukemia or brain neoplasm taken separately. Thus, this study found a significantly increased risk of all neoplasms in children with higher-than-median MW exposure to MPBS.

Carlberg and Hardell analyzed the survival of 1678 glioma patients from Sweden with respect to their usage of mobile phones. The longest use of wireless phones for >20 years latency group (time since first use) yielded a maximal increased hazard ratio (HR) of 1.7, 95% confidence interval (CI) = 1.2–2.3 for all gliomas. For the most dangerous brain tumor, glioblastoma multiforme (astrocytoma grade IV), mobile phone use resulted in maximum HR = 2.0, 95% CI = 1.4–2.9, and cordless phone use resulted in HR = 3.4, 95% CI = 1.04–11 for >20 years latency group. The HR for astrocytoma grade IV increased statistically significantly per year of latency for wireless phones, HR = 1.020, 95% CI = 1.007–1.033, but not per 100 h cumulative use, HR = 1.002, 95% CI = 0.999–1.005. The HR increase was not statistically significant for other types of glioma.

In their recent study, Hardell and Carlberg analyzed a pooled data from two case–control studies on malignant brain tumors with patients diagnosed during 1997–2003 and 2007–2009 in Sweden (Hardell and Carlberg, 2015). In total, 1498 cases and 3530 control subjects participated. Mobile phone use increased the risk of glioma, the highest threefold increased risk being observed in the longest latency group >20 years. The risk for 3G users increased by 4.7% per 100 h cumulative use and by 15.7% per year of latency. The risk increased linearly up to 10,000 h cumulative use and up to 28 years of latency.

The authors concluded that the risk for mobile and cordless phone use increased significantly with both cumulative use and latency.

9.5 Discussion

Many parameters including duration and PD of exposure are important for the NT MW effects to be observed. The reason for variability in the NT MW effects is their dependence on a number of biological and physical parameters, which varied between studies and have been previously reviewed (Belyaev, 2010). Interestingly, the analysis of the duration of the exposure showed that exposure was on average twice as high in the case of studies showing MW effects than in studies with no significant effects (Cucurachi et al., 2013). This chapter focused on the role of duration and dose of NT MW exposure in appearance of biological and health effects. Thus, the studies reporting no effects at any exposure condition, which could be comparable between different doses and durations of exposure, were not considered.

Overall, available studies show that the NT MW biological effects depend on duration of exposure. Importantly, prolonged exposure to NT MW from mobile communication correlated with increased cancer risk. There is no doubt that the duration of exposure plays a significant role in health effects of NT MW exposure and the SAR concept is not applicable for guidelines in preventing the NT MW effects. Thus, available data strongly suggest that duration of exposure should be applied for assessing health effects and be considered in safety limits. Adaptation to prolonged exposure was also reported, which should be taken into account in the assessment of risks.

On the whole, a strong variation in dose response for different endpoints was reported. Few *in vitro*, *in vivo*, and human studies show dose-dependent effects, while other studies did not. So far, there is limited evidence that the dose may be a feasible parameter for assessment of cancer risks.

Russia was the first country in the world to introduce safety standards for RF/MW exposure. These limits were based on around 30-years of research performed in several Soviet institutions. Most data were never published. On the basis of the obtained data, the safety standards (SanPiN) were issued. According to the SanPiN study (SanPiN, 1996), MW (300 MHz–300 GHz) exposure was limited by exposure duration and PD. For example, the PD of whole-body exposure was limited to $25\,\mu W/cm^2$ for durations >8 h, while exposure at $1000\,\mu W/cm^2$ was limited by a duration ≤0.2 h. According to 40 Soviet studies published in Russian and selected by the Russian National Committee of Non-Ionizing Radiation Protection (RNCNIRP) based on standard quality criteria, the unfavorable bioeffects were observed in animals under chronic MW exposures (Grigoriev et al., 2003). These studies have been performed in the 1960s–1990s in five research institutions of Moscow, St. Petersburg, Tomsk, and Kiev. Animals (mice, rats, rabbits, guinea pigs) were chronically, up to 4 months, exposed to MW at different frequencies and various modulations with PD from 1 to $1000\,\mu W/cm^2$ continuously or intermittently. The endpoints included the analysis of weight of animal, histological analysis and weight of tissues, central nervous system, arterial pressure, blood and hormonal studies, immune system, metabolism and enzymatic activity, reproductive system, teratogenic and genetic effects. RNCNIRP concluded that: (1) data on chronic exposure should be considered during development of the radiation guidelines, (2) application of the SAR concept at NT PD less than $100\,\mu W/cm^2$ is questionable, (3) the role of other parameters such as modulation and duration of exposure should be taken into account, (4) further development of the MW safety guidelines would greatly benefit from the knowledge of the biophysical mechanisms behind the NT MW effects.

During recent years, significant progress has been made in understanding the NT MW bioeffects (Belyaev, 2015a). These NT effects occurred because of oxidative stress, modification in transmembrane processes, changes in gene/protein expression, cell metabolism, transmembrane signal transduction, cell cycle progression, intracellular signaling cascades, and conformational changes (Belyaev, 2015b).

Current safety standards are insufficient to protect the public from chronic exposures to NT MW. The SAR concept, which has been widely adopted for protection against acute thermal effects of MW, is not useful for the evaluation of health risks from NT MW of mobile communication. New safety standards should be developed based on knowledge of mechanisms for NT MW effects. Precautionary principles should be implemented while new standards, which would be feasible for protection against NT MW effects, are in progress. It should be anticipated that a definite part of the human population, such as hypersensitive persons, which constitute about 1–10% of the general population in economically developed countries (Belpomme et al., 2015), children, and pregnant women could be especially vulnerable to NT MW exposures.

9.6 Conclusion

While the dose rate/SAR concept is adequate for description of acute thermal effects, it is not applicable for chronic exposures to NT MW. Such parameters as PD, dose, and duration of exposure should be further analyzed for development of reliable safety standards, which would protect against detrimental effects of chronic exposures to MW at NT intensities. Especially informative are those studies where both power level and duration of exposure are varied. Taken together, these studies indicate that response to NT MW depend on both PD and duration exposure. Importantly, the same response is observed with lower PD but prolonged exposure as at higher PD and shorter exposure. While some studies indicated dose–response relationships, these effects seem to be endpoint dependent. In some cases, the effects observed after short exposures can be restored due to adaptation during prolonged exposure. Thus, the dose-dependent response has not been established universally. So far, a combination of exposure duration with PD seems to be the most appropriate value for setting the safety standards for exposure to MW, e.g., from mobile communication systems such as mobile phones, base stations, and WiFi.

Acknowledgments

Financial support from the National Scholarship Program of the Slovak Republic, the Russian Foundation for Basic Research, the Slovak Research and Development Agency (APVV-15–0250), and the VEGA Grant Agency (2/0109/15) of the Slovak Republic is gratefully acknowledged.

References

Adang, D., Remacle, C. and Vorst, A. V. (2009) Results of a long-term low-level microwave exposure of rats, *IEEE Trans Microwave Theory Tech* 57(10): 2488–97.

Agarwal, A., Deepinder, F., Sharma, R. K., Ranga, G. and Li, J. (2008) Effect of cell phone usage on semen analysis in men attending infertility clinic: an observational study, *Fertil Steril* 89(1): 124–8.

Agarwal, A., Desai, N. R., Makker, K., Varghese, A., Mouradi, R., Sabanegh, E. and Sharma, R. (2009) Effects of radiofrequency electromagnetic waves (RF-EMW) from cellular phones on human ejaculated semen: an in vitro pilot study, *Fertil Steril* 92(4): 1318–25.

Belpomme, D., Campagnac, C. and Irigaray, P. (2015) Reliable disease biomarkers characterizing and identifying electrohypersensitivity and multiple chemical sensitivity as two etiopathogenic aspects of a unique pathological disorder, *Rev Environ Health* 30(4): 251–71.

Belyaev, I. (2010) Dependence of non-thermal biological effects of microwaves on physical and biological variables: Implications for reproducibility and safety standards. In L. Giuliani and M. Soffritti (eds.) *Non-thermal Effects and Mechanisms of Interaction between Electromagnetic Fields and Living Matter*. An ICEMS Monograph, Vol. 5. Bologna, Italy: Ramazzini Institute, European Journal of Oncology – Library, www.icems.eu/papers.htm?f=/c/a/2009/12/15/MNHJ1B49KH.DTL.

Belyaev, I. (2015a) Biophysical mechanisms for nonthermal microwave effects. In M. Markov (ed.) *Electromagnetic Fields in Biology and Medicine*. Boca Raton, FL: CRC Press.

Belyaev, I. (2015b) Electromagnetic field effects on cells and cancer risks from mobile communication. In P. J. Rosch (ed.) *Bioelectromagnetic and Subtle Energy Medicine*, 2nd edition. Boca Raton, FL: CRC Press.

Belyaev, I. Y. and Kravchenko, V. G. (1994) Resonance effect of low-intensity millimeter waves on the chromatin conformational state of rat thymocytes, *Z Naturforsch [C]* 49(5–6): 352–8.

Belyaev, I. Y., Alipov, Y. D., Shcheglov, V. S. and Lystsov, V. N. (1992a) Resonance effect of microwaves on the genome conformational state of *E. coli* cells, *Z Naturforsch [C]* 47(7–8): 621–7.

Belyaev, I. Y., Shcheglov, V. S. and Alipov, Y. D. (1992b) Existence of selection rules on helicity during discrete transitions of the genome conformational state of *E. coli* cells exposed to low-level millimetre radiation, *Bioelectrochem Bioenerg* 27(3): 405–11.

Belyaev, I. Y., Alipov, Y. D., Shcheglov, V. S., Polunin, V. A. and Aizenberg, O. A. (1994) Cooperative response of *Escherichia coli* cells to the resonance effect of millimeter waves at super low-intensity, *Electro and Magnetobiol* 13(1): 53–66.

Belyaev, I. Y., Hillert, L., Protopopova, M., Tamm, C., Malmgren, L. O. G., Persson, B. R. R., Selivanova, G. and Harms-Ringdahl, M. (2005) 915 MHz microwaves and 50 Hz magnetic field affect chromatin conformation and 53BP1 foci in human lymphocytes from hypersensitive and healthy persons, *Bioelectromagnetics* 26(3): 173–84.

Binhi, V. N. (2011) *Principles of Electromagnetic Biophysics* (in Russian). Moscow: Fizmatlit.

Bozhanova, T. P., Bryukhova, A. K. and Golant, M. B. (1987) About possibility to use coherent radiation of extremely high frequency for searching differences in the state of living cells. In N. D. Devyatkov (ed.) *Medical and Biological Aspects of Millimeter Wave Radiation of Low Intensity*, Vol. 280. Fryazino, USSR: IRE, Academy of Science.

Buchner, K. and Eger, H. (2011) Changes of clinically important neurotransmitters under the influence of modulated RF fields—A long-term study under real-life conditions. Original study in German, *Umwelt-Medizin-Gesellschaft* 24(1): 44–57.

Cam, S. T. and Seyhan, N. (2012) Single-strand DNA breaks in human hair root cells exposed to mobile phone radiation, *Int J Radiat Biol* 88(5): 420–4.

Campisi, A., Gulino, M., Acquaviva, R., Bellia, P., Raciti, G., Grasso, R., Musumeci, F., Vanella, A. and Triglia, A. (2010) Reactive oxygen species levels and DNA fragmentation on astrocytes in primary culture after acute exposure to low intensity microwave electromagnetic field, *Neurosci Lett* 473(1): 52–5.

Cardis, E., Armstrong, B. K., Bowman, J. D., Giles, G. G., Hours, M., Krewski, D., McBride, M., Parent, M. E., Sadetzki, S., Woodward, A. et al. (2011) Risk of brain tumours in relation to estimated RF dose from mobile phones: Results from five Interphone countries, *Occup Environ Med* 68(9): 631–40.

Croft, R. J., Chandler, J. S., Burgess, A. P., Barry, R. J., Williams, J. D. and Clarke, A. R. (2002) Acute mobile phone operation affects neural function in humans, *Clin Neurophysiol* 113(10): 1623–32.

Cucurachi, S., Tamis, W. L., Vijver, M. G., Peijnenburg, W. J., Bolte, J. F. and de Snoo, G. R. (2013) A review of the ecological effects of radiofrequency electromagnetic fields (RF-EMF), *Environ Int* 51: 116–40.

Duan, Y., Zhang, H. Z. and Bu, R. F. (2011) Correlation between cellular phone use and epithelial parotid gland malignancies, *Int J Oral Maxillofac Surg* 40(9): 966–972.

Esmekaya, M. A., Aytekin, E., Ozgur, E., Guler, G., Ergun, M. A., Omeroglu, S. and Seyhan, N. (2011) Mutagenic and morphologic impacts of 1.8 GHz radiofrequency radiation on human peripheral blood lymphocytes (hPBLs) and possible protective role of pre-treatment with *Ginkgo biloba* (EGb 761), *Sci Total Environ* 410(411): 59–64.

Frey, A. H. (1993) Electromagnetic field interactions with biological systems, *FASEB J* 7(2): 272–81.

Gapeyev, A. B., Mikhailik, E. N. and Chemeris, N. K. (2008) Anti-inflammatory effects of low-intensity extremely high-frequency electromagnetic radiation: Frequency and power dependence, *Bioelectromagnetics* 29(3): 197–206.

Gerner, C., Haudek, V., Schandl, U., Bayer, E., Gundacker, N., Hutter, H. P. and Mosgoeller, W. (2010) Increased protein synthesis by cells exposed to a 1,800-MHz radio-frequency mobile phone electromagnetic field, detected by proteome profiling, *Int Arch Occup Environ Health* 83(6): 691–702.

Grigoriev, Y. G., Stepanov, V. S., Nikitina, V. N., Rubtcova, N. B., Shafirkin, A. V. and Vasin, A. L. (2003) ISTC Report. Biological effects of radiofrequency electromagnetic fields and the radiation guidelines. Results of experiments performed in Russia/Soviet Union. Moscow: Institute of Biophysics, Ministry of Health, Russian Federation.

Hardell, L. and Carlberg, M. (2015) Mobile phone and cordless phone use and the risk for glioma – Analysis of pooled case-control studies in Sweden, 1997–2003 and 2007–2009, *Pathophysiology* 22(1): 1–13.

Hardell, L., Carlberg, M. and Hansson Mild, K. (2012) Use of mobile phones and cordless phones is associated with increased risk for glioma and acoustic neuroma, *Pathophysiology* 20(2): 85–110.

Hou, Q., Wang, M., Wu, S., Ma, X., An, G., Liu, H. and Xie, F. (2015) Oxidative changes and apoptosis induced by 1800-MHz electromagnetic radiation in NIH/3T3 cells, *Electromagn Biol Med* 34(1): 85–92.

IARC (2013) *Non-ionizing Radiation, Part 2: Radiofrequency Electromagnetic Fields*, IARC Monographs on the Evaluation of Carcinogenic Risks to Humans, Vol. 102. Lyon, France: IARC Press, monographs.iarc.fr/ENG/Monographs/vol102/mono102.pdf.

ICNIRP (1998) ICNIRP Guidelines. Guidelines for limiting exposure to time-varying electric, magnetic, and electromagnetic fields (up to 300 GHz), *Health Phys* 74: 494–522.

Jorge-Mora, T., Misa-Agustino, M. J., Rodriguez-Gonzalez, J. A., Jorge-Barreiro, F. J., Ares-Pena, F. J. and Lopez-Martin, E. (2011) The effects of single and repeated exposure to 2.45 GHz radiofrequency fields on c-Fos protein expression in the paraventricular nucleus of rat hypothalamus, *Neurochem Res* 36(12): 2322–32.

Kim, H. S., Paik, M. J., Lee, Y. H., Lee, Y. S., Choi, H. D., Pack, J. K., Kim, N. and Ahn, Y. H. (2015) Eight hours of nocturnal 915 MHz radiofrequency identification (RFID) exposure reduces urinary levels of melatonin and its metabolite via pineal arylalkylamine N-acetyltransferase activity in male rats, *Int J Radiat Biol* 91(11): 898–907.

Koveshnikova, I. V. and Antipenko, E. N. (1991) On the quantitative regularities of the cytogenic effect of microwaves, *Radiobiologiya* 31(1): 149–51.

Kucer, N. and Pamukcu, T. (2014) Self-reported symptoms associated with exposure to electromagnetic fields: a questionnaire study, *Electromagn Biol Med* 33(1): 15–7.

Kwee, S. and Raskmark, P. (1998) Changes in cell proliferation due to environmental non-ionizing radiation. 2. Microwave radiation, *Bioelectrochem Bioenerg* 44: 251–5.

Li, C. Y., Liu, C. C., Chang, Y. H., Chou, L. P. and Ko, M. C. (2012) A population-based case-control study of radiofrequency exposure in relation to childhood neoplasm, *Sci Total Environ* 435–436: 472–8.

Liu, K., Zhang, G., Wang, Z., Liu, Y., Dong, J., Dong, X., Liu, J., Cao, J., Ao, L. and Zhang, S. (2014) The protective effect of autophagy on mouse spermatocyte derived cells exposure to 1800 MHz radiofrequency electromagnetic radiation, *Toxicol Lett* 228(3): 216–24.

Lu, Y. S., Huang, B. T. and Huang, Y. X. (2012) Reactive oxygen species formation and apoptosis in human peripheral blood mononuclear cell induced by 900 MHz mobile phone radiation, *Oxid Med Cell Longev* 2012: 740280.

Manta, A. K., Stravopodis, D. J., Papassideri, I. S. and Margaritis, L. H. (2014) Reactive oxygen species elevation and recovery in *Drosophila* bodies and ovaries following short-term and long-term exposure to DECT base EMF, *Electromagn Biol Med* 33(2): 118–31.

Markova, E., Hillert, L., Malmgren, L., Persson, B. R. and Belyaev, I. Y. (2005) Microwaves from GSM mobile telephones affect 53BP1 and γ-H2AX foci in human lymphocytes from hypersensitive and healthy persons, *Environ Health Perspect* 113(9): 1172–7.

Markova, E., Malmgren, L. O. G. and Belyaev, I. Y. (2010) Microwaves from mobile phones inhibit 53BP1 focus formation in human stem cells more strongly than in differentiated cells: Possible mechanistic link to cancer risk, *Environ Health Perspect* 118(3): 394–9.

Morgan, L. L., Miller, A. B., Sasco, A. and Davis, D. L. (2015) Mobile phone radiation causes brain tumors and should be classified as a probable human carcinogen (2A) (review), *Int J Oncol* 46(5): 1865–71.

Naziroglu, M., Cig, B., Dogan, S., Uguz, A. C., Dilek, S. and Faouzi, D. (2012) 2.45-Gz wireless devices induce oxidative stress and proliferation through cytosolic Ca^{2+} influx in human leukemia cancer cells, *Int J Radiat Biol* 88(6): 449–56.

Nikolova, T., Czyz, J., Rolletschek, A., Blyszczuk, P., Fuchs, J., Jovtchev, G., Schuderer, J., Kuster, N. and Wobus, A. M. (2005) Electromagnetic fields affect transcript levels of apoptosis-related genes in embryonic stem cell-derived neural progenitor cells, *FASEB J* 19(12): 1686–8.

Panagopoulos, D. J. and Margaritis, L. H. (2010) The effect of exposure duration on the biological activity of mobile telephony radiation, *Mutat Res* 699(1–2): 17–22.

Peinnequin, A., Piriou, A., Mathieu, J., Dabouis, V., Sebbah, C., Malabiau, R. and Debouzy, J. C. (2000) Non-thermal effects of continuous 2.45 GHz microwaves on Fas-induced apoptosis in human Jurkat T-cell line, *Bioelectrochemistry* 51(2): 157–61.

Persson, B. R. R., Salford, L. G. and Brun, A. (1997) Blood-brain barrier permeability in rats exposed to electromagnetic fields used in wireless communication, *Wirel Netw* 3: 455–61.

SanPiN 2.2.4/2.1.8. (1996) Radiofrequency electromagnetic radiation (RF EMR) under occupational and living conditions, Moscow: Minzdrav.

Sarapultseva, E. I., Igolkina, J. V., Tikhonov, V. N. and Dubrova, Y. E. (2014) The in vivo effects of low-intensity radiofrequency fields on the motor activity of protozoa, *Int J Radiat Biol* 90(3): 262–7.

Schrader, T., Munter, K., Kleine-Ostmann, T. and Schmid, E. (2008) Spindle disturbances in human-hamster hybrid (AL) cells induced by mobile communication frequency range signals, *Bioelectromagnetics* 29(8): 626–39.

Sefidbakht, Y., Moosavi-Movahedi, A. A., Hosseinkhani, S., Khodagholi, F., Torkzadeh-Mahani, M., Foolad, F. and Faraji-Dana, R. (2014) Effects of 940 MHz EMF on bioluminescence and oxidative response of stable luciferase producing HEK cells, *Photochem Photobiol Sci* 13(7): 1082–92.

Tkalec, M., Malaric, K. and Pevalek-Kozlina, B. (2007) Exposure to radiofrequency radiation induces oxidative stress in duckweed *Lemna minor* L, *Sci Total Environ* 388(1–3): 78–89.

Trivino Pardo, J. C., Grimaldi, S., Taranta, M., Naldi, I. and Cinti, C. (2012) Microwave electromagnetic field regulates gene expression in T-lymphoblastoid leukemia CCRF-CEM cell line exposed to 900 MHz, *Electromagn Biol Med* 31(1): 1–18.

Valbonesi, P., Franzellitti, S., Bersani, F., Contin, A. and Fabbri, E. (2014) Effects of the exposure to intermittent 1.8 GHz radio frequency electromagnetic fields on HSP70 expression and MAPK signaling pathways in PC12 cells, *Int J Radiat Biol* 90(5): 382–91.

10

Practical Principles of Dosimetry in Studying Biological Effects of Extremely High-Frequency Electromagnetic Fields

Andrew B. Gapeyev
Russian Academy of Sciences

This chapter considers some questions of dosimetry for which solutions are necessary by studying the mechanisms of biological effects of extremely high-frequency electromagnetic radiation (EHF EMR). The purpose of dosimetry is to establish a relationship between parameters of external electromagnetic fields (EMF) influencing a biological object and parameters of the fields penetrating to or arising in the exposed system, leading to primary biological effects. In spite of the large number of studies in the field of electromagnetic biology and therapy, the correct dosimetric supply is an exception rather than a rule. This results in both irreproducibility and inaccurate interpretation of experimental results. On the basis of our own results from our laboratory and data available in the literature, the practical principles of dosimetry are demonstrated by choosing a radiating system, providing the matching of a radiator with a feeding tract and irradiated biological object, measuring the radiation pattern of an antenna, determining an incident power density, measuring the distribution of the specific absorption rate (SAR) in an irradiated object, and assessing the degree of heterogeneity of SAR, dependence on radiation frequency, and geometry of the irradiated object. Examples of experimental, theoretical, and numerical methods of determining the electromagnetic energy absorption are considered for *in vitro* and *in vivo* irradiation of biological systems. Specific absorption rate in the skin of laboratory animals was determined on the basis of both microthermometric measurements of initial rates of temperature rise in the skin induced by exposure and microcalorimetric measurements of specific heat of the skin. Theoretical calculations of SAR in the skin were performed taking into account dielectric parameters of the skin, which were obtained from standing wave ratio measurements on reflection of electromagnetic waves from the skin surface, and considering the effective area of stationary overheating measured by a method of infrared thermography. The numerical method was developed to determine electromagnetic energy reflected, absorbed, and transmitted in the model of flat layers. SAR values obtained from experimental measurements, theoretical calculations, and numerical analysis show good concordance. These results can be used for dosimetric supply of experiments on studying the mechanisms of biological effects of EHF EMR. Despite the complexity in implementation of dosimetry in a range of extremely high frequencies, the knowledge of its features and methods of measurement as well as an assessment of the physical values characterizing electric, magnetic, and electromagnetic fields should become mandatory. Without solutions to dosimetry problems, the investigation into the mechanisms of biological effects of EMR becomes incorrect. Inconsistent or incomplete dosimetric supply and absence of necessary controls result in the occurrence of artifacts, difficulties in the analysis of experimental results, and undervaluation of even the most interesting and important effects of EMR.

10.1 Introduction

The problem of research into the mechanisms of biological effects of extremely high-frequency electromagnetic radiation (EHF EMR) has a special place in bioelectromagnetics in connection with the widespread use of this range of electromagnetic waves (EMW) in clinical practice for the prevention, diagnosis, and treatment of various diseases (Rojavin and Ziskin, 1998; Pakhomov and Murphy, 2000; Betskii and Lebedeva, 2007). Setting the goal of studying the mechanisms of biological effects of EMR, a researcher or clinician should, initially, have a clear idea of the characteristics of the influencing physical factors such as the intensity [characterized by incident power density (IPD) or specific absorption rate (SAR)], carrier frequency (or the corresponding wavelength), the electromagnetic field (EMF) structure (near- or far-field radiation zones), field gradient, the presence or absence of modulation and the type of modulation (pulse-, amplitude-, frequency-, phase-, or complex-modulated field), exposure duration and exposure mode (acute or chronic exposure, continuous, or fractionated), polarization (linear or circular), localization of exposure, etc. Therefore, one of the most pressing problems in this area is to ensure the correct dosimetric supply of the experiments required for their careful planning and implementation. The purpose of dosimetry is to establish a relationship between parameters of external EMF influencing a biological object and parameters of the fields penetrating into or arising in the exposed system, which lead to primary biological effects. In spite of the large number of works in the field of electromagnetic biology and therapy, the correct dosimetric supply is an exception rather than a rule. This results in both irreproducibility and inaccurate interpretation of experimental results.

Currently, based on the results of dosimetry, epidemiological, and experimental data on the biological effects of EMR, hygienic standards were designed, which are both mandatory and recommended (European Prestandard ENV 50166:1995; IEEE Std C95.1-2005; Russian Sanitary Regulations and Standards 2.1.8/2.2.4.1383-03). It should be noted that the Russian standards for electromagnetic safety are the most stringent in the world. In Russia, the maximum permissible level of Incident Power Density (IPD) for the population is $10\,\mu W/cm^2$ for irradiation by stationary radiofacilities (Russian Sanitary Regulations and Standards 2.1.8/2.2.4.1383-03) and $100\,\mu W/cm^2$ for irradiation by mobile radiofacilities (Sanitary Epidemiological Regulations and Standards 2.1.8/2.2.4.1190-03), while the international maximum permissible levels (EU, USA, Japan) are 10 times higher. Certain issues relating to the development of science-based standards, and, in particular, the valuation of EHF EMR having certain features of action on biological objects, continue to be actively discussed (ICNIRP, 1998; D'Andrea et al., 2007; Roy, 2007; Hardell and Sage, 2008; Grandolfo, 2009; Dhungel et al., 2015).

In certain experimental conditions, the study of biological effects of EHF EMR raises the following issues: choosing a radiating system, matching of a radiator with feeding tract and exposed biological object, measuring the radiation pattern of an antenna, determining an IPD, measuring distribution of SAR in the irradiated object, and assessing a degree of heterogeneity of SAR, dependence on radiation frequency, and geometry of the irradiated object (Gapeyev and Chemeris, 2010).

This chapter aims to summarize the results of works devoted to dosimetric supply of experiments directed to studying the mechanisms of EHF EMR effects at different levels of organization of biological objects. To provide methodological assistance to interested researchers, special attention is given to the phase-solving fundamental issues of dosimetry.

10.2 Choosing a Radiating System

The radiating system used should not only expose the object to EMR with selected parameters but also allow the control of these parameters in the system in the presence of the exposed object, as its presence can change the values of the selected parameters of the radiation. It is always easier and more expedient to use standard EMR emitters that are widely used in radiophysics and radioengineering (Betskii et al., 2000). These radiators are divided into free-space (horn antennae of different configurations, dielectric antennae, and antennae of special forms) and close-ended (different sections of transmission lines, e.g., strip lines, waveguides, resonators, etc.) systems.

Electromagnetic waves in waveguides have only one field component, electric or magnetic, which lies in planes perpendicular to the direction of wave propagation. The other component of the field in general has a longitudinal component. However, in spite of the disparity of these waves to the transverse wave, different waveguide sections are often used in practice as a close-ended exposure system. In this case, a thin plastic or quartz capillary filled with a physiological solution or cell suspension may be put into the waveguide through a hole in the nonradiating broad wall. The open side of the waveguide is connected with an efficient absorber (matched load) to eliminate the reflected wave. The absence of such an absorber can lead to severe distortion of the field in the waveguide, and at characteristic dimensions of the waveguide section and the wavelength of the excited EMF, it can turn the section into a resonator. Another disadvantage of the waveguide exposure is that the proximity of the object to the walls of the waveguide can strongly distort the field distribution pattern. When the dimensions of the capillary, the waveguide, and the wavelength reach a certain value, strong absorption of the radiation power in a narrow frequency band may be observed due to the formation of a cylindrical dielectric resonator (capillary effect). This phenomenon is due to the conversion of the fundamental mode to other wave types and vibration effects (Belyakov, 1987; Harlanov, 2012).

When horn irradiation systems are used, biological objects are often placed in the near-field zone of the antenna. In this case, due to the interference of incident waves and waves reflected from the sample and the inner surface of the horn, absorption maxima of EMR may occur on the surface of the irradiated object, the availability and distribution of which are highly dependent on the distance to the object, the size and shape of the irradiated object, its orientation, and the frequency of the radiation (Betskii et al., 1989; Khizhnyak and Ziskin, 1994). In the near-field zone of the horn antennae, broadband matching of irradiated object with the radiator is practically impossible. The main type of waves through the horn falls on the irradiated object, which is, in an electrodynamic sense, a strong disturbance in the near-field zone of the horn. In such a system, there is a conversion into the higher wave types, which are in turn reflected from critical sections of the horn corresponding to this wave type. Thus, a multimode interference pattern appears in the near-field zone of the horn, forming a fraction of the power absorbed by an object, reradiated in sides, and reflected into waveguide. Changing the frequency of the radiation leads to changes in the interference pattern, which results in the appearance of the resonance-like dependence of absorption energy on EMR frequency. Changes in the frequency by 200 MHz can lead to qualitative and quantitative transformation of energy absorption pattern of EHF EMR, with the equivalent quality factor reaching 500 or more and the heating rate in the areas of the absorption peaks being 10-fold higher than average over the exposure zone. Similar effects were found by observing the absorption of EHF EMR in water samples (Polnikov and Putvinskii, 1988; Betskii et al., 1989; Khizhnyak and Ziskin, 1994), and these were reproduced in model calculations (Balantsev et al., 1991). Multiwave processes can be eliminated by several means such as the following: using a quasi-optical system, irradiating objects in the far-field zone of the antenna, and eliminating the heterogeneity of dielectric permittivity (ε) throughout the irradiated area by means of effective selection of type and thickness of the absorber. Obviously, the simplest of these methods is to conduct the irradiation of biological objects in the far-field zone of standard radiators. The standard radiators of EHF EMR have good directivity and ensure uniform distribution of SAR in the plane of the irradiated object in the far-field zone at a substantially greater area than in the near-field zone of radiation.

Open-ended waveguides can be considered as the simplest types of antennae. At the open end of the waveguide, the same field structure is preserved as in the waveguide of infinite length. The wave front is flat, and the field is in phase. However, at the open end of the waveguide, there is a reflection of the fundamental wave. Except for the reflected wave, higher types of waves arise due to the appearance of currents on the outer surfaces of the waveguide walls that cause distortion of the field pattern. Directional pattern is relatively wide because of the small cross-sectional size of the waveguide. Disadvantages of waveguide radiators connected with a broad directional pattern and the presence of the reflected wave can be substantially eliminated if a portion of the waveguide at the radiating end is made the divergent. This results in the so-called pyramidal horn (Figure 10.1a). However, the shape of the wavefront of the pyramidal horn is spherical, so the field in the aperture plane is outphasing (the outphasing may be eliminated by the use of different types of lenses).

The analysis of available radiation systems shows the need to consider the specifics of interaction of the antennae with the exposed object. As for the free-space radiating systems (sections of waveguides and horn antennae), an optimal matching of the radiation with the object can be obtained when it is placed in the far-field zone of radiators, where the influence of the object on the radiating system is absent. The current level of knowledge in the field of EHF radiotechnical devices allows for the development and construction of the antenna systems, ensuring good matching of the radiation with the irradiated biological object of a given geometry in the given frequency range. We used a special form of radiator called "channel radiator" (Figure 10.1c) (Gapeyev and Chemeris, 1999). The channel radiator is a section of the channel waveguide with a cross-section of 17.5×12.5 mm. The principles underlying the design of the channel radiator allow for the elimination of the disadvantages of the waveguide and horn antennae. When a plane wavefront and thus the equiphase condition of EMF at the antenna output are saved, the directivity of the radiator is improved due to the transition to the larger cross-section. The critical wavelength in the waveguide increases as the broad side of the cross-section, which means that the wave impedance approaches the characteristic impedance of a vacuum ($120\pi\ \Omega$), thereby improving the matching of the waveguide to the free space.

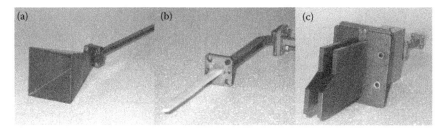

FIGURE 10.1   Radiators of EHF EMR. (a) Pyramidal horn antenna, (b) dielectric antenna, and (c) channel radiator. (Adapted from Gapeyev, AB and Chemeris, NK., *Biomed. Radioelectron.*, 1, 13–36, 2010.)

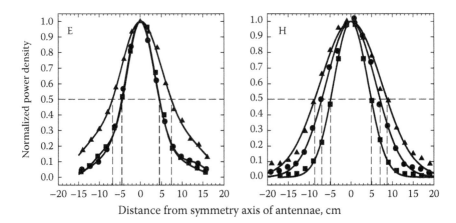

FIGURE 10.2 Profiles of directional patterns in *E* and *H* planes for pyramidal horn (■) and dielectric (▲) antennae and for channel radiator (●). The sensor antenna was at a distance of 500 mm from radiating antennae. The directivities were 260, 100, and 190 for the pyramidal horn, dielectric, and channel antennae, respectively. (Adapted from Gapeyev, A.B. and Chemeris, N.K., *Biomed. Radioelectron.*, 1, 13–36, 2010.)

The currents on the outer surfaces of the waveguide walls are smaller because the broad walls of the channel waveguide are not closed, and so the result is the minimization of EMF distortion in the near-field zone. Reflected waves from the open end of the wave-guide are minimized due to the fact that the radiating end of the antenna is smoothly sloped at an angle of 25°. Figure 10.2 shows the profiles of directional patterns for pyramidal horn and dielectric antenna (Figure 10.1b) as well as for the channel radiator. As can be seen from the figure, the pyramidal horn antenna shows the best directivity and the dielectric antenna the worst. The directivity of the channel radiator is not much worse than that of the horn antenna.

10.3 Matching a Radiator with Feeding Tract and Exposed Biological Object

The use of standard horn or dielectric antennae for exposure of biological objects in the near-field zone of the emitters can lead to heterogeneous distribution of EMF in the object plane. Spatial heterogeneity of EMF in the near-field zone of horn antennae is unacceptably high (Betskii et al., 1989; Khizhnyak and Ziskin, 1994), which may lead to local frequency-dependent overheating of the irradiated object. Local SAR in these overheating spots may exceed the average SAR throughout the exposed area by several orders. As for dielectric antennae, the matching of the antenna with feeding tract is strongly dependent on the shape and configuration of the antenna, and standing waves usually occur in the waveguide tract due to poor matching. As a result, the standing wave ratio (SWR) is nonuniformly dependent on the radiation frequency. Changing the frequency even in the range of 100–200 MHz can lead to SWR variation by 2–3 times, which is directly reflected on the value of EHF energy absorbed by the object. Therefore, when the object is exposed using dielectric antenna, one should have a clear idea of SAR value in the real object, its spatial distribution, and dependence on frequency of the radiation (Gapeyev and Chemeris, 1999).

For most of the setups, SWR < 1.5 can be considered as satisfactory, and it is at this point that the reflection losses are about 4% of the power. However, the quality of matching is characterized not only by the SWR in a frequency range, but its slope versus frequency. If a broadband matching is needed, SWR value should have a smooth dependence on the frequency while remaining below a predetermined limit, for example, SWR < 1.5 in the band $\Delta\nu/\nu$ at least ± 5%–10% of the average frequency (Korbanskii, 1973).

Thus, there is a need for the development of special, fundamentally new radiating systems that will enable artifact-free exposure of biological objects in the near-field zone of radiation. As detailed experimental and theoretical studies of EMF patterns in the near-field zone of radiators are difficult to implement, it is optimal to conduct the irradiation of biological objects in the far-field zones of standard antennae. In this case, the exposure conditions are defined more clearly:

1. Wave front is formed in the far-field zone, and the wave is a transverse plane;
2. Vectors E and H and the direction of wave propagation are mutually perpendicular. In contrast to the near-field zone, in the far field-field zone, there is only the traveling wave;
3. The components of vectors E and H in the far-field zone are decreased inversely to the first power of the distance from the antenna and their ratio is constant, while the electric field energy predominates in the near-field zone (proportion of reactive power of EMF abruptly increases as coming to the antenna); and
4. A matching of the radiation with a load in the far-field zone is mainly determined by parameters of load, whereas an antenna greatly affects the matching in the near-field zone.

10.4 Determining Parameters of EHF EMR in Exposed Objects

Dosimetry in EHF EMR range is often done by determining the incident power density. However, due to reflection and interference of EMW on interaction with the substance, the correspondence between IPD and an energy absorbed by the biological environment is affected. Moreover, taking into account the specific characteristics of the irradiated object, frequency dependence of EMF energy absorption is possible due to geometric resonances, which can be interpreted as resonance-like biological effects. Therefore, local or average SAR, as the value uniquely associated with EMF parameters in the environment by means of the electrical characteristics of the substance, is a more indicative parameter characterizing the interaction between the EMF and a sample (Kuznetsov, 1994).

Knowledge of the spatial distribution of the electrical characteristics of biological tissues, in principle, allows using Maxwell's equations to calculate the field strength inside the irradiated object and to determine charges and currents occurring in it and on its surface. Next, from the obtained values, the full picture of the distribution of the field in the object can be obtained, and so local values of SAR can be determined. However, identifying a solution for such problems for the whole variety of possible objects, fields, and exposure conditions is impossible, even if numerical calculation methods will be used.

In the near-field (also called the Rayleigh's induction zone) and intermediate (called the Fresnel's diffraction zone) radiation zones, i.e., at distances $R < 2D^2/\lambda$ (where D is the maximum size of the antenna aperture, λ is wavelength) from the radiating end of the antenna, vectors E and H are nonperpendicular to each other and their amplitudes even in a vacuum are not linked via the characteristic wave impedance ς (Kuznetsov, 1994). Therefore, the total EMF distribution in an object would require the analysis of the two components of the waves. This, at least in general, is extremely challenging. However, some common patterns may be identified, for example, by the analysis of interaction of model objects with fields of simplest types of radiators. EMR absorption by the object in the vicinity of one of two types of point emitters, electric or magnetic dipoles, has been calculated (Iskander et al., 1980; Lachtakia et al., 1981). The proportion of reactive energy of the field, in contrast to the active part of the total energy, is not transmitted to the external environment and sharply increases when approaching the radiator. Thus, a complicated situation arises in the near-field zone since the maximum and minimum values of E and H amplitudes are not achieved at the same points along the direction of EMW propagation. EMF can be highly heterogeneous in the near-field zone; there may be significant variations in the E/H ratio from the value of the wave resistance $120\pi \cdot \Omega$, which is characteristic for a plane wave.

In the far-field zone of the radiator, also called Fraunhofer's diffraction zone, the distance from the object to the antenna is $R > 2D^2/\lambda$. In this case, the transverse waves, corresponding to the free space, have already been formed, and the incident waves can be regarded as plane waves, and the interaction of EMR with the object can be considered excluding exact source structure. However, in this situation, conducting theoretical analysis is a difficult task, and for each particular case, it requires the use of approximate calculation methods when the original form of Maxwell's equations and their solutions are greatly simplified. For example, the components of vectors E and H of EMF can be explicitly defined from Maxwell's equations for the electric Hertz's dipole specified by retarded scalar and vector potentials (Aizenberg, 1957):

$$
\left\{
\begin{aligned}
H_\phi &= \frac{I_0 l}{4\pi}\left(\frac{2\pi}{\lambda}\right)^2\left[i\left(\frac{\lambda}{2\pi R}\right) + \left(\frac{\lambda}{2\pi R}\right)^2\right]\sin\theta e^{-ikR} \\
E_r &= \frac{I_0 l}{2\pi}\left(\frac{2\pi}{\lambda}\right)^2\left[\left(\frac{\lambda}{2\pi R}\right)^2 - i\left(\frac{\lambda}{2\pi R}\right)^3\right]\zeta\cos\theta e^{-ikR} \\
E_\theta &= \frac{I_0 l}{4}\left(\frac{2\pi}{\lambda}\right)^2\left[i\left(\frac{\lambda}{2\pi R}\right) + \left(\frac{\lambda}{2\pi R}\right)^2 - i\left(\frac{\lambda}{2\pi R}\right)^3\right]\zeta\sin\theta e^{-ikR} \\
H_r &= H_\theta = E_\phi = 0
\end{aligned}
\right. ,
$$

where λ is the wavelength, l is the length of the vibrator, $k = \omega/\upsilon = 2\pi/\lambda$ is the wave number, and $\zeta = \sqrt{\mu/\varepsilon}$ is the wave impedance.

From these expressions, it is clear that there are E_r and E_θ components of the electric field of the dipole in the near-field zone ($R \ll \lambda$); vector E surpasses vector H by $\pi/2$, which indicates the presence of a standing wave. In the far-field zone ($R \gg \lambda$), the phase shift is absent and $E \perp H$, because there is no E_r component. However, the analysis is simple only in the case of an elementary dipole. If we consider the multipole radiation or a real antenna, the receipt and analysis of the expressions for the EMF components in the near-field radiation zone are significantly impeded.

In the far-field zone, EMW propagation can be described using a plane wave model. In this case, the wave front is flat, vectors E and H and direction of wave propagation are mutually perpendicular, electric and magnetic fields have the same phase, and the amplitude ratio E/H is constant in the space. The power flux density, i.e., EMR power passing through a unit area perpendicular to the direction of wave propagation, is associated with magnetic and electric fields by the following ratio:

$$P_0 = \frac{EH}{2} = \frac{E^2}{2\zeta} = \frac{\zeta H^2}{2},$$

where ς is the wave impedance.

The electric component of EHF EMR in various biological objects can be estimated from the following equation (Kuznetsov, 1994):

$$E = \sqrt{\frac{2P_0\zeta}{n}} = \sqrt{\frac{2\sqrt{2}P_0}{\varepsilon_0 c\sqrt{\sqrt{(\varepsilon')^2 + (\varepsilon'')^2} + \varepsilon'}}},$$

where P_0 is IPD, ζ is the wave impedance of vacuum, n is the refractive index of the medium, ε_0 is the dielectric constant of vacuum (8.85×10^{-12} F/m), c is the speed of light (3×10^8 m/s), and ε' and ε'' are real and imaginary parts of the complex dielectric permittivity ε, respectively, depending on the frequency of the radiation. For example, when IPD is 10 mW/cm^2 and radiation frequency is 42 GHz, electric field strength in the human skin at 36°C–37°C is about 132 V/m, while in water at 20°C it is about 125 V/m. It should be noted that these are upper bounds of the electric field strength because the electric field strength will be exponentially decreased as the radiation penetrates into the absorbing medium. For comparison, the intensity of the electrostatic field of the biological membrane at a direction perpendicular to its surface is about 10^7 V/m.

The abovementioned formula does not take into consideration the magnetic permeability μ of the medium, because μ with high accuracy (up to 10^{-5}) is equal to unit for biological tissues, while ε may be significantly larger than unity (Kuznetsov, 1994). It can be also shown from the Lorentz force equation that the ratio of the magnetic force to the electric force is equal to v^2/c^2, where v is the velocity of a charged particle, c is the speed of light. Since $v \ll c$ for ions or electrons in biological tissues (Ismailov, 1987), considering the influence of EHF EMR on a moving charged particle, the magnetic component of the radiation can be neglected accurately to within 1%. The interaction energy of the magnetic field with the magnetic moment of the electron has the same order. Thus, from an energy point of view, the magnetic component of EHF EMR when considering its interaction with moving charged particles and their magnetic moments is neglected. An exception is the consideration of the cases of anomalous Zeeman's splitting levels (Belyaev et al., 1996). The magnitude of the magnetic induction at IPD = 10 mW/cm^2 and a frequency of 50 GHz in water at 20°C is equal to 1.93 mT (1.53 A/m), which is much less than the Earth's geomagnetic field (about 48 mT; 40 A/m).

10.5 Determining the Specific Absorption Rate

In spite of all methodological procedures that are used to prevent the distortion of the field distribution in the irradiated objects, the real energy absorption by the object is always different from that calculated theoretically. In this regard, in specific experimental conditions when studying biological effects of EHF EMR, different experimental methods of dosimetry need to be used to determine the SAR distribution in the object, the dependence of SAR on radiation frequency, and relative position of the object and the antenna.

Upon irradiation of biological objects in close-ended systems, it is possible to determine the total power absorbed by an object from the difference between the power of the incident wave and the power of waves reflected from an object and transmitted to the other end of the exposure system. However, using the power difference method, absorption of the radiation energy by the system walls (and possibly by other objects contained in it) allows only an approximate determination of the true power absorbed by the object itself, as EMR absorption by walls is changed with the placement of the object and, thus, cannot be strictly taken into account when measuring the unperturbed system. For example, it was shown that the apparent value of average SAR calculated without taking into account the absorption of EMR by waveguide walls exceeds the maximal value of the average SAR by 1.4 times (Chou et al., 1984).

The EMF energy absorption in biological objects is complicated because of the layered structure (aqueous solutions or cell suspensions in dielectric cuvettes, Petri dishes, etc., and cutaneous and hair covering). In these cases, the mode of standing waves must be considered for an objective assessment of the results of experimental studies. As a result of standing waves, absorption of EMF energy by surface layers can be significantly increased compared to the mode of running wave propagation (King and Smith, 1981). The results of calculations for a model system of the layered structure showed that the SAR in surface layers of the irradiated object can be very high (Ryakovskaya and Shtemler, 1983). The outer dielectric layer acts as an impedance matcher between air and the aqueous layer, partially compensating the energy reflected from the absorbing layer. When ε < 8 and thickness of the used material corresponds to the thickness of the walls and bottom of the standard experimental cuvettes, maximal changes in SAR can reach 250% at certain wavelengths. Using EMR with frequencies of 0.7–2.5 GHz, it was shown that the SAR distribution on the surface or through the depth of the irradiated cell suspension in standard Petri dishes is heterogeneous (Burkhardt et al., 1996; Schönborn et al., 2001). The degree of the heterogeneity strongly depends on the ratio of the diameter of the Petri dish to the height of cell suspension layer. Maximal heterogeneity of SAR was observed near the walls of the Petri dish. Similar results were obtained when the power density distribution was calculated in the layers of water exposed to EHF EMR in the dielectric polystyrene plates (Alekseev and Ziskin, 2001b).

The absorption of EMR power can be determined and monitored by the method of acoustic detection proposed by Polnikov and Putvinskii (1988). The method is based on the registration of thermoelastic vibrations caused by the absorption of modulated EHF EMR (square wave modulation with frequencies of 2–1000 Hz). Thermoelastic vibrations are generated in the object and in the air layer adjacent to the absorbing surface, and the amplitude of the acoustic signal is proportional to the absorbed radiation power. The sensitive piezoelectric sensors allow measuring acoustic vibrations at low power of EMR absorbed by biological object. However, this method is not widely used in experimental practice since it does not allow for correct determination of the spatial temperature distribution in the irradiated object. Sound travel time (about 300 ns) in the heating area (about 0.5 mm) is significantly shorter than the duration of the electromagnetic pulse of 500 μs (at modulation frequency of 1 kHz). Therefore, rising acoustic pulses have time to leave the absorption zone for the duration of the pulse of EHF EMR.

In carrying out dosimetry *in vivo* using open-ended systems, the experimenter usually faced additional difficulties caused by the complex geometry of the irradiated object, nonuniform layer structure of the exposed skin, and the presence of blood flow in the irradiated tissues. The methodology of dosimetry was demonstrated as an example of EHF EMR energy absorption in the skin of laboratory animals (mice and rats) using experimental and theoretical methods (Gapeyev et al., 2002). The following tasks were solved in determining SAR in the skin of laboratory animals: (1) determining of SAR using microthermometry method, (2) determining of dielectric parameters of the skin, (3) calculating SAR considering obtained values of dielectric parameters of the skin of laboratory animals, (4) determining SAR by numerical methods in the model of plane layers, and (5) analyzing the applicability and accuracy of the approaches for determining SAR.

In these experiments, waveguide radiator, with an open segment of the waveguide having a section of 5.2 × 2.6 mm without flange, was used as the radiator of EHF EMR. The use of such an antenna was due to the experimental need for local exposure of rat skin in the near-field zone. Earlier, it was experimentally demonstrated that the use of the open segment of the waveguide as a radiator, in the absence of other special antennae, is the optimal solution when objects should be irradiated in the near-field zone (at a distance of 1–5 mm from the radiating end of an antenna) (Alekseev and Ziskin, 2003). Near the open end of the waveguide, the same field structure as in the waveguide of infinite length is preserved, with a flat wave front and in-phase field. If the radiating end of the waveguide is prepared in a special way, then the interference pattern of EMF distribution throughout the near-field zone is absent, and, in contrast to the horn antennae, there is a unique heating spot on the irradiated object (Alekseev and Ziskin, 2003). In our experimental conditions, EHF dielectrometry was possible only under direct contact of the object and the antenna, which is ideally implemented in the near field of the waveguide radiator.

Since the exposure of biological objects to intense EHF EMR is always accompanied by the heating of the object, the desired value of SAR can be obtained from measurements of the initial rate of temperature rise following the procedure developed in the work by Allis et al. (1977). The use of temperature microsensors is justified because they do not make a large error in determining the temperature growth rate and steady-state levels of overheating (Dunscombe et al., 1988; Alekseev and Ziskin, 2001a). The kinetics of the temperature increase in exposed objects is usually well approximated by a two-exponential dependence:

$$\Delta T(t) = c_1[1 - e^{-t/\tau_1}] + c_2[1 - e^{-t/\tau_2}],$$

where τ_1 and τ_2 are time constants, c_1 and c_2 are coefficients. Then the steady-state level of overheating can be determined using the formula

$$\Delta T(\infty) = c_1 + c_2,$$

and the rate of temperature rise and corresponding SAR can be calculated using:

$$\left.\frac{dT(t)}{dt}\right|_{t \to 0} = \frac{c_1}{\tau_1} + \frac{c_2}{\tau_2}, \quad SAR = C\left.\frac{dT(t)}{dt}\right|_{t \to 0},$$

where C (J/kg·°C) is specific heat capacity of substance.

When measuring the initial rate of temperature rise, some features should be taken into account that can introduce systematic error in the determination of SAR. Theoretical calculations show that the initial phase of temperature rise upon the exposure of the skin to EHF EMR is nonlinear and cannot be described by a simple exponential expression. Therefore, precise determination of the initial rate of temperature rise is very difficult. Determination of $dT/dt|_{t \to 0}$ in a short period of time does not improve the situation because the increase in temperature during this time is very small and is comparable to the noise. This is further discussed by Alekseev and Ziskin (2003). When the human skin was exposed to sufficiently high IPD (200 mW/cm²), the temperature rise during irradiation for 0.1 s exceeded 0.05°C. For high-precision measurements of the initial rate of temperature rise in the irradiated sample, measurement time should be very small (about 2–3 ms), but then the temperature rise during this time will be <0.002°C. Given that most of the temperature sensors (including good infrared cameras) have a thermal noise level at 0.05 K, the correct measurement of such low levels of temperature changes, which are within the noise level of the instrument, is difficult to measure. Due to the high nonlinearity of the initial phase of temperature rise, an error can reach more than 30% even for kinetics with minimal noise.

In our experiments to measure the temperature, we used temperature microsensors with a relative error of about 0.05°C (the microthermocouples were placed in the center of the exposed surface perpendicular to the E-vector of the electromagnetic field), approximated the temperature kinetics by two-exponential dependence, and the final value of the initial rate of temperature rise in the irradiated sample was calculated as the average over a sufficiently large number of measurements (20 series, 3–5 measurements for each) (Gapeyev et al., 2002). It was found that for the irradiation of the rat skin from the open end of the waveguide at a distance of about 1 mm and at IPD of 17 mW/cm², the average rate of temperature rise was $0.074 \pm 0.005°C/s$ and the average level of steady-state overheating was $0.61 \pm 0.02°C$ (Gapeyev et al., 2002).

To calculate SAR based on microthermometric measurements, the specific heat capacity of rat skin was determined by microcalorimetry. The average heat capacity of rat skin was about $3100 \pm 100 J/(kg·°C)$. From these data, SAR was found to be $230 \pm 20 W/kg$ upon exposure of the rat skin to EHF EMR with IPD of about 17 mW/cm². Taking into account the linear dependence of the rate of temperature rise on IPD in the wide power range (at least up to 50 mW/cm²), this value can be converted to any other level of IPD using the linear regression coefficient.

It should be noted that in specific experimental conditions, the accuracy of calculations of SAR based on microthermometric measurements can strongly depend on the quality of matching of the radiator with the object, the object's geometry, the accuracy of the temperature measurements, the conditions of heat exchange with the environment, blood flow, etc. Kinetic parameters of heating of the human skin exposed to EHF EMR at a frequency of 42.25 GHz with the use of the waveguide open-ended radiator and a pyramidal horn antenna were studied in previous studies (Alekseev and Ziskin, 2003; Alekseev et al., 2005). Using a thermographic method (Guy, 1971), it has been shown that the SAR distribution on the exposed surface of the skin is described by circularly symmetrical Gaussian type function. A microthermocouple with a diameter <0.1 mm inserted into the exposed area did not significantly disturb the SAR distribution and kinetic parameters of the temperature change if the thermocouple was perpendicular to E-vector of the EMF (Alekseev and Ziskin, 2001a). The temperature changes of the irradiated skin were well fit by the kinetic dependences, which are solutions of the 3-D bio-heat transfer equation for homogeneous tissue:

$$\rho C \frac{\partial T}{\partial t} = k \cdot \left(\frac{\partial^2 T}{\partial z^2} + \frac{1}{r} \frac{\partial T}{\partial r} + \frac{\partial^2 T}{\partial r^2} \right) - V_s (T - T_b) + Q(z, r, t),$$

where ρ is the density of tissue (kg/m³), C is the specific heat (J/kg °C), k is the coefficient of tissue heat conduction (W/m·°C), T is tissue temperature (°C), T_b is arterial blood temperature (°C), $V_s = f_b \cdot \rho_b \cdot C_b$ (W/m³ °C), where f_b is the specific blood flow rate (mL/s/mL), ρ_b (kg/m³) and C_b (J/kg·°C) are the density and specific heat of blood, respectively; $Q(z, r, t)$ is heat depositions from exposure (W/m³), where t is time (s), r is the radial distance from the beam axis, and z is distance from tissue surface in the direction of the beam axis. The equation was solved numerically by Alekseev and Ziskin (2003).

The authors showed that the initial phase of temperature rise in the skin exposed to EHF EMR is weakly dependent on the changes of blood flow rate or the conditions of heat exchange with the environment. Thus, the solution of the bioheat transfer equation, which is close to the experimentally obtained temperature kinetic, allows determining accurately the heating parameters and the corresponding SAR. The authors note that as a result of the presence of thermal noise, the exact definition of SAR only by temperature measurements is not possible since it is impossible to define the exact initial rate of temperature rise (Alekseev and Ziskin, 2003). In turn, changes in blood flow rate or the conditions of heat exchange with the environment can strongly influence the steady-state overheating of the skin during exposure to EHF EMR (Foster et al., 1978; Walters et al., 2004). It was found that exposure of the forearm and middle finger at the same IPD (208 mW/cm²) produced different kinetics of temperature changes, and the average steady-state overheating was 4.7 ± 0.4°C and 2.5 ± 0.3°C, respectively (Alekseev et al., 2005). The authors explained that the different levels of overheating at the same IPD were due to different blood flow rates in the forearm and the fingers. It was directly shown that the decline of blood flow rate with occlusion increases the level of overheating during exposure to EHF EMR and that the use of a vasodilator (nonivamide/ nicoboxil) reduces the level of overheating. Thus, the significant impact of blood flow on the level of tissue overheating upon exposure to EHF EMR was demonstrated, and it was experimentally and in model calculations shown that the higher the initial temperature of the skin (i.e., the higher blood flow rate), the lower the level of steady-state overheating during irradiation with EHF EMR (Alekseev et al., 2005). With the use of different models, it was found that the blood flow in the dermis and muscle layers causes the greatest influence, and the blood flow in the fat layer and the thickness of the epidermis causes less influence on the level of overheating of the skin during EHF EMR exposure (Alekseev and Ziskin, 2009).

Using dielectric parameters of absorbing medium and IPD, SAR can be determined from theoretical calculations. It is known that IPD (P_0) is connected to the electric field strength (E) as follows (Kuznetsov, 1994):

$$P_0 = \frac{nE^2}{2\xi},$$

where ξ is the wave resistance of vacuum, n is the refractive index of the medium, and

$$SAR = \frac{\sigma E^2}{2\rho},$$

where $\sigma = \varepsilon_0 \varepsilon'' \omega$ is the conductivity, ρ is the density of the medium, $\varepsilon_0 = 8.85 \times 10^{-12}$ F/m is the dielectric constant of vacuum, ε'' is the imaginary part of the complex permittivity, and ω is the circular frequency of the radiation.

Since only part of the power passes through the interface due to reflection, SAR on the surface of the irradiated object can be calculated by the formula:

$$SAR = \frac{\sigma \xi (1 - R) P_0}{n \rho},$$

where R is the reflection coefficient.

Thus, determining SAR based on IPD requires knowledge on parameters of EMW interaction with the irradiated surface (distribution of absorbed energy on the irradiated surface and reflection coefficient) and some characteristics of the medium such as the dielectric constant and density. To determine the distribution of the energy absorbed by the exposed surface, the distribution of temperature fields on the rat skin were recorded by infrared thermography (Gapeyev et al., 2002). The overheating spot during irradiation of the rat skin using waveguide radiator had the form of an ellipse with a square of about 1.2 cm², which corresponded to IPD of about 17 mW/cm² at an output power of the generator of 20 mW.

To determine the dependence of the reflection coefficient of EMW on the thickness of dielectric layer, samples of different thickness were placed on the open end of the waveguide. It was quite possible, since it was shown that the use of the open end of the waveguide does not make a fundamental distortion in the observed pattern, that errors associated with it are much less than errors in determining SWR and the complex refractive index from the experimental dependence of the reflection coefficient on layer thickness (Alekseev and Ziskin, 2001c). To check the correctness of the method and to evaluate the degree of deviation from the true values, before measurements were taken on the skin, dependence of SWR on the layer thickness for water was obtained. The water layers were simulated by using water-soaked filter papers. Then, experimental points obtained were fit by a given function and values of the complex refractive index and the dielectric constant for water and the skin determined (Gapeyev et al., 2002).

The values of the complex refractive index and the dielectric constant for water are in good agreement with both theoretical values obtained by the Debye equation and experimental results obtained previously (Alekseev and Ziskin, 2001c). The values of the complex refractive index and the dielectric constant of the skin coincide with values for human skin (Kislyakov, 1994; Kuznetsov, 1994). Figure 10.3 shows the experimental data and curves obtained by using the experimental model calculations for the parameters

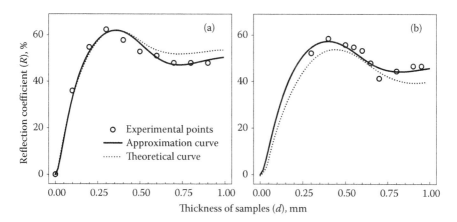

FIGURE 10.3 Dependencies of reflection coefficients on the thickness of water layer (a) and the thickness of the rat skin (b) at a fixed frequency of 42.25 GHz. (Adapted from Gapeyev, A.B. et al., *Biophysics*, 47(4), 706–15, 2002.)

and the parameters obtained by the Debye equation for water and for the skin (Kislyakov, 1994; Kuznetsov, 1994). Knowing the limits of the complex permittivity, which conditionally met the static case and corresponded to a field infinitely fast changed over time, as well as the characteristic relaxation frequency included into the Debye equation, allows theoretical determination of the dependence of the complex permittivity on the frequency (Kuznetsov, 1994) and calculation of corresponding SAR considering the reflection coefficient for a fixed frequency, IPD, and density of the absorbed material.

Thus, upon exposure of the rat skin through the waveguide radiator at an output power of 20 mW, corresponding IPD of 170 W/m^2, reflection coefficient R of 0.47, refractive index n of 4.2, and skin density ρ of 1150 kg/m^3, calculated SAR was 290 W/kg, wherein skin conductivity (σ) = 41.3 (m·Ω)$^{-1}$ at the frequency of 42.25 GHz.

An important role in dosimetry is played by mathematical models that can be used to calculate the absorption, reflection, and absorbed energy distribution in the irradiated object with the help of numerical methods [for example, the impedance method (Orcutt and Gandhi, 1988), the recursive method (Vinogradov and Pirogov, 1983), the method of impedance characteristics (Born and Wolf, 1970), and matrix method (Kozar et al., 1978)]. The model of flat layers is the simplest and most commonly used. This model allows determining the energy distribution of EMW upon the interaction with biological objects with layered structure. To date, a lot of calculation methods for flat-layer model were developed both in stationary (Kozar et al., 1978; Ryakovskaya and Shtemler, 1983; Vinogradov and Pirogov, 1983; Orcutt and Gandhi, 1988; Alekseev and Ziskin, 2001b,c) and in nonstationary (Vinogradov et al., 1984) exposure conditions. However, existing methods are either highly specialized and only suitable for special conditions or are poorly applied to systems consisting of thick absorbing layers.

To determine the distribution of EMW energy upon the interaction with matter in terms of the model of flat layers, a calculation method based on the phased calculation of the amplitudes of the reflected and transmitted waves in the course of propagation of EMW in a multilayer system was developed (Gapeyev et al., 2002). The basis behind the method is that the complex amplitude of EMW undergoes two types of changes: a division into two (reflected and transmitted) by reaction with the interface between two layers with different dielectric constants, and attenuation when passing through the layer (Figure 10.4). As the result of the numerical experiment in a medium with dielectric constant equal to that obtained experimentally for the rat skin, it was found that in the surface layer of the skin with a thickness up to 0.2 mm, SAR was about 280 W/kg at IPD of 170 W/m^2.

Thus, with the use of different approaches to determining SAR in the skin of laboratory animals exposed to EHF EMR with a frequency of 42.25 GHz at IPD of 17 mW/cm^2, the following values were obtained (Gapeyev et al., 2002). Calculation of SAR by the initial rate of temperature rise recorded with microthermometric method, and taking into account the value of the specific heat capacity of the skin, gives SAR of about 230 W/kg. The calculation of SAR based on IPD, permittivity, and density of the skin gives a SAR of about 290 W/kg. Numerical simulation in the model of flat layers in a medium with dielectric constant equal to the value for the rat skin gives a SAR of about 280 W/kg. The error of determining SAR from theoretical calculations and numerical experiments is connected mainly with an error in determining the complex dielectric permittivity

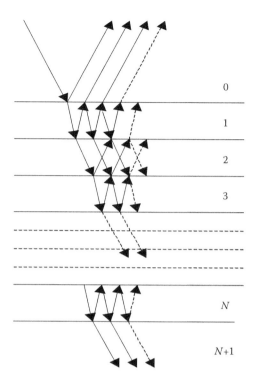

FIGURE 10.4 Illustration of the method of phased calculation of the amplitudes of reflected and transmitted waves in the course of propagation of EMW in a multilayer system. (Adapted from Gapeyev, A.B. et al., *Biophysics*, 47(4), 706–15, 2002.)

of the skin. The relative error in this case may reach 20%, which is due to the accuracy in determining the reflection coefficient of EMW. Thus, it was shown that taking into account the actual errors, SAR values obtained from experimental measurements, theoretical calculations, and numerical simulations are in good concordance.

To study the mechanisms of biomedical effects of EHF EMR, an important task is to develop the correct methods of assessment of radiation absorption in biological systems and the penetration depth of waves into the skin and underlying tissues. Studies have been undertaken on human skin to determine the frequency dependence of EHF EMR reflection in a range of 37–74 GHz, to develop homogeneous and flat-layered models for calculation of the dielectric constant of various skin layers, their content of free water, and the penetration depth of the radiation (Alekseev and Ziskin, 2007). Four types of models were used: (a) homogeneous model of a semi-infinite layer of skin tissue; (b) a two-layer model including the stratum corneum and underlying homogeneous skin tissue; (c) a three-layer model containing the stratum corneum, the rest of the epidermis plus dermis, and a semi-infinite layer of fat; and (d) a four-layer model with the outer and inner layers of the stratum corneum, with the rest made up of epidermis and plus dermis, and a semi-infinite layer of fat. It was shown that the best agreement between the experimental and calculated results was obtained when multilayer models were used for calculation, especially when body areas with a sufficiently thick epidermis layer were irradiated, which in this case acts as a matching element reducing radiation reflection (Alekseev et al., 2008b). The total water content in the superficial layers of the epidermis calculated by the authors was 38%–43% (12% of free water), while in inner layers of epidermis and dermis it was 65%–70% (about 40% of free water). The skin depths at the frequency of 42 GHz were about 420 and 620 microns for water and the skin tissue, respectively. The method of EHF EMR reflectometry has been used by the authors for the noninvasive determination of the water content in living skin (Alekseev et al., 2008c).

Calculations performed for multilayer skin models showed that EHF EMR is able to penetrate deep enough into the skin (Alekseev et al., 2008a,b). EHF EMR is almost completely absorbed in the epidermis and dermis, however, reaching all skin structures. The targets for the radiation may be keratinocytes, melanocytes, collagen fibers, Langerhans cells, Merkel cells, free nerve endings, immune cells, microcirculatory bed of the skin, etc., transforming the primary recepted signals into physiological responses.

10.6 Conclusion

In spite of complexity in implementing dosimetry in a range of extremely high frequencies, the knowledge of its features and methods of measurement and an assessment of the physical values characterizing electric, magnetic, and electromagnetic fields should become mandatory. Without the solution to dosimetry problems, the investigation into the mechanisms of biological effects of electromagnetic radiation becomes incorrect. It is necessary to determine and specify all parameters of the electromagnetic field and exposure conditions since a change in the majority of them can cardinally modify the response of a biological system. Inconsistent or incomplete dosimetric supply and absence of necessary controls imply an occurrence of artifacts, difficulties for the analysis of experimental results, and undervaluation of even the most interesting and important effects of electromagnetic radiation.

The author is very grateful to Sokolov, P.A. and Chemeris, N.K. for their participation in certain studies.

References

Aizenberg G.Z. *Antennae of Ultrashort Waves*. Svyaz'izdat, Moscow, USSR, 1957 (in Russian).

Alekseev S.I., Gordiienko O.V., Ziskin M.C. Reflection and penetration depth of millimeter waves in murine skin. *Bioelectromagnetics* 29(5): 340–4, 2008a.

Alekseev S.I., Radzievsky A.A., Logani M.K., Ziskin M.C. Millimeter wave dosimetry of human skin. *Bioelectromagnetics* 29(1): 65–70, 2008b.

Alekseev S.I., Radzievsky A.A., Szabo I., Ziskin M.C. Local heating of human skin by millimeter waves: Effect of Blood Flow. *Bioelectromagnetics* 26: 489–501, 2005.

Alekseev S.I., Szabo I., Ziskin M.C. Millimeter wave reflectivity used for measurement of skin hydration with different moisturizers. *Skin Res. Technol.* 14: 390–6, 2008c.

Alekseev S.I., Ziskin M.C. Distortion of millimeter-wave absorption in biological media due to presence of thermocouples and other objects. *IEEE Trans. Biomed. Eng.* 48: 1013–9, 2001a.

Alekseev S.I., Ziskin M.C. Human skin permittivity determined by millimeter wave reflection measurements. *Bioelectromagnetics* 28(5): 331–9, 2007.

Alekseev S.I., Ziskin M.C. Influence of blood flow and millimeter wave exposure on skin temperature in different thermal models. *Bioelectromagnetics* 30(1): 52–8, 2009.

Alekseev S.I., Ziskin M.C. Local heating of human skin by millimeter waves: A Kinetics Study. *Bioelectromagnetics* 24: 571–81, 2003.

Alekseev S.I., Ziskin M.C. Millimeter wave power density in aqueous biological samples. *Bioelectromagnetics* 22: 288–91, 2001b.

Alekseev S.I., Ziskin M.C. Reflection and absorption of millimeter waves by thin absorbing layers. *Bioelectromagnetics* 21: 264–71, 2001c.

Allis J.W., Blackman C.F., Fromme M.L., Benane S.G. Measuring of microwave radiation absorbed by biological systems. 1. Analysis of heating and cooling data. *Radio Sci.* 12(6S): 1–8, 1977.

Balantsev V.N., Lebedev A.M., Permyakov V.A., Plotnikov S.A., Sevastyanov V.V. Numerical study of distribution of specific absorption rate in two-dimensional models of horn antennae with biological objects, Chap. 3. In *Reports Collection of International Symposium on Low-Intensity Millimeter Waves in Medicine.* Institute of Radio Electronics, Moscow, USSR, 660–4, 1991 (in Russian).

Belyaev I.Y., Shcheglov V.S., Alipov Y.D., Polunin V.A. Resonance effect of millimeter waves in the power range of 10^{-19} to 3×10^{-3} W/cm^2 on *Escherichia coli* cells at different concentrations. *Bioelectromagnetics* 17(4): 312–21, 1996.

Belyakov E.V. High-Q resonance in the waveguide with a strongly absorbing dielectric electronic technique. Series 1. *SHF Electron.* 7(401): 51–3, 1987 (in Russian).

Betskii O.V., Devyatkov N.D., Lebedeva N.N. Treating with electromagnetic fields, Chapter 1. Sources and properties of electromagnetic waves. *Biomed. Radioelectron.* (7): 3–9, 2000 (in Russian).

Betskii O.V., Lebedeva N.N. Application of low-intensity millimeter waves in biology and medicine. *Biomed. Radioelectron.* (8–9): 6–25, 2007 (in Russian).

Betskii O.V., Petrov I.Y., Tyazhelov V.V., Khizhnyak E.P., Yaremenko Y.G. Distribution of millimeter wave electromagnetic fields in model and biological tissues under exposure in the near-field zone of emitters. *Doklady USSR Acad. Sci.* 309(1): 230–3, 1989 (in Russian).

Born M., Wolf E. *Basics of Optics.* Nauka, Moscow, USSR, 1970 (in Russian).

Burkhardt M., Poković K., Gnos M., Schmid T., Kuster N. Numerical and experimental dosimetry of Petri dish exposure setups. *Bioelectromagnetics* 17(6): 483–93, 1996.

Chou C.K., Guy A.W., Johnson P.B. SAR in rats exposed in 2450 MHz circularly polarized waveguides. *Bioelectromagnetics* 5(4): 389–98, 1984.

D'Andrea J.A., Ziriax J.M., Adair E.R. Radio frequency electromagnetic fields: Mild hyperthermia and safety standards. *Prog. Brain Res.* 162: 107–35, 2007.

Dhungel A., Zmirou-Navier D., van Deventer E. Risk management policies and practices regarding radio frequency electromagnetic fields: Results from a WHO survey. *Radiat. Prot. Dosim.* 164(1–2): 22–7, 2015.

Dunscombe P.B., Constable R.T., McLellan J. Minimizing the self-heating artifacts due to the microwave irradiation of thermocouples. *Int. J. Hyperthermia* 4(4): 437–45, 1988.

European Prestandard ENV 50166:1995. Human exposure to electromagnetic fields: two parts: ENV 50166-1: Low frequencies (0 to 10 kHz); ENV 50166-2: High frequencies (10 kHz to 300 GHz). CENELEC, Brussels, Belgium, 1995.

Foster K.R., Kritikos H.N., Schwan H.P. Effect of surface cooling and blood flow on the microwave heating of tissue. *IEEE Trans. Biomed. Eng.* 25: 313–16, 1978.

Gapeyev A.B., Chemeris N.K. Dosimetry questions at studying biological effects of extremely high-frequency electromagnetic radiation. *Biomed. Radioelectron.* 1: 13–36, 2010 (in Russian).

Gapeyev A.B., Chemeris N.K. The influence of continuous and modulated EHF EMR on animal cells. Part II. Problems and methods of EHF EMR dosimetry. *Herald New Med. Technol.* 6(2): 39–45, 1999 (in Russian).

Gapeyev A.B., Sokolov P.A., Chemeris N.K. A study of absorption of energy of the extremely high frequency electromagnetic radiation in the rat skin by various dosimetric methods and approaches. *Biophysics* 47(4): 706–15, 2002.

Grandolfo M. Worldwide standards on exposure to electromagnetic fields: An overview. *Environmentalist* 29: 100–17, 2009.

Guy A.W. Analysis of electromagnetic fields induced in biological tissue by thermographic studies on equivalent phantom models. *IEEE Trans. Microw. Theory Tech.* 19(2): 205–14, 1971.

Hardell L., Sage C. Biological effects from electromagnetic field exposure and public exposure standards. *Biomed. Pharmacother.* 62(2): 104–9, 2008.

Harlanov A.V. Electroosmosis and influence of electromagnetic waves on liquid course through a capillary. *Biomed. Radioelectron.* (9): 22–9, 2012 (in Russian).

IEEE Std C95.1-2005. *IEEE Standard for Safety Levels with Respect to Human Exposure to Radio Frequency Electromagnetic Fields, 3 kHz to 300 GHz.* The Institute of Electrical and Electronics Engineers, New York, 2006.

International Commission on Non-Ionizing Radiation Protection (ICNIRP). ICNIRP guidelines for limiting exposure to time-varying electric, magnetic and electromagnetic fields (up to 300 GHz). *Health Phys.* 74(4): 494–522, 1998.

Iskander M.F., Barber P.W., Durney C.H., Massoudi H. Irradiation of Prolate Spheroidal Models of Humans in the Near Field of a Short Electric Dipole. *IEEE Trans. Microw. Theory Tech.* 28(7): 801–7, 1980.

Ismailov E.S. *Biophysical Influence of SHF EMR.* Energoatomizdat, Moscow, USSR, 1987 (in Russian).

Khizhnyak E.P., Ziskin M.C. Heating patterns in biological tissue phantoms caused by millimeter wave electromagnetic irradiation. *IEEE Trans. Biomed. Eng.* 41(9): 865–73, 1994.

King R.W.P., Smith G.S. *Antennas in Matter: Fundamentals, Theory and Applications.* MIT Press, Cambridge, MA and London, 1981.

Kislyakov A.G. The penetration depth of millimeter waves in the human skin. *Radio Technol. Electron.* 39(11):1852–58, 1994.

Korbanskii I.N. *Antennae.* Energiya, Moscow, USSR, 1973.

Kozar A.V., Kolesnikov V.S., Pirogov Y.A. Distribution of electric field in multilayer resonance-type systems. Herald of Moscow University. Series 3. *Phys. Astron.* 19(1): 78–86, 1978 (in Russian).

Kuznetsov A.N. *Biophysics of Electromagnetic Impact.* Energoatomizdat, Moscow, USSR, 1994 (in Russian).

Lachtakia A., Iskander M.F., Durney C.H., Massoudi H. Near-field absorption in prolate spheroidal models of humans exposed to a small loop antenna of arbitrary orientation. *IEEE Trans. Microw. Theory Tech.* 29(6): 588–94, 1981.

Orcutt N., Gandhi O.P. A 3-D Impedance method to calculate power deposition in biological bodies subjected to time varying magnetic fields. *IEEE Trans. Biomed. Eng.* 35(8): 577–83, 1988.

Pakhomov A.G., Murphy M.R. Low-intensity millimeter waves as a novel therapeutic modality. *IEEE Trans. Plasma Sci.* 28(1): 34–40, 2000.

Polnikov I.G., Putvinskii A.V. Acoustic detection of absorption of electromagnetic radiation of millimeter range in biological objects. *Biophysics* 32(5): 893–5, 1988 (in Russian).

Rojavin M.A., Ziskin M.C. Medical application of millimetre waves. *Q. J. Med.* 91: 57–66, 1998.

Roy C.R. Rapporteur Report: ICNIRP international workshop on EMF dosimetry and biophysical aspects relevant to setting exposure guidelines. *Health Phys.* 92(6): 658–67, 2007.

Ryakovskaya M.L., Shtemler V.M. Absorption of electromagnetic waves of millimeter range in biological preparations with a plane layer structure. In (Devyatkov N.D. ed.) *Effect of Nonthermal Action of Millimeter Radiation on Biological Subjects.* Moscow, USSR, Institute of Radioelectronics of USSR Academy of Sciences, 172–81, 1983 (in Russian).

Sanitary Epidemiological Regulations and Standards 2.1.8/2.2.4.1190-03. Hygienic requirements for the placement and operation of land mobile radio. Russian Ministry of Health, Moscow, USSR, 2003.

Sanitary Epidemiological Regulations and Standards 2.1.8/2.2.4.1383-03. Hygienic requirements for the placement and operation of radio transmitting facilities. Russian Ministry of Health, Moscow, USSR, 2003 (in Russian).

Schönborn F., Poković K., Burkhardt M., Kuster N. Basis for optimization of *in vitro* exposure apparatus for health hazard evaluations of mobile communications. *Bioelectromagnetics* 22(8): 547–59, 2001.

Vinogradov S.V., Ekzhanov A.E., Pirogov Y.A. Recursive method for studying nonstationary modes of multilayer interference systems. Herald of Moscow University. Series 3. *Phys. Astron.* 25(3): 62–4, 1984 (in Russian).

Vinogradov S.V., Pirogov Y.A. Calculation of distribution of electric field in multilayer resonance-type systems. Herald of Moscow University. Series 3. *Phys. Astron.* 24(4): 50–2, 1983 (in Russian).

Walters T.J., Ryan K.L., Nelson D.A., Blick D.W., Mason P.A. Effects of blood flow on skin heating induced by millimeter wave irradiation in humans. *Health Phys.* 86: 115–20, 2004.

11

Experimental Approaches for Local Heating and Absorption Measurements

Andrei G.
Pakhomov
Old Dominion University

In bioelectromagnetics, accurate dosimetry is the key that enables comparison of results generated under diverse exposure conditions and in different laboratories. Conversely, inadequate or erroneous dosimetry renders any biological results irreproducible and useless, as they cannot be fit in a large knowledge database. It has been widely accepted in the bioelectromagnetics community that no biological study should be published unless accompanied by dosimetry detailed enough for independent replication of findings. However, even genuine attempts to document dosimetry may suffer from improper use of experimental techniques and instrumentation in electromagnetic fields (EMFs), and the variety of approaches being employed by different research groups may complicate the situation further.

The paramount importance of experimental and theoretical dosimetry has made it a topic of hundreds of original studies and of several editions of *Radiofrequency Radiation Dosimetry Handbook* [1,2]. *This chapter is an abridged version of the material presented earlier in one of these Handbooks [3].*

11.1 Definitions

In any biological experiment involving EMFs, and regardless of the particular mechanisms of how bioeffects are produced, the "extent" of EMF interaction with the living object has to be characterized by certain metrics. The specific absorption rate (SAR, W/kg) is a convenient and universal metric for many types of EMF exposure, and it is defined as:

$$SAR = \frac{\sigma E^2}{\rho_m},$$ (11.1)

where E is the induced electric field in tissue (V/m); σ is the electrical conductivity of the tissue (S/m), and ρ is the mass density of the tissue (kg/m³).

Being a measure of the rate of energy transfer from the field into the tissue, SAR can also be conveniently estimated indirectly, from the rate of temperature rise in the tissue under exposure. With the assumption that all the absorbed field energy is transformed into heat, and disregarding any heat dissipation, SAR (W/kg) at any location in the tissue is measured as:

$$SAR = \frac{c\Delta T}{\Delta t},$$ (11.2)

where ΔT is the temperature increment (°C), c is the specific heat capacity of the tissue [J/(kg × °C)], and Δt is the duration (s) over which ΔT is measured. In practice, ΔT has to be measured over as short as possible a time interval immediately after the onset of exposure, when the heat dissipation is minimal and can be disregarded.

Over time, SAR has been widely accepted as the principal exposure metric for many (although not all) types of EMFs. Nonetheless, several misconceptions exist. First, the close connection between SAR and heating may be misinterpreted as a synonymy of these two parameters. Proponents of this position tend to disregard the fact of heat dissipation and the obvious fact that the temperature can be intentionally kept the same or even lowered at very high SAR values. The "opposite" misconception is interpretation of any temperature rise as resulting directly from EMF energy absorption, whereas it could also be a result of conventional heat transfer from adjacent areas where the EMF absorption actually occurred.

11.2 Local Temperature Measurements in Intense Pulsed EMFs

11.2.1 General Requirements of Temperature Probes

One critical feature of thermometers intended for measurements during EMF exposure is their EMF compatibility. First, thermometers that have an electrical circuitry may be vulnerable to EMF pickup: the electronic measurement circuitry, wire leads, and temperature probes may function as antennas, and erroneously report the detected electrical signal as a change in temperature. These artifacts can be minimized or eliminated by using thermometers equipped with remote sensors and by moving the electronic measurement circuitry to the outside of the field; by aligning the wire leads perpendicular to the E-vector of the field and by shortening them; by using high-resistance wires; by reducing the physical dimensions of temperature probes; or by using nonmetal temperature probes and connectors, as in fiber-optic thermometers.

Another problem of EMF compatibility results from the fact that temperature sensors have dielectric properties different from those of biological tissues and samples. The insertion of a probe into a sample may change EMF absorption and heating. It is common knowledge that metal probes may alter the local field and therefore should be avoided. However, it is much less known or appreciated that dielectric nonuniformities (e.g., plastic probes or air bubbles in biological samples) can also cause profound local-field distortion [4], leading to erroneous measurements (an example of such behavior is shown in Figure 11.5 and will be discussed later). Perhaps the most practical way to reduce artifacts from introduction of temperature probes into EMF-exposed samples is reducing the size of the probes.

Hence, reducing the size of the temperature probes improves their EMF compatibility by decreasing both the EMF pickup (by metal probes) and the field distortion (by both metal and nonmetal probes). In addition, smaller probes can be positioned in biological samples with less damage to surrounding tissues; they have lower heat capacity, which means more accurate and faster reading of temperature changes; and they have better spatial resolution for temperature and SAR "mapping."

11.2.2 Selection and Availability of Temperature Probes

Several miniature temperature probes that are (or have been) available on the market are shown in Figure 11.1.

Vitek model 101 Electrothermia Monitor is a thermistor probe that was manufactured by BSD Medical Devices, Salt Lake City, UT. On the basis of the original design by Bowman [5], it became the first commercially available "non-field–perturbing" temperature probe. With a sensor size of about 1 mm, it has a sensitivity of about 5 mK and characteristic response time of about 240 ms [6]. This inexpensive probe has been employed widely in biological studies [7–14] and remains in use in some laboratories, although its production has been discontinued.

Vitek   Luxtron   TP21   HERO   Thermocouples

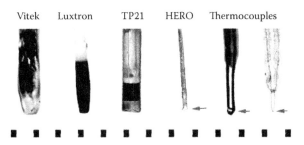

FIGURE 11.1 The appearance and comparative dimensions of some commercially available temperature probes. The ruler divisions are 1 mm. For the smaller probes, the temperature-sensing element is identified by an arrow. See text for more information.

Different models of fluoroptic thermometers have been manufactured since 1978 by Luxtron corporation (Santa Clara, CA). Fluoroptic temperature sensing is based on the fluorescence decay time of a special thermosensitive phosphorescent (phosphor) sensor located at the end of a fiber-optic cable. Luxtron fluoroptic thermometers have also been popular among experimenters [15–21]. The depicted temperature probe is from a Luxtron 850 model thermometer. With this device, the maximum measurement rate was about 30 samples/s; however, at this rate, sequential measurements of a constant temperature could vary by as much as 0.5°C–1°C [22]. Averaging of sequential measurements reduced the "noise," but also reduced the temporal resolution. Other authors reported a noise level of 100–250 mK and a response time of 250 ms for a 700-series Luxtron thermometer [23].

TP-21-MO2 is a different type of fiber-optic sensor that is utilized in the Reflex thermometer that was made by Nortech Fibronic Inc. (Quebec, QC, Canada). The manufacturer's specified temperature resolution is 0.1°C, and the probe's response time is 0.3 s. The exact principle of operation of this probe has not been disclosed. Same as Luxtron and HERO temperature probes, TP-21 is completely immune to EMF pickup. The highest acquisition rate by the Reflex thermometer is limited to only 4 samples/s; however, in our experience, the stability of readings over time and reproducibility of measurements were notably better than in the other tested fiber-optic thermometers (Luxtron 850 and VELOCE). Reflex thermometers are no longer available on the market.

The FOT-HERO fiber-optic probe manufactured by FISO Technologies (Quebec, QC, Canada) is by far the smallest and the fastest on the market. It utilizes a new technology of Fabry–Perot white light interferometry [24]. The heat-sensitive element is just a 5-μm thick film of SiH at the tip of the fiber-optic cable, and the absolute change in the thickness of this film due to thermal expansion is measured with the help of a VELOCE signal conditioner (up to 200 kHz sampling rate). The HERO probe has an outstanding response time of less than 1 ms and a resolution of 0.03°C at 1-ms averaging time (see www.fiso.com). This probe is indispensable for unique high-speed temperature measurements; its advantages, however, are offset by the high cost, fragileness, and appreciable sensitivity of temperature readings to the positioning and bending of the fiber-optic cable.

Premanufactured thermocouples of various sizes are available, for example, from Omega Engineering Inc. (www.omega.com). The thermocouples shown in the figure are made of 125-μm (left) and 25-μm (right) copper and constantan wires (T-type thermocouples); the smaller one will be referred to as a microthermocouple (MTC). Thermocouples, as well as leads connecting them to amplifiers or other devices, are not "non-field–perturbing": they may affect local field distribution just by the virtue of their presence in the field, and the field-induced currents may heat the thermocouple junction or affect the amplifier directly, producing erroneous readings. The response time in the smallest thermocouples is comparable to that of the HERO probe (e.g., it is 2 ms in MTC).

When proper measures are taken to minimize possible EMF pickup artifacts, thermocouples may be a convenient and inexpensive alternative to fiber-optic thermometry, and they have often been employed in EMF bioeffect studies [22,25–33].

11.2.3 Comparative Evaluation of Temperature Probes

11.2.3.1 Microwave Exposure Setup

We utilized a setup that was employed previously [31,34] for *in vitro* exposure of small biological objects to extremely high peak power microwave pulses (EHPP). EHPP were generated by either an MH300 system with an MF-IM65-01 RF plug-in (Epsco Inc., Itasca, IL), or by a model 337X magnetron transmitter (Applied Systems Engineering, Inc.). The following EHPP exposure parameters were used in this study: carrier frequency between 9 and 10 GHz (typically 9.6 GHz), peak output power 75–115 kW (MH300) or 230–260 kW (337X), pulse width from 0.5 to 2 μs, and isolated pulses or brief pulse trains with up to 1-kHz repetition rate.

A separate transmitter (HP 8690A sweep oscillator with Hughes a 8020H amplifier) was used to generate microwave pulses of the matching carrier frequency, but of longer duration (1–100 ms) and lower peak power (<20 W). Microwave pulses were transmitted via a waveguide (WR90, 22.86 × 10.16 mm) to a custom-made exposure chamber mounted atop the waveguide flange and filled with saline. The waveguide opening was sealed with a sapphire matching plate, flush with the flange (Figure 11.2). Incident and reflected powers in the waveguide were measured via directional couplers by an HP 438A power meter using HP 8481A power sensors. Reflection from the exposure chamber into the waveguide was less than 3%. EHPP shape was nearly rectangular, as monitored via an HP 432 detector on a TEK 2430A digital oscilloscope.

11.2.3.2 Exposure Chamber Design and Analytical Evaluation of Local SAR

Initially, the design of the exposure cell was similar to the one proposed by Chou and Guy [35], Figure 11.2a. A vertical section at the end of the waveguide was separated by a quarter-wave matching plate and filled with saline. The waveguide walls in the cell were covered with a lacquer to prevent its electrical contact with the solution and possible corrosion. The complex dielectric constant of this solution at 9.3–9.7 GHz and at various temperatures was calculated using the equations given by Stogryn [36]. With the solution normality of 0.12–0.14, its relative permittivity (ε_r) and conductivity (σ) at 25°C were calculated as 63.4 and 15.9 S/m, respectively. To verify the calculations, the dielectric properties of the solution were also obtained experimentally using an HP

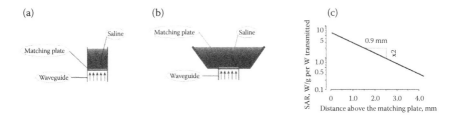

FIGURE 11.2 Basic design of the exposure chamber that was employed in microdosimetry and thermometry experiments. (a) The original design after Chou and Guy [35], and (b) the modified design with improved access to the area immediately above the matching plate. (c) Calculated absorption in the saline along the axis of the waveguide as a function of the distance above the matching plate. Calculations were performed using Equation 11.3 for 9.5 GHz at 25°C. Note that local SAR decreased exponentially with the distance above the plate, at a rate of about twofold per 1 mm. With the saline depth of 10–20 mm, the microwave radiation intensity at the surface was attenuated by a factor of 2^{10}–2^{20}, so any microwave leakage to the outside of the exposure chamber and backward reflection from the saline–air interface could be disregarded.

8510 measurement system (see [37] for detailed procedures). Measured values ($\varepsilon_r = 63$, $\sigma = 16$ S/m) were virtually the same as the calculated ones. The respective linear loss coefficient of the solution was 3.71 Np/cm.

SAR in the solution along the axis of the waveguide decreased exponentially with increasing distance above the matching plate, and this could be calculated according to [35] as:

$$SAR = (1/\rho)\frac{2(P_i - P_r)}{S} 2\alpha e^{-2\alpha z} \text{ (W/g)}, \tag{11.3}$$

where ρ is the saline density (g/cm³). P_i and P_r are the incident and reflected power values (W; their difference will be referred to in later sections as a "transmitted power"), S is the cross section of the waveguide (2.32 cm²), α is the linear loss coefficient in the saline (Np/cm), and z is the distance above the matching plate (cm). On the basis of this equation, a power of 100 kW transmitted to the exposure cell would produce SAR values of 610, 304, 145, and 69 kW/g at distances of at 0, 1, 2, and 3 mm above the plate, respectively (calculated for 9.5 GHz and 25°C). In other words, SAR in the saline decreased approximately twofold per millimeter.

In the later experiments, the waveguide walls beyond the matching plate were removed to form a larger reservoir (Figure 11.2b) with better access from the sides to the area where the SAR was the highest. The entire reservoir was filled with saline; because of shallow microwave penetration into the saline at the employed microwave frequencies, the removal of waveguide walls had no measurable effect on SAR in the chamber along the waveguide axis.

For testing of temperature probes, they were positioned 0.5–2 mm above the center of the matching plate. Peak SAR at this location reached 0.3–0.8 MW/g for EHPP and 5–50 W/g for millisecond-range pulses. Accurate measurement of local temperature and SAR in such a high-intensity pulsed microwave field and within a steep field gradient is perhaps among the most challenging tasks in practical microdosimetry. This is exactly the reason why we chose such "unfavorable" conditions to compare the performance of different temperature probes and to evaluate their applicability for EMF microdosimetry.

11.2.3.3 Temperature Measurements

Three probes were selected for comparative evaluation of their performance under the extreme conditions of microwave exposure described earlier. The probes were (1) TP-21 with a Reflex thermometer, (2) HERO with a Veloce signal conditioner, and (3) a copper–constantan MTC (25-µm wires), connected to a DAM50 DC amplifier (WPI) via a cold junction compensator (Omega). Zero DC level of the amplifier output was regularly checked with a model 181 nanovoltmeter (Keithley Instruments, Inc. Cleveland, OH) and was adjusted whenever a measurable drift was detected. Within a narrow range of temperature changes employed in this study (<10°C), the inherent nonlinearity of the thermocouple readings was negligible.

MTC was fixed in a custom-made glass holder. Bare wires ending with the thermocouple junction extended from the holder by 1–2 mm; they were covered with a lacquer for electrical insulation from the saline.

The three probes were positioned in the exposure chamber such that they were touching each other at the same level, 1–1.5 mm above the center of the matching plate. The precise and well-reproducible positioning was accomplished by means of three separate micromanipulators and was controlled via a stereo microscope. To allow for this visual control, the probes had to be inserted at about a 45° angle to the field propagation vector.

The outputs of the Reflex thermometer, Veloce signal conditioner, and DAM50 amplifier were connected to the analog inputs of a BIOPAC MP100 data acquisition system (WPI). The calibration factor in each channel was fine-tuned to obtain exactly identical temperature readings from the three probes (Figure 11.3).

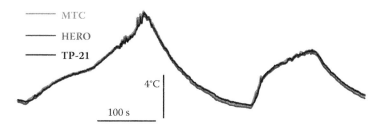

FIGURE 11.3 Concurrent recording of temperature changes by the MTC, HERO, and TP-21 temperature probes. The probes are brought together in the exposure chamber by means of three micromanipulators. The temperature changes were caused by arbitrary additions of hot and cold saline into the exposure chamber (no EMF exposure) and were recorded using a BIOPAC data acquisition system. The goal of this experiment was to verify the identical calibration of the temperature probes and thermometers employed. See text for more detail.

11.2.3.4 Performance of Probes under "Regular" Microwave Exposure

These experiments were performed using relatively long pulses (10–100 ms) at a transmitter output power of less than 20 W. Calculated peak SAR values at the location of the temperature sensors were within 10–40 W/g; the time-average SAR values varied widely depending on the pulse repetition rate and duty cycle.

The ability of the probes to produce correct temperature readings during microwave irradiation was evaluated in two ways: (1) the probe's reading should match the concurrent readings of the other probes and (2) the presence of the probe in the exposure chamber in the vicinity of the other probes should not affect readings of these probes.

When MTC and HERO were positioned in the exposure chamber together (in the absence of the TP-21 probe), their response to microwave heating was almost identical (Figure 11.4). Removal of the MTC from the exposure area had no effect on the HERO readings, indicating that possible field distortion and additional microwave heating in the vicinity of the MTC were negligible.

Peak heating reading by the MTC could be slightly higher (Figure 11.4) or slightly lower (e.g., the 5th and the 7th exposures Figure 11.5) than respective concurrent readings of the HERO readings probe. The reason for these small differences has not been fully understood; they could be caused by such factors as slightly uneven positioning of the probes relative to the irradiator; different size, shape, and material of the probes, leading to unequal field distortion and absorption changes in their immediate vicinity; different heat dissipation via the leads; or a combination thereof.

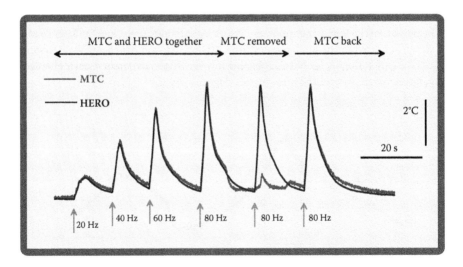

FIGURE 11.4 Readings of the temperature rise by a HERO probe and by an MTC positioned together in the exposure chamber. Exposures (shown by gray arrows): 2-s trains of 10-ms pulses at different repetition rates, 15.5-W peak power output (peak SAR ~35 W/g). Note that withdrawal of the MTC had no effect on HERO readings in response to a microwave exposure. (When withdrawn, the MTC was at an arbitrary remote location within the exposure chamber; its readings during the "removed" time interval can be disregarded.)

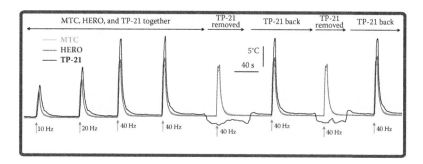

FIGURE 11.5 Concurrent recordings of microwave heating by the MTC, HERO, and TP-21 probes. Exposures (shown by arrows): 5-s trains of 20-ms pulses at different repetition rates, 10-W peak output (peak SAR ~25 W/g). Note that TP-21 readings of the microwave heating are substantially higher than that of the other two probes. Note also that readings of the HERO and MTC probes reversibly decreased when the TP-21 probe was withdrawn. See text and Figure 11.4 for more explanation.

The situation changed dramatically when the TP-21 probe was positioned in the immediate vicinity of the two smaller probes (Figure 11.5). Not only did the TP-21 probe report much higher temperature rise than the two other probes, but it also affected the readings of the other probes: The temperature rise reported by HERO and MTC in the presence of TP-21 was notably higher compared to the same exposure in the absence of TP-21. The difference of TP-21 readings from those of HERO and MTC could be as high as 5°C, and the effects of the withdrawal and reintroduction of the TP-21 probe on the readings of the two other probes were well-reproducible.

In other experiments, the TP-21 probe was deliberately positioned 0.5–1 mm above the other two probes, i.e., farther away from the matching plate. Despite considerable additional attenuation of the field at these locations, readings of TP-21 were higher or just equal to those of the other probes.

The TP-21 probe is immune to microwave radiation and has no metal elements. It is probably safe to assume that this probe is capable of reporting correct temperature even when it is exposed to microwaves. This conjecture is well confirmed by the fact that we observed no pulse microwave pickup or temperature changes when a TP-21 probe was exposed to air. Hence, the only explanation for the higher temperature readings taken by this probe in saline is the local-field distortion (as described in Section 11.2.1), resulting in higher local field intensity than it would be at the same location in the absence of the probe. In turn, higher local field intensity resulted in a profoundly greater heating in the immediate vicinity of the TP-21 probe. As a consequence, the smaller HERO and MTC probes also displayed higher temperature readings when they were situated in the area of field distortion near the TP-21 probe.

On the basis of these findings, only the smallest probes (HERO and MTC) were used in subsequent experiments.

11.2.3.5 Performance of Probes under EHPP Microwave Exposure

These experiments employed 1- and 2-μs pulses at the transmitter peak output power of up to 250 kW; peak SAR at the location of the temperature probes reached 600–800 kW/g.

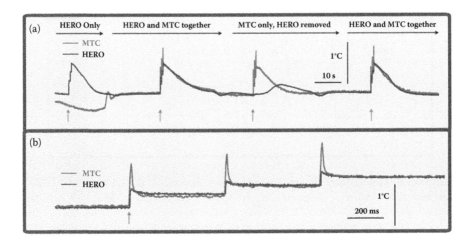

FIGURE 11.6 Concurrent temperature readings by HERO and MTC probes during exposures to trains of extremely high-power microwave pulses. All exposures were identical trains of 3 EHPPs: 2-µs pulse width, 500-ms inter-pulse interval, 250-kW peak output, 700-kW/g peak SAR. The onset of each exposure is identified by an arrow. (a) Alternate readings by both probes situated in the exposure area together and when one of the probes was withdrawn. Note that the presence of either probe in the exposure area had no apparent effect on readings of the other probe. (b) The second exposure from (a), shown at an expanded timescale. Note transient EHPP pickup artifacts, which affected MTC readings during the first 20–30 ms after each EHPP.

An example of simultaneous temperature readings with HERO and MTC probes exposed to EHPP is shown in Figure 11.6a. Exposure to a train of 3 EHPPs (2-µs pulse width, 500 ms inter pulse interval, 250-kW peak output, 700 kW/g peak SAR) caused a distinct stepwise temperature rise. In contrast to "regular" microwave pulses (Section 11.2.3.4), EHPP irradiation could induce transient artifacts in MTC measurements. These artifacts are probably caused by EMF pickup and instant overheating of MTC, MTC wires, and their immediate microenvironment. Due to the tiny volume and negligible heat capacity of the MTC, the excessive heat is fully dissipated in just 20–40 ms after exposure, without any measurable impact on the temperature of the medium. Indeed, temperatures measured by the HERO probe were practically unaffected by the presence or absence of MTC, and vice versa. When the probes were situated in the exposure area together, their readings became identical shortly after the exposure. In other words, the MTC was capable of measuring correct temperature before the EHPP(s) and shortly (although not instantly) after it. The fact that the presence of the MTC did not affect heating of the surrounding medium was additionally confirmed in a different type of experiment where the MTC-recorded heating curves served as "self-control" [32].

11.2.3.6 Summary: Applicability of Different Probes for Microwave Thermometry and Dosimetry

The records presented in Figure 11.5 provide an excellent example of how a fiber-optic probe (made only of dielectric materials) may produce erroneous readings when exposed in saline to microwave pulses (25 W/g peak SAR and 5–20 W/g time-average SAR). Although this probe is commonly regarded as "non-field–perturbing" or as "artifact-free," these definitions are not entirely correct. When exposed in a conductive medium, TP-21 can profoundly distort the local field in its vicinity, resulting in erroneous temperature readings. On the contrary, smaller probes made solely of dielectric materials (HERO) or solely conductive materials (MTC) displayed almost identical readings. These results indicate once again that the size of the temperature probe, especially in intense microwave fields, may be a more important parameter than the materials it is made of.

At the same time, it is worth mentioning that the error in temperature readings by the TP-21 probe was logarithmically proportional to the time-average SAR. For the case illustrated in Figure 11.5, the readings of TP-21 exceeded those of two other probes by 4.5%, 10.5%, and 16.5% at the time-average SAR of 5, 10, and 20 W/g, respectively (Figure 11.7). One may anticipate that at low enough SAR, the error becomes negligible, and TP-21 probe can be safely employed for temperature measurements. For the case illustrated in Figures 11.2 and 11.5, the apparent "safety limit" will be at 2 W/g; however, this value may vary greatly depending on specific exposure conditions (frequency, surrounding medium, exact probe position, etc.).

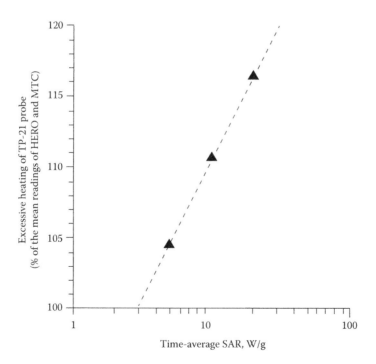

FIGURE 11.7 The error in temperature readings by TP-21 probe as a function of time-average SAR during exposure. Temperature readings of TP-21 are shown as a percentage of concurrent readings of HERO and MTC probes, measured in °C. The graph is based on the data illustrated in Figure 11.6.

Overall, the experimental data suggest that relatively large "non-field–perturbing" temperature probes (e.g., TP-21, Luxtron, VITEK), which are undoubtedly useful for measurements in weak EMFs, may produce large errors in high-intensity, high-gradient fields and should be used with caution.

Since these probes appeared on the market, they have been a popular choice to measure local SAR from the rate of local microwave heating. Such measurements render accurate results when all of the following conditions apply: (1) The presence of the probe during exposure does not introduce large measurement errors, as shown in Figure 11.5; (2) The field and heating gradients are relatively low, meaning that the change in these parameters within a distance comparable with the size of the probe can be neglected; (3) Microwave heating should be slow enough to allow for adequate data sampling by a particular thermometer (sampling rates of different thermometers are largely in the range from 1 to 30 Hz); (4) Microwave heating should be fast enough to enable the reliable detection of the initial linear portion of the heating curve, prior to any considerable heat dissipation. Alternatively, the rate of heat dissipation can be accurately measured and included in the equation for calculating SAR (e.g., [38]).

The above procedure of local SAR measurements has been widely used and addressed in published literature. Hence, this method will not be further discussed in this chapter, with the exception of one underappreciated problem. This problem arises from the fact that local heating in tissues during exposure can be produced directly by the EMF absorption at the given location, and indirectly, by heat flow from neighboring areas that might have absorbed more radiation and thereby reached a higher temperature. In an artificial "extreme" situation, as shown in Figure 11.8a (two absorptive layers separated by an aluminum foil), a temperature sensor placed in the EMF-shielded absorptive layer will still record a heating curve that can be erroneously used for SAR calculation. Note that SAR definition is based on the local E-field, not on the heating rate (Section 11.1); hence, in the absorptive layer shielded by the foil, SAR is equal to zero by definition, irrespective of the local heating rate during exposure. The indirect heating component can be recognized by a delayed start of the heating curve after the onset of EMF exposure, typically leading to S-shaped heating curves. However, this delay can be easily missed if the sampling rate of the temperature probe is slow (Figure 11.8b); or it can be misinterpreted by the experimenter as a meter "noise" and disregarded.

In real-life situations, SAR measurement errors introduced by indirect heating can be very significant and perhaps sometimes approach the 100% level, like in the hypothetical setting shown in Figure 11.8. Large errors may be expected, for example, when skin is exposed to millimeter-waves, and local SAR under the skin is measured with a subcutaneous temperature probe (millimeter-waves are mostly absorbed within skin, so subcutaneous heating will have a large indirect component). This type of error in local SAR determination is very difficult to account or correct for, and it is inherent for methods that require collection of thermometry data over extended time intervals. In contrast, the method of "instant" SAR determination with fast temperature probes as described later in this chapter is completely immune to this error.

Overall, the size, the slow response, and the potential errors in readings of the larger temperature probes limit their utility for high-resolution thermometry and dosimetry, especially in high-gradient and high-intensity fields. When possible, preference should

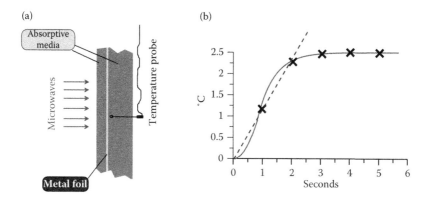

FIGURE 11.8 Illustration of local SAR measurement error resulting from indirect heating. (a) A hypothetical exposure setting where two layers of an absorptive material are separated by a metal foil. Microwaves will only be absorbed in front of the foil. However, temperature may also rise in the area shielded by the foil, because of heat flow from the exposed area. A sample heating curve for a location in the shielded area is shown by the curve in fragment (b); all numbers are arbitrary. If the heating rate in the shielded area is measured with a thermometer having a discrete sampling rate of 1 Hz (black symbols), the S-shaped onset of the curve will not be detected. An experimenter who is not aware that the area was shielded will mistakenly calculate local SAR from the slope of the apparently linear portion of the curve (dashed line). If the absorptive media is water [$C = 4.2$ J/ (°C × g)], the calculated local SAR will be about 5 W/g. Actually, SAR in the shielded area is zero by definition.

be given to the smaller probes, of which only two types can be found on the market, namely, MTC and HERO.

These smaller probes performed equally well with "regular" microwave pulses, but not with EHPP. EMF pickup artifacts were clearly present in MTC readings, but they lasted no more than 20–40 ms after each microwave pulse. In practice, the repetition rate of EHPP in biological experiments is usually kept low, e.g., 0.5–2 Hz or less (otherwise, heating will exceed limits of biological function). In these cases, the time-average temperature readings by MTC will not be affected by the short-lived pickup artifacts. Indeed, in the experiments described in [32], slow concurrent recording of EHPP heating by MTC and by a Luxtron probe looked virtually identical. The EMF pickup transients can also be removed by low-pass filtering of the MTC signal, although it must be understood that this procedure is purely cosmetic.

EMF pickup artifacts are an inherent property of "artifact-prone" probes such as thermocouples. Proper positioning of the probe leads and their shielding are the most efficient ways to minimize these artifacts, although the results are sometimes unpredictable. In our setup, the layer of saline ensured very efficient shielding of the leads, but, depending on specific exposure conditions and settings, the magnitude of possible artifacts may be many times greater than reported here. In general, thermocouples, as well as other "artifact-prone" probes should not be used for temperature measurements unless they are proven to perform properly under specific exposure settings. Interestingly, this rule does not apply to the precision local microdosimetry technique using MTC (described later), as the temperature readings are taken before and after the exposure (not during it) and therefore are not distorted by the artifacts.

The uniqueness of the HERO probe used with the Veloce signal conditioner is in combining the "artifact-free" performance features of a fiber-optic sensor with the minimum thermal mass and virtually instant response typical of the MTC. The HERO probe outperforms other temperature sensors and can potentially be considered a "gold standard." However, based on our experience with a HERO/Veloce thermometer (manufactured in 2002), the device was tricky to calibrate and the calibration values could be rather unstable. For example, moderate bending or simple repositioning of the fiber-optic cable, as well as subtle movement of the cable plug-in (such as touching it with a finger) could shift the output zero and calibration values, requiring recalibration. The Veloce conditioner itself has no indicators, and any temperature (or output voltage) readout can only be obtained using external devices. Although Veloce was supplied with specific guidelines and software to communicate with a computer, we found it greatly advantageous to employ an MP100 data acquisition system with Acknowledge™ software instead. Overall, the HERO/Veloce thermometer was an indispensable tool for unique, high-speed, high-resolution, artifact-free, one-time measurements; however, it may be troublesome to use it for routine everyday measurements. (Note: We have not assessed the latest modification of this thermometer, which might have already been improved by the manufacturer.)

11.3 Pulsed EMF Microdosimetry Using the MTC and HERO Temperature Probes

11.3.1 General Considerations

As shown in Section 11.2, MTC and HERO are the most suitable probes for high-resolution, quick temperature measurements with minimum local-field distortion. In this section, we will show how these probes can be utilized for precision microdosimetry in high-intensity pulsed EMF. Although a HERO/Veloce thermometer would be a top choice for this task, the cost of this device may be prohibitively high for most experimenters. Therefore, we will largely focus on the use and performance of MTC to demonstrate that it can be reliably used in many microdosimetry applications, despite possible EMF pickup artifacts.

A conventional way to calculate SAR from temperature rise is based on the measurement of the initial slope of the heating curve during EMF exposure. When MTC is used for this purpose, EMF pickup transients may obscure the initial slope, making accurate measurements impossible (e.g., see the top record in Figure 11.9). Therefore, MTC could not be employed in the conventional manner for measuring local SAR from EHPP exposure, and an alternative procedure had to be developed.

11.3.2 A Key to Microdosimetry in Intense Pulsed Fields: The Temperature Plateau Phenomenon

An unexpected observation from the temperature curves measured with the MTC and HERO probes (but not with any other probes) was the presence of an extended

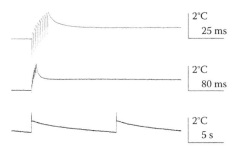

FIGURE 11.9 The shape of the heating/cooling curves recorded by MTC at different timescales. Heating was produced by identical trains of 10 EHPPs (110-kW peak transmitted power, 1-μs width, 1-ms inter-pulse interval, 1 train/12 s). Note the EHPP pickup artifacts (they also mark the time when the EHPP train was applied), the brief temperature decline after the exposure is over, and the long-lasting temperature plateau afterward. Exposures were performed in the chamber shown in Figure 11.2, A; MTC was positioned in saline about 1 mm above the center of the matching plate. (Adapted from Pakhomov, A.G., et al., High-resolution microwave dosimetry in lossy media. In: Klauenberg B, Miklavcic D, editors. *Radio Frequency Radiation Dosimetry*. Dordrecht, the Netherlands: Kluwer Academic Publishers, 187–197. With kind permission of Springer Science and Business Media.)

temperature plateau after a short microwave pulse or a brief train of pulses (Figures 11.6B, 11.9, and 11.10). It should be emphasized that this was an actual plateau, and not a part of the declining slope that was artificially expanded to make it appear flat. The presence of this plateau was a distinguishing feature of fast MTC and HERO readings and typically could not be detected when using larger and slower probes. Perhaps this fact explains why, despite numerous studies in the areas of EMF thermometry and dosimetry, the plateau phenomenon has not been described until recently [32].

At first glance, the existence of a temperature plateau after an exposure is somewhat counterintuitive: if an object's temperature has been increased above the background during EMF heating, one may expect it to go down immediately once the external influx of energy is discontinued. However, measurements taken with the smallest and fastest probes show it unambiguously that the temperature decrease after cessation of exposure may occur after a short delay, forming a postexposure temperature plateau. Potential reasons for plateau formation, at a conceptual level, are explained later and illustrated in Figure 11.11.

Figure 11.11a depicts a hypothetical situation when a temperature probe is placed inside a sphere made of an absorptive material. It is also assumed that the presence of the temperature probe causes no field distortion, does not affect heating in the sphere, and that some sort of an EMF exposure causes uniform heating of the sphere above its environment. In this case, cessation of EMF exposure will have no immediate effect on the temperature readings by the probe. First, the surface layers of the sphere will begin to cool down to the environment temperature; next, the intermediate-depth layers will start to cool down; ultimately, the process will reach the center of the sphere in the vicinity of the probe, which will then display a temperature decline. However, this process takes some finite time, and there will be a distinct delay between turning off the EMF field and the onset of the temperature decline, displayed as a postexposure plateau on the

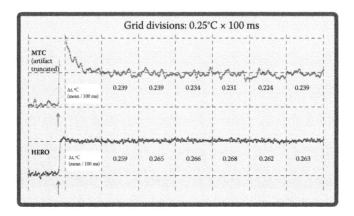

FIGURE 11.10 Examples of the temperature plateau following exposure to a single extremely high-peak power microwave pulse (pulse duration 2 μs, peak transmitted power about 240 kW). Exposure setting is the same as in Figure 11.9, except for using the chamber shown in Figure 11.2b. The same temperature recording was made using the MTC (top) or the HERO probe (bottom). The time when the EHPP was delivered is shown by arrows. Numbers in each square show the temperature difference from the pre-exposure level (ΔT, °C), as averaged over a respective 100-ms interval identified by grid divisions. Note that the temperature stayed practically constant (within 3 digits of precision!) for at least 700 ms after the microwave pulse, meaning the detection of a true postexposure temperature plateau. Note also that the initial EMF pickup artifact is not present when using the HERO probe, which is artifact-free. Slightly higher readings of the HERO probe compared to the MTC, similar to those seen in Figure 11.5, could be a result of minor field distortion in the immediate vicinity of this probe.

temperature curve. The time duration of this plateau will depend on many parameters (e.g., the size of the sphere, the rate of heat conduction), but the principal conclusion from this hypothetical example is that the presence of a postexposure temperature plateau is indeed expected and makes no surprise in such exposure setting.

Figure 11.11b depicts a more realistic situation, when an absorptive object has been nonuniformly heated by an EMF exposure. The object is assumed to have semi-infinite structure, meaning that it has one flat surface and the other surfaces are far away, so they need not be accounted for. We assume again that the presence of the temperature probe had no effect, and expect that the irradiated surface of the cylinder experiences maximum heating, whereas deeper areas will be heated less and less, due to gradual power loss when the field propagates through the object. Hence, irradiation will produce a temperature gradient in this object; for simplicity, we will consider the situation when the attained temperature is the same at any given depth from the surface and that the temperature change with the distance from the surface is linear (although in real-life exposure conditions it will usually be closer to exponential).

In this setting, once the EMF radiation is turned off, the temperature sensor will be instantly subjected to two opposite processes: it will be cooled down by heat conduction to the deeper (cooler) layers, but concurrently it will also be warmed up by heat conduction from the upper (warmer) layers. In the idealized situation, two processes completely

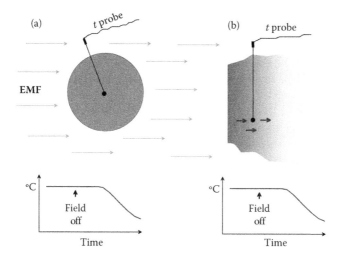

FIGURE 11.11 Hypothetical exposure settings that illustrate situations when local temperature inside of an EMF-heated object may remain unchanged for a finite time interval after cessation of exposure. In (a), the temperature probe is located within a sphere that has been heated to a uniform temperature by some modality of EMF exposure. In (b), the probe is placed into a medium that has attained a uniform temperature gradient during the EMF exposure; the exposed surface is assumed to be warmer (grey) than deeper layers (light grey). Dark grey arrows show the direction of heat flow once the exposure is terminated. In both (a) and (b) settings, the temperature probe may record a short-lived temperature plateau after the exposure is turned off. See text for more explanation.

balance each other, resulting in a temperature plateau that will last for a certain time after the irradiation is discontinued.

One may argue that real-life EMF exposures will seldom, if ever, produce field and temperature distribution patterns such as those described earlier. However, if we consider only a very small volume within a large exposed object, these patterns may be indistinguishable from the idealized situation. Hence, with a microscopic temperature probe that measures instant temperature in its immediate vicinity, the idealized assumptions may indeed be true for real-life situations.

As far as practical applications are concerned, the plateau can be readily and routinely detected if (1) the temperature probe is small enough, such as MTC or HERO, (2) the temperature data are acquired at a rate of no less than 100 Hz (1000 Hz preferred), so that a plateau that lasts only a few tens of milliseconds will not be missed, (3) the high-frequency cutoff filter of a signal amplifier, if one is used, should be no less than 50–100 Hz; any low-frequency filters must be disabled to allow DC recording, (4) the overall duration of EMF exposure (a pulse or a train of pulses) should not exceed 10–100 ms, (5) the EMF intensity is high enough to produce measurable heating during this brief exposure, and (6) *the probe is positioned so that it is surrounded by a uniform absorptive medium, not at the interface of media with different absorptive and heat conduction properties.* Failure to meet the above conditions does not necessarily prevent one from the detection of the postexposure temperature plateau, but the chances will be decreased.

With EHPP exposure and temperature recording using MTC, the plateau is not instant after the exposure is turned off. Typically, the plateau occurs after a brief (~10–20 ms) decline of the reported temperature (Figures 11.9, 11.10, and 11.12). The reasons for this artifact may be one or both of the following: (1) relaxation of the EMF pickup signal in the electronic circuitry of the amplifier, and (2) additional heating of the MTC by EMF-induced currents in the MTC itself and in its leads. In the latter case, the brief decline of temperature truly reflects dissipation of this "additional" heat from the MTC into the surrounding medium. Fortunately, the size and the heat capacity of the MTC are so small that the additional heating of the medium by the presence of MTC typically does not affect the level of the following plateau. For example, in Figure 11.11A, the plateau levels were identical after EHPP trains that ended with different sizes of the temperature artifact (the largest artifact occurred with the shortest (1-ms) interval between EHPP). This brief initial artifact was not present when the temperature changes were recorded with the HERO probe (which is artifact-free), but the plateau level was practically the same as with the MTC probe that picked up a large artifact (Figure 11.10).

11.3.3 Postexposure Temperature Plateau, Specific Absorbed Dose, and Specific Absorption Rate

On the basis of the previous discussion, the presence of the postexposure plateau means that, at the given location, the dissipation of heat generated by EMF absorption was negligible, or that the opposite heat flows were balanced and cancelled each other. In either case, heat dissipation can be disregarded, so the plateau level (i.e., the difference between the temperature before the exposure and the plateau) is linearly proportional to the absorbed energy (which, in turn, is proportional to the delivered energy). Indeed, when the delivered energy is kept constant, the level of the plateau does not change, regardless of the particular pulse repetition rate or overall train duration (Figure 11.12a). Likewise, equal increments in the amount of delivered energy (by adding an extra four pulses to the train, Figure 11.12b) produced proportional and identical increments in the measured plateau level.

Hence, the local specific absorbed dose [SAD; (J/g)] can be accurately measured from the postexposure temperature plateau level as:

$$SAD = C\,\Delta T, \tag{11.4}$$

where C is the specific heat capacity of the medium [J/(kg × °C)] and ΔT is the plateau level (°C).

Once the SAD is known, we no longer need to measure the initial slope of the heating curve to calculate SAR. Instead, SAR (W/g) can be calculated as a ratio of SAD and known duration of the EMF pulse (Δt, s):

$$SAR = \frac{SAD}{\Delta t} = \frac{C\,\Delta T}{\Delta t}. \tag{11.5}$$

In case heating is produced by a train of pulses, Δt is the product of the pulse number and individual pulse duration (s). This "alternative" method of fast and accurate measurement of local SAR is illustrated in Figure 11.13.

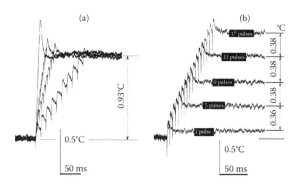

FIGURE 11.12 The postexposure plateau level does not depend on the pulse repetition rate (a) and is proportional to the delivered (and absorbed) energy (b). Heating was produced by EHPP trains (1-μs pulse width, ~100-kW peak transmitted power); pickup artifacts show the exact time when microwave pulses were applied. Temperature was measured as a mean value for a 40-ms period. (a) Trains of 10 pulses (therefore, delivering the same dose) were applied with interpulse intervals of 1, 3, 5, and 13 ms. The postexposure plateau level (0.93°C) did not depend on the interpulse interval or the amplitude of the pickup artifacts. (b) Changing the absorbed dose by varying the number of pulses in the EHPP changes the level of postexposure temperature plateau in a linear manner. Heating by a single pulse was about 0.09°C. (Reproduced from Pakhomov AG, et al., High-resolution microwave dosimetry in lossy media. In: Klauenberg B, Miklavčič D, editors. *Radio Frequency Radiation Dosimetry*. Dordrecht, the Netherlands: Kluwer Academic Publishers, 187–197. With kind permission of Springer Science and Business Media.)

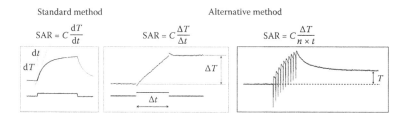

FIGURE 11.13 Standard and alternative methods of specific absorption rate (SAR) measurement from heating dynamics. Shown are the microwave pulse duration (lower traces) and sample heating curves (upper traces). The standard method uses the initial slope of the heating curve (C is the specific heat capacity of medium). The alternative method calculates SAR from the level of the after-pulse plateau and the duration of the pulse, or of all pulses in a train (in the right fragment, *n* is the number of pulses and *t* is the known duration of each pulse). Both methods can be used with the same result when the plateau is present and no artifacts are recorded. However, the alternative method is the only choice when the initial slope cannot be measured accurately. (Reproduced from Pakhomov AG, et al., High-resolution microwave dosimetry in lossy media. In: Klauenberg B, Miklavčič D, editors. *Radio Frequency Radiation Dosimetry*. Dordrecht, the Netherlands: Kluwer Academic Publishers, 187–197. With kind permission of Springer Science and Business Media.)

The final step to justify the alternative method was to compare the measured SAR values with those predicted theoretically using the Equation 11.3. SAR measurements were performed with MTC in well-defined exposure conditions (Figure 11.2a), and the comparison results are shown in Figures 11.14 and 11.15.

In Figure 11.14a, heating of 0.1°C was produced by a single high-power pulse. The calculated and measured SARs were 585 and 420 kW/g, respectively. As stated by Bassen and Babij [39], "an absolute accuracy of ±3 dB is the best case measurement uncertainty that can be achieved when attempting to determine the maximum and minimum SARs within an irradiated biological body." In our exposure system and with the use of MTC, the difference between calculated and measured SARs was only 1.4 dB.

The match was even better for exposure to relatively low-power, longer microwave pulses (Figure 11.14b). Measured values were only 0.4–0.5 dB apart from the calculated ones. This small error could result from the combined inaccuracy of power meters, field probes, and other devices, or even to some imperfection of the analytical equation itself (the equation is for TE_{10} wave propagation mode, but other modes can also arise in the waveguide after the matching plate). Figure 11.14 also demonstrates that the presence of the temperature plateau after a microwave pulse is not a specific attribute of exposure to extremely high-power pulses. The plateau can be as well pronounced with reasonably low-amplitude pulses of ms-range duration, and SAR measurements with MTC are accurate as long as the plateau is present.

Figure 11.15 gives an example of the "local-field mapping" with MTC and illustrates the spatial resolution of the method. Heating curves recorded in spots as close as 0.5 mm

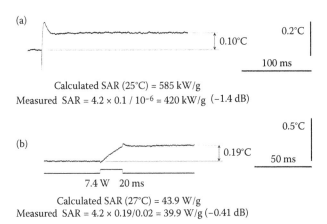

(a)

0.2°C

0.10°C

100 ms

Calculated SAR (25°C) = 585 kW/g

Measured SAR = 4.2 × 0.1 / 10⁻⁶ = 420 kW/g (−1.4 dB)

(b)

0.5°C

0.19°C   50 ms

7.4 W   20 ms

Calculated SAR (27°C) = 43.9 W/g

Measured SAR = 4.2 × 0.19/0.02 = 39.9 W/g (−0.41 dB)

FIGURE 11.14 Comparison of calculated and MTC-measured local SAR values. (a) Exposure to a single extremely high-power pulse (96-kW transmitted power, 1-μs width); the moment of exposure coincides with the artifact on the heating curve. (b) A moderate-power pulse (7.4 W, 20-ms width, indicated on the lower trace). The temperature before the pulse and during the plateau was measured as a mean of a 25-ms period. Calculated SAR values were obtained using Equation 11.3; see text. The difference of the measured and calculated SAR was −1.4 dB (a) and −0.41 dB (b). (Reproduced from Pakhomov, A.G., et al., High-resolution microwave dosimetry in lossy media. In: Klauenberg B, Miklavcic D, editors. *Radio Frequency Radiation Dosimetry*. Dordrecht, the Netherlands: Kluwer Academic Publishers, 187–197. With kind permission of Springer Science and Business Media.)

A train of 10 high-power pulses (110 kW, 1-μs width, 2-ms interval)

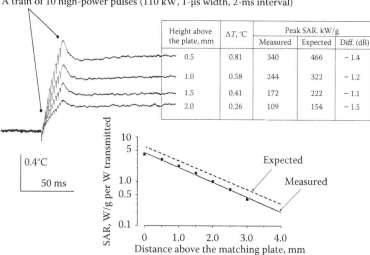

Height above the plate, mm	ΔT, °C	Peak SAR. kW/g		
		Measured	Expected	Diff. (dB)
0.5	0.81	340	466	−1.4
1.0	0.58	244	322	−1.2
1.5	0.41	172	222	−1.1
2.0	0.26	109	154	−1.5

0.4°C

50 ms

SAR, W/g per W transmitted

10
5

1.0
0.5

0.1

Expected

Measured

0   1.0   2.0   3.0   4.0
Distance above the matching plate, mm

FIGURE 11.15 Local SAR in the exposure chamber as a function of the distance above the matching plate, along the axis of the waveguide. Heating curves (left) produced by a train of 10 high-power pulses (110-kW output, 1-μs width) were recorded at the heights from 0 to 3mm above the matching plate (for clarity, only four curves are shown). The respective heights, plateau levels, and local SAR values are provided in the table. The "expected" local SAR values at the same distances above the matching plate were calculated using Equation 11.3. In the graph below, the expected and measured local SAR have been normalized to the transmitted power and plotted against the distance above the matching plate. (Reproduced from Pakhomov, A.G., et al., *High-resolution microwave dosimetry in lossy media.* In: Klauenberg B, Miklavcic D, editors. *Radio Frequency Radiation Dosimetry.* Dordrecht, the Netherlands: Kluwer Academic Publishers, 187–197. With kind permission of Springer Science and Business Media.)

to each other were substantially different, and measured SAR values were always close to the calculated ones.

Similarly, MTC was used for horizontal field mapping in a plane parallel to the matching plate. Horizontal field distribution was calculated according to standard waveguide equations [40] for TE_{10} mode. Again, the measured and calculated values were within ±1.5 dB apart.

11.3.4 Verification of the Local SAR Measurement Technique in a Different Exposure Setting

Explanations for the formation of the postexposure temperature plateau, as discussed in Section 11.3.2, did not rely on a particular type of radiation or a specific exposure setup. The same is also true for Equations 11.4 and 11.5, which are employed to calculate SAR from the temperature plateau level. In principle, this method of local SAR measurement should be valid for various radiation modalities (not limited to microwaves or even to EMFs) and various exposure settings. The only requirements that enable the use of this method are: (1) Irradiation should be performed as a single pulse or a brief train of pulses, (2) The energy deposited by this pulse (or the train) is sufficient to produce a measurable temperature rise, and (3) The pulse (or the train) duration is short enough not to obscure the temperature plateau.

Hence, this method was additionally tested in exposure conditions very different from those described earlier. A volume of 0.45 mL of a mammalian cell culture medium (RPMI 1640 supplemented with 10% fetal bovine serum, obtained from American Type Tissue collection, Manassas, VA) was placed in an electroporation cuvette with 2-mm gap between two conductive walls (Bio-Rad Laboratories, Hercules, CA); the cuvette is shown in the inset of Figure 11.16. The conductive walls were connected to a Blumlein line pulse generator (custom built at the Old Dominion University, Norfolk, VA, see [41] for design details) to produce a high-voltage, 10-ns duration electrical pulse into the sample. The Blumlein line was charged from a high-voltage DC power supply (HR-100, American High Voltage, Elko, NV) until a breakdown voltage was reached across a spark gap in a pressurized switch chamber. The breakdown voltage (and, consequentially, the voltage of the 10-ns pulse delivered to the sample) was varied from 10 to 25–30 kV by increasing the pressure of SF_6 gas in the switch chamber. The shape and voltage of the pulse delivered to the sample were monitored on a 500-MHz digital oscilloscope (TDS3052B, Tektronix, Wilsonville, OR) via a custom-made high-voltage probe. The pulse shape was close to rectangular, with some additional oscillations caused by impedance mismatch and the capacitance of the load.

Assuming that the pulse shape is perfectly rectangular, the power dissipation in the sample during the pulse can be calculated straightforwardly as U^2/R, where U is the applied voltage (V) and R is the ohmic resistance of the sample (in our case, it was

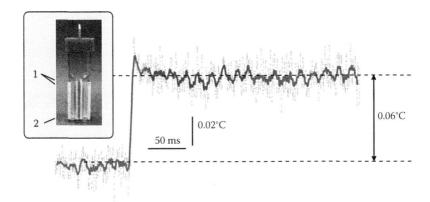

FIGURE 11.16 Heating of the medium in an electroporation cuvette by a 10-ns electrical pulse. Individual data points (grey) show the temperature readings sampled at 10 kHz using the HERO/Veloce thermometer system. The solid line is the "running average" of 49 data points. Heating produced by this single pulse of 10.4 kV (52 kV/cm) was approximately 0.06°C, which corresponded to the SAR value of 24 MW/g. The expected SAR value, calculated from the pulse voltage and electrical resistance of the sample (8 Ω), was 30 MW/g. See text for more detail. The inset shows the electroporation cuvette with a 2-mm gap between the conductive walls (1) filled with the sample medium (2).

about 8 Ω). With the parallel-plate design of the cuvette, the E-field in the sample is close to uniform, so SAR (W/g) can be calculated as:

$$SAR = \frac{U^2}{(RM)},$$ (11.6)

where M is the mass of the sample (0.45 g). In our experiments, these "expected" SAR values were compared with those measured using a HERO probe and a VELOCE signal conditioner. The probe was placed in the medium in the gap between the conductive walls, approximately in the geometrical center of the electroporation cuvette. Temperature data were filtered at 1 kHz and sampled at 10 kHz. To improve the signal-to-noise ratio, up to 12–15 independent runs were averaged off-line.

An example of such temperature recording, with a clear postexposure temperature plateau, is shown in Figure 11.16. The voltage applied across the sample was 10.4 kV, yielding a SAR of 30 MW/g using Equation 11.6. At the same time, the plateau level detected by the HERO probe was about 0.06°C, which corresponds to a SAR of 24 MW/g, using Equation 11.5. Thus, the "expected" and HERO-measured SAR values were remarkably close; in other similar experiments using various voltages, the difference between the expected and measured SAR did not exceed 1.5 dB. Although this difference can probably be explained by a combined inaccuracy of measurement or less than perfect shape of the electric pulse, it may also be indicative of the fact that the pulse energy was not entirely transferred into heat. For example, some of this energy could be spent in an electrochemical reaction, e.g., between the aluminum walls of the cuvette and water or other constituents of the medium.

Regardless of these details, the experiments clearly established that the postexposure temperature plateau is not unique for the exposure system shown in Figure 11.2; the plateau can be found in totally different exposure settings and using a different type of pulsed energy. Furthermore, the procedure of local SAR calculation from the plateau level remains valid and accurate in these different exposure conditions.

Note also that with pulse duration as small as 10 ns, the slope of the heating curve obviously cannot be measured by any conventional means, even if the EMF pickup artifacts are entirely absent. In such cases, measuring SAR from the postexposure temperature plateau is probably the only possible approach, aside from theoretical calculations.

11.4 Summary

In this chapter, we have reviewed practical methods and techniques for high-resolution thermometry and microdosimetry in high-intensity pulsed RF fields. The methods were first reported in [22,32], and, over the years, were repeatedly employed in varied biological studies [31,33,34,42,43]. The described method of local SAR measurement has proven to be fast, accurate, and reliable; it is likely to produce valid data whenever a temperature plateau after a pulsed exposure can be reliably detected (regardless of the radiation wavelength, incident power, absorptive media properties, or exposure setup). When biological experiments are performed at radiation intensities too low to produce a measurable temperature plateau after a pulse, the intensity can be cranked up just for dosimetry purposes. In this case, SAR values measured at the increased field intensity can be linearly scaled down (proportionally to the output power) to determine actual local SAR in the biological experiments.

References

1. Durney CH, Massoudi H, Iskander MF, editors (1986) *Radiofrequency Radiation Dosimetry Handbook*, 4th edition. San Antonio, TX: Armstrong Laboratory (AFMC), Occupational and Environmental Health Directorate, Brooks Air Force Base.
2. Roach WP, editor (2009) *Radiofrequency Radiation Dosimetry Handbook*, 5th edition. Directed Energy Bioeffects Division, Radio Frequency Radiation Branch, Brooks City-Base, TX 78235-5147.
3. Pakhomov AG (2009) Practical guide to high-resolution thermometry and microdosimetry in pulsed electromagnetic fields. In: Roach WP, editor. *Radiofrequency Radiation Dosimetry Handbook*, 5th edition. Directed Energy Bioeffects Division, Radio Frequency Radiation Branch, Brooks City-Base, TX 78235-5147, 69–100.
4. Alekseev SI, Ziskin MC (2001) Distortion of millimeter-wave absorption in biological media due to presence of thermocouples and other objects. *IEEE Trans Biomed Eng* 48: 1013–1019.
5. Bowman RR (1976) A probe for measuring temperature in radio-frequency heated material. *IEEE Trans Microw Theory Tech* 24: 43–45.
6. Burkhardt M, Poković K, Gnos M, Schmid T, Kuster N (1996) Numerical and experimental dosimetry of Petri dish exposure setups. *Bioelectromagnetics* 17: 483–493.
7. Adey WR, Byus CV, Cain CD, Higgins RJ, Jones RA, Kean CJ, Kuster N, MacMurray A, Stagg RB, Zimmerman G. (2000) Spontaneous and nitrosourea-induced primary tumors of the central nervous system in Fischer 344 rats exposed to frequency-modulated microwave fields. *Cancer Res* 60: 1857–1863.
8. Galvin MJ, McRee DI (1986) Cardiovascular, hematologic, and biochemical effects of acute ventral exposure of conscious rats to 2450-MHz (CW) microwave radiation. *Bioelectromagnetics* 7: 223–233.
9. McRee DI, Davis HG (1984) Whole-body and local dosimetry in rats exposed to 2.45-GHz microwave radiation. *Health Phys* 46: 315–320.
10. Nawrot PS, McRee DI, Galvin MJ (1985) Teratogenic, biochemical, and histological studies with mice prenatally exposed to 2.45-GHz microwave radiation. *Radiat Res* 102: 35–45.
11. Adair ER, Adams BW, Akel GM (1984) Minimal changes in hypothalamic temperature accompany microwave-induced alteration of thermoregulatory behavior. *Bioelectromagnetics* 5: 13–30.
12. Burr JG, Krupp JH (1980) Real-time measurement of RFR energy distribution in the *Macaca mulatta* head. *Bioelectromagnetics* 1: 21–34.
13. Chou CK, Guy AW, McDougall JA, Lai H (1985) Specific absorption rate in rats exposed to 2,450-MHz microwaves under seven exposure conditions. *Bioelectromagnetics* 6: 73–88.

14. Spiers DE, Adair ER, Baummer SC (1989) Acute thermoregulatory responses of the immature rat to warming by low-level 2,450-MHz microwave radiation. *Biol Neonate* 56: 48–56.

15. Braithwaite L, Morrison W, Otten L, Pei D (1991) Exposure of fertile chicken eggs to microwave radiation (2.45 GHz, CW) during incubation: Technique and evaluation. *J Microw Power Electromagn Energy* 26: 206–214.

16. Tattersall JE, Scott IR, Wood SJ, Nettell JJ, Bevir MK, Wang Z, Somasiri NP, Chen X. (2001) Effects of low intensity radiofrequency electromagnetic fields on electrical activity in rat hippocampal slices. *Brain Res* 904: 43–53.

17. Kojima M, Hata I, Wake K, Watanabe S, Yamanaka Y, Kamimura Y, Taki M, Sasaki K. (2004) Influence of anesthesia on ocular effects and temperature in rabbit eyes exposed to microwaves. *Bioelectromagnetics* 25: 228–233.

18. Adair ER, Adams BW, Hartman SK (1992) Physiological interaction processes and radio-frequency energy absorption. *Bioelectromagnetics* 13: 497–512.

19. Moriyama E, Salcman M, Broadwell RD (1991) Blood-brain barrier alteration after microwave-induced hyperthermia is purely a thermal effect: I. Temperature and power measurements. *Surg Neurol* 35: 177–182.

20. Saalman E, Nordén B, Arvidsson L, Hamnerius Y, Höjevik P, Connell KE, Kurucsev T. (1991) Effect of 2.45 GHz microwave radiation on permeability of unilamellar liposomes to 5(6)-carboxyfluorescein. Evidence of non-thermal leakage. *Biochim Biophys Acta* 1064: 124–130.

21. Shellock FG, Schaefer DJ, Crues JV (1989) Alterations in body and skin temperatures caused by magnetic resonance imaging: Is the recommended exposure for radiofrequency radiation too conservative? *Br J Radiol* 62: 904–909.

22. Pakhomov AG, Mathur SP, Murphy MR (2003) High-resolution temperature and SAR measurement using different sensors. June 22–27, 2003, Wailea, Maui, HI, 224–225.

23. Schuderer J, Schmid T, Urban G, Samaras T, Kuster N (2004) Novel high-resolution temperature probe for radiofrequency dosimetry. *Phys Med Biol* 49: N83–N92.

24. Choqueta P, Juneaua F, Bessetteb J (2000) New generation of Fabry-Perot fiber optic sensors for monitoring of structures, Newport Beach, CA, 1–9.

25. Dunscombe PB, Constable RT, McLellan J (1988) Minimizing the self-heating artefacts due to the microwave irradiation of thermocouples. *Int J Hyperthermia* 4: 437–445.

26. Dunscombe PB, McLellan J, Malaker K (1986) Heat production in microwave-irradiated thermocouples. *Med Phys* 13: 457–461.

27. Gammampila K, Dunscombe PB, Southcott BM, Stacey AJ (1981) Thermocouple thermometry in microwave fields. *Clin Phys Physiol Meas* 2: 285–292.

28. Gerig LH, Szanto J, Raaphorst GP (1992) The clinical use of thermocouple thermometry. *Front Med Biol Eng* 4: 105–117.

29. Haveman J, Smals OA, Rodermond HM (2003) Effects of hyperthermia on the rat bladder: A pre-clinical study on thermometry and functional damage after treatment. *Int J Hyperthermia* 19: 45–57.

30. Moriyama E, Matsumi N, Shiraishi T, Tamiya T, Satoh T, Matsumoto K, Furuta T, Nishimoto A. (1988) Hyperthermia for brain tumors: Improved delivery with a new cooling system. *Neurosurgery* 23: 189–195.

31. Pakhomov AG, Du X, Doyle J, Ashmore J, Murphy MR (2002) Patch-clamp analysis of the effect of high-peak power and CW microwaves on calcium channels. In: Kostarakis P, editor. *Biological Effects of EMFs*, Proceedings of the 2nd International Workshop on Biological Effects of EMFs, Rodos, Greece, Oct. 2002, 281–288.

32. Pakhomov AG, Mathur SP, Akyel Y, Kiel JL, Murphy MR (2000) High-resolution microwave dosimetry in lossy media. In: Klauenberg B, Miklavčič D, editors. *Radio Frequency Radiation Dosimetry*. Dordrecht, the Netherlands: Kluwer Academic Publishers, 187–197.

33. Pakhomov AG, Mathur SP, Doyle J, Stuck BE, Kiel JL, Murphy MR. (2000) Comparative effects of extremely high power microwave pulses and a brief CW irradiation on pacemaker function in isolated frog heart slices. *Bioelectromagnetics* 21: 245–254.

34. Pakhomov AG, Doyle J, Stuck BE, Murphy MR (2003) Effects of high power microwave pulses on synaptic transmission and long term potentiation in hippocampus. *Bioelectromagnetics* 24: 174–181.

35. Chou CK, Guy AW (1978) Effects of electromagnetic-fields on isolated nerve and muscle preparations. *IEEE Trans Microw Theory Tech* 26: 141–147.

36. Stogryn A (1971) Equations for calculating the dielectric constant of saline water. *IEEE Trans Microw Theory Tech* 19: 733–736.

37. Bao JZ, Swicord M, Davis C (1996) Microwave dielectric characterization of binary mixtures of water, methanol, and ethanol. *J Chem Phys* 104: 4441–4450.

38. Chemeris NK, Gapeyev AB, Sirota NP, Gudkova OY, Tankanag AV, Konovalov IV, Buzoverya ME, Suvorov VG, Logunov VA. (2006) Lack of direct DNA damage in human blood leukocytes and lymphocytes after in vitro exposure to high power microwave pulses. *Bioelectromagnetics* 27: 197–203.

39. Bassen HI, Babij TM (1990) Experimental techniques and instrumentation. In: Gandhi OP, editor. *Biological Effects and Medical Applications of Electromagnetic Energy*. Englewood Cliffs, NJ: Prentice Hall, 141–173.

40. Paris DT, Hurd FK (1969) *Basic Electromagnetic Theory*. New York: McGraw-Hill Companies.

41. Kolb JF, Kono S, Schoenbach KH (2006) Nanosecond pulsed electric field generators for the study of subcellular effects. *Bioelectromagnetics* 27: 172–187.

42. Pakhomov AG, Gajsek P, Allen L, Stuck BE, Murphy MR (2002) Comparison of dose dependences for bioeffects of continuous-wave and high-peak power microwave emissions using gel-suspended cell cultures. *Bioelectromagnetics* 23: 158–167.

43. Walker K, 3rd, Pakhomova ON, Kolb J, Schoenbach KS, Stuck BE, Murphy MR, Pakhomov AG. (2006) Oxygen enhances lethal effect of high-intensity, ultrashort electrical pulses. *Bioelectromagnetics* 27: 221–225.

12

Dosimetry in Electroporation-Based Technologies and Treatments

Eva Pirc, Matej
Reberšek, and
Damijan Miklavčič
University of Ljubljana

12.1 Introduction 233
    Electroporation—The Phenomenon • Physical
    Dosimetry in Electroporation • Biophysical
    Dosimetry in Electroporation

12.2 Applications..245
    Electrochemotherapy • Irreversible
    Electroporation • Gene Transfection

12.3 Conclusion 250

To measure is to know.
If you cannot measure it, you cannot improve it.

Lord Kelvin

12.1 Introduction

Electroporation is a platform technology that is already established in medicine and food processing (Haberl et al., 2013a). It is based on increased cell membrane permeability due to exposure to electric pulses (Weaver, 1993; Kotnik et al., 2012). If the cell is able to fully recover afterwards, we call it reversible electroporation; when the damage is too great and the cell dies, we call it irreversible electroporation (IRE). Electrochemotherapy (ECT) is an antitumor therapy in which locally applied high-voltage (HV) pulses trigger a transient permeabilization of tumor cells. Diffusion of a chemotherapeutic drug (bleomycin or cisplatin) is enabled, resulting in higher cytotoxicity. Effectiveness is accomplished if sufficient drug concentration and electric field in the tumor are achieved. ECT is a highly efficient treatment, with complete response rates of between 60% and 70% on a single treatment and with objective response rates up to 80% (Mali et al., 2013),

233

and is comparable to, if not more efficient than, other similar skin-directed technologies (Spratt et al., 2014). It is used in the treatment of cutaneous and subcutaneous tumors, following standard operating procedures (SOPs; Mir et al., 2006).

If the exposure of the cell to electric field is too excessive, it dies, presumably due to a loss of homeostasis. IRE is used as an ablation method for normal and tumor tissues. It is called "nonthermal" ablation, because cells primarily die due to membrane permeabilization and not due to the increase in the temperature of the tissues. However, we should not overlook local temperature increases around the electrodes, which can be significant at higher amplitudes, increased duration, or number of pulses (Garcia et al., 2014; Kos et al., 2015). Considerable research has also been undertaken in the area of gene transfection and biopharmaceutical drugs stimulating an immune response. Gene electrotransfer (GET) is a nonviral method for delivering DNA molecules into cells. DNA vaccination using electric pulses and clinical trials of GET of DNA with interleukin-12 in patients with metastatic melanoma also has shown great promise in clinical practice (Heller et al., 2001; Haberl et al., 2013a; Lambricht et al., 2016).

Many of its biotechnological applications such as inactivation of microorganisms and extraction of biomolecules have only recently started to emerge, while nonthermal food pasteurization is already being used in the industry (Toepfl et al., 2006; Kotnik et al., 2015). Electroporation is more commonly termed as pulsed electric field (PEF) treatment in food technology. Food preserved by PEF maintains color and flavor, and the antioxidant levels also stay unaffected (Haberl et al., 2013). It is efficient for increasing the shelf life of liquid food (Toepfl et al., 2006). A combination of mild preheating to 60°C and subsequent electroporation reduces the energy needed for efficient disinfection to 40 kJ/L (Gusbeth and Frey, 2009).

Microalgae are currently the most intensely investigated feedstock for biomass production with electroporation; they are getting implemented in biofuel applications (Golberg et al., 2016; Postma et al., 2016). A combination of grape fermentation and electroporation led to an increased content of polyphenolic compounds and less acidity, thereby resulting in a slightly smoother taste and color intensity in wine (Mahnič-Kalamiza et al., 2014). Overall, it is a fast-growing field with great potential.

12.1.1 Electroporation—The Phenomenon

Electroporation is a phenomenon in which cells that are exposed to a high enough electric field increase permeability and conductivity of their membranes. Each biological cell is surrounded by a membrane that mainly consists of phospholipids. Lipids in aqueous conditions spontaneously form a two-molecule thick layer as a result of their dielectric properties. Water and water-soluble molecules cannot pass the entirely intact barrier only by diffusion (Deamer and Bramhall, 1986). In addition, biological membrane also contains glycolipids, cholesterol, and various proteins, which enable selective transport of some molecules from intracellular space to the cell interior and vice versa (Kotnik et al., 2012).

Several theoretical descriptions of the electroporation phenomenon have been proposed. The most established and likely one is that electroporation is based on the aqueous pores formation in the lipid bilayer (Freeman et al., 1994; Weaver and Chizmadzhev, 1996; Kotnik et al., 2012). When cells are exposed to a high pulsed electric field, voltage is induced across their membranes. This results in the rearrangement of their membrane components, leading to the formation of hydrophilic pores in the bilayer, the presence of which increases the ionic and molecular transport to otherwise impermeable membranes (Pucihar et al., 2008; Kotnik et al., 2012). Experimental observation of the pore formation was not successful with known techniques, but molecular dynamic (MD) simulation provides convincing corroboration. From the electrical point of view, a cell can be modelled as an electrolyte (conductive media), surrounded by an electrically insulated/dielectric shell. Each cell under physiological conditions has a resting transmembrane voltage in the range of -90 to $-40\,mV$ (Kotnik et al., 2010). This is a result of ion imbalance in the cytoplasm, controlled by Na^+–K^+ pumps and K^+ leak channels. Na^+–K^+ pumps export Na^+ ions out of the cell and simultaneously import K^+ ions; meanwhile, K^+ ions can freely cross the membrane through K^+ leak channels, to achieve electrical and concentration equilibrium. Applied electric pulses cause local field distortion in the cell and their surroundings. Due to low-membrane conductivity in the vicinity of the cell, the electrical field concentrates mainly in the cell membrane, resulting in electrical potential difference across the membrane. The induced transmembrane voltage superimposes to the resting potential. It can affect transport through the membrane, stimulate cells, and if high enough, lead to the electroporation of cell membrane. Increased cell permeabilization is observed with electric field increase; induced transmembrane voltage is dependent on position, cell shape, and orientation. Delay between external and inducted voltage is in the microsecond range and is determined by the membrane time constant τ_m (Isokawa, 1997). If cells are exposed to electric field in low conductivity medium, delay significantly increases (Kotnik et al., 2010). When short, intense electric pulses (nanosecond pulses, tens of kV/cm, with a period similar to τ_m or shorter) are applied, the outer membrane acts as a short circuit because of cell frequency response, and the applied voltage also appears across the interior of the cell (Kotnik and Miklavčič, 2006). Nanosecond pulses can induce a high enough voltage to cause electroporation of internal organelles (Batista Napotnik et al., 2016). Because organelle interior is electrically more conductive then cytosol, and organelle membrane dielectric permittivity is lower than a cell membrane permittivity, a voltage induced on organelle membrane can exceed the one induced on the cell membrane, resulting in increase of induced voltage amplitude (Kotnik et al., 2010; Retelj et al., 2013). But at the same time, pulses also cause plasma membrane permeabilization (Kotnik et al., 2006; Batista Napotnik et al., 2010).

12.1.2 Physical Dosimetry in Electroporation

The local electric field, i.e., the one "felt" by the cell is the one that leads to membrane electroporation. Applicator/electrode characteristic and applied pulse characteristics define the electric field distribution and intensity. For various applications, different pulse shapes, voltages, duration, repetition frequencies, and sequences are needed. Therefore, special pulse generators have been designed called electroporators. Because biological load characteristics vary considerably, and in addition their conductivity changes due to electroporation during the pulse delivery, development of these devices is challenging.

12.1.2.1 Dosimetry of Pulse Delivery

The electroporation signal is, as mentioned before, characterized by pulse amplitude, shape and duration, number of pulses, pulse repetition frequency, and pulse orientation sequence. Most common pulse shapes that are used in electroporation are square wave (also bipolar), exponential decay, and bell-shaped pulses. When designing an electroporation device, one should always keep in mind that a biological sample as a load has a resistive–capacitive nature and can vary from sample to sample, and in addition the impedance of a biological sample decreases during pulse delivery (Pliquett et al., 2000; Pavlin et al., 2005). The most simple and inexpensive way to generate pulses is by a capacitor discharge circuit (Figure 12.1). When we are dealing with higher voltages, it is more efficient and easier to use smaller capacitors and connect them in to a Marx generator (Figure 12.1c). The main problem here becomes simultaneous switching; the switching element must be chosen with respect to their maximum operating voltage and response rate. The generated pulses have typical capacitor discharge-exponentially decay shape (Reberšek and Miklavčič, 2011; Reberšek et al., 2014). Micro- and millisecond square pulses are usually generated by an HV power supply switching circuit (Figure 12.2a), with fast-power MOSFET (Metal-Oxide-Semiconductor Field-Effect Transistor) or IGBT (Isolated-Gate Bipolar Transistor) used as the switch. All the required energy must be generated and stored in the capacitor before delivery. To minimize a voltage drop, a very large capacitor is needed, resulting in difficult voltage modification. We also have

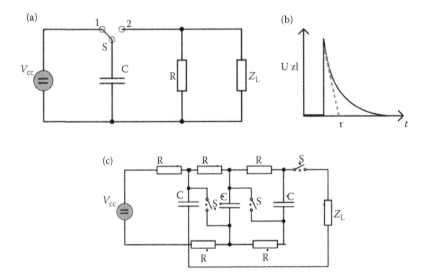

FIGURE 12.1   Panel (a) is capacitor discharge circuit, a built-in resistance R is added to limit the decrease of time constant $\tau = C * |Z_L| \cdot I_F * |Z_L| \geq 10R, \tau \approx RC$, resulting in 90% energy dissipation through R. Panel (c) is a proposed Marx bank circuit. Capacitors C are charged in parallel through resistor R and then switched to series building up the voltage to $n * U$ and discharged through the load Z_L, by switching all switches simultaneously. The maximum applied voltage is equal to the load power supply voltage multiplied by the number of capacitors and time constant. In panel (b), a generated pulse shape is presented, the discharge time is time constant dependent.

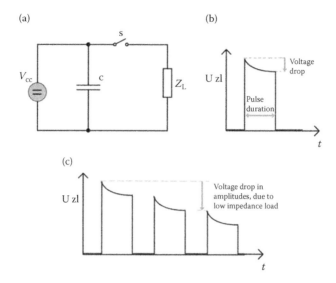

FIGURE 12.2 High-voltage power supply switching circuit (a). The variable power supply V_{CC} defines the amplitude of the output pulse. Switches control pulse duration and pulse repetition frequency. The voltage drop that occurs during pulse is proportional to load impedance $\left(\text{voltage drop} = \text{pulse duration}/\left(C * |Z_L|\right)\right)$ (b). In (c) C an example of reduced amplitude is shown, that can occur in a case of low impedance load that requires too high current.

some time limitations; pulses can be generated after the capacitor is recharged to the preset voltage. Low impedance of a load (tissue or cell suspension) requires large power/currents, which quite often leads to a significant voltage drop. Protocols in which a larger number of pulses are delivered can result in reduced amplitude of pulses (Figure 12.2). This is one of the main reasons why we need to measure when using devices for electroporation. Furthermore, we focus on measuring and quality control.

To achieve an efficient electric field that enables electroporation, we are dealing with HV and currents; therefore, generator construction can be challenging. The shorter the pulses, the more complicated the circuit designs that are required; it is really challenging to generate high-power and short-duration pulses. HV switches with short rise times are needed; spark gaps, UV lasers, SiC MOSFETs, or IGBTs can be used, depending on the application. With nano- and picosecond pulses, pulse-forming networks are a common solution, e.g., a transmission line (Figure 12.3a). Transmission lines operate in both charging and discharging phases. Generated energy should be stored in a large capacitor and then discharged to the load. After pulse generation, a new pulse can be delivered when the capacitor is recharged, resulting in repetition frequency limitation (Bertacchini et al., 2007; Syamsiana and Putri, 2011; Reberšek et al., 2014).

Electrodes, together with the biological sample, present the load for the pulse generator. The main problem that we encounter here is that electrodes get polarized where they get in touch with the sample due to water molecules and hydrated ions that are present in the surrounding area. It is a frequency-dependent phenomenon, which can be modeled as a capacitor in series with a resistor (Chafai et al., 2015). In a cell suspension, a counterion

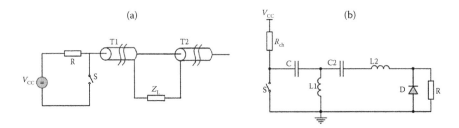

FIGURE 12.3 Concept of Blumlein transmission line (a) and diode opening switch (DOS) circuit (b), commonly used in nanosecond pulse generator designs. The Blumlein transmission line generators have a variable high-voltage power supply *V*, a charging resistor *R*, and two transmission lines. T1 and T2 are charged when the switch is turned off and then discharged through the load when switched. The pulse duration cannot be modified as it equals twice the electrical length of the transmission line, if the impedance of the load is twice the impedance of the transmission line. The DOS generators can be composed of more accessible electrical components than Blumlein generators. Pulse is formed by a diode that must be forward and reverse pumped with adequate sinusoidal current. Diode should stop conducting when the majority of the total energy is stored in L2. That means a current must be maximized at the time of switching. High voltage is switched by diodes, which means MOSFET-s does not need to withstand the whole output amplitude, and does not need to be faster than the output pulse. Pulse duration is determent with diode reverse recovery time. But finding the appropriate matching of capacitors and inductors values in LC oscillator for optimal switching can be challenging. (From Reberšek, M., et al., *IEEE Electr. Insul. Mag.*, 30(3), 8–18, 2014; Sanders, J. M., et al., *IEEE Trans. Dielectr. Electr. Insul.*, 16(4), 1048–1054, 2009.)

layer is formed at each electrode and electric field driving charge transport is reduced, resulting in lower suspension conductivity. At the contacts with tissue, electrodes stimulate the release of electrolytes, resulting in the development of a poorly conductive region where wounds can occur. Luckily, polarization decreases with increasing frequency.

Electrodes must be user friendly; the wires connecting them should be long enough to enable easy handling and smooth application. But each additional wire/connection has some parasitic properties resulting in route losses—the higher the frequencies, the more the parasitic characteristics are manifested. In worst cases, when dealing with nanosecond pulses, the generated pulses at the end of electrodes can completely differ from the ones at the output stage of the electroporator. At high frequencies, reflection on the lines must also be taken into account. The thickness of the wire must be compatible with the output current (Kolb et al., 2006; Batista Napotnik et al., 2016).

Different electrodes are available on the market, and they need to be chosen considering the targeted load and pulse generator restrictions. Electrode geometry and position also determine electric field distribution. *In vitro*, four main groups of electrodes are present: single-cell chambers, macroelectrodes (two-plate electrodes separated for at least 1 mm), microelectrodes (glued onto cover glass, with separation of 100 μm), and flow-through chambers (polyethylene or polypropylene used as insulating materials, combined with stainless steel electrodes) (Reberšek et al., 2014). When using nano- or picosecond pulses, impedance matching must be ensured. For *in vivo* use macroplate and needle electrodes are commonly used. Electrochemotherapy standard operating procedure (ESOP) (Mir et al., 2006) describes three different types of electrodes that were developed within the Cliniporator project and are compatible with the Cliniporator generator (Table 12.1). According to SOP, plate electrodes are recommended for treatment of small and superficial tumor nodules. For treatment of thicker and deeper-seated tumor nodes, needle electrodes are more suitable. During the development of Cliniporator, voltage characteristics—sequences—were determined and acquired for better efficacy (Ongaro et al., 2016). Sometimes, large volumes need to be treated; lately, a new type of electrode that covers larger surfaces has become a subject of study (Ongaro et al., 2016).

12.1.2.2 Measuring

When dealing with electroporation, measuring is crucial for achieving effective electroporation. Quality assurance can only be provided by appropriate measurements. Unfortunately, failed efforts to confirm other group's published work are increasing (Kaiser, 2016). The description of the used equipment and the process are flawed. In many papers describing/using electroporation, insufficient detail is reported, and quite often measurements are not reported (Batista Napotnik et al., 2016; Campana et al., 2016). The electroporation field needs to promote research reproducibility, and to improve this, we further need to try and answer the following questions: "Why measuring is necessary?", "How to measure?", "Which data are significant for researches?", and "Which electroporator to choose?"

For electroporation, you need an electroporator. A considerable number of electroporation devices can be found in the market, some are designed for specific applications and some are multifunctional. Most often, they are compatible with different electrodes. It seems however that as the market grows, manufacturers tell us less about their devices. We have already reached a critical point in the field of gene transfection, where preprogramed electroporation procedures are most commonly used so that quite often the researcher does not even know basic pulse parameters, such as pulse shape, repetition rate, and even less about applied voltage amplitude/electric field strength (see Tables 12.1 and 12.2). The researchers are only aware of the program number they used on their device. This limited information availability restricts and hampers the development of new knowledge. In this case, preengineering of electroporation devices limits researchers and hampers the sharing and comparing of results or protocols and even the further development of gene transfection field.

The most complicated of all are nanosecond electroporation systems. They usually consist of nanosecond pulse generator, transmission line or delivery system, and electroporation chamber/electrodes (Pakhomov et al., 2009; Ibey et al., 2010; Batista Napotnik et al., 2016). When using nanosecond pulses, it should always be taken into account that pulses reflect on impedance change and lose power in the transmission line. If the impedance matching is not guaranteed, reflections are present and load dependent. When load impedance is higher, reflections are positive and add to amplitude that would be present on a matched load. In case of lower load impedance, amplitude on the load will be lower. Pulses can become bipolar and cancelation effect can occur. Nanosecond pulses travel approximately 20 cm/ns in coaxial cable (Batista Napotnik et al., 2016).

12.1.2.3 Why Is Measuring Necessary?

For quality assurance! To be sure that the pulses are delivered and the device operates according to its specifications.

The first and most logical answer is because we want to know if our pulses were successfully delivered and we need to know what was delivered. If the current flows through a load, a delivery was more or less successful. But we still do not know anything about the pulse shape and voltage amplitude, or how many pulses were actually delivered. Due to the nature of electric discharge circuits that are commonly used in electroporation devices, amplitude of successive pulses can be lower with each successive pulse delivered (Figure 12.4), if the pulse repetition rate is in the higher half of device operation range. Low-conductivity media in cuvettes is used for two reasons— to reduce heating of the sample due to the current flow and to facilitate pulse generation. With lower resistant loads, problems can occur because the pulse generators cannot deliver "what they promise," i.e., high currents. A typical cuvette resistance with a low conductivity media is somewhere between 100 and 50 Ω. But high conductive media can have a resistance between 10 and 15 Ω, and even lower, requiring even up to 10 × higher currents. This large variation of load characteristics represents a great challenge in electroporator design resulting in different solutions. Because of software or hardware errors, devices can have an unexpected delay during generation, one or more pulses may be omitted, a voltage amplitude may deviate from the expected value, etc. In the case of clinical medical devices with CE certification, these errors should not occur, or if they do, an alarm must be triggered. But when we are dealing with self-developed or commercial devices not classified as medical devices, monitoring is necessary. Also, electronic components and, consequently, devices are aging, and their characteristics may change with time. Built-in measurement systems are usually comfortable solutions, but some manufacturers are taking shortcuts. If the measurement system is a part of the device, it needs periodic calibration; so if this is not part of the unit maintenance, the measurement system is questionable.

The second reason why measuring is necessary is the reproducibility of research results. For research reproducibility, at least similar, if not exactly the same, pulses are needed. Different research groups have different electroporation devices whose output pulses may derogate from specified shapes/parameters. If we know exactly what kind of pulse is needed, a custom setting might lead us towards better matching. Not reporting pulse parameters hinders the comparison of results and hinders progress of research.

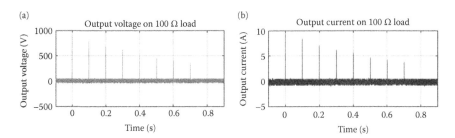

FIGURE 12.4 An example of output pulse measurements, we used a SENNEX electroporator with surface pin electrodes. The biological load was simulated with a 100 Ω resistor. Amplitude of each successive pulse is lower, due to improper operation. The last pulse does not even reach 50% of the preset amplitude. Panel (a) shows output voltage measured on 100 Ω load and (b) output current.

12.1.2.4 How and What to Measure and How to Report

Only correct measurement will upgrade our research, reduce resources used, and help advance the field by contributing to enhancing/improving reproducibility. The easiest way is to measure applied voltage and current. We need an adequate measurement from our oscilloscope and probes. What is appropriate depends on what we want to measure. First, we need to know what pulses are expected—at least amplitude, repetition frequency, and pulse duration. All oscilloscopes have limitations; a bandwidth tells us how fast it follows the signal changes, or more theoretically said, the maximum frequency range that it can measure (Figure 12.8). Closely related to frequency bandwidth is rise time specification. The specified rise time of an oscilloscope defines the fastest rising pulse it can measure. If not specified, it can be calculated as Rise time = 0.35/Bandwidth. For most applications, micro- or millisecond pulses are used; oscilloscopes and probes with a few MHz bandwidth are thus suitable. Measurement gets complicated when we reach the nanosecond HV pulses, where GHz range bandwidths and high rise times are needed. To minimize stray inductance and capacitances and reflections on lines, probes should be located as close to the electrodes as possible, with no additional connecting wires.

For the adequate presentation of pulses delivered, we propose to attach at least two measurements to your publications. One of a single pulse zoomed and another with reduced time scale, where all delivered pulses are displayed (if the number of pulses is low enough to keep a measurement meaningful) (Figure 12.5). If attaching the measurement does not appear to be suitable, an adequate description of a pulse, common notations, and pulse parameters are required. But in some cases when pulses strongly deviate from classical forms, an image tells us a lot more. An example of pulse characteristics determination is given in Figure 12.7.

Exponential decay pulses (Figure 12.6a) are best described by their maximum value, A_{MAX}, and time constant, τ. The value of time constant depends on circuit output stage characteristics. It is defined as the time maximum amplitude A_{MAX}, which drops to 37% of A_{MAX}. Square wave pulses are described with amplitude at high stage (that is choose to best fit the high level) and time t_{FWHM} (Full Width at Half Maximum-FWHM). t_{FWHM} is best described as the time passed between when the pulse reaches 50% of maximal amplitude at the rising and falling phase. Other pulse shapes are best described if we

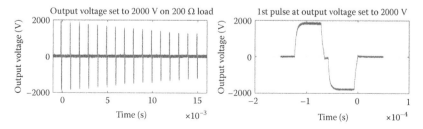

FIGURE 12.5 An example of proposed measurements to accompany the report. For the adequate presentation of pulses delivered, we propose that you attach at least two measurements to your publications. One of a single pulse zoomed and another with reduced time scale, where all delivered pulses are displayed (if the number of pulses is low enough to keep a measurement meaningful) The example measurements were made with the help of a CHEMIPULSE IV electroporator.

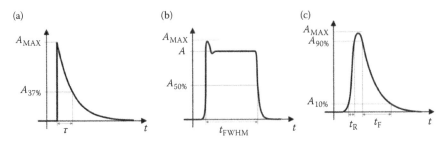

FIGURE 12.6   (a) Exponential decay pulse, (b) square wave pulse, and (c) Gaussian or bell-shaped pulse.

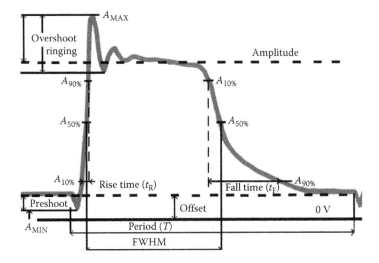

FIGURE 12.7   Pulse characteristics: a presentation of useful terms for description.

define their rise (t_R) and fall times (t_F) and maximum amplitude A_{MAX}. Rise time is time required for a pulse to rise from 10% to 90% of its steady value. Similarly, fall time is the time taken for the amplitude of a pulse to decrease from a specified value (usually 90% of the peak value exclusive of overshoot or undershoot) to another specified value (usually 10% of the maximum value exclusive of overshoot or undershoot) (Reberšek et al., 2014).

12.1.3 Biophysical Dosimetry in Electroporation

Biological cells can be electroporated in suspension, attached, or in tissue. We distinguish *in vitro* and *in vivo* electroporation; the connection between them is not taken for granted. New applications of medical electroporation are first demonstrated *in vitro*, if their efficacy is shown also *in vivo* in an appropriate animal model, human clinical studies can be done (Hofmann., 2000). Electric pulse parameters for effective *in vivo* application can be determined from *in vitro* experiments considering application specifications (Maček-Lebar et al., 2002).

12.1.3.1 Electroporation *In Vitro*

Cell membrane electroporation and consequently increased membrane permeability is controlled by the electric field strength. Because cells are in suspension and we usually work with low cell volume fractions, we can assume the surrounding field is homogenous and uniform throughout all the conducting media (Susil et al., 1998; Kotnik et al., 2010). But induced transmembrane voltage is not uniform on the cell surface, it is dependent on cell size, membrane characteristics and orientation to the field, frequency, and time and space (Teissié and Rols, 1993). *In vitro,* we are dealing with dilute cell suspensions, where the local field outside cells does not affect other cells. If volume fractions are higher than 10%, the induced transmembrane voltage cannot be easily estimated by Schwan's equation; local cell fields influence each other, and therefore approximate analytical or numerical calculations are needed (Pavlin et al., 2005). Even if we increase cell volume fraction of cells, there is still a big difference between tissue and a dense suspension. Plated cells are permeabilized with lower electroporation parameters than when in suspension (Towhidi et al., 2008). In tissues, cells form specific structures and are in contact with each other (Kotnik et al., 2010). *In vitro* experiments can be performed in electroporation chambers, especially with short pulse durations (nano- and picosecond pulses); chamber characteristics such as frequency responses can have a great impact on the results. Results are different when using different cuvettes; pulses are usually applied through two electrodes; the field delivered in is consequently different. From *in vitro* to *in vivo*, one needs to keep in mind that electric pulses are much larger compared to diameters of the cells (Maček-Lebar et al., 2002).

12.1.3.2 Electroporation *In Vivo*

In case of *ex vivo* electroporation of tissues, or *in vivo* electroporation, the electric field can no longer be considered homogenous because tissue is a highly inhomogeneous conductor. Some biological materials are also anisotropic, and therefore electric field orientation must also be considered. Tissue inhomogeneity is frequency dependent, it varies from tissue to tissue and is smaller at higher frequencies. Tumors mostly have a higher water content as a result of cellular necrosis (Miklavčič et al., 2006). In preclinical and clinical studies a few years ago, authors often considered the treated tissues as being linear electric conductors (i.e., with constant tissue conductivities) (Čorović et al., 2013).

Cell membranes have low electric conductivity in comparison to cytoplasm and extracellular medium. Electroporation changes the conductivity of cells, and thereby the field distribution is changed (Sel et al., 2005). To analyze tissue electroporation, we need to know the characteristics of the treated tissue. A macroscopic description is the most common and is described by specific conductivity and relative permittivity. Applied voltage rests among the most resistive tissue, which in the case of external electroporation is the skin. Skin conductivity is 10–100 times lower than the tissue underneath. Restive heating occurs and should be considered so as to avoid damaging healthy cells (Lacković et al., 2009; Kos et al., 2012). Numerical methods are used to define the local electric field distribution within the tumors (Miklavčič et al., 2010; Edhemovic et al., 2011). For deep-seated tumors and tumors in internal organs, which are surrounded by tissues with different electric properties, individualized patient-specific treatment planning is required (Miklavčič and Davalos, 2015). Tumors vary in shape, size, and location. The shape and position of the used long-needle electrodes and even the applied voltage (Hjouj and Rubinsky, 2010; Edhemovic et al., 2011; Pavliha et al., 2012) are analyzed and optimized for each tumor; coverage of the whole tumor with a sufficiently high electric field (which is one of the two prerequisites for successful treatment) (Miklavčič et al., 1998, 2006) can currently only be assured by means of numerical modeling of electric field distribution (Pavliha and Kos, 2013). Electric field calculations based on real input data are performed. Image-guided

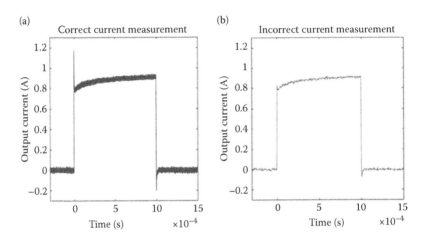

FIGURE 12.8 Current measurement examples, in (b) an oscilloscope with a too low bandwidth was used, consequently current spike was not detected. The spike is clearly visible in (a), that was captured by a faster oscilloscope.

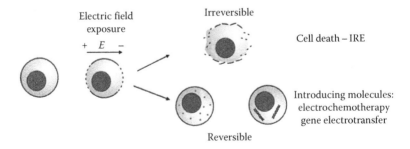

FIGURE 12.9 Symbolic representation of different electroporation applications. When externally applied electric field reaches the cell membrane threshold value, the cell gets permeabilized. We distinguish reversible and irreversible electroporation. The result of irreversible electroporation is cell death, which we exploit for nonthermal ablation, so-called IRE. In case of reversible electroporation, cell membrane can fully recover after the electroporation process. During the electroporation process, molecules are introduced into the cell at electrochemotherapy and gene electrotransfer.

insertion of electrodes is used (Kos et al., 2010; Miklavčič et al., 2010; Grošelj et al., 2015). Nonthermal IRE is also an electroporation-based application that is used for ablation of pathological tissue, it similarly requires a specific treatment planning. But in that case, calculations of temperature increase should also be considered (Županič and Miklavčič, 2011) as the conductivity increases with temperature. It is also necessary to measure *in vivo,* the voltage and current measurement of applied pulses and electrical impedance tomography (EIT) are common, but they do not tell us, how the electrical conductivity of tissue affects the electric field distribution. MR-EIT (Magnetic Resonance-Electrical Impedance Tomography) enables reconstruction of electric field distribution by measuring the electric current density distribution and electric conductivity during electroporation by using MR imaging and numeric algorithms (Kranjc et al., 2015).

12.2 Applications

Further on We will focus on three main applications (Figure 12.9). The most established ECT, IRE, is used for tissue ablation and gene electrotransfection. In each section, we review pulses used, their characteristics, and main principles. For each application, we tried to discover if researches report adequate data. At the end of the chapter, a review of commercially available electroporation devices can be found, including their characteristics. We have summarized all the important parameters, so as to help researchers select the appropriate device for their application. Table 12.2 describes their specifications and limitations. We focused on devices available for ECT, IRE, and gene electrotransfer. After a detailed review of the manufacturer's Internet pages and literature, we wrote to all the listed producers and asked them to kindly review the data we found in literature, device specifications, and on the Internet and update if necessary. We wrote to 25 producers: 13 were pleased to cooperate, the data they provided can be found in Tables 12.1 and 12.2. Three producers (LONZA, MaxCyte, and Ichor Medical Systems) replied to our email and informed us that the requested information was confidential. The data collection lasted for 1 month, with one reminder email. We contacted producers through their official emails published on their homepages. In addition, we also wrote directly to employees, whose contact information we have in our database. There have been no responses from BIO-RAD, Eppendorf, NEPA GENE, Oncosec, Scientz, Sigma-Aldrich, Thermo Fischer, and Tritech Research. Overall, the ones that do not specify their pulse parameters did not change their mind, only Inovio is excluded. We could not find any technical specifications or generated pulse descriptions of their devices, but in the end they provided all the missing data. Most of the producers who did not cooperate sell devices that are mainly used in the biopharmaceutical drug industry. When we are working with a device that has a CE mark, a small derogation from the specification is allowed. But due to aging and the huge diversity in biological load characteristics, control with an external measuring system is required for quality assurance. Noncertified devices can generate pulses that highly deviate from the preset values, so the use of an external measuring system is necessary.

12.2.1 Electrochemotherapy

One of the leading applications on the electroporation field is ECT. It is highly effective, with complete response rates between 60% and 70% and objective response rates of about 80% (Mali et al., 2013). It is suitable for treatment of cutaneous and subcutaneous tumors of different histotypes, both skin and nonskin cancers, as well as metastases. European Standard Operating Procedures of Electrochemotherapy (ESOPE) have been established in 2006; it increased reproducibility and improved clinical practice results (Campana et al., 2016). SOP however only defines ECT for smaller skin tumors (<3 cm in diameter); we still do not have any guidelines for internal or larger tumor ECT. National Institute for Health and Care Excellence (NICE) has recognized ECT as an integral part of the multidisciplinary treatment for patients with skin metastases of nonskin origin and melanoma (NICE interventional procedure guidance IPG 446, http://www.nice. org.uk/guidance/ipg446) (Campana et al., 2016). Lately, ECT has been introduced into the treatment of deep-seated tumors, it is really suitable for treatment of liver metastases, especially when they are located close to major blood vessels and consequently not manageable with surgery (Edhemovic et al., 2014). Recently, recommendations for improving the quality of reporting clinical ECT studies has been released (Campana et al., 2016), on initiative of the Steering Committee of the COST TD 1104 Action. Really good guidelines could raise the level of research even higher. That is a good example for other, high-quality applications. Standardized reporting enables faster and greater progress (Miklavčič et al., 2014).

But the main challenge is still the successful use of the application, the presence of a cytotoxic agent within tumor tissue, and adequate coverage of the tumor with electric pulses above the threshold of reversible membrane electroporation are crucial. Some studies have been conducted that have introduced the method for the determination of effective electrical parameters for ECT from a systematic *in vitro* study performed on cells in culture (Maček-Lebar et al., 2002; Larkin et al., 2007). It has been proven that ECT parameters optimized *in vitro* are applicable *in vivo*. Currently, eight or two groups of four 100 μs square wave pulses with a repetition frequency of 1 Hz or 5 kHz are most commonly used. In the ESOPE clinical study, a 5-kHz electric pulse repetition frequency was used based on preliminary data assuming that higher electric pulse frequency has a comparable effect as lower pulse repetition frequency in ECT (Marty et al., 2006). The advantages of a higher frequency are shorter duration of electroporation, the sensation of only one application of electric pulses and also muscle contraction is obtained only right after the electric pulses, delivery, therefore an electrode displacement due to muscle contraction during pulse' delivery is avoided. Patients report less pain is associated with 5 kHz than with 1-Hz repletion frequency electroporation (Županič et al., 2007; Serša et al., 2010). Pulse voltage amplitude is most commonly somewhere between 200 and 1000 V. It is electrode and target tissue dependent, which means it should be set to the value that ensures the electric field between the electrodes is higher than 400 V/cm. From Table 12.1 various implementations of electrodes and associated voltage amplitudes can be observed. Within a Cliniporator project, a clinical electroporator was designed, which is classified as a medical device and it is for now the only one with a medical device CE mark. SOP bases on the use of Cliniporator with associated electrodes, but new electric pulse generators are coming to the market, with new electrodes that might have a completely different design, we always need to keep in mind that the voltage amplitude must be optimized specifically for each electrode configuration.

As with all treatments ECT also has some side effects. Transient lesions and some localized pain can appear in areas that are in direct contact with the electrodes (Mir and Orlowski, 1999). A problem can also occur if electroporation pulses interfere with heart muscle rhythm. There is very little chance for this phenomena when the application is used for skin treatment (Mali et al., 2008). But deep-seated tumors, which can be located close to the heart, are also treated with ECT, and even in an open surgery the probability of electroporation pulses interfering with the heart is increasing. The most dangerous possible interference is induction of ventricular fibrillation (Wiggers and Wegria, 1940; Han, 1973; Reilly, 1998). Fibrillation can occur if the amplitude of the applied electric pulses is greater than the threshold level for fibrillation, and if electrical pulses are delivered during late atrial or ventricular systole. The vulnerable period for ventricles is near the peak of the T wave, and for atria in the S wave (Mali et al., 2005). The delivery of electric pulses must be synchronized with the ECG so as to reduce the risk (Mali et al., 2015).

12.2.2 Irreversible Electroporation

IRE as a nonthermal tissue ablation is a promising application for ablation of tumors tissue that is located near bile ducts or blood vessels (Scheffer et al., 2014). IRE causes cell death due to cell membrane electroporation and not due to tissue's temperature increase; however, a local temperature increase occurs around the electrodes, when a greater number of pulses are administered. IRE has almost the same main challenges as ECT, the tumor tissue should be covered with an adequate electric field, but in case of IRE, the electric field should be above the IRE threshold (Rubinsky et al., 2007). In addition, the magnitude should be selected to minimize the electroporation of healthy tissue, so as to avoid significant thermal damage (Shafiee et al., 2009; Županič and Miklavčič, 2011). It is mainly used for the treatment of deep-seated tumors either during open surgery or percutaneously in liver, pancreas, kidney, lung, and other organs.

IRE does not have an SOP, treatment protocols vary with research groups, tumor types, and stages of development. An individual treatment plan required for each specific tumor and is crucial for successful outcome. IRE can be minimally invasive in combination with ultrasound, computed tomography guidance, or magnetic resonance imaging (Jourabchi et al., 2014). In comparison to ECT, safety is even more important, because with IRE we are ablating about $50\,cm^3$ of tissue and the number of applied pulses is at least 90 (Bertacchini et al., 2007). To achieve IRE threshold, applied electric fields should be higher, i.e., delivered pulses should have amplitudes up to 3000 V and currents up to 50 A (Bertacchini et al., 2007). A Cliniporator VITAE or NanoKnife electroporator is used (their specifications can be found in Table 12.2). Higher electric fields, open surgery, and proximity of the heart raises the risk, delivered pulses might interfere with cardiac activity if delivered at inappropriate heart rhythm phase (Thomson et al., 2011). Pulses should be synchronized with the refractory period of the cardiac rhythm. The overall time for the procedure is extremely short in comparison to benchmark treatments. It lasts only a few minutes, actual time can be calculated from the number of delivered pulses and average heart rate, because pulse delivery is coupled to the heart rate (Davalos et al., 2005; Bertacchini et al., 2007; Rubinsky et al., 2008).

Traditional IRE is based on the use of a series of unipolar electric pulses, normally accompanied by a significant muscle contraction; therefore, general anesthesia and neuromuscular blocking agents are necessary to prevent muscle contraction (Rubinsky et al., 2008). Researchers are investigating different techniques to minimize the contractions. According to the Golberg and Rubinsky approach, surrounding a central energized electrode with a series of grounded electrodes reduces the volume of tissue exposed to electric fields with the potential to induce contraction (Golberg and Rubinsky, 2012). This procedure requires that at least 16 grounded electrodes be surrounded by one superficially inserted, energized electrode. Arena (Arena et al., 2011) uses high-frequency IRE named H-FIRE. H-FIRE utilizes high frequency, bipolar bursts to eliminate muscle contraction, without sacrificing the efficiency of cell death due to nonthermal electroporation. He showed that H-FIRE at 250 or 500 kHz has the same ablation and precision outcomes as traditional IRE (Arena et al., 2011)

Treatment plans have been developed that can help clinicians. Electrode configuration and pulse parameters are proposed, but a proper electrode placement can be in some cases really challenging (Edd and Davalos, 2007; Kos et al., 2015). Clinical studies are going on all around the world trying to specify the optimal parameters for specific cancer types. Each set of pulse specifications, number of pulses, voltage amplitude, and pulse duration have an effect on IRE outcomes. Pulse length is responsible for thermal effects in tissue. The maximal duration can be calculated for each electric field that would not induce thermal effects or at least minimize them. Typical IRE pulses consist of a series of 100-µs pulses separated by at least 100 µs. The pause between pulses enables a cooldown. Davalos (Davalos et al., 2005) showed that the threshold for IRE in most cell types is at least 800 V/cm. Rubinsky (Rubinsky et al., 2008) proposed that for prostate cancer cells a field of 250 V/cm is sufficient with use of 90 pulses to ensure complete ablation of that region. Raffa demonstrated that performing IRE in the presence of fibril boron nitride nanotubes lowers the necessary voltage threshold required to cause tumor cell death. IRE at 800 V/cm was 2.2 times more effective at causing cell destruction when performed in the presence of fibril boron nitride nanotubes (Raffa et al., 2012). Contradictory to recent guidelines for ECT, there are no specific guidelines on how to report clinical cases and studies.

12.2.3 Gene Transfection

Gene electrotransfer is a promising non-viral gene delivery method (Kandušer et al., 2009). It is used for treatment of cancer and other diseases (Shibata et al., 2006; Daud et al., 2008), for DNA vaccination (Chiarella et al., 2010; Sardesai and Weiner, 2011), and genetic modification of organisms (Golden et al., 2007; Grewal et al., 2009). "Cancer immunoediting" is a process combining the immune system and tumors. The immune system can protect the host against tumor growth, or promote cancer development by selection of tumor variants with reduced immunogenicity (Zou et al., 2005). Immunotherapy can include cancer vaccines based on plasmid DNA (pDNA) vectors (Serša et al., 2015).

Electroporation is used to promote antigen, oligonucleotides, and immunomodulatory molecule delivery in to tumor tissue. They can stimulate the immune system or act on immunosuppressor genes (Serša et al., 2015). *In vitro* electric pulses are frequently used for the transfection of bacterial and eukaryotic cells. *In vivo* the technique is termed DNA electrotransfer, electrogenetherapy, or also gene electrotherapy. It has been successfully used since 1998. However, exact molecular mechanisms of DNA transport are unknown (Kandušer et al., 2009; Serša et al., 2015). DNA transfer can only be achieved by reversible electroporation, because dead cells are not able to express transferred genes (Andre et al., 2010). The DNA must be injected before electroporation; the application requires sufficiently intense electric fields, which means sufficiently long pulses should be applied, but we also need to ensure reversible electroporation. Permeabalized cell membrane should interact with the plasmid; thus a DNA–membrane complex is formed. DNA, then, with an as yet unknown process, is transferred into the cytoplasm and transported to the nucleus. In cases where the application is successful, the process is followed by gene expression (Golzio et al., 2002; Faurie et al., 2010).

Initially, we thought one of the most important mechanisms for efficient gene electrotransfer was electrophoretic movement of DNA during the pulse. A long millisecond, square, or exponential decaying pulses were used with 400–600 V/cm and up to 20 ms long (Bettan et al., 2000). Some studies showed that DNA transfection is enhanced in a combination of short HV and long low-voltage (LV) pulses. It was suggested that HV pulses are crucial for permeabilization of the cell membrane and pore formation, while LV pulses electrophoretically drag negatively charged DNA into the cell. Eight HV pulses 100 μs with amplitude 1300 V/cm followed by one longer 100 ms LV pulse 100 V/cm (Šatkauskas et al., 2002) were proposed. Further, Miklavčič's group showed that short HV pulses are not only crucial but also sufficient for successful DNA delivery, at optimal plasmid concentrations. They suggested that electrophoretic force of LV pulses is crucial in *in vivo* conditions where suboptimal plasmid concentration is the limiting factor for efficient transfection (Kandušer et al., 2009). As induced electric field is tissue dependent, it is important to define targeted tissue. In comparison to ECT, where targeted tissue is always tumor, a definition in the case of gene electrotransfer is more complicated. Electroporation parameters depend on the type of tumor antigen and target tissues, and the target cells in specific tissue are different (Serša et al., 2015). Electric pulse parameters have to be experimentally or numerically optimized for given electrodes' positions and geometry (Županič et al., 2010). Gene electrotransfer efficiency is electroporation media dependent, divalent cations such as Ca^{2+} and Mg^{2+} are necessary for the formation of DNA–membrane complex during the pulses. They act as a bridge between negatively charged DNA and the negatively charged cell plasma membrane, and thus improve DNA–membrane binding (Haberl et al., 2013a).

Classical gene electrotransfection parameters are hard to define; for example, 8×5 ms, 700 V/cm, 1 Hz are efficient *in vitro*. Studies show that more than 30% of cells can express the gene coded by plasmid DNA, while preserving cell viability to a large extent (Golzio et al., 2002; Chopinet et al., 2012). In the case of skin tumors, rate significantly decreases *in vivo* (Rols and Teissié, 1998; Čemazar et al., 2009). A lot of studies have been performed and parameters described to enable better gene transfer. In other studies, electric field direction and orientation changes during the pulse delivery have been shown to increase the area, making DNA entry into the cell more competent. The introduction of DNA only occurs in the part of the membrane facing the cathode. It was shown that the percentage of cells expressing genes increases when electric field direction and orientation change (Pavlin et al., 2011). Also, a new prospect was presented, involving nanosecond electric pulses (pulse duration: 4–600 ns). Very short HV pulses (several tens of kV/cm) are able to disturb membranes of internal organelles, due to cell membranes charging time. We can conclude there is an option short nanosecond pulses can effect the nuclear envelope. Combination of medium or long electrical pulses with short HV nanosecond pulses enhance gene expression by increasing the number of plasmids entering the nucleus (Beebe et al., 2003; Chopinet et al., 2013; Guo et al., 2014).

Overall, we can say that the field of gene electrotransfer is complex and many known and as yet unknown factors mutually affect the process.

Longer electric pulses are optimal for higher transfection efficiency, but they reduce viability. Shorter pulses enable lower transfection efficiency and preserve viability. The number of studies is increasing, fields and applications are spreading, with insufficient and incomplete data and pulse parameters. There are no standard procedures or reporting guidelines defined that would in the future enable a proof of concept. In comparison to ECT, a greater variety of pulses are being used, for each tissue specific pulse parameters are optimal, for each application a specific procedure seems to be required. One of the problems is that electroporation device manufacturers produce devices with preprogramed procedures. Users only select the "appropriate program" and the device ensures optimal transfection. Pulse parameters and characteristics stay unknown due to the device patent. The field is getting more and more chaotic and is in need of a more consistent, explicit, and well-defined research with guidelines on reporting; including electric pulse protocols.

12.3 Conclusion

Electroporation is a platform technology, which is already established in medicine and food processing. When we are dealing with electroporation, measuring is crucial for achieving effective electroporation. Quality assurance can only be provided by appropriate measurements, i.e., measuring the voltage and current using an oscilloscope.

Due to the huge variation in biological load characteristics, delivered pulses may significantly deviate from the pre-set. Low impedance of a load (tissue or cell suspension) requires large power/currents, which quite often leads to significant voltage drop. Protocols in which a larger number of pulses (or long pulses) are delivered can result in reduced amplitude of pulses. The delivered pulse shape repetition frequency, pulse duration, and amplitude must be always monitored. The measurement probes should be located as close to the load as possible, the oscilloscope bandwidth should be high enough, and in the case of nanosecond application, the reflections and losses must be considered.

One of the leading applications in the electroporation field is ECT that already has well-established protocols, reporting guidelines, and good research reproducibility. Unfortunately, failed efforts to confirm other published paperwork are increasing (Kaiser, 2016). We believe the main reason for this situation is flawed descriptions of the

TABLE 12.1 Various Implementations of Electrodes, with Associated Voltage Amplitudes, Provided by the Manufacturers

Producer	Device	Type of electrodes	Number of output channels	Number of electrodes	Electrodes' geometric description		Pulse number	Pulses amplitude
Angiodynamics								
	NanoKnife	Needle	/	1–6 Probe outputs	Probes spaced 1.5 cm apart with the active electrode length set at 2 cm		90 (Pulses for each pair of electrodes)	(100–3000) V
Bionmed Technologies								
	SENNEX	Needle/pin surface	/	4	Linear layout for small tumors and pentagon layout for larger, more extensive tumors. Pin electrodes are 3 mm thick at the top and needle 0.3 mm.		8	1000 V
IGEA								
	Cliniporator EPS02	Needle	7	7	Hexagonal configuration (diameter: 0.7 mm/length: 10 mm/20 mm/30 mm)		HV: 4	HV: 730 V
		Needle	2	8 (2 × 4)	Linear configuration (diameter: 0.7 mm/length: 10 mm/20 mm/30 mm)		HV: 8	HV: 400 V
		Plate	2	2	Linear configuration (10 mm × 30 mm × 0.8 mm)		HV: 8	HV: 960 V
		Needle	2	6 (2 × 3)	Finger configuration with orthogonal linear needles (diameter: 0.7 mm/ length: 5 mm/10 mm)		HV: 8	HV: 400 V

(Continued)

TABLE 12.1 (*Continued*)   Various Implementations of Electrodes, with Associated Voltage Amplitudes, Provided by the Manufacturers

Producer	Device	Type of electrodes	Number of output channels	Number of electrodes	Electrodes geometric description		Pulse number	Pulses amplitude
		Needle	2	6 (2 × 3)	Finger configuration with longitudinal linear needles (diameter: 0.7 mm/length: 5 mm/10 mm)		HV: 8	HV: 400 V
		Partially isolated needles	7	7	Hexagonal configuration (diameter: 0.7 mm/length: 40 mm)		HV: 4	HV: 730 V
	Cliniporator VITAE	Needle	2–6	2–6	Single long needle/diameter: 1.2 mm/active part: 1–4 cm; soft tissues custom geometry ECT		HV: 4 + 4 (polarity exchange)	HV: (500–3000) V
		Needle	2–6	2–6	Single long needle/diameter: 1.8 mm/active part 1–4 cm; bones custom geometry ECT		HV: 4 + 4 (polarity exchange)	HV: (500–3000) V
Intracel	TSS20 Ovodyne	Silver, tungsten, and platinum electrodes	/	2	Electrodes. Silver electrodes are available made from 0.8 mm diameter silicon rubber insulated silver wire, the exposed pole being flattened into a "paddle" shape approximately 2 × 1 mm. Silver wire length extending beyond the electrode holder is 40 mm. Tungsten electrodes are produced from 0.5 mm o.d. tungsten rod		/	/

(*Continued*)

TABLE 12.1 (Continued) Various Implementations of Electrodes, with Associated Voltage Amplitudes, Provided by the Manufacturers

Producer	Device	Type of electrodes	Number of output channels	Number of electrodes	Electrodes geometric description	Pulse number	Pulses amplitude
Inovio							
	CELLECTRA 5PSP	Needle electrodes, intramuscular	/	5	An array consisting of 5 needle electrodes, adjustable from 13, 19, and 25 mm in length depending on BMI, forming a pentagon on a 1-cm circle	3	Max 200 V
	CELLECTRA 2000–5P	Needle electrodes, intramuscular	/	5	An array consisting of 5 needle electrodes, adjustable from 13, 19, and 25 mm in length depending on BMI, forming a pentagon on a 1-cm circle	3	Max 200 V
	CELLECTRA 2000–3P	Needle electrodes, intradermal	/	5	An array consisting of 3 needle electrodes, 3 mm in length, forming an isosceles triangle, 3-mm spacing (short side) and 5-mm spacing (long sides)	2 Sets of 2 pulses	Max 200 V
Leroy BIOTECH							
	ELECTROvet S13		/		All are compatible with plate and needle electrodes		
	ELECTROvet EZ ELECTRO cell B10		/	8 (2 × 4)	Needle electrodes are specially designed and manufactured for the treatment of subcutaneous tumors. 8 mm between the two rows of four needles. Each needle spaced 2 mm apart		
	ELECTRO cell S20 MILLIPULSES		/	2	10-mm length/10-mm spacing between electrodes centers/3 mm		

TABLE 12.2   A Review of Commercially Available Electroporation Devices, Including Their Capabilities and Characteristics

Producer	Device	Pulse description	Pulse number	Pulses amplitude	Pulses duration	Pulses repetition frequency	Pulse sequences	Maximum voltage	Maximum current
Angiodynamics									
	NanoKnife	Square wave	90 (Pulses for each pair of electrodes)	(100–3000) V	(20–100) µs	ECG synchronized, 90, 120, 240 ppm	/	3000 V	50 A
BEX Co., Ltd.									
	CUY21EDIT	Square wave	/	(1–500) V	T = 0.1–999.9 ms	/	/	/	I < 5.0 A (1–125 V) I < 2.2 A (126–250 V) I < 1.0 A (251–500 V)
	CUY21EDIT II	Square wave exponential	/	(1–200) V (Square) (1–400) V (Exponential; PP) (1–350) V (Exponential; DP)	T = 0.05–1000 ms (Square) T = 0.01–99.9 ms (Exponential)	/	/	/	I < 1A (Square) I < 10 A (Exponential)
	Genome editor	Square wave	/	(1–200) V	T = 0.10–1000 ms	/	/	/	
	CUY21Vitro-EX	Exponential	/	1–900 V (PP) 1–500 V (DP)	T = 0.01–99.9 ms	/	/	/	I < 50 A
	LF301	Square wave sinus (AC)	/	0–1200 (Square wave) 0–75 V$_{rms}$ (AC)	T = 0–100 µs (Square) fAC = 1 MHz; T = 0–100 s (Prefusion)/0–10 s (Post-fusion)	/	/	/	R > 50 Ω

(Continued)

TABLE 12.2 (*Continued*) A Review of Commercially Available Electroporation Devices, Including Their Capabilities and Characteristics

Producer	Device	Pulse description	Pulse number	Pulses amplitude	Pulses duration	Pulses repetition frequency	Pulse sequences	Maximum voltage	Maximum current
Bionmed Technologies	SENNEX	Square wave	8	1000 V	100 μs	100 ms	10 Imp/s	1000 V	/
BTX-Harvard Apparatus	AgilePulse *in vivo* system	/	3 Groups of pulses: from 1 to 10 pulses in each group	(50–1000) V	(0.050–10) ms	(0.200–1000) ms	/	1000 V	At max voltage and minimum resistance: 1000 V/10 Ω = 100 A
	AgilePulse MAX system	/	3 Groups of pulses: from 1 to 10 pulses in each group	(50–1200) V	HV: (0.050–10) ms	(0.200–1000) ms	(1–5000) Hz	1200 V	At max voltage and minimum resistance: 1200 V/10 Ω = 120 A
	ECM 2001	Square wave AC	1–9	HV: (10–3000) V LV: (10–500) V (0–150) V (Vpp)	HV: (1–99) μs LV: (0.01–0.99) ms Duration: (0–99) s	(0.01–0.99) ms (1–99) ms Post fusion—ramp: 1–9 s	1 mHz	3000 V	/
	ECM 830	Square wave	1–99	HV: (505–3000) V LV: (5–500) V	HV: (10–600) μs LV: (10–999) μs; (1–999) ms; (1–10) s	100 ms–10 s	/	3000 V	500 A limit at 100 μs
	ECM 630	Exponential decay wave	1–99	HV: (50–2500) V LV: (10–500) V	10 μs–10 s	/	/	2500 V	6000 A in LV mode
	Gemini SC2	Square waves and exponential decay waves	LV: 1–10 HV: 1–2 Exponential decay: 1	(10–3000) V	50 μs–100 ms	100 ms–30 s	/	3000 V	/

(*Continued*)

TABLE 12.2 (Continued) A Review of Commercially Available Electroporation Devices, Including Their Capabilities and Characteristics

Producer	Device	Pulse description	Pulse number	Pulses amplitude	Pulses duration	Pulses repetition frequency	Pulse sequences	Maximum voltage	Maximum current
	Gemini X2	Square waves and exponential decay waves	Square wave: LV mode-1-120 (10 per sample) HV mode-1-36 (3 per sample); Exponential decay: 1-12 (R internal <100Ω) and 1-24 (R internal >100Ω)	(5–3000) V	10 µs–1s	100 ms–30 s	/	3000 V	/
	ECM 399	Exponential decay waves	1	(2–2500) V	Max. at 500 V: 125 ms; Max. at 2500 V: 5 ms	100 ms–10 s	/	2500 V	/
Cyto Pulse Science, Inc.									
	OncoVet	/	/	(50–1000) V	(0.05–10) ms	(0.2–1000) ms	(1–5000) Hz	/	/
IGEA									
	Cliniporator EPS02	Square wave	LV: 1–10 HV: 1–10	LV: (20–200) V HV: (100–1000) V	LV: (1–200) ms HV: (50–1000) µs	LV: (0.45–500) Hz HV: (1–5000) Hz	24 Configurations	1000 V	LV: 5 A HV: 20 A
	Cliniporator VITAE	Square wave	HV: 4 + 4 (polarity exchange); 4–8	HV: (500–3000) V	100 µs	HV: (1–5000) Hz	Costum	3000 V	50 A
Inovio									
	CELLECTRA 5PSP	Square wave	3	Max 200 V	52 ms	1 Hz	/	Max 200 V	0.5 A Constant current

(Continued)

TABLE 12.2 (*Continued*) A Review of Commercially Available Electroporation Devices, Including Their Capabilities and Characteristics

Producer	Device	Pulse description	Pulse number	Pulses amplitude	Pulses duration	Pulses repetition frequency	Pulse sequences	Maximum voltage	Maximum current
	CELLECTRA 2000–5P	Square wave	3	Max 200 V	52 ms	1 Hz	/	Max 200 V	0.5 A Constant current
	CELLECTRA 2000–3P	Square wave	2 Sets of 2 pulses	Max 200 V	52 ms	3 Hz	2 Sets seperated by 3 s	Max 200 V	0.2 A Constant current
Intracel									
	TSS20 Ovodyne	Square wave	1–999	(0.1–99.9) V	(1–999) ms	(10–9990) ms	1–999	99.9 V	100 mA TSS20 1000 mA EP21
Leroy Biotech									
	ELECTROvet Sl3	Square wave pulse generator	1–10,000 or infinite	0–1350 V	5–5000 µs	0.1–10,000 ms	0.1–10,000 Hz	1300 V	10 A
	ELECTROvet EZ	Square wave pulse generator	1–10,000 or infinite	0–1500 V	5–5000 µs	0.1–10,000 ms	0.1–10,000 Hz	1500 V	25 A
	ELECTRO cell B10	Square wave bipolar pulse generator—high voltage low voltage	1–10,000 or infinite	0–1000 V	5–5000 µs	0.1–10,000 ms	0.1–10,000 Hz	1000 V	10 A
	ELECTRO cell S20	Square wave pulse generator	1–10,000 or infinite	0–2000 V	5–5000 µs	0.1–10,000 ms	0.1–10,000 Hz	2000 V	25 A

(*Continued*)

TABLE 12.2 (*Continued*) A Review of Commercially Available Electroporation Devices, Including Their Capabilities and Characteristics

Producer	Device	Pulse description	Pulse number	Pulses amplitude	Pulses duration	Pulses repetition frequency	Pulse sequences	Maximum voltage	Maximum current
Molecular Devices									
	Axoporator 800 A	Square and bi-level pulses with positive and negative polarity, as well as bipolar	Train duration: 10 ms–100 s	±(1–100) V	MONOPOLAR: 200 µs–1 s BI-POLAR: 400 µs–1 s BI-LEVEL: 10 ms–20 s	MONOPOLAR ftrain = (1–2000) Hz BI-POLAR ftrain = (1–2000) Hz BI-LEVEL ftrain = (0.024–50) Hz	Rectangular pulse Bipolar pulse Positive bi-level pulse	5 V peak to peak	±10.0 µA
NPI									
	ELP-01D	Square wave	/	(0–110) V	(0–9999.9) ms	(0–9999.9) ms	/	/	/
OnkoDisruptor									
	Onkodisruptor Electroporator	/	/	/	/	50 + 50 µs Pause: 10 µs s	8 Biphasic	1500 V	5 A
Supertech Instruments									
	SP-4a	With RC time constant of the exponential decay of wave	0–99	HV: (200–400) V LV: (0–200) V	(1–250) ms	(10–60,000) ms	Single pulse mode and burst mode (up to 99 pulses)	400 V	11.9 A

Sources: Angiodynamics, http://www.angiodynamics.com/; BEX Co., Ltd., http://www.bexnet.co.jp/; Bionmed Technologies, http://bionmed.de/; BTX-Harvard Apparatus, http://www.btxonline.com; Cyto Pulse Science, Inc., http://www.cytopulse.com/; IGEA, http://www.igea.it/; Inovio, http://www.inovio.com/; Intracel, http://intracel.co.uk/; Leroy Biotech, https://www.leroybiotech.com; Molecular Devices, http://www.moleculardevices.com/; NPI http://www.npielectronic.de/; OnkoDisruptor, https://www.onkodisruptor.com; Supertech Instruments, http://www.superte.ch/Electroporator.html

Note: We wrote to all listed producers and kindly asked them to review the data we found in literature and on the Internet and update if necessary. While most of the manufacturers were pleased to cooperate, we did not get information about pulse characteristics from others. We are still missing all the information about pulse characteristics from, Lonza, TriTech, Ichor Medical Systems, Thermo Fischer, Eppendorf, Maxcyte, and Oncosec.

equipment used and the process. Many papers describing/using electroporation have reported insufficient details, and quite often measurements are not reported (Batista Napotnik et al., 2016; Campana et al., 2016). The field of electroporation is in need of promoting reproducible research that can only be achieved by adequate measurements, standardized reports, and proper use of electroporators and electrodes.

Acknowledgments

This study was supported by the Slovenian Research Agency (ARRS) and conducted within the scope of the European Associated Laboratory for Electroporation in Biology and Medicine (LEA-EBAM). We also thank all the manufacturers who kindly provided data about their devices.

References

Andre, F. M. and L. M. Mir. 2010. Nucleic acids electrotransfer in vivo: Mechanisms and practical aspects. *Current Gene Therapy* 10, no. 4 (August 1, 2010): 267–80. doi:10.2174/156652310791823380.

Arena, C. B., M. B. Sano, J. H. Rossmeisl, J. L. Caldwell, P. A. Garcia, M. N. Rylander, and R. V. Davalos. 2011. High-frequency irreversible electroporation (H-FIRE) for non-thermal ablation without muscle contraction. *Biomedical Engineering Online* 10 (2011): 102. doi:10.1186/1475-925X-10–102.

Batista Napotnik, T., M. Reberšek, T. Kotnik, E. Lebrasseur, G. Cabodevila, and D. Miklavčič. 2010. Electropermeabilization of endocytotic vesicles in B16 F1 mouse melanoma cells. *Medical & Biological Engineering & Computing* 48, no. 5 (May 2010): 407–13. doi:10.1007/s11517-010-0599-9.

Batista Napotnik, T., M. Reberšek, P. T. Vernier, B. Mali, and D. Miklavčič. 2016. Effects of high voltage nanosecond electric pulses on eukaryotic cells (in vitro): A systematic review. *Bioelectrochemistry* (Amsterdam, Netherlands) 110 (February 27, 2016): 1–12. doi:10.1016/j.bioelechem.2016.02.011.

Beebe, S. J., J. White, P. F. Blackmore, Y. Deng, K. Somers, and K. H. Schoenbach. 2003. Diverse effects of nanosecond pulsed electric fields on cells and tissues. *DNA and Cell Biology* 22, no. 12 (December 2003): 785–96. doi:10.1089/104454903322624993.

Bertacchini, C., P. M. Margotti, E. Bergamini, A. Lodi, M. Ronchetti, and R. Cadossi. 2007. Design of an irreversible electroporation system for clinical use. *Technology in Cancer Research & Treatment* 6, no. 4 (August 2007): 313–20.

Bettan, M., M. A. Ivanov, L. M. Mir, F. Boissière, P. Delaere, and D. Scherman. 2000. Efficient DNA electrotransfer into tumors. *Bioelectrochemistry (Amsterdam, Netherlands)* 52, no. 1 (September 2000): 83–90.

Campana, L. G., A. J. P. Clover, S. Valpione, P. Quaglino, J. Gehl, C. Kunte, M. Snoj, M. Čemažar, C. R. Rossi, D. Miklavčič, and G. Serša. 2016. Recommendations for improving the quality of reporting clinical electrochemotherapy studies based on qualitative systematic review. *Radiology and Oncology* 50, no. 1 (March 1, 2016): 1–13. doi:10.1515/raon-2016-0006.

Čemazar, M., M. Golzio, G. Serša, P. Hojman, S. Kranjc, S. Mesojednik, M.-P. Rols, and J. Teissié. 2009. Control by pulse parameters of DNA electrotransfer into solid tumors in mice. *Gene Therapy* 16, no. 5 (May 2009): 635–44. doi:10.1038/gt.2009.10.

Chafai, D. E., A. Mehle, A. Tilmatine, B. Maouche, and D. Miklavčič. 2015. Assessment of the electrochemical effects of pulsed electric fields in a biological cell suspension. *Bioelectrochemistry* 106, Part B (December 2015): 249–57. doi:10.1016/j.bioelechem.2015.08.002.

Chiarella, P., V. M. Fazio, and E. Signori. 2010. Application of electroporation in DNA vaccination protocols. *Current Gene Therapy* 10, no. 4 (August 2010): 281–86.

Chopinet, L., L. Wasungu, and M.-P. Rols. 2012. First explanations for differences in electrotransfection efficiency *in vitro* and *in vivo* using spheroid model. *International Journal of Pharmaceutics*, Special Issue : Drug Delivery and Imaging in Cancer, 423, no. 1 (February 14, 2012): 7–15. doi:10.1016/j.ijpharm.2011.04.054.

Chopinet, L., T. Batista Napotnik, A. Montigny, M. Reberšek, J. Teissié, M.-P. Rols, and D. Miklavčič. 2013. Nanosecond electric pulse effects on gene expression. *The Journal of Membrane Biology* 246, no. 11 (November 2013): 851–59. doi:10.1007/s00232-013-9579-y.

Čorović, S., I. Lacković, P. Sustaric, T. Sustar, T. Rodic, and D. Miklavčič. 2013. Modeling of electric field distribution in tissues during electroporation. *Biomedical Engineering Online* 12 (2013): 16. doi:10.1186/1475-925X-12-16.

Daud, A. I., R. C. DeConti, S. Andrews, P. Urbas, A. I. Riker, V. K. Sondak, P. N. Munster, D. M. Sullivan, K. E. Ugen, J. L. Messina, and R. Heller. 2008. Phase I trial of interleukin-12 plasmid electroporation in patients with metastatic melanoma. *Journal of Clinical Oncology: Official Journal of the American Society of Clinical Oncology* 26, no. 36 (December 20, 2008): 5896–903. doi:10.1200/JCO.2007.15.6794.

Davalos, R. V., L. M. Mir, and B. Rubinsky. 2005. Tissue ablation with irreversible electroporation. *Annals of Biomedical Engineering* 33, no. 2 (February 2005): 223–31. doi:10.1007/s10439-005-8981-8.

Deamer, D. W. and J. Bramhall. 1986. Special issue: Liposomes permeability of lipid bilayers to water and ionic solutes. *Chemistry and Physics of Lipids* 40, no. 2 (June 1, 1986): 167–88. doi:10.1016/00093084(86)90069-1.

Edd, J. F. and R. V. Davalos. 2007. Mathematical modeling of irreversible electroporation for treatment planning. *Technology in Cancer Research & Treatment* 6, no. 4 (August 1, 2007): 275–86. doi:10.1177/153303460700600403.

Edhemovic, I., E. M. Gadzijev, E. Brecelj, D. Miklavčič, B. Kos, A. Županič, B. Mali, T. Jarm, D. Pavliha, M. Marčan, G. Gasljevic, V. Gorjup, M. Music, T. P. Vavpotic, M. Čemažar, M. Snoj, and G. Serša. 2011. Electrochemotherapy: A new technological approach in treatment of metastases in the liver. *Technology in Cancer Research & Treatment* 10, no. 5 (October 2011): 475–85.

Edhemovic, I., E. Brecelj, G. Gasljevic, M. M. Music, V. Gorjup, B. Mali, T. Jarm, B. Kos, D. Pavliha, B. Grcar Kuzmanov, M. Čemažar, M. Snoj, D. Miklavčič, E. M. Gadzijev, and G. Serša. 2014. Intraoperative electrochemotherapy of colorectal liver metastases. *Journal of Surgical Oncology* 110, no. 3 (September 2014): 320–27. doi:10.1002/jso.23625.

Faurie, C., M. Reberšek, M. Golzio, M. Kandušer, J.-M. Escoffre, M. Pavlin, J. Teissié, D. Miklavčič, and M. P. Rols. 2010. Electro-mediated gene transfer and expression are controlled by the life-time of DNA/membrane complex formation. *The Journal of Gene Medicine* 12, no. 1 (January 2010): 117–25. doi:10.1002/jgm.1414.

Freeman, S. A., M. A. Wang, and J. C. Weaver. 1994. Theory of electroporation of planar bilayer membranes: Predictions of the aqueous area, change in capacitance, and pore-pore separation. *Biophysical Journal* 67, no. 1 (July 1994): 42–56.

Garcia, P. A., R. V. Davalos, and D. Miklavčič. 2014. A numerical investigation of the electric and thermal cell kill distributions in electroporation-based therapies in tissue. *PloS One* 9, no. 8 (2014): e103083. doi:10.1371/journal.pone.0103083.

Golberg, A. and B. Rubinsky. 2012. Towards electroporation based treatment planning considering electric field induced muscle contractions. *Technology in Cancer Research & Treatment* 11, no. 2 (April 1, 2012): 189–201. doi:10.7785/tcrt.2012.500249.

Golberg, A., M. Sack, J. Teissié, G. Pataro, U. Pliquett, G. Saulis, T. Stefan, D. Miklavčič, E. Vorobiev, and W. Frey. 2016. Energy-efficient biomass processing with pulsed electric fields for bioeconomy and sustainable development. *Biotechnology for Biofuels* 9, no. 1 (April 27, 2016): 1–22. doi:10.1186/s13068-016-0508-z.

Golden, K., V. Sagi, N. Markwarth, B. Chen, and A. Monteiro. 2007. In vivo electroporation of DNA into the wing epidermis of the butterfly, *Bicyclus anynana*. *Journal of Insect Science* 7, no. 53 (October 1, 2007): 1–8. doi:10.1673/031.007.5301.

Golzio, M., J. Teissié, and M.-P. Rols. 2002. Direct visualization at the single-cell level of electrically mediated gene delivery. *Proceedings of the National Academy of Sciences* 99, no. 3 (February 5, 2002): 1292–97. doi:10.1073/pnas.022646499.

Grewal, R. K., M. Lulsdorf, J. Croser, S. Ochatt, A. Vandenberg, and T. D. Warkentin. 2009. Doubled-haploid production in chickpea (*Cicer arietinum* l.): Role of stress treatments. *Plant Cell Reports* 28, no. 8 (June 19, 2009): 1289–99. doi:10.1007/s00299-009-0731-1.

Grošelj, A., B. Kos, M. Čemažar, J. Urbancic, G. Kragelj, M. Bosnjak, B. Veberic, P. Strojan, D. Miklavčič, and G. Serša. 2015. Coupling treatment planning with navigation system: A new technological approach in treatment of head and neck tumors by electrochemotherapy. *Biomedical Engineering Online* 14, no. Suppl 3 (2015): S2. doi:10.1186/1475-925X14-S3-S2.

Guo, S., D. L. Jackson, N. I. Burcus, Y.-J. Chen, S. Xiao, and R. Heller. 2014. Gene electrotransfer enhanced by nanosecond pulsed electric fields. *Molecular Therapy—Methods & Clinical Development* 1 (September 17, 2014): 14043. doi:10.1038/mtm.2014.43.

Gusbeth, C. A. and W. Frey. 2009. Critical comparison between the pulsed electric field and thermal decontamination methods of hospital wastewater. *Acta Physica Polonica A* 115, No. 6: 1092-1093 (2009).

Haberl, S., D. Miklavčič, G. Serša, W. Frey, and B. Rubinsky. 2013a. Cell membrane electroporation—Part 2: The applications. *IEEE Electrical Insulation Magazine* 29, no. 1 (January 2013): 29–37. doi:10.1109/MEI.2013.6410537.

Haberl, S., M. Kandušer, K. Flisar, D. Hodžić, V. B. Bregar, D. Miklavčič, J.-M. Escoffre, M. Rols, and M. Pavlin. 2013b. Effect of different parameters used for *in vitro* gene electrotransfer on gene expression efficiency, cell viability and visualization of plasmid DNA at the membrane level. *The Journal of Gene Medicine* 15, no. 5 (May 2013): 169–81. doi:10.1002/jgm.2706.

Han, J. 1973. Ventricular vulnerability to fibrillation. In Dreifus, L. S., Likoff, W. (eds): *Cardiac Arrhythmias*. New York: Grune and Stratton, Inc., pp. 87–95.

Heller, R., J. Schultz, M. L. Lucas, M. J. Jaroszeski, L. C. Heller, R. A. Gilbert, K. Moelling, and C. Nicolau. 2001. Intradermal delivery of interleukin-12 plasmid DNA by *in vivo* electroporation. *DNA and Cell Biology* 20, no. 1 (January 2001): 21–26. doi:10.1089/10445490150504666.

Hjouj, M. and B. Rubinsky. 2010. Magnetic resonance imaging characteristics of non-thermal irreversible electroporation in vegetable tissue. *The Journal of Membrane Biology* 236, no. 1 (July 15, 2010): 137–46. doi:10.1007/s00232-010-9281-2.

Hofmann, G. A. 2000. Instrumentation and electrodes for in vivo electroporation. *Methods in Molecular Medicine* 37 (2000): 37–61. doi:10.1385/1-59259-080-2:37.

Ibey, B. L., A. G. Pakhomov, B. W. Gregory, V. A. Khorokhorina, C. C. Roth, M. A. Rassokhin, J. A. Bernhard, G. J. Wilmink, and O. N. Pakhomova. 2010. Selective cytotoxicity of intense nanosecond-duration electric pulses in mammalian cells. *Biochimica Et Biophysica Acta* 1800, no. 11 (November 2010): 1210–19. doi:10.1016/j. bbagen.2010.07.008.

Isokawa, M. 1997. Membrane time constant as a tool to assess cell degeneration. *Brain Research Protocols* 1, no. 2 (May 1997): 114–16.

Jourabchi, N., K. Beroukhim, B. A. Tafti, S. T. Kee, and E. W. Lee. 2014. Irreversible electroporation (nanoknife) in cancer treatment. *Gastrointestinal Intervention* 3, no. 1 (June 2014): 8–18. doi:10.1016/j.gii.2014.02.002.

Kaiser, J. 2016. Calling all failed replication experiments. *Science* 351, no. 6273 (February 5, 2016): 548–548. doi:10.1126/science.351.6273.548.

Kandušer, M., D. Miklavčič, and M. Pavlin. 2009. Mechanisms involved in gene electrotransfer using high- and low-voltage pulses—An in vitro study. *Bioelectrochemistry (Amsterdam, Netherlands)* 74, no. 2 (February 2009): 265–71. doi:10.1016/j. bioelechem.2008.09.002.

Kolb, J. F., S. Kono, and K. H. Schoenbach. 2006. Nanosecond pulsed electric field generators for the study of subcellular effects. *Bioelectromagnetics* 27, no. 3 (April 2006): 172–87. doi:10.1002/bem.20185.

Kos, B., A. Županič, T. Kotnik, M. Snoj, G. Serša, and D. Miklavčič. 2010. Robustness of treatment planning for electrochemotherapy of deep-seated tumors. *The Journal of Membrane Biology* 236, no. 1 (July 2010): 147–53. doi:10.1007/ s00232-010-9274-1.

Kos, B., B. Valič, T. Kotnik, and P. Gajšek. 2012. Occupational exposure assessment of magnetic fields generated by induction heating equipment—The role of spatial averaging. *Physics in Medicine and Biology* 57, no. 19 (October 7, 2012): 5943–53. doi:10.1088/0031-9155/57/19/5943.

Kos, B., P. Voigt, D. Miklavčič, and M. Moche. 2015. Careful treatment planning enables safe ablation of liver tumors adjacent to major blood vessels by percutaneous irreversible electroporation (IRE). *Radiology and Oncology* 49, no. 3 (September 2015): 234–41. doi:10.1515/raon-2015-0031.

Kotnik, T. and D. Miklavčič. 2006. Theoretical evaluation of voltage inducement on internal membranes of biological cells exposed to electric fields. *Biophysical Journal* 90, no. 2 (January 15, 2006): 480–91. doi:10.1529/biophysj.105.070771.

Kotnik, T., G. Pucihar, and D. Miklavčič. 2010. Induced transmembrane voltage and its correlation with electroporation-mediated molecular transport. *The Journal of Membrane Biology* 236, no. 1 (July 2010): 3–13. doi:10.1007/s00232-010-9279-9.

Kotnik, T., P. Kramar, G. Pucihar, D. Miklavčič, and M. Tarek. 2012. Cell membrane electroporation—part 1: The phenomenon. *IEEE Electrical Insulation Magazine* 28, no. 5 (September 2012): 14–23. doi:10.1109/MEI.2012.6268438.

Kotnik, T., W. Frey, M. Sack, S. H. Meglič, M. Peterka, and D. Miklavčič. 2015. Electroporation-based applications in biotechnology. *Trends in Biotechnology* 33, no. 8 (August 2015): 480–88. doi:10.1016/j.tibtech.2015.06.002.

Kranjc, M., B. Markelc, F. Bajd, M. Čemažar, I. Serša, T. Blagus, and D. Miklavčič. 2015. In situ monitoring of electric field distribution in mouse tumor during electroporation. *Radiology* 274, no. 1 (August 19, 2014): 115–23. doi:10.1148/radiol.14140311.

Lacković, I., R. Magjarević, and D. Miklavčič. 2009. Three-dimensional finite-element analysis of joule heating in electrochemotherapy and in vivo gene electrotransfer. *IEEE Transactions on Dielectrics and Electrical Insulation* 16, no. 5 (October 2009): 1338–47. doi:10.1109/TDEI.2009.5293947.

Lambricht, L., A. Lopes, S. Kos, G. Serša, V. Préat, and G. Vandermeulen. 2016. Clinical potential of electroporation for gene therapy and DNA vaccine delivery. *Expert Opinion on Drug Delivery* 13, no. 2 (February 2016): 295–310. doi:10.1517/17425247.2016.1121990.

Larkin, J. O., C. G. Collins, S. Aarons, M. Tangney, M. Whelan, S. O'Reily, O. Breathnach, D. M. Soden, and G. O'Sullivan. 2007. Electrochemotherapy: Aspects of preclinical development and early clinical experience. *Annals of Surgery* 245, no. 3 (March 2007): 469–79. doi:10.1097/01.sla.0000250419.36053.33.

Maček-Lebar, A., G. Serša, S. Kranjc, A. Groselj, and D. Miklavčič. 2002. Optimisation of pulse parameters *in vitro* for *in vivo* electrochemotherapy. *Anticancer Research* 22, no. 3 (June 2002): 1731–36.

Mahnič-Kalamiza, S., E. Vorobiev, and D. Miklavčič. 2014. Electroporation in food processing and biorefinery. *The Journal of Membrane Biology* 247, no. 12 (December 2014): 1279–1304. doi:10.1007/s00232-014-9737-x.

Mali, B., T. Jarm, F. Jager, and D. Miklavčič. 2005. An algorithm for synchronization of *in vivo* electroporation with ECG. *Journal of Medical Engineering & Technology* 29, no. 6 (December 2005): 288–96. doi:10.1080/03091900512331332591.

Mali, B., T. Jarm, S. Čorovič, M. S. Paulin-Kosir, M. Čemažar, G. Serša, and D. Miklavčič. 2008. The effect of electroporation pulses on functioning of the heart. *Medical & Biological Engineering & Computing* 46, no. 8 (August 2008): 745–57. doi:10.1007/s11517-008-03467.

Mali, B., T. Jarm, M. Snoj, G. Serša, and D. Miklavčič. 2013. Antitumor effectiveness of electrochemotherapy: A systematic review and meta-analysis. *European Journal of Surgical Oncology: The Journal of the European Society of Surgical Oncology and the British Association of Surgical Oncology* 39, no. 1 (January 2013): 4–16. doi:10.1016/j. ejso.2012.08.016.

Mali, B., V. Gorjup, I. Edhemovic, E. Brecelj, M. Čemažar, G. Serša, B. Strazisar, D. Miklavčič, and T. Jarm. 2015. Electrochemotherapy of colorectal liver metastases— An observational study of its effects on the electrocardiogram. *Biomedical Engineering Online* 14, no. Suppl 3 (2015): S5. doi:10.1186/1475-925X-14-S3-S5.

Marty, M., G. Serša, J. R. Garbay, J. Gehl, C. G. Collins, M. Snoj, V. Billard, P. F. Geertsen, J. O. Larkin, D. Miklavčič, I. Pavlovic, S. M. Paulin-Kosir, M. Čemažar, N. Morsli, D. M. Soden, Z. Rudolf, C. Robert, G. C. O'Sullivan, and L. M. Mir. 2006. Electrochemotherapy—An easy, highly effective and safe treatment of cutaneous and subcutaneous metastases: Results of ESOPE (European standard operating procedures of electrochemotherapy) study. *European Journal of Cancer Supplements* 4, no. 11 (November 2006): 3–13. doi:10.1016/j. ejcsup.2006.08.002.

Miklavčič, D., K. Beravs, D. Semrov, M. Čemažar, F. Demsar, and G. Serša. 1998. The importance of electric field distribution for effective in vivo electroporation of tissues. *Biophysical Journal* 74, no. 5 (May 1998): 2152–58.

Miklavčič, D., S. Čorovič, G. Pucihar, and N. Pavselj. 2006. Importance of tumour coverage by sufficiently high local electric field for effective electrochemotherapy. *European Journal of Cancer Supplements* 4, no. 11 (November 2006): 45–51. doi:10.1016/j.ejcsup.2006.08.006.

Miklavčič, D., M. Snoj, A. Županič, B. Kos, M. Čemažar, M. Kropivnik, M. Bracko, T. Pecnik, E. Gadzijev, and G. Serša. 2010. Towards treatment planning and treatment of deep-seated solid tumors by electrochemotherapy. *Biomedical Engineering Online* 9 (February 23, 2010): 10. doi:10.1186/1475-925X-9-10.

Miklavčič, D., B. Mali, B. Kos, R. Heller, and G. Serša. 2014. Electrochemotherapy: From the drawing board into medical practice. *Biomedical Engineering Online* 13 (2014): 29. doi:10.1186/1475-925X-13-29.

Miklavčič, D. and R. V. Davalos. 2015. Electrochemotherapy (ECT) and irreversible electroporation (IRE)-advanced techniques for treating deep-seated tumors based on electroporation. *Biomedical Engineering Online* 14, no. Suppl 3 (August 27, 2015): I1. doi:10.1186/1475-925X-14-S3-I1.

Mir, L. M. and S. Orlowski. 1999. Mechanisms of electrochemotherapy. *Advanced Drug Delivery Reviews* 35, no. 1 (January 4, 1999): 107–18.

Mir, L. M., J. Gehl, G. Serša, C. G. Collins, J.-R. Garbay, V. Billard, P. F. Geertsen, Z. Rudolf, G. C. O'Sullivan, and M. Martya. 2006. Standard operating procedures of the electrochemotherapy: Instructions for the use of bleomycin or cisplatin administered either systemically or locally and electric pulses delivered by the cliniporator™ by means of invasive or non-invasive electrodes. *EJC Supplements* 4, no. 11 (November 1, 2006): 14–25. doi:10.1016/j.ejcsup.2006.08.003.

Ongaro, A., L. G. Campana, M. De Mattei, F. Dughiero, M. Forzan, A. Pellati, C. R. Rossi, and E. Sieni. 2016. Evaluation of the electroporation efficiency of a grid electrode for electrochemotherapy from numerical model to in vitro tests. *Technology in Cancer Research & Treatment* 15, no. 2 (April 1, 2016): 296–307. doi:10.1177/1533034615582350.

Pakhomov, A. G., B. L. Ibey, A. M. Bowman, F. M. Andre, and O. N. Pakhomova. 2009. Nanosecond-duration electric pulses open nanometer-size pores in cell plasma membrane. In *World Congress on Medical Physics and Biomedical Engineering, September 7–12, 2009, Munich, Germany*, edited by O. Dössel and W. C. Schlegel. IFMBE Proceedings 25/13. Springer, Berlin-Heidelberg, pp. 17–20. http://link.springer.com/chapter/10.1007/978-3-642-03895-2_6.

Pavliha, D., B. Kos, A. Županič, M. Marčan, G. Serša, and D. Miklavčič. 2012. Patient-specific treatment planning of electrochemotherapy: Procedure design and possible pitfalls. *Bioelectrochemistry* 87 (October 2012): 265–73. doi:10.1016/j.bioelechem.2012.01.007.

Pavliha, D. and B. Kos. 2013. Planning of electroporation-based treatments using web-based treatment-planning software. *The Journal of Membrane Biology* 246, no. 11 (2013): 833–42. doi:10.1007/s00232-0139567-2.

Pavlin, M., M. Kandušer, M. Reberšek, G. Pucihar, F. X. Hart, R. Magjarevićcacute, and D. Miklavčič. 2005. Effect of cell electroporation on the conductivity of a cell suspension. *Biophysical Journal* 88, no. 6 (June 2005): 4378–90. doi:10.1529/biophysj.104.048975.

Pavlin, M., S. Haberl, M. Reberšek, D. Miklavčič, and M. Kandušer. 2011. Changing the direction and orientation of electric field during electric pulses application improves plasmid gene transfer in vitro. *Journal of Visualized Experiments: JoVE*, no. 55 (September 12, 2011): 3309. doi:10.3791/3309.

Pliquett, U., E. Gersing, and F. Pliquett. 2000. Evaluation of fast time-domain based impedance measurements on biological tissue. *Biomedizinische Technik* 45, nos. 1–2 (January 1, 2000): 6–13.

Postma, P. R., G. Pataro, M. Capitoli, M. J. Barbosa, R. H. Wijffels, M. H. M. Eppink, G. Olivieri, and G. Ferrari. 2016. Selective extraction of intracellular components from the microalga *Chlorella vulgaris* by combined pulsed electric field–temperature treatment. *Bioresource Technology* 203 (March 2016): 80–88. doi:10.1016/j.biortech.2015.12.012.

Pucihar, G., T. Kotnik, D. Miklavčič, and J. Teissié. 2008. Kinetics of transmembrane transport of small molecules into electropermeabilized cells. *Biophysical Journal* 95, no. 6 (September 15, 2008): 2837–48. doi:10.1529/biophysj.108.135541.

Raffa, V., C. Riggio, M. W. Smith, K. C. Jordan, W. Cao, and A. Cuschieri. 2012. Bnnt-mediated irreversible electroporation: Its potential on cancer cells. *Technology in Cancer Research & Treatment* 11, no. 5 (October 1, 2012): 459–65. doi:10.7785/tcrt.2012.500258.

Reberšek, M. and D. Miklavčič. 2011. Advantages and disadvantages of different concepts of electroporation pulse generation. *Automatika (Zagreb)* 52: 12–19.

Reberšek, M., D. Miklavčič, C. Bertacchini, and M. Sack. 2014. Cell membrane electroporation—part 3: The equipment. *IEEE Electrical Insulation Magazine* 30, no. 3 (May 2014): 8–18. doi:10.1109/MEI.2014.6804737.

Reilly, J. P. 1998. *Applied Bioelectricity*. New York, NY: Springer New York, 1998. http://link.springer.com/10.1007/978-1-4612-1664-3.

Retelj, L., G. Pucihar, and D. Miklavčič. 2013. Electroporation of intracellular liposomes using nanosecond electric pulses—A theoretical study. *IEEE Transactions on Biomedical Engineering* 60, no. 9 (September 2013): 2624–35. doi:10.1109/TBME.2013.2262177.

Rols, M. P. and J. Teissié. 1998. Electropermeabilization of mammalian cells to macromolecules: Control by pulse duration. *Biophysical Journal* 75, no. 3 (September 1998): 1415–23.

Rubinsky, B., G. Onik, and P. Mikus. 2007. Irreversible electroporation: A new ablation modality—clinical implications. *Technology in Cancer Research & Treatment* 6, no. 1 (February 1, 2007): 37–48. doi:10.1177/153303460700600106.

Rubinsky, J., G. Onik, P. Mikus, and B. Rubinsky. 2008. Optimal parameters for the destruction of prostate cancer using irreversible electroporation. *The Journal of Urology* 180, no. 6 (December 2008): 2668–74. doi:10.1016/j.juro.2008.08.003.

Sanders, J. M., A. Kuthi, Y. H. Wu, P. T. Vernier, and M. A. Gundersen. 2009. A linear, single-stage, nanosecond pulse generator for delivering intense electric fields to biological loads. *IEEE Transactions on Dielectrics and Electrical Insulation* 16, no. 4 (August 2009): 1048–54. doi:10.1109/TDEI.2009.5211853.

Sardesai, N. Y. and D. B. Weiner. 2011. Electroporation delivery of DNA vaccines: Prospects for success. *Current Opinion in Immunology* 23, no. 3 (June 2011): 421–29. doi:10.1016/j.coi.2011.03.008.

Šatkauskas, S., M. F. Bureau, M. Puc, A. Mahfoudi, D. Scherman, D. Miklavčič, and L. M. Mir. 2002. Mechanisms of *in vivo* DNA electrotransfer: Respective contributions of cell electropermeabilization and DNA electrophoresis. *Molecular Therapy: The Journal of the American Society of Gene Therapy* 5, no. 2 (February 2002): 133–40. doi:10.1006/mthe.2002.0526.

Scheffer, H. J., K. Nielsen, M. C. de Jong, A. A. J. M. van Tilborg, J. M. Vieveen, A. (R. A.) Bouwman, S. Meijer, C. van Kuijk, P. M. van den Tol, and M. R. Meijerink. 2014. Irreversible electroporation for nonthermal tumor ablation in the clinical setting: A systematic review of safety and efficacy. *Journal of Vascular and Interventional Radiology*, Special Issue: Interventional Oncology 25, no. 7 (July 2014): 997–1011. doi:10.1016/j.jvir.2014.01.028.

Sel, D., D. Cukjati, D. Batiuskaite, T. Slivnik, L. M. Mir, and D. Miklavčič. 2005. Sequential finite element model of tissue electropermeabilization. *IEEE Transactions on Biomedical Engineering* 52, no. 5 (May 2005): 816–27. doi:10.1109/TBME.2005.845212.

Serša, G., S. Kranjc, J. Scancar, M. Krzan, M. Cemažar. 2010. Electrochemotherapy of mouse sarcoma tumors using electric pulse trains with repetition frequencies of 1 Hz and 5 kHz. *The Journal of Membrane Biology* 236, no. 1 (July 2010):155–62. doi:10.1007/s00232-010-9268-z.

Serša, G., J. Teissié, M. Čemažar, E. Signori, U. Kamensek, G. Marshall, and D. Miklavčič. 2015. Electrochemotherapy of tumors as in situ vaccination boosted by immunogene electrotransfer. *Cancer Immunology, Immunotherapy* 64 (2015): 1315–27. doi:10.1007/s00262-015-1724-2.

Shafiee, H., P. A. Garcia, and R. V. Davalos. 2009. A preliminary study to delineate irreversible electroporation from thermal damage using the arrhenius equation. *Journal of Biomechanical Engineering* 131, no. 7 (June 12, 2009): 074509. doi:10.1115/1.3143027.

Shibata, M.-A., Y. Ito, J. Morimoto, K. Kusakabe, R. Yoshinaka, and Y. Otsuki. 2006. In vivo electrogene transfer of interleukin-12 inhibits tumor growth and lymph node and lung metastases in mouse mammary carcinomas. *The Journal of Gene Medicine* 8, no. 3 (March 2006): 335–52. doi:10.1002/jgm.854.

Spratt, D. E., E. A. Gordon Spratt, S. Wu, A. DeRosa, N. Y. Lee, M. E. Lacouture, and C. A. Barker. 2014. Efficacy of skin-directed therapy for cutaneous metastases from advanced cancer: A meta-analysis. *Journal of Clinical Oncology: Official Journal of the American Society of Clinical Oncology* 32, no. 28 (October 1, 2014): 3144–55. doi:10.1200/JCO.2014.55.4634.

Susil, R., D. Šemrov, and D. Miklavčič. 1998. Electric field-induced transmembrane potential depends on cell density and organization. *Electro- and Magnetobiology* 17, no. 3 (January 1, 1998): 391–99. doi:10.3109/15368379809030739.

Syamsiana, I. N. and R. I. Putri. 2011. High voltage pulse generator design with voltage control for pulse electric field (PEF) pasteurization. In *2011 International Conference on Electrical Engineering and Informatics (ICEEI)*, 1–5, 2011. doi:10.1109/ICEEI.2011.6021712.

Teissié, J. and M. P. Rols. 1993. An experimental evaluation of the critical potential difference inducing cell membrane electropermeabilization. *Biophysical Journal* 65, no. 1 (July 1993): 409–13.

Thomson, K. R., W. Cheung, S. J. Ellis, D. Federman, H. Kavnoudias, D. Loader-Oliver, S. Roberts, P. Evans, C. Ball, and A. Haydon. 2011. Investigation of the safety of irreversible electroporation in humans. *Journal of Vascular and Interventional Radiology* 22, no. 5 (May 2011): 611–21. doi:10.1016/j.jvir.2010.12.014.

Toepfl, S., A. Mathys, V. Heinz, and D. Knorr. 2006. Review: Potential of high hydrostatic pressure and pulsed electric fields for energy efficient and environmentally friendly food processing. *Food Reviews International* 22, no. 4 (December 1, 2006): 405–23. doi:10.1080/87559120600865164.

Towhidi, L., T. Kotnik, G. Pucihar, S. M. P. Firoozabadi, H. Mozdarani, and D. Miklavčič. 2008. Variability of the minimal transmembrane voltage resulting in detectable membrane electroporation. *Electromagnetic Biology and Medicine* 27, no. 4 (2008): 372–85. doi:10.1080/15368370802394644.

Weaver, J. C. 1993. Electroporation: A general phenomenon for manipulating cells and tissues. *Journal of Cellular Biochemistry* 51, no. 4 (April 1, 1993): 426–35. doi:10.1002/jcb.2400510407.

Weaver, J. C. and Yu. A. Chizmadzhev. 1996. Theory of electroporation: A review. *Bioelectrochemistry and Bioenergetics* 41, no. 2 (December 1996): 135–60. doi:10.1016/S0302-4598(96)05062-3.

Wiggers, C. J. and R. Wegria. 1940. Ventricular fibrillation due to single, localized induction and condenser shocks applied during the vulnerable phase of ventricular systole. *American Heart Journal* 19, no. 3 (March 1, 1940): 372. doi:10.1016/S0002-8703(40)90014-X.

Zou, P., W. Liu, and Y.-H. Chen. 2005. The epitope recognized by a monoclonal antibody in influenza a virus M2 protein is immunogenic and confers immune protection. *International Immunopharmacology* 5, no. 4 (April 2005): 631–35. doi:10.1016/j.intimp.2004.12.005.

Županič, A. and D. Miklavčič. 2011. Tissue heating during tumor ablation with irreversible electroporation. *Elektrotehniški vestnik* 78, no. (1–2): 42–47.

Županič, A., S. Ribaric, and D. Miklavčič. 2007. Increasing the repetition frequency of electric pulse delivery reduces unpleasant sensations that occur in electrochemotherapy. *Neoplasma* 54, no. 3 (2007): 246–50.

Županič, A., S. Čorović, D. Miklavčič, and M. Pavlin. 2010. Numerical optimization of gene electrotransfer into muscle tissue. *Biomedical Engineering Online* 9 (November 4, 2010): 66. doi:10.1186/1475-925X-9-66.

13

Quantifying in Bioelectromagnetics

Pawel Bienkowski
and Hubert Trzaska
*Wroclaw University
of Technology*

13.1 Introduction

A couple of years ago, a known specialist in electro- and magnetotherapy was invited to present his works at a meeting of electronic engineers at a special session devoted to possible bioeffects caused by exposure to an electromagnetic fields (EMFs) and the necessity of protection of workers and general public against uncontrolled exposure. His therapeutic results were fascinating—almost 60%–80% of positive effects in selected therapies. In the other cases, the results were less optimistic. The main problems with the wider application of these therapies were conservatism of the medical sector and the role of the pharmaceutical industry. After the presentation, he was asked a question about the technique and accuracy of the exposure parameters selection, i.e., frequency of stimuli, its intensity, duration, character, etc. The answer was they were selected on the basis of our experience and the accuracy. What an accuracy, I push a button, that's all. Of course the latter was a kind of a metaphor; however, it reflects, in some sense, an approach of a medic to the question of accuracy.

Lately, in many presentations devoted to magnetotherapy applications, methods using "electromagnetic therapy" are presented. In these methods, very low frequencies and coil applicators are in use and the units of exposure are tesla or gauss. Apart from this, there are no quantities that would characterize EMF; all of this indicates that in these therapies, the magnetic field is of concern. A repeated question to the presenters regarding why they called it electromagnetic therapy rather than magnetic was answered: this is indicated in the devices manual.

Both these anecdotes illustrate a medic's relation to technical problems as "an approach with trepidation." And inversely, engineers have a similar problem with biomedicine.

A further illustration of this: The authors, although involved in different experiments in the field of bioelectromagnetics, never formulate any suggestions or ideas in the field of biology and medicine so as to not frustrate people involved in these fields. In some sense, this is the result of a discussion, which happened years ago, with a known doctor involved in the field of bioelectromagnetics. He suggested that the best way of quantifying the exposure, especially at a workplace, was by a phantom replacing a person(s) of concern. The phantom should be equipped with a number of wideband antennas (no comment). The antennas should be in the form of spiral antennas (oh God!).

All of the abovementioned points show a gap between the two worlds of medicine and engineering. Fortunately, the problem is being better understood, and steps are being taken to increase the understanding of these two worlds (i.e., medicine and engineering), for instance by organizing "schools of understanding" through the European program, Cooperation in Science and Technology (COST).

In this chapter, selected comments related to the problem of EMF quantification are presented. The comments are formulated on the grounds of well-known rules of EMF metrology and long-time experience of the authors and their colleagues. This chapter focuses mainly upon accuracy of procedures applied in bioelectromagnetics, especially those of an electromagnetic nature.

13.2 Accuracy

Every single measuring procedure is subject error. Moreover, the real value of a measured quantity is never known. The real value may be approached if the measurements are repeated in series and a mean value is calculated. The resultant error consists of a number of partial errors. These errors are usually presented as follows:

a. Random errors

These errors are usually difficult to indicate. They include variations of measurements conditions (temperature, humidity, local circumstances), alternations of measured field (modulation, propagation) and its interference during measurements, presence of other radiations, change of personnel performing the measurements, etc. Presence of the errors and their scale may be identified by way of repeated measurements. If we assume constant measuring conditions, a dispersion of meter readouts would indicate the presence of errors. Thus, the use of statistics may limit the occurrence of errors.

b. Systematic errors

Contrary to the previous one, here readouts of a meter, in constant conditions, during repeated measurements, should remain unchanged (similar). The main role here is played by the accuracy of the meter calibration (accuracy of the meter: the accuracy cannot exceed that of a calibration tool). Other accuracy limiting factors are specificity of measured field (polarization, modulation, curvature, spectrum), mutual interactions between the probe (meter) and the surroundings, presence and role of other objects in the area, and so on.

c. Gross errors (mistakes)

These errors are usually characterized by the presence of meter indications that are considerably different from other ones. In the majority of physical measurements, indications are omitted during the final analysis. That approach is not suggested in our case. Here, unbelievably low levels of fields may appear close to big power sources and vice versa, and so a measuring team may be disappointed by it. The team experience may be helpful here in judging the correctness of the indications or, at least a meters tester [1] should be used to check a meter during measurements (Figure 13.1). Such a procedure is required by Polish standards from accredited labs.

In far-field EMF measurements, errors are estimated at a level not significantly exceeding 10%. In our case, in the near field, the errors were observed to reach 20% or more. The error budget, in the latter case, was estimated for E- and H-field measurements of a stable, monochromatic field. However, while magnitudes related to energy absorption, proportional to E^2 and H^2, are of concern, the error is twice as much. In the case of modulated field measurements, the error increases due to nonlinear dynamic characteristics of a meter.

Also, while nonstationary fields are of concern, for instance near a rotating radar antenna, the error reaches levels that allow for the conclusion that the measurements are of a qualitative rather than a quantitative nature [2]. In order to illustrate these considerations, let us assume that a tool of class estimated at the level of 20% is applied for

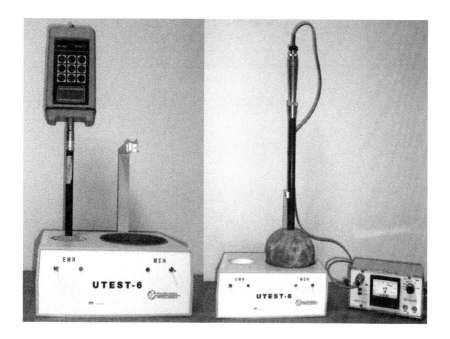

FIGURE 13.1 EMR and MEH meters testing in UTEST-6.

man size measurements (Figure 13.2). The result of the measurement: the high ranges between a dwarf and a basketball player. However, judging this visually would be much more accurate. But this is a reality in bioelectromagnetics.

Also, previous studies quantifying EMF in laboratory experiments and undertaking measurements of EMF for protection purposes included well-prepared statistical analysis. However, only in very few cases was any accuracy even taken into consideration. This makes it impossible to evaluate the correctness of the procedures. The statistics of results and the accuracy have nothing in common with each other. Figure 13.3 presents a situation where the statistic mean value (smv) and measured values (*v*) are contained within accuracy frames of a meter (±δ) and expected value (ev). However, when the class of the meter is not taken into consideration and the meter is applied in unacceptable conditions, the (smv) and (*v*) are outside the frames of ±δ.

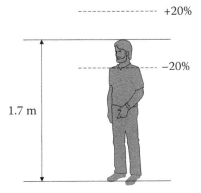

FIGURE 13.2   A size of a man measurement with 20% uncertainty.

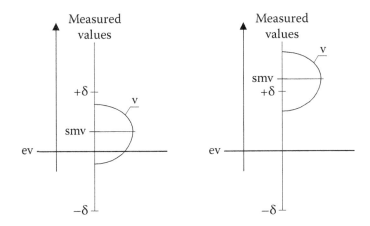

FIGURE 13.3   Believable (left) and unbelievable (right) results of an experiment.

13.3 Quantities

13.3.1 *E&H* Measurements

A variety of quantities were considered and proposed for bioelectromagnetic purposes [3,4]. We cannot discuss all of them, and so limit our discussion to measurable quantities. With some exceptions (for instance a Hall cell application for H-field measurement), these are H-field measurement using small loop antennas (probes) and E-field measurement using short dipole antennas (probes). The expressions "small" and "short" indicate the size of an antenna in relation to the shortest wavelength within the measuring frequency band at which the device is used. Let us briefly analyze the measurable quantities in the aspect of factors limiting the measurement accuracy. The factors occur mainly due to specificity of the near field.

A schematic diagram of E-field measurement is shown in Figure 13.4. A symmetric dipole antenna is placed at distance R to a source (primary or secondary) of EMF. The field intensity is inversely proportional to R^{-3}–R^{-1} depending upon the source structure and size and the distance R. Due to "mirror reflection" effect, the input impedance of the dipole is affected by a coupling with its reflection. Both factors quickly disappear with increase in R.

Usually the antenna(s) are placed in a dielectric casing (random), which remarkably limits the errors. The main reason for this randomness is to protect the antenna against mechanic or/and electric (overloading) damage. However, it simultaneously limits the possibility to measure in close vicinity of the source, and the presented analyses try to show possible maximal value of the errors. A remarkable role may be played by external field(s) (EF) reflection(s) (rr—reflected ray), and diffraction(s) and by attributes of the person performing the measurement and other objects in the area. In the case of the H-field, an important role is played by the so-called "antenna effect," i.e.,

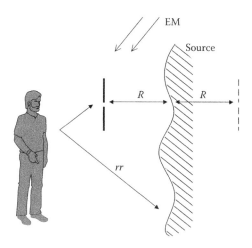

FIGURE 13.4   E-field measurement.

mutual sensitivity of a loop to E- and H-field. The effect may play a dominating role while H-field is measured near the source of the dominating E-field, for instance near and within a flat capacitor, which is used in experiments as an exposure system, and in diathermy or dielectric heating.

The abovementioned factors are specific to a limited number of cases, only to monochromatic EMF. This is the case in laboratory experiments. In the EMF measurements for labor safety and/or protection purposes (again: in the near field), due to modulated, achromatic field measurement no negligible role is played by the probe frequency response and its dynamic characteristics.

Till now, the level (class) of standards applied in EMF meter calibration has not been mentioned anywhere. The EMF standards are one of the least accurate standards of physical quantities [5]. The class is on the level of several percent. In the case of exposure systems used both for therapeutic purposes and in laboratory experiments, due to their loading or coupling to the exposed object, the class is dramatically reduced. The accuracy of standards and exposure systems is the "Achilles heel" of the electromagnetic side of bioelectromagnetics.

Now let us briefly discuss examples of derived quantities, i.e., quantities calculated on the ground of E- and/or H-field measurement. They are proportional to E^2 or H^2 and, as mentioned above, the error of these measurable quantities is doubled, as compared to E and H measurement. Apart from this introductory error, other factors limiting accuracy, specific for separate quantities, should be considered:

a. Power density (S)

   Earlier, it was mentioned that there is no magnitude that would characterize EMF. Now, we have to remember that such a quantity exists and may be played by the Poynting vector, which characterizes power transportation by an electromagnetic wave and in practical applications is called "power density" (S). The power density is measurable using surface antennas (horn, opened waveguide), and this is given as a ratio of power measured at the input of the antenna to its surface. Other methods include resonant dipoles, Yagi-Uda antennas, or their combination. In the latter cases, these antennas require a calibration, and then S is calculated by dividing E^2 by Z_0 (where Z_0 is the intrinsic impedance of free space; $Z_0 = 120\ \pi$). Here, voltage is measured at the input of the antenna and E is then read from calibration curves.

   The methods are fully correct in the far field. Unfortunately, they are useless in the near field because of resonant sizes of antennas and their directional properties. The near field conditions possibly require small antennas (sensors). Moreover, the H-field dominates close to magnetic sources (solenoid), while the E-field dominates near electric sources (short wire, plates of a capacitor). Thus, a measurement of the power density, by way of the H-field measurement near "electric" sources (and inversely), may be loaded with errors of any amplitude. This is illustrated while S is measured by E-field measurement in Figure 13.5, which presents a run of the S measurement error (δ) near a $\lambda/2$ symmetric dipole in a direction perpendicular to the dipole as a function of kR (where: k is propagation constant $k = 2\pi/\lambda$).

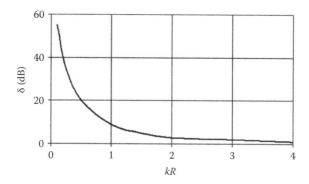

FIGURE 13.5   Error δ of S measurement by E measurements versus kR.

Results of estimations presented in Figure 13.5 allow several conclusions to be drawn:

- The error vanishes for $kR > 2$, i.e., for $R > \lambda/\pi$. The boundary of the near field (R_{nf}) may by estimated using the formula $R_{nf} = 2D^2/\lambda$ (where D: size of the radiator). In the discussed example, $R_{nf} = \lambda/2$. It means that the error in the case vanishes nearer the source rather than at the boundary.
- If we would like to reduce the error and measure S at a distance, say one foot, the frequency should be limited to the maximum 300 MHz.
- If the measurements are performed nearer, as above, and/or the frequency is lower, the error δ is positive. It means that the measured S majorizes the real one. It is useful for measurements that are performed for protection purposes; however, the approach is not suggested for any kind of bioelectromagnetics studies.

b. Specific absorption rate (SAR)

The SAR, from the very beginning, has created doubts and controversies [6–10]. The discussion was devoted to the biomedical aspect of the SAR. Justesen [7] even calls it "scientifically illicit." Regardless of aspect, let us look at it from a technical point of view.

The SAR, with regard to its technical aspect, is directly taken from EMF energy application for dielectric heating. Including any simplifications and assumptions of theoretical estimations of energy absorbed in a heated body, the estimations were performed for a flat capacitor in which a homogeneous object is immersed. Due to this, a homogeneous E-field distribution within the object was evident. The aim of these estimations was to assess necessary time of the procedure (t) to get a temperature increase (ΔT), which is given by Equation 13.1:

$$t = \frac{\rho c \Delta T}{\varepsilon_0 \varepsilon_r \omega E^2 tg\delta},$$

(13.1)

where ρ is mass density,
c is specific heat,

ε_0 is electric permittivity,
ε_r is relative permittivity,
ω is angular frequency, and
δ is dielectric loss angle.

Starting from both the SAE definitions [9], the formula given below can be derived:

$$\frac{c\Delta T}{t} = \text{SAR} = \frac{\sigma E^2}{\rho}, \tag{13.2}$$

where σ is conductivity.

It confirms that in the case of the SAR definitions, the same assumptions were taken into account as in the simplest technological approach. Contrary to the latter, here penetration of EM energy into the body is much more complex. The penetration is a function of the body size and its electric structure, frequency and polarization of the field, and reflection and refraction of the energy, not to mention energy dispersion due to radiation and thermoregulation *in vivo*. Thus, the procedure of SAR estimation and its spatial distribution within an exposed body, using external E measurement, may be acceptable (in the light of accuracy) only in a limited number of cases. A similar conclusion may be drawn for T measurements as the latter is, in practice, measured on the surface of the body.

13.3.2 Current Measurements

Although the majority of protection standards require the current (I) measurements, the quantity is, in some sense, forgotten. As presented above, the $E\&H$ measurements are loaded with remarkable errors, and in the case of I, the measurement is even worse. Finally, the eddy currents are not measurable at all. The current is measured, in general, in two cases: on direct contact of a body with a charged body (source) and while exposed to an EMF. The cases are illustrated in Figure 13.6.

The touch current measurement looks conceptually simpler. It may be done, for instance, using a thermocouple (TC) connected between a source and a coil. The main sources of measuring errors include problems in accurate calibration of the measuring device, unknown parameters of the person involved (including his/her clothing and shoes), ground impedance, frequency dependence, temperature, humidity, etc. Even very approximate estimations with use of the simplest assumptions show that errors here would well exceed what was estimated earlier. Disregarding the errors, the authors studied the problem in the case of a counterpoise, played by an operator of a transceiver (TRX). It was shown that at frequencies below, say 300 MHz, the role of hands' (mouths) current is important, even while low power devices are of concern, and increases for lower frequencies. The currents may exceed values indicated in protection standards. However, studies of the role of current are not included in biomedical investigations, while bioeffects caused by mobile phones, for instance, are studied.

While a person is exposed to external EMF, it plays the role of a vertical (ground plane) antenna. Apart from the abovementioned factors limiting the current measurement accuracy, here the essential role is played by field polarization. The maximal value of the current depends on the parallelism of the E-field vector and the main axis of the body. Disregarding the height of the person and its behavior in relation to the fields, it is worth mentioning that the horizontal component of the current remains immeasurable.

It seems that another manner of current induction was observed by the authors during measurements at an antenna tower (Figure 13.7). A hypothesis was formulated that the human body and a metallic ladder create something similar to a loop antenna.

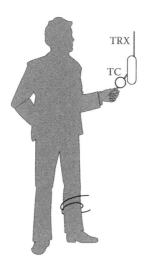

FIGURE 13.6   Two cases of the current measurement.

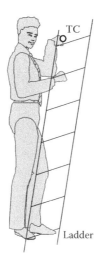

FIGURE 13.7   A person at a ladder.

The current measured at the hands and feet exceeded 100 mA close to FM radio transmitting antennas.

13.3.3 Temperature Measurements

While thermal effects are of concern, the temperature measurement, its value, and distribution is of importance. The temperature rise allows for calculating SAR. The main advantage of the temperature measurement is its accuracy, which is much better as compared to the abovementioned errors in the E, H, and I measurements. However, these quantities may be measured outside of the body, and then on the basis of those values (with better or worse accuracy), may be assessed to identify their role within the body. The temperature measurement may be performed at the surface of a body or just below the surface, *in vivo*, and within a phantom model that allows for temperature distribution assessment. The main disadvantage of the measurements is the sensitivity of the method. Even if we assume measurable $\Delta T \approx 0.01$ K in real time, the temperature rise at level of minutes require above megahertz frequencies and $E > 100$ V/m (compare with Equation 13.1). Another disadvantage, especially important in diathermy and noninvasive experiments *in vivo*, is the possibility of estimating the temperature distribution within a body on the basis of E (H, S, I) measurements outside it, and then modeling calculations. A similar procedure, although very uncertain, in the case of E, H, S, I allows for finding their distribution within the body (with an assumption of temperature independent parameters of the body); in the case of temperature rise estimations, the accuracy of the procedure is additionally degraded by the heat transfer, cooling, radiation, and the thermoregulation mechanisms. Fortunately, in therapeutic applications and in laboratory experiments, the source of exposure is well known. In the case of exposure to less determined field(s), the considerations reach the level of predicting from dregs.

Studies in the field were started by Guy [11]. He exposed homogeneous models of regular shape and then investigated their temperature using thermographic methods and confirmed the presence and role of thermal effects.

13.4 Measured Values RMS, Peak

In practice, two values of a measured quantity, i.e., effective value (root mean square, RMS) and amplitude (peak), are considered. Sometimes a mean value is of concern. The effective value (RMS) represents energy transfer, while the amplitude illustrates maximal value of the measured quantity. Most of the available instruments are calibrated in RMS, while more advanced ones allow readings of any value.

A detection device for RMS measurement possesses square law characteristics. Such characteristics are seen in a thermistor and a diode detector; however, in the latter the characteristic is nonlinear only in its initial part owing to its dynamic characteristics. As far as the harmonic signal is concerned, there are no problems as all the values are interchangeable. Again, problems start when achromatic EMF is to be measured, in particular for pulsed or nonstationary fields.

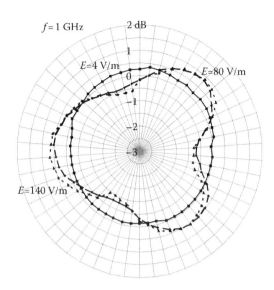

FIGURE 13.8 Measured pattern of an omnidirectional probe.

Because of many advantages (thermal detectors have "short" dynamic characteristics and are sensitive to damage while overloading), the diode detectors are the most popular solution in meters offered on the market. Thus, as a result of their square law characteristics initially, which then transform to a linear one, the value that is measured depends on the level of measured quantity. Let us illustrate the issue with an example.

The synthesis of an omnidirectional pattern of a measuring probe requires a summation, with square, output voltages from three mutually perpendicular sensors. If the summation is linear, in the pattern it would appear minima. Results of measurements of the pattern of a spherical probe are shown in Figure 13.8. The pattern is omnidirectional while measured in $E = 4$ V/m field. For stronger fields, minima may be as deep as 3 dB, indicating linear summation of the voltages.

The experiment may be repeated with any omnidirectional probe. It is enough to put a probe at a distance from a source of EMF and turn it. Then the procedure should be repeated nearer to the source, where the field is stronger. If better accuracy is expected, the field intensity should be measured, for instance as indicated in Figure 13.8.

In the case of the pattern, the difference is "only" 3 dB, and the question remains which indication of the meter should be read. What value is indicated by such a device, while fields of time-alternating amplitudes are measured, remains unanswered.

13.5 Meters

The first doubts, regarding meters, were formulated by Mild [12]. He found that even devices offered by known manufacturers did not always fulfill parameters declared in their manuals.

Daily work in the authors' laboratory includes, amongst other tasks, periodic testing of EMF meters used by Polish control institutions. This involvement has made it possible to investigate different types of meters, their parameters, and their possible faults and damages. The damages are, most often, the result of intense working conditions during measurements, transportation, and more so, inappropriate use of the meters. However, the most surprising are faults and imperfections of newly bought devices. Several examples of these are listed below:

- Probes with a diode detector assure quite correct readouts while constant, or slowly alternating, harmonic EMF is measured. The conditions are similar to that of calibration and respond to the probes' dynamic characteristics. Remarkable errors, while achromatic and/or AM modulated fields are measured, were mentioned above; it includes a deformation of a probe directional pattern.
- The calibration is performed in laboratory conditions where temperature is around 20°C. In practice, the measurements may be performed in temperatures ranging from, say, −20°C to +40°C. Results of measurements show that Schottky barrier diodes are the most temperature sensitive, and because of this, the results of measurements may be more that 50% less while measurements are performed in temperatures just below 0°C and the effect is the strongest while the lowest EMF levels are of concern.
- The sources of errors listed thus far are evident and characterize any type of meter. Now let us remember that the measurements are usually performed in quite strong fields and that the measured spectrum may be unknown *a priori*. Thus, the field may penetrate a monitor (indication part of a meter) by any possible route and affect its work. In order to illustrate the phenomenon, Figure 13.9 shows a set of screens and filters, in a MEH-type meter, which protect against the penetration

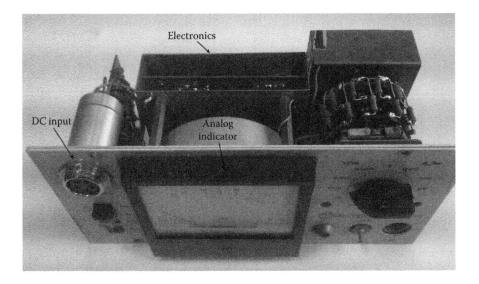

FIGURE 13.9   Screens and filters in a MEH type meter.

through an analog indicator and DC voltage input from a probe. Electronic circuitry is enclosed in a metallic compartment with low-pass filters at any input or output. All these techniques are necessary to ensure correct working of the device even in EM environments with high intensity.

The susceptibility of a meter may be proven by its user in several simple ways. For instance:

- A meter, with disconnected probe, is exposed to EMF. Presence of any indications and their dependence from the field frequency and amplitude changes would confirm the susceptibility presence and its range.
- No device would work without a probe. In this case, it is suggested to cover the probe with conducting foil and then start the experiments.

The susceptibility is not a specificity of considered area of measurements. It characterizes any electric and electronic device and is a subject of involvement of electromagnetic compatibility. In general, the phenomenon usually increases with frequency due to increase of electric sizes of devices and their elements. In the case considered, e.g., in the case of handheld meters, an additional effect may be created by the body of a measuring person, which plays the role of an additional EMF receptor and, thereby, a device stimulator. Moreover, the most susceptible are the small, inexpensive meters in a plastic casing.

13.6 Exposure Systems

EMF energy absorbed in an exposed object may be estimated in different ways. It is most often done by:

Enmf measurement around (close) the object, and then calculations of the absorbed energy and its distribution within the object.

Excitation measurement of a radiator (antenna) on the basis of the radiator parameter (radiation pattern) calculations of EMF distribution around the object and absorbed energy.

Energy losses measurement in a transmission line due to any object presented(s) in it.

In the two former cases, factors limiting accuracy of measurements and absorption estimations are identical to that previously discussed. An important role is played here by propagation effects (reflection, refraction, penetration); shape, size, and parameters of exposed object; other objects in the area; and frequency.

The latter case was discussed in Ref. [13]. It was shown that such approximation as uniform EMF distribution, especially in a cell of length comparable with the wavelength, is inacceptable. In these conditions, an assumption of identical energy absorption, calculated by dividing energy loss in the cell by number of exposed objects, is wrong. Moreover, mutual interactions (couplings) between objects and between the objects and the walls of the cell, and their role in energy quantity absorbed by a separate object, are not taken into account. Thus, apart from quite accurate measurement of the energy losses in the cell, the energy absorbed in any object is unknown, not to mention the energy distribution within these objects.

13.7 Nonthermal Effects

Presence of low-level effects finds its confirmation and interpretation in many aspects. Let us briefly present three of them:

a. The market offers a lot of devices that generate EMF, and these devices are able to cure diseases of the body and mind and also provide protection against any possible invasive factor. For instance, an ELF generator that sets up "a protective field" able to "combat jet lag" or is helpful while "making crucial decisions" [14].

   Such a device usually includes a carrier wave generator working at a stable frequency within frequency range between low frequencies and microwaves or jumping from one frequency to another. An essential role is played by a pulsed generator(s), whose amplitude, frequency, and duration time vary in any possible combinations. The combinations may be selected using a mode switch (headache, stomach diseases, emotional problems). These pulses modulate the carrier wave generator or act independently, in solutions without the generator. Usually these are small sized, battery powered devices for personal use. Thus, a patient's body may be stimulated by direct coupling.

   A number of different devices were tested by the authors technically, and results of their use were observed by the authors themselves, but no significant remarkable results were obtained. A similar conclusion was drawn following medical tests performed by medical doctors. Although it is impossible to totally exclude any therapeutic abilities of these devices, including placebo, the majority of them seem to be gadgets and nothing more. Regardless, all of them have a medical certificate of effectiveness.

b. Hypersensitivity to EMF is the objective of many serious studies showing possible allergy in a part of the population exposed to relatively small levels of EMF, well below levels mentioned in protection standards. An example of the approach to the issue was a COST-244 WG meeting in Graz [15]. Apart from all these studies and investigations, there are persons complaining of different diseases of the body and mind caused by exposure to EMF. Nobody helps them as they exhibit no evident symptoms of the problems (medicine), as the measured fields do not exceed those permitted by the standards (sanitary inspection), and as nobody was murdered (police). These people are often advised to visit our lab. *Ad hoc* tests performed in a screened chamber, using any available frequency and modulation, have never shown any sensitivity to these stimuli. It may be concluded that in the majority of cases, these were as a result of phobias, mental illness, or other reasons. However, it is impossible to exclude that some cases do show enhanced sensitivity. A question that remains unanswered is who should investigate and help them [16]?

c. Therapeutic applications using fields and currents are currently becoming more popular in medicine and assure results in the treatment of many diseases. Levels of the stimuli usually are well below that required for temperature increase. Disregarding the presence of the saddle effects (nonlinear dependence between the stimuli intensity and its results), the stimuli amplitude choice is not of primary importance. The amplitude and the stimuli shape are chosen in an experimental way.

In the case of electro- and magnetostimulation, the intensity of the stimuli is lower than in therapeutic applications. However, there may be a problem with frequency choice due to resonant phenomena. The authors also assist in medical tests with teleelectrostimulation. The concept is primarily based upon nonlinear properties of cells walls [17]. From an engineer's point of view, if such nonlinearity exists it should be measurable not only in bioexperiments on the walls but also at the macroscopic scale. Thus, experiments with a possibility of detection of AM modulated signals in tissues *in vivo* and *in vitro* were performed, unfortunately no results were obtained. Then, similar experiments performed using frequency mixing gave positive results. So, it was assumed that a carrier wave modulated in amplitude with the stimulating signal should give stimulation results similar to that while using invasive methods (electrodes). Teleelectrostimulation is being intensively studied now. However, problems with the carrier wave frequency routinely arise. It was found that the most intensive results of the stimulation are frequency dependent and appear at 1420 MHz. This "mysterious" frequency is protected by the Radio Regulations. It has nothing in common with biomedicine, but was established for cosmic investigations protection. It includes hydrogen spectrum studies, which were done by observations of the Zeeman split of energy levels of hydrogen, the most common element in the universe. The same may be said regarding a living body. Here was a reason to prove telestimulation effects using this frequency. Although a large number of effects were found in a relatively narrow frequency band, one question remains unanswered: which mechanism here is more important—nonlinearity or hydrogen or both of them?

13.8 Final Comments

Selected problems related to accuracy of EMF quantitation in bioelectromagnetic studies have been briefly mentioned here. More detailed studies are found in the literature [18,19]. And it can be concluded that the accuracy is not revelational, and sometimes the results are of a qualitative rather than a quantitative character. It is not a result of bad will or dilettantism, but the result of conditions and factors that limits the accuracy. Regardless, doubts may be formulated upon very precise conclusions based on such approximate introductory data; for instance, while levels in protection standards are sometimes given with accuracy to four significant figures—what a way to measure—it still remains an unanswered question.

A slightly more optimistic conclusion may be drawn: even using inaccurate tools and methods, as discussed earlier in this chapter, more accurate results can be obtained. It requires better understanding of EMF problems and closer cooperation between biomedical and technical worlds.

This chapter is addressed mainly to biologists and medical doctors involved in bioelectromagnetics so as to direct their attention toward the physical and technical side of their experiments. Contrarily, in Section 13.7, as an illustration, are presented doubts and problems faced by engineers regarding an interpretation of biomedical, legal, or physical phenomena.

References

1. Bienkowski P., Trzaska H. 2015. EMF meters validation in situ. *Bulg. J. Public. Health* Vol. 7, No. 2(1), pp. 94–98.

2. Bienkowski P., Trzaska H. 2010. Non-stationary EMF standard. *Proceedings of 9-th International Symposium On EMC Europe 2010*, pp. 386–389.

3. Wacker P.F., Bowman R.R. 1971. Quantifying hazardous electromagnetic fields: Scientific basis and practical considerations. *IEEE Trans. Microw. Theory Tech.* Vol. 19, No. 2, pp. 178–187.

4. Youmans H.D., Ho H.S. 1975. Development of dosimetry for RF and microwave radiation-I: Dosimetric quantities for RF and microwave electromagnetic fields. *Health Phys.* Vol. 29, No. 2, pp. 313–316.

5. Kanda M. et al. 2000. International comparison GT/RF 86-1 electric field strengths: 27 MHz to 10 GHz. *IEEE Trans. Electromag. Comp.* Vol. 42, No. 2, pp. 190–205.

6. Frey A. 1970. On the SAR. *Bioelectromag. Soc. Newsl.* November, No. 10, pp. 1–3.

7. Justesen D.R. 1980. Response to Frey on SAR. *Bioelectromag. Soc. Newsl.* January, No. 20, pp. 1–6.

8. Frazer J.W. 1980. Letter to the editor. *Bioelectromag. Soc. Newsl.* January, No. 20, pp. 1–8.

9. Neil W.C. 1981. Proposed terminology system for radiofrequency dosimetry. *Bioelectromag. Soc. Newsl.* October, No. 26, pp. 1–6.

10. Trzaska H. 2005. Limitations in the SAR use. *The Environmentalist.* Vol. 25, No. 2, pp. 181–185.

11. Guy A.W. 1971. Analyses of electromagnetic fields induced in biological tissues by thermographic studies on equivalent phantom models. *IEEE Trans. Microw. Theory Tech.* Vol. 19, No. 2, pp. 205–214.

12. Bestrode R., Mild K.H., Nilsson G. 1986. Calibration of commercial power density meters at RF and microwave frequencies. *IEEE Trans. Instrum. Meas.* Vol. IM-35, No. 2, pp. 111–115.

13. Trzaska H. 2015. Engineering problems in bioelectromagnetics. In: *Biology and Medicine*, M. Markov Ed., CRC Press, pp. 69–78

14. 1982. *Bioelectromag. Soc. Newsl.* No. 30, p. 6. ELF Wave Antidote (after the April 1982 issue of OMNI).

15. 1994. *Proceedings of the COST 244 meeting on Electromagnetic Hypersensitivity*, Technische Universität Graz, Graz, September 26–27.

16. Trzaska H. 1994. Hypersensitivity? *Ibid.* pp. 106–117.

17. Arberg S.L., Adgimolaev T.A. 1980. Is this possible a detection of high frequency electromagnetic fields by nerve cell membranes (in Russian). *Elektronnaja Obrabotka Materialov* No. 1/1980, pp. 74–75.

18. Bienkowski P., Trzaska H. 2012. *Electromagnetic Measurements in the Near Field.* SciTech Publishing, Raleigh, NC.

19. Grudzinski G., Trzaska H. 2014. *Electromagnetic Field Standards and Exposure Systems.* ScitTech Publishing, Edison, NJ.

14

Blood and Vascular Targets for Magnetic Field Dosing

Harvey N.
Mayrovitz
*Nova Southeastern
University*

14.1 Introduction

14.1.1 Dose as a Factor in the Early Days

Horatio Donkin, a senior assistant physician with East London Hospital for Children at the time, writing in the *British Medical Journal* in 1878 (Donkin 1878), provides a number of remarks on magnetic therapeutics. These remarks were chiefly in response to claims made regarding magnet use by an English professor, Gamgee (Gamgee 1878), of the ability of metals and magnets to alter various neurological functions described by the famous French physician M. Charcot and others of the time. Among several targeted critiques offered by Donkin, the one that captures the essence of the issue was put by him in part as follows:

> ... the physiologist, being somewhat conversant with the infinite permutations and combinations of nervous and so called "functional" phenomena in the human organism, is, as a rule, aware of the complexity of the matter in hand, when these phenomena become the subject of a scientific investigation, and recognises, as physicians strive to do, the necessity of applying the most rigorous methods of research before arriving at any conclusion in so difficult a field of observation.

In the third edition of his text "Medical Electricity," Roberts Bartholow (1887) wrote about the "Physiological effects of magnetic applications" in which he takes the view that with respect to the application of a magnet to the skin "*The production of physiological effects, that can be recognized, is merely a matter of magnetic intensity*" or as we would state—a matter of dose—with the intrinsic—but unsubstantiated thought that bigger was better. In 1892, F. Peterson, a physician, and A. E. Kennelly, an engineer, combined their efforts to test this concept with experiments performed at the Edison Laboratory in Orange, NJ (Peterson and Kennelly 1892). They built a large iron electromagnet, reported to require two strong men to lift, that produced a field strength of 5000 G within the confines of its poles separated by 1.2 cm. Introduction of various substances within the pole space and the observation of the effects was the mode of investigation used. Human and frog blood placed on a slide within the pole space and observed microscopically "*failed to show the feeblest traces of polarization, movement or vibration.*" Similar observations of the frog's web blood circulation at field strength of 1500 G also showed no visible effects of the field. Contrastingly, when a 1–2 mA current was passed from toe to ankle, the circulation was arrested and restored when the current was removed. On the basis of these and a series of other experiments, the authors concluded "*... that neither direct nor reversed magnetism exerts any perceptible influence upon the iron contained in the blood, upon the circulation, upon....*" It was their view that "*The ordinary magnets used in medicine have a purely suggestive or psychic effect, and would in all probability be quite as useful if made of wood,*" a view that is echoed by many today. Despite this, they had an open mind as evidenced by the closing statement which stated that "*... we do not deny the possibility of there being invented someday magnets enormously more powerful than any yet known to us, which may produce effects upon the nervous system...*" this being a clear foreshadowing of the potential importance of dose at the target site as a factor. Indeed, others of the time had put forward the use of what we would call double-masked experiments which showed that magnets so applied could alter skin temperature and muscular contraction (Columbia and Richet 1880).

14.1.2 Blood and Vascular Targets

When considering possible blood and vascular impacts of electromagnetic field (EMF) stimulation, as is the main topic herein, there is good reason to believe that effects on cellular membranes and associated ion transport are involved since it is these processes that ultimately regulate function. The possibility of membrane transport modulation that is dependent on frequency, amplitude, or duration "windows" makes the search for such EMF–membrane transport interactions elusive, but this has still been considered by several investigators (Berg 1999). To provide a suitable and manageable template for discussion, it is our plan to mainly discuss blood and vascular targets that relate to potential EMF impacts on blood flow, with the components of Figure 14.1 useful in illustrating some major target components under consideration. When considering possible dose-dependent aspects (Markov and Colbert 2000; Colbert et al. 2008), it should be recognized that for static magnetic fields (SMF), dose mainly relates to field intensity effects at the target, whereas for pulsed electromagnetic fields (PEMF) it also includes frequency parameters and induced currents. Thus, we have chosen to consider and discuss separately the SMF and PEMF effects on targets that impact blood flow. Further, to help provide a contextual basis, brief overviews of some physiological basis and blood flow control involvement are provided as appropriate, for which Figure 14.1a serves well as an initial canvas.

Aside from EMF-related forces that act on red blood cells (RBCs), which are discussed in Section 14.3.1, blood velocity (cm/s) and flow (mL/s) changes are expected to result from EMF actions that cause changes in blood vessel diameters (D), since single vessel vascular resistance depends on $1/D^4$.

Physiologically, changes in diameter occur when ion channels within the membranes of vascular smooth muscle (VSM) or endothelial cells (ECs) change state to allow more or less entry or exit of selective ions that control the relaxation state of the VSM. Often, this state change depends on the level of the transmembrane potential that can "gate" open more or close more selective channels. Such interactions may involve a host of intermediary components and processes, but an important one is the flux of Ca^{++} through selective ion channels in which an inward Ca^{++} flux is facilitated by a partial depolarization of the membrane potential. When this occurs in VSM, the result is an increase in VSM contraction force reducing the vessel's diameter (vasoconstriction), as simply schematized in Figure 14.1b. Physiologically, these events can be caused by changes in the impulse rate of the nerves that innervate the blood vessel walls, since the amount of vasoactive substance (e.g., norepinephrione, NE; Figure 14.1b) released by the nerve terminals depend on the impulse rate. Contrastingly, if Ca^{++} influx is increased in ECs, biochemical processes are triggered that include an increase in the release of nitric oxide (NO) that diffuses to VSM causing these cells to be less contracted (more relaxed). This generally results in an increase in the vessel's diameter (vasodilation), as schematized in Figure 14.1c. This EC-initiated effect depends on the EC synthesis and release of vasoactive substances such as NO. NO release is induced by chemical and physical stimuli including acetylcholine (Ach) and adenosine triphosphate (ATP) and mechanical events such as wall shear stress. NO is synthesized from L-arginine in the presence of constitutive NO synthase (cNOS), which is Ca^{++} dependent, and inducible NO synthase (iNOS), which is dependent on the presence of cytokines and other substances. NO diffuses to VSM where it activates cytoplasmic guanosine triphosphate (GTP) that then increases cyclic guanosine monophosphate [cGMP], causing a decrease in Ca^{++} entry and a decrease in cytosolic free Ca^{++} in VSM, thereby resulting in relaxation.

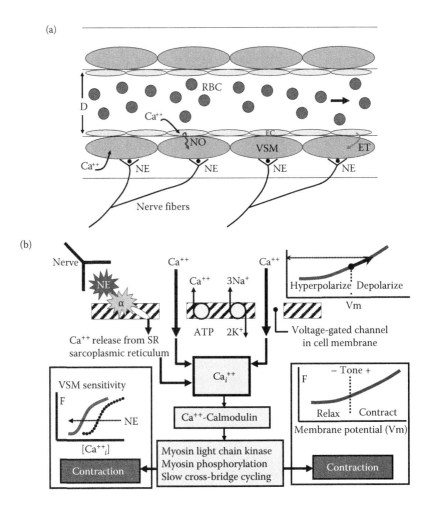

FIGURE 14.1 (a) Simplified schema of potential vascular-related targets. RBC, red blood cells; *D*, vessel diameter; VSM, vascular smooth muscle; EC, endothelial cells; NO, nitric oxide (a potent vasodilator); ET, endothelium (a potent vasoconstrictor); NE, norepinephrine (released from many sympathetic nerves, main peripheral vessel action is vasoconstriction). (b) Overview of some calcium-related contraction mechanisms as potential targets. VSM resting membrane potential (Vm) is mainly determined by high membrane K^+ permeability and outward K^+ current. Internal $[K^+]$ is maintained by active $3Na^+$-$2K^+$-ATPase pump. More Ca^{++} enters through voltage-gated channels if cell membrane depolarization occurs. Additional Ca^{++} entry is also triggered by ligand-gated channels such as that caused by the release of NE from nerve terminals acting on α-receptors. Ca^{++} release from SR stores makes more Ca^{++} available and promotes vasoconstriction, whereas Ca^{++} removal by the SR and membrane pumps reduces Ca^{++} availability promoting less vasoconstriction (vasodilation). Sensitivity of VSM contractile machinery to Ca^{++} concentration $[Ca^{++}]$ increases with increased [NE] enhancing VSM contraction force F.

(Continued)

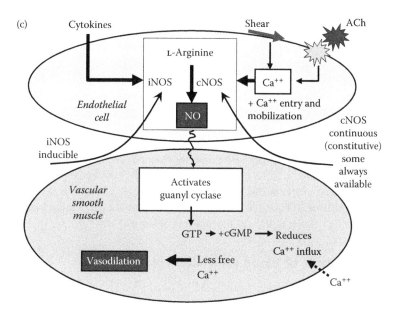

FIGURE 14.1 (CONTINUED) (c) Overview of some EC-related mechanisms as potential targets. If Ca⁺⁺ influx is increased in ECs, biochemical processes are triggered that include an increase in the release of NO that diffuses to VSM causing these cells to be less contracted (more relaxed), which generally results in an increase in the vessel's diameter (vasodilation). This effect depends on EC synthesis and release of NO induced by chemical and physical stimuli including Ach and wall shear stress. NO is synthesized from L-arginine in the presence of constitutive nitric oxide synthase (cNOS) that is Ca⁺⁺ dependent and inducible nitric oxide synthase (iNOS) that depends on cytokines and other substances. In VSM, NO activates cytoplasmic GTP, which then increases cGMP, causing a decrease in Ca⁺⁺ entry and a decrease in cytosolic free Ca⁺⁺ in VSM, resulting in relaxation.

14.2 Blood Flow Modulation by Electromagnetic Fields: General Considerations

Blood flow changes associated with EMF stimulation may occur in two generalized ways: one is related to potential blood property changes and the other to vascular and hemodynamic changes.

14.2.1 Blood Property Changes

These would include modification of blood's properties that subsequently decrease or increase blood flow under the same hemodynamic driving forces. Examples of such changes would be alterations in blood viscosity, RBC orientation, or RBC physical or chemical properties.

14.2.2 Vascular and Hemodynamic Changes

These would include modification of blood vessel diameters and other properties that either directly affect blood flow via alterations in vascular resistance to blood flow and/or cause changes in keep the full text blood pressure and not abbreviate that directly affect blood flow. With reference to Figure 14.1a, examples include changes in VSM membrane potentials that cause alterations in channel ionic currents (e.g., Ca^{++}), resulting in vessel diameter (*D*) changes. Increased amounts of Ca^{++} entry cause vasoconstriction. Another example would be EMF-related changes of nerve membrane potentials that cause changes in vasoactive substance release, for example, NE, thereby affecting vessel diameter and blood flow. A third example would be EMF-related changes in EC membrane potentials and currents that would alter their release of vasoactive compounds such as the vasodilator NO and the vasoconstrictor endothelin (ET). These processes in part depend on Ca^{++} fluxes in ECs, as discussed in Section 14.1.2.

14.2.3 Magnitude of Blood Flow Change

In assessing potential cellular or vascular EMF-related effects on blood flow, a key question is if and how much change (increase or decrease) occurs with an applied SMF or PEMF. It is not useful, in our opinion, to consider or search for mechanisms or dosing issues in which such blood flow changes have not been demonstrated. So, our approach is to first examine what is known regarding blood flow changes. Further, because "healthy" or "sick" targets may show different outcomes depending on dose or other factors, EMF-related impacts that are present or not present in healthy tissue may not apply to "sick" or otherwise traumatized tissues. This is a concept that should always be borne in mind.

14.3 Blood Flow Modulation by Electromagnetic Fields: What Do We Know?

14.3.1 Direct Force-Related Effects in Blood Vessels

Blood flow is partly the movement of iron-containing RBCs and various ions within and together with electrically conducting blood plasma. The electrically conducting blood flowing in a magnetic field experiences an electromotive force that induces currents in the blood. In turn, these currents produce a magnetic field that perturbs the originally applied field and also causes an electromotive force that affects the blood flow itself. If the applied field is an SMF applied at right angles to the flow, then the effect is reduced blood flow. One physical aspect thought to be causing this is an alteration in the velocity profile caused by the SMF. In earlier theoretical and experimental work, effort was directed to characterizing changes in steady laminar flow for which velocity profiles were caused to shift from parabolic to a more blunt profile (Shercliff 1953). The profile shift and thereby the effect on flow depends on a parameter denoted as $M = \mu H r (\sigma/\eta)^{1/2}$, in which μ is permeability, *H* is the applied field, *r* is half the flow channel width, and σ and η are the fluid conductivity and viscosity. Calculations and some experimental data indicate an increase in flow resistance with increasing values of *M*. The magnitude of this effect has been further explored theoretically and experimentally for a range of SMF doses. For example, a magnetic field (**B**) applied 90° to the velocity (**v**) of flowing blood generates an electric field (**E**) that is orthogonal to both **v** and **B** with a value determined as $E = v \times B$. So, potential differences (φ) between opposite walls of a blood vessel with diameter *D* ($\varphi = ED$) are directly related to **v**, and this forms a basis of a blood flow measurement (Kolin et al. 1957). The relationship $v \times B$ also shows that the blood-flow-reducing force is dependent on the magnitude of the blood velocity, which in humans is greatest in the aorta. Theoretical analyses using Navier–Stokes equations have predicted a 10% reduction in aortic blood flow due to a 5 T SMF (Chen and Saha 1984), whereas simulated fields in an arterial model (Sud and Sekhon 1989) predicted a magnetic field-related reduction in ascending aorta blood flow of 1.3% for a 12.5 T field at the target. Blood flow simulations with 15% saline solutions (Keltner et al. 1990) exposed to SMF also showed reduced flow, but by <1% if exposed to 2.35 T and 2%–3% if 4.7 T.

A theoretical analysis that takes into account the nonzero electrical conductivity of vessel walls and tissues (Kinouchi et al. 1996) indicates a dose-dependent reduction in aortic blood flow (*Q*) approximately as the square of the applied magnetic field which, based on analysis of their data, can be represented as %*Q* reduction = $0.052B^{1.973}$ in which *B* is in Tesla. It is noteworthy that when wall and tissue conductivity is included in a more complete analysis, the calculated SMF flow reduction is somewhat greater than if these conductivities are considered to be zero. This is in part due to a greater magnetically induced force tending to retard blood flow. On the basis of some of these findings, it appears that only in blood vessels with the highest blood velocities will there be magnetic forces sufficient to reduce blood flow and then only when sufficiently high SMF are applied. Thresholds for 5% and 10% reductions in aortic blood flow appear to be in the range of 10.1 and 14.4 T, respectively (Kinouchi et al. 1996), as would be determined from the aforementioned equation.

A component not specifically included in the earlier theoretical models and analyses were RBC magnetic properties. These arise in part due to the hemoglobin–iron–oxygen complex and electronic spin configurations (Nakano et al. 1971), which in the presence of an applied magnetic field (*H*) result in an induced RBC magnetic dipole per unit volume (*m*). This is associated with an RBC magnetic susceptibility ($\chi = H/m$) that is diamagnetic ($\chi < 0$) if hemoglobin is oxygenated and paramagnetic ($\chi > 0$) if deoxygenated. One consequence is that an RBC experiences a force/volume (**F**) that affects its motion in the presence of applied magnetic fields. This force can be expressed in various forms (Watarai and Namba 2002; Bor Fuh et al. 2004; Han and Frazier 2004; Haverkort et al. 2009), one being $F = \chi H \nabla B$ in which χ is most accurately expressed as the difference between χ_{RBC} and χ_{PLASMA} (Zborowski et al. 2003). If there was spatial variation only in the *z*-direction, then this simplifies to $F = \chi H \, \partial B / \partial z = (\chi/\mu) B \, \partial B / \partial z$. Such magnetically based forces cause RBCs to experience a torque ($\tau = m \times B$), which tends to align the disk face plane parallel to the applied field direction. This rotational movement results in relative motion between RBCs and plasma that causes frictional energy loss and thereby an increase in the effective viscosity of the blood in an apparent dose-dependent manner based on a novel electrohydraulic discharge test (El-Aragi 2013). These RBC rotational features may not be exclusively dependent on hemoglobin presence or oxygenation state since experiments indicate that RBCs devoid of hemoglobin when suspended in solution orient themselves with the plane of their discoid axis aligned with the applied field (Suda and Ueno 1994). The alignment is dose-dependent over reported ranges from 2 to 8 T (Suda and Ueno 1994) and between 1 and 4 T (Higashi et al. 1993), with the rotational torque accounting for this alignment theorized to be caused by a membrane-dependent directional anisotropy of χ that favors the observed orientation (Higashi et al. 1993; Suda and Ueno 1994; Takeuchi et al. 1995; Higashi et al. 1997). However, as opposed to these effects observed when RBCs were suspended in phosphate buffer solutions, when whole blood was used and similarly exposed to fields ranging from 2 to 8 T, essentially no orientation with the applied field was observed (Suda and Ueno 1996). Despite these somewhat divergent observations, perhaps partly due to oppositely directed influences of transmembrane proteins and the lipid bilayer (Higashi et al. 1993), the fact that SMFs do increase blood's effective viscosity and decrease blood flow has been experimentally demonstrated (Haik et al. 2001; Yamamoto et al. 2004; El-Aragi 2013). The viscosity increase is dose-dependent, reported as 3.7% for 1.5 T exposure (Yamamoto et al. 2004), about 8% for 3.0 T (Haik et al. 2001), and near 20% for 5 T (Haik et al. 2001). However, the magnetic effect on viscosity, at least at 1.5 T, appears to depend on the shear rate with which the blood is flowing with only a 0.3% increase observed at a shear rate of 130 s^{-1} (Yamamoto et al. 2004) with 2.6% and 3.7% increases at shear rates of 49.4 and 18.3 s^{-1}. Average shear rate (γ) in a circular blood vessel with laminar steady flow is calculable from average blood velocity (V) divided by the blood vessel diameter (D) as $\gamma = 2V/D$ or in terms of volumetric blood flow (Q) as $\gamma = (8/\pi D^3)\,Q$. Thus, the magnetic effect depends on both the magnitude of the flow and the diameter of the vessel in which the flow is occurring. Applying this concept to the human vasculature, it is the veins that normally experience the lower shear rates and would most likely be vulnerable to the abovementioned magnetic effects. Calculations based on measured blood velocities and vessel lumen areas in superior vena cava (Wexler et al. 1968), femoral vein (Willenberg et al. 2011), inferior vena cava (Wexler et al. 1968), jugular and vertebral veins (Ciuti et al. 2013), and saphenous vein (Abraham et al. 1994) reveal the following calculated average shear rates: 8.2, 13.2, 16.5, 35.2, 53.9, and 56.6 s^{-1}, respectively. Thus, the low venous shear rates combined with the deoxygenated state of venous blood suggest that the greatest magnetic effect of the type described would be detectable in the postcapillary vasculature. As far as this author knows, this avenue of research has yet to be specifically undertaken.

14.3.2 Effects in Smaller Blood Vessels of Animal Models

Normal blood flow conditions in smaller blood vessels, including the microcirculation, differ from that in larger vessels. The net flow direction in a specific artery or vein is generally known. Contrastingly, smaller arterioles, capillaries, and venules within tissues and organs tend to exist as networks with variable and possibly near-random orientations among flow directions (Mayrovitz et al. 1975, 1976, 1987, 1990; Mayrovitz 1992). Thus, it is unlikely that magnetically related vector forces as discussed in the previous section would have much of a net impact on the flow. However, if applied fields affect blood or tissue components that directly or indirectly alter vascular diameters, then the variable directional aspect is no longer very important. So with this in mind, we ask two broad questions: (1) what is the specific evidence of EMF-related changes in small blood vessels or tissue blood flow and (2) what might be mechanisms involved?

Placement of rats into the bore of a superconducting SMF of 8T for 20 min was used to assess pre- and postexposure effects on blood velocity in small venules (20–40 μm) of the skin (Ichioka et al. 1998) via microscopic observations using the skinfold chamber with velocity measured using the dual-slit method (Intaglietta et al. 1975). Compared to preexposure, the first minute of postexposure showed a 17% increase, which the author speculated was due to an exposure-related blood flow reduction with a subsequent post-exposure transient hyperemia. The apparent hyperemia lasted about 10 min and was not present in sham-exposed animals. The interpretation of this data is complicated by the fact that the entire animal was exposed to the field, so the effect, if real, could be due to flow reduction in the large vessels, with the observation in small venules being a reflection of such an upstream event.

Possibly arguing for a direct effect is the fact that the venules were observed to be dilated after exposure. But even this might be explained as a response to a prior blood flow decrease. Subsequent measurements by these authors (Ichioka et al. 2000) using laser Doppler methods showed a slight reduction in blood perfusion during the 20-min exposure but no associated hyperemia, thereby leaving the question still open. An associated skin temperature decrease during exposure may be related to factors other than blood flow reduction (Ichioka et al. 2003) and in fact may have been the cause of the small blood flow reduction previously reported (Ichioka et al. 2000). However, later work using a similar dorsal skinfold chamber to observe the skin skeletal muscle microcirculation demonstrated an apparent dose-dependent blood flow effect (Brix et al. 2008) when only the chamber, and not the whole body, was exposed to varying SMF intensities. Analysis of video recordings of 24 capillaries for each of the 10 awake hamsters obtained during SMF exposure showed a dose-dependent decrease in capillary RBCs velocity with an apparent threshold for change at near 500 mT at the target. This velocity reduction occurred within 1 min of exposure and was associated with no measured change in capillary diameter, functional capillary density, or systemic blood pressure. Quantitatively, an average preexposure capillary blood velocity of about 0.37 mm/s was reduced to 0.22 mm/s (40%) at a target field intensity of 587 mT. This reduction could be reversed by reducing or eliminating the field intensity. In interpreting these important findings, we suggest that the absence of a diameter change would be expected since capillaries do not have VSM. With no capillary diameter change, the already mentioned velocity change represents a hefty 40% flow reduction. Subsequent measurements of diameters of arterioles feeding the capillary network showed no significant SMF-related change. This finding would seem to negate the possibility that the observed decrease in capillary flow could be attributed to SMF-related arteriolar vasoconstriction. However, there is a caveat to this that depends on the accuracy with which such arteriolar diameters are measured, given the inverse fourth power dependence of vascular resistance on diameter. Whether the observed reduction is due to SMF-related increases in blood viscosity or orientations as previously described or due to RBCs displacement in inhomogeneous fields (Okazaki et al. 1987) is as yet unknown. A mathematical model analysis that includes a drag force due to interactions between a central core of flowing RBCs and a cell-free periphery plasma indicates a dose-dependent effective viscosity increase and corresponding flow decrease in small vessels that increases with increasing field gradient (Bali and Awasthi 2011). This analysis requires a number of assumptions; so whatever the mechanism turns out to be, it appears to not occur in a measurable way for field intensities less than about 500 mT (Brix et al. 2008).

A direct effect of SMF on skin microcirculation might be inferred from studies on rabbit ear microvessel vasomotion (Okano et al. 1999). The rabbit ear circulation was exposed to an SMF of about 1 mT for 10 min and effects of infused NE or Ach, alone or with the magnetic field present, were evaluated. It was found that in the absence of the SMF, NE caused vasoconstriction and Ach caused vasodilation, but with the SMF, the prior NE vasoconstriction was blunted or blocked and Ach vasodilation was blunted or blocked. The authors called this a biphasic response. But it was not possible to assess the direct effect of the SMF on microvascular blood flow without infusing vasoactive substances that modified vascular tone. Hence it is not known if the SMF by itself causes a change (increase or decrease) or not. Without pharmacological modification, an SMF with doses of 1, 5, and 10 mT caused both increased and decreased vasomotion within 10 s of SMF activation (Ohkubo and Xu 1997) independent of dose. On the basis of these and further experimental findings in animals (Xu et al. 1998; Ohkubo et al. 2007), an interpretation is that there may be an SMF-induced vasodilation if a high vascular tone is present and vasoconstriction if a low vascular tone is present. An important note on the above observations is that they are based on a measurement that depends on the relative optical density of the microscopic field being viewed. The method, known as microphotoelectric plethysmography (MPPG) (Asano et al. 1965) senses optical density changes of the entire microscopic field being viewed and uses the magnitude of these as an index of vasodilation and the standard deviation of these changes as an index of vasomotion. It does not directly measure blood flow nor is it certain whether the vasodilation is due to the venous or arteriolar vessels. Application of this method to assess changes in the rabbit ear circulation when exposed to a 0.25 T SMF indicated an increase in MPPG that occurred about 20 min after field application (Gmitrov et al. 2002). This was interpreted as an increase in blood flow. Whether or not these changes actually indicate a blood flow change, the MPPG measurement that was made on tissue exposed to a much higher B-field dose than used in earlier studies (Ohkubo and Xu 1997; Okano et al. 1999) revealed the absence of a prior-reported vasomotion-like feature and a delay of about 20 min in the B-field response, further pointing out the relevance of dose in a potential physiological response. Bidirectional changes in arteriolar tone have also been noted to occur almost immediately in rat spinotrapezius microvessels when an exteriorized preparation was exposed to a 70 mT SMF (Morris and Skalak 2005). Further discussions on electromagnetic differential field effects on microcirculation and with respect to wound healing processes have recently become available (Mayrovitz 2015; Ohkubo and Okano 2015).

14.3.3 Effects on Vascular Cells

Two main vascular cell types to be considered as potential targets for SMF-related blood flow modulation are ECs and VSM cells. As schematized in Figure 14.1, ECs pave the inner surface of all blood vessels and are secretors of various vasoactive substances that may cause vasodilation or vasoconstriction. For example, vasodilation may be caused by release of NO and vasoconstriction is caused by the release of ET. VSM is present in most blood vessel types, and when VSM is caused to contract there is generally a reduction in blood flow, whereas if it relaxes vasodilation occurs and blood flow increases. The state of VSM is determined by multiple factors, but one of the common denominators is Ca^{++} entry or exit from EC and VSM via ion channels. For example, release of NO from EC and its vasodilatory effects on VSM is mediated by Ca^{++} entry into EC. Conversely, Ca^{++} entry into VSM is associated with increased vasoconstriction and blood flow reduction. In both cell types, transcellular Ca^{++} flux is largely determined by the cell's membrane potential that in turn controls the ion channel's gate state. Shifts in the membrane gate state depend on conformational changes in membrane-spanning proteins that require force changes that may in part depend on the interaction between ion drift currents and an applied field (St. Pierre and Dobson 2000; Hughes et al. 2005). Given this factual physiological scenario, we may ask the following question: To what extent is there a connection between applied SMF and altered states of either cell type that would suggest an SMF-related blood flow modulation?

14.3.3.1 Endothelial Cells

Low-dose SMF (0.03–0.12 mT) affects proliferation rates of cultured human umbilical vein EC when administered for 24 h, but has no apparent effect on NO concentration (Martino et al. 2010; Martino 2011). Pulsed magnetic fields (PMF; 15 Hz and 1.8 mT) acting on isolated cardiac ECs also indicated increased cell proliferation (Li et al. 2015), and SMF may play a role in arteriogenesis (Okano et al. 2007, 2008), but there appears to be no experimental work directly showing a link between an applied SMF and altered vasoactive substance release from ECs. Given the current availability of a variety of EC types it would seem that focused experimental work directed toward this area would be useful.

14.3.3.2 Vascular Smooth Muscle Cells

Virtually no experimental work has been systematically carried out in which the responses of cultured VSM cells to either SMF or PMF have been evaluated and reported. In view of the important role played by VSM in the control of blood circulation, this appears to be an important target for future research. However, it has been shown that gradient SMF can alter frog sciatic nerve c-fiber conduction in a dose-dependent manner (Okano et al. 2012), with a slight but significant reduction at an exposure of 0.7 T but not at 0.21 T. If translatable to sympathetic nerves that innervate blood vessels, this could imply a potential vascular effect. However, more work in this area is importantly needed. Other vascular effects are potentially subsequent to alterations in circulating vasoactive substances that in turn may impact VSM. An example of this is the decreased circulating levels of angiotensin II and aldosterone of hypertensive rats exposed to whole body SMF. The normal progression of hypertension with age was blunted when they were continuously exposed to SMF of 10 and 25 mT apparently due to reduced levels of both hormones (Okano and Ohkubo 2003). Both of these hormones cause vasoconstriction mediated in part through actions on VSM. Additional vascular-related processes appear to be involved when baroreceptors (BR) in the carotid sinus (CS) of rabbits are exposed to 0.35 T SMF. Bilateral CS exposure was associated with a decrease in mean blood pressure and an increase in an index of blood flow in the rabbit ear (Gmitrov and Ohkubo 2002). These changes occurred in the absence and presence of intravenously administered Ca^{++} channel blocker verapamil. However, an SMF-induced increase in BR reflex sensitivity present without intervention was absent in the presence of verapamil administration. This suggests that the SMF effect on blood pressure and BR sensitivity, which is a measure of heart rate change caused by pharmacologically induced small blood pressure changes, are dependent on Ca^{++} flux processes with multiple possibilities at mechanisms and sites of action.

14.3.4 Effects on Human Circulation

There are two broad aspects to consider when addressing the issue of SMF effects on human circulation. One relates to the exposure of the whole body or specific functional targets such as the CS area and the other relates to assessing direct effects on circulation of specific tissues such as skin or muscle.

Whole body 30-min exposure within an MRI in which only the SMF was active (1.5 T) increased NO metabolites (nitrite, NO_2^- + nitrate NO_3^-) by about 18% in 33 young men with no change when the field was not activated in a separate group (Sirmatel et al. 2007). The possible significance of this finding goes to the expansive role of NO as a cardiovascular modulator. For example, animal experiments indicate that an SMF produced by a small magnet implanted near the CS (180 mT) for 6 weeks adds to the expected increase in plasma NO that would be produced by a calcium channel blocker (Okano and Ohkubo 2006). Although the SMF resulted in a slightly reduced blood pressure in comparison to sham-treated rats, the blood pressure reduction was greatest with the combined SMF and Ca^{++} blocker. Since blood pressure reductions were experienced without changes in skin blood flow, the blood pressure reduction in this case may not be due to peripheral vasodilation. An absence of a skin blood flow effect despite a blood pressure reduction was also seen when spontaneously hypertensive rabbits (SHRs) were housed in cages for 12 weeks while being exposed to a 5-mT SMF (Okano et al. 2005). Contrastingly, it has been argued that, at least in rabbits, there is an increase in peripheral vessel sensitivity to NO associated with SMF actions on carotid BRs (Gmitrov 2013). Concentric magnets appear to have no significant effect on forearm blood flow in young adults (Martel et al. 2002).

14.3.5 Human Skin Circulation as a Target

Despite the ability to noninvasively assess skin blood flow with the potential to systematically expose this tissue to various field intensity doses, human skin circulation as a target has been systematically studied by only a few investigators using SMF (Mayrovitz et al. 2001, 2002, 2005; Mayrovitz and Groseclose 2005; Li et al. 2007; Yan et al. 2011) and PMFs (Ueno et al. 1986; Mayrovitz and Larsen 1992, 1995; Mayrovitz et al. 2002; Wenzel et al. 2005; McNamee et al. 2011; Kwan et al. 2015). The experimental data and concepts that have emerged from these and related investigations is worthy of further in-depth characterization in that it may serve to well illustrate and amplify various aspects not previously reported to a similar level of detail. In the following subsections these are categorized by static and pulsed magnetic fields separately.

14.3.5.1 Static Magnetic Fields and Human Skin Circulation

A possible skin blood flow effect due to an SMF of a ceramic magnet placed on forearm skin was assessed in a group of young persons as depicted in Figure 14.2a. The magnet, which

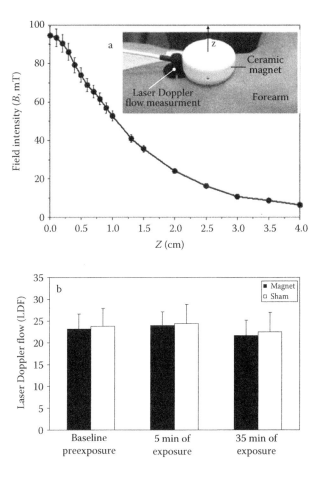

FIGURE 14.2 Measurement of forearm skin blood flow with and without an SMF. (a) A ceramic magnet with a B-field intensity as shown in the figure or a sham magnet was placed on the forearm of a group of 12 healthy subjects for 35 min and the laser Doppler flow (LDF) within 5 mm distal to the magnet edge monitored continuously. (b) Results show mean and SEM for LDF that demonstrated no difference between sham and magnet exposed.

had a surface B-field intensity of 95 mT, was in place for 35 min and was compared to the effects of a sham similarly placed (Mayrovitz et al. 2002). As shown in Figure 14.2b, the net result of this evaluation indicated no significant difference between LDF values associated with exposure to sham or to magnet. Another method to assess skin blood flow changes is by using laser Doppler imaging (LDI). In this method, instead of placing a probe directly on skin, a laser beam scans skin areas of interest to determine LDI blood flow (Mayrovitz and Carta 1996; Mayrovitz and Leedham 2001). This method was used to assess possible SMF blood flow changes when one index finger was exposed to a magnet with B-field properties as shown in Figure 14.3, while the other index finger was exposed to a sham and various LDI scans run (Mayrovitz et al. 2001), as illustrated in Figure 14.4a through c. Over an exposure

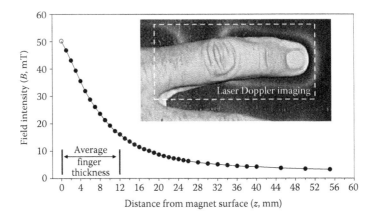

FIGURE 14.3 Laser Doppler imaging setup of finger dorsum exposed to magnetic pad. Index finger is shown lying on a rectangular magnet with the B-field intensity as shown in the figure. The finger dorsum blood flow is measured to a depth of about 1.5–2.0 mm via scanning the finger surface over a time span of about 4 min per scan. In practice, one index finger lies on a magnet and the other on a sham, which are both scanned simultaneously as shown in the figure. Over the blood flow measurement depth, the B-field gradient is ~4 mT/mm.

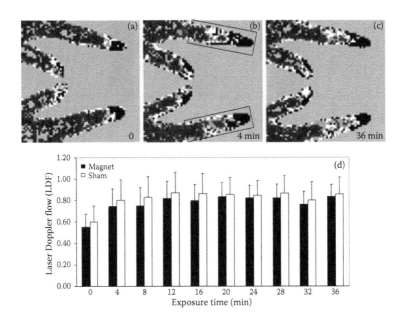

FIGURE 14.4 LDI scan results for magnet- and sham-exposed index fingers. (a) through (c) show an example scan for baseline (a), after 4 min of exposure (b), and after 36 min of exposure (c). In this example, the upper finger is exposed to the magnet. In part (d), the summarized results (mean + SEM) are shown for the evaluated group, indicating no significant difference in LDI flow between sham or magnet exposed.

interval of 35 min there could be demonstrated no significant difference when exposed to sham or magnet with a surface B-field intensity of about 50 mT and a field gradient within the depth at which flow was measured d*B*/d*z* of about 4 mT/mm (Figure 14.4d).

An extension of this work was to test the effect of an increased field intensity dose using neodymium magnets with surface B-fields near 0.4 T (Mayrovitz and Groseclose 2005). For this purpose, finger dorsum LDF flow was measured simultaneously in finger 2 (F2) and finger 4 (F4) of the nondominant hand while F2 rested on a magnet and F4 rested on a sham as shown in Figure 14.5a. Exposure was for 30 min with an average finger thickness at the LDF flow site being 12 ± 1 mm (mean ± SD), which resulted in an average B-field dose at the F2 measurement site of 0.88 ± 0.05 T. Evaluations made in a group of 12 young healthy adults indicated a statistically significant decrease in LDF flow in the SMF-exposed finger as compared to the sham exposed finger with a pattern as shown in Figure 14.5b. The overall percentage reduction in LDF flow at F2 compared to pre-magnet exposure was near 18% as assessed after 30 min of exposure. Parenthetically, an LDF flow reduction occurred independent of the pole (north or south) facing the skin. This appears to be the first reported finding of an SMF-related reduction in average skin blood flow.

It may be reasoned that if an SMF can affect average flow then it might also have an impact on vascular responses to perturbations. This aspect was assessed in several ways

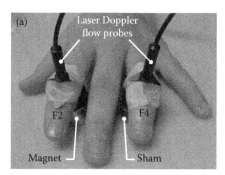

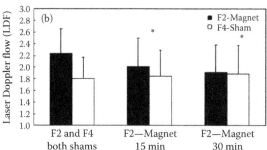

FIGURE 14.5 Simultaneous LDF measurements with neodymium magnet exposure. (a) Setup for measuring finger dorsum LDF flow simultaneously in fingers F2 and F4 of the nondominant hand. F2 was supported by a magnet with surface field of 0.4 T. Average measured B-field at F2 dorsum target was 0.88 T. (b) Results (mean + EM) for 15- and 30-min SMF exposure in a group of 12 young adults indicate a statistically significant decrease in LDF flow as compared to the sham-exposed finger.

by evaluating the SMF effect on the LDF flow responses to inspiratory gasps (IGs) and to vascular occlusions. The inspiratory gasp vascular response (IGVR) relates to the transient reduction in skin blood flow that accompanies a rapid inspiration (IG). It is believed that this response is triggered by afferent sympathetic impulses to skin arterioles during rapid thorax and lung enlargement (Mayrovitz and Groseclose 2002). The concept to be tested using this perturbation is that an SMF alters the vasoconstriction response either by interfering with the afferent nerve traffic to the arteriole or by somehow inhibiting the VSM's response.

This was tested using an experimental setup as shown in Figure 14.6 in which LDF flow was measured simultaneously on the third finger dorsum of one hand that was exposed to a magnet while the other hand's third finger was exposed to a sham magnet. After an interval of 20-min exposure of both fingers to sham, a series of IGs were performed. Thereafter, one finger was exposed to the SMF of a magnet and the other to the sham for an additional 20 min and a series of IGs were again done. An example of the typical response for a single IG is shown in Figure 14.7. The characteristic feature of the LDF flow response is a substantial reduction in flow in both fingers with a return to pre-IG levels. The response is quantified by calculating the percentage reduction in flow and this is termed the IGVR. The net effect of this procedure on a group of 24 subjects (Mayrovitz et al. 2005) demonstrated essentially no difference between IGVR whether exposed to sham or magnet as illustrated in Figure 14.8. In retrospect, the inability to demonstrate a significant impact of the SMF on IGVR may in part have been due to the intensity of B-field at the finger dorsum LDF-measuring site, which for the previously described

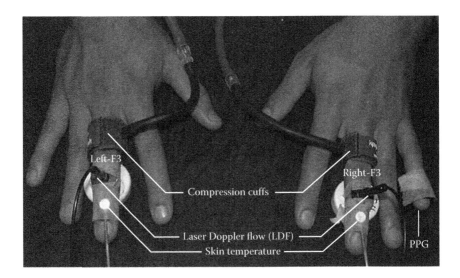

FIGURE 14.6 Experimental setup to evaluate SMF affects on vascular responses. LDF is measured simultaneously on the third finger dorsum while fingers rest on either a magnet or a sham. The pulse is continuously monitored with the photoplethysmography (PPG), and finger skin temperatures are monitored with the thermocouple placed just distal to the flow probes. The compression cuffs around the base of the fingers are used to produce a full blood flow occlusion.

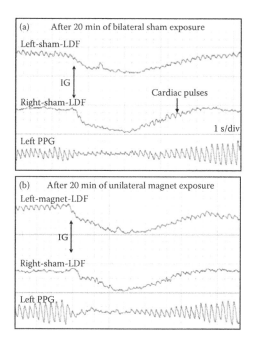

FIGURE 14.7 Example blood flow responses to IGs. (a) Bilateral IG responses after both middle fingers exposed to shams for 20 min. (b) Bilateral IG responses after left finger exposed to magnet and right exposed to sham for 20 min. There is essentially no difference between sham and magnet exposed in this subject. IG, inspiratory gasps; LDF, low Doppler flow.

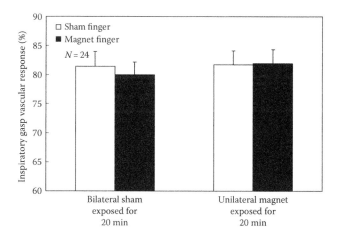

FIGURE 14.8 Composite results for inspiratory gasp vascular responses (IGVR). The IGVR is the percentage reduction of the mean LDF flow caused by a rapid inspiration. The figure shows the overall mean + SEM IGVR for 24 healthy subjects first exposed bilaterally to sham and then unilaterally to a magnet with the field properties similar to that shown in Figure 14.1a. There was no difference in vascular responses whether exposed to sham or to magnet.

study averaged 31.5 mT. This author believes that the use of now available higher intensity magnets to clarify this issue would be a worthwhile future endeavor.

In contrast to a possible inhibition of a vasoconstriction response of IG, the question as to the effects of an SMF on transient vasodilatory responses subsequent to an ischemic event was investigated using the experimental setup also as shown in Figure 14.6, but now simultaneously inflating both finger cuffs to produce a 5-min full occlusion as illustrated with the response shown in Figure 14.9. Ordinarily, when tissue experiences an interval of blood flow and oxygen deprivation, which in this case was for 5 min, there is a subsequent immediate transient hyperemia that depends on the vasodilatory state of the tissue when the occlusion is released. The results of a series of assessments, of which Figure 14.9 is typical, suggest no significant effect on this hyperemia response as a result of 24 min of SMF exposure to the B-field intensity used.

However, experiments studying the effects of an SMF on spectral content of LDF blood flow may indicate subtle changes in properties without measurable alterations in average flow. Such was the result reported in some experimental animal studies (Li et al. 2007; Xu et al. 2013) and has been further investigated in human skin as depicted in Figure 14.10. This represents an experimental setup to assess LDF spectral content changes when the middle part of the third finger of a hand rests first on a sham magnet for 15 min and then on a 0.4-T surface field magnet for 30 min. Such measurements suggest a change in spectral content with an indication of a trend for decreasing mean LDF flow associated with the magnet as illustrated in Figure 14.11. Such SMF-dependent changes in the spectral content of skin blood flow represent an interesting parameter for further study, but at this time the presence of a significant SMF effect remains an open question.

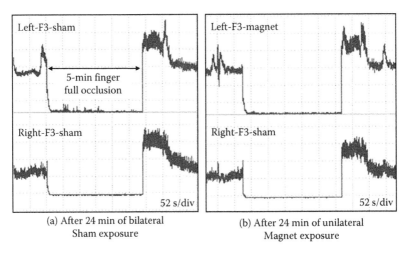

(a) After 24 min of bilateral   (b) After 24 min of unilateral
Sham exposure   Magnet exposure

FIGURE 14.9   Blood flow responses after full vascular occlusions using the experimental setup of Figure 14.6 (a) Both fingers exposed to sham magnets (b) Left finger exposed to magnet for 24 minutes prior to the 5-minute occlusion and right finger simultaneously exposed to sham magnet. The hyperemic response is little affected by the magnet exposure. This example response was similar to that found in a group of 20 subjects.

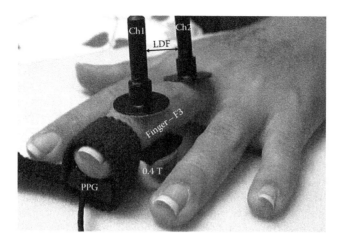

FIGURE 14.10 Experimental setup to assess LDF spectral content changes. Middle part of the third finger of nondominant hand rests on either a 0.4-T magnet or a sham while laser Doppler blood flow and PPG are continuously recorded at the magnet exposed finger dorsum (ch1) and on the same finger proximal to it (ch2). Resultant LDF signal is subjected to spectral analysis as illustrated in Figure 14.11. LDF, Laser Doppler Flow.

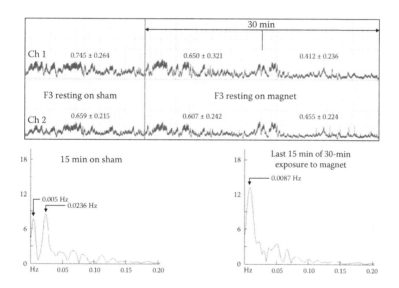

FIGURE 14.11 Example LDF flow recording and associated frequency spectrum. F3 initially rests on sham for 15 min, and then the sham is replaced by a 0.4-T magnet, which in this example resulted in a measured 0.09 T at the dorsum LDF measurement site. The similarity in Ch 1 and Ch 2 LDF temporal patterns is evident. The substitution of the sham for the magnet seems to indicate a shift in the spectral content. Numbers in Ch 1 and Ch 2 indicate the LDF average during each of the consecutive 15 min intervals (mean ± SD).

One aspect to be considered in such investigations is the possibility of spectral components linked to specific physiological processes. For example, applications of spectral and wavelet analyses suggest that there are at least five bands that present when recording skin blood flow with LDF (Kvernmo et al. 1998; Kvandal et al. 2006). These bands include those due to EC activity (0.0095–0.02 Hz), neurogenic activity (0.02–0.06 Hz), myogenic activity (0.06–0.15 Hz), and respiratory activity (0.15–0.4 Hz). There is also usually a band related to the heart rate (0.6–1.6 Hz). In an initial attempt to probe the possible impact of magnetic fields on the skin blood flow spectrum, LDF was measured on the dorsum of fingers 3 and 4 while being exposed to an SMF of 0.12 T produced by magnets secured to a turntable as shown in Figure 14.12. LDF flow was generally recorded for 8 continuous minutes before altering the number of magnets placed on the rotating turntable with each 8-min sample subjected to spectral analysis. The concept under test with such an arrangement would be that if there were any impact of the magnet on LDF flow, then there should be a spectral component associated with that of the impulse rate. Some sample spectra are shown in Figure 14.13, which appears to indicate

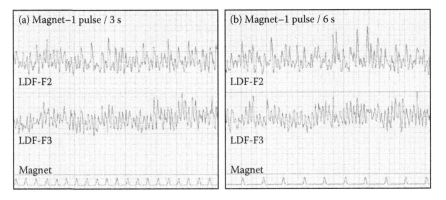

FIGURE 14.12 Setup and example for rotating magnet exposure (a). The subject's hand rests on the wood block while the turntable rotates exposing fingers 2 and 3 to the fields of the rotating magnets. Average field experienced at the finger dorsum was 0.12 T, which was produced by each of the magnets. The rotation speed (Ω) was fixed and the rate of exposure was determined by the number of magnets that were placed on the turntable. Parts (b) and (c) are example recordings each of 60 s duration showing LDF on F2 and F3 with 4 magnets placed (b) and two magnets placed (c).

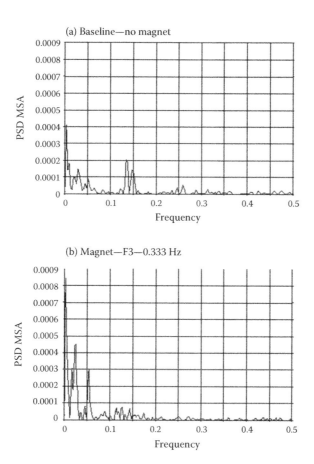

FIGURE 14.13 Power spectral distribution of LDF without and with rotating magnets. (a) Power spectral density (PSD) of baseline LDF for finger F3. (b) PSD of 8-min exposure to rotating field (1 impulse/3 s = 0.333 Hz) as exemplified by the sample of Figure 14.11a. The measured peak B-field measured at the finger dorsum was 0.125 T MSA is the mean squared amplitude.

a shift in the spectrum during magnet exposure at a rotation frequency of 0.333 Hz, but the full interpretation must await further data accumulation and analysis. However, the spectral analysis approach to the question appears to be a useful tool and is currently being investigated.

14.3.5.2 Pulsed Magnetic Fields and Human Skin Circulation

One major difference to consider when discussing PEMF is the fact that the time-changing component gives rise to induced currents within target tissues and their surroundings. A possible coupling of various forms of PEMF to circulation-related parameters and processes has been the subject of considerable discussion (McKay et al. 2007; Pilla 2015) with various applications (Markov 2007) and some evidence of a PEMF-induced VSM relaxation possibly involving a calcium-dependent NO process

(Pilla 2012), although no such process has been specifically shown to effect skin circulation. However, there is evidence that skin circulation can be altered by exposure to PEMF even if the involved mechanisms are unclear. Early workers used a form of PEMF in which a baseband time-varying signal modulated a radio frequency carrier typically at 27.12–27.33 MHz, which are diathermy frequencies (Guy et al. 1974). Application of such signals when at a sufficiently low intensity was by some termed "athermic" (Liebesny 1938) to suggest that effects observed were not attributable to tissue heating, but this may not be confirmed in all cases (Silverman 1964; Silverman and Pendleton 1968) even if short duty cycles are used. Thus some early work using PEMF may in part be related to mild heating. Nonetheless, there are reports of increased lower extremity blood perfusion indices in persons free of vascular disease (Erdman 1960) and in persons with peripheral arterial disease (Hedenius et al. 1966; Valtonen et al. 1973). Further, direct measurements of skin blood perfusion in arms of healthy persons (Mayrovitz and Larsen 1992) and in persons with lymphedematous arms (Mayrovitz et al. 2002) and with lower extremity ulcers (Mayrovitz and Larsen 1995) indicate a PEMF-related increase in blood flow.

An important contribution to clarifying the question of heat versus EMF effects on vascular targets was addressed, not in human skin, but by exposing blood vessels in the web of the frog to PEMF of 0.1, 1, and 10 MHz at burst rates of 10 kHz at 50% duty cycles for up to 60 min at magnetic and electric field strengths at the target calculated to not raise tissue temperature by more than 0.036°C (Miura and Okada 1991). When topical NE was used to pre-constrict microscopically observed arterioles to about half their resting diameter, exposure to PEMF at each frequency caused diameters to increase after 10–20 min of exposure, with a peak vasodilation at about 60 min. The 10-MHz excitation produced the largest vasodilation (to about 68% of pre-constriction) but was not significantly different than that caused by 1-MHz excitation. The amount of vasodilation was dose-dependent for pre-constricted arterioles and for non-pretreated arterioles. Percentage increases in arteriole diameters for non-pretreated arterioles were on average 12% and 26% for target magnetic field intensities reported as 7.3 and 30.1 mG, respectively, and electric field intensities of 2.19 and 9.00 v/cm, respectively, with preconstricted arterioles dilating more for the same dose.

On the basis of the manipulations of the suffusing fluid content, the authors concluded that the PEMF vasodilation depended on Ca^{++} efflux from VSM and/or influx into VSM sarcoplasmic reticulum, both well-known processes that alter VSM tone, as discussed in earlier sections of this chapter. On the basis of the prior work (Miura and Okada 1988; Okada and Miura 1990) they hypothesized that the PEMF facilitates activation of the Ca^{++}-ATPase calcium pump, thereby modulating VSM tone. Subsequent work provided evidence of a strong involvement of NO via a PEMF-induced cyclic GMP activation when cerebella tissue was exposed to a similar 10 MHz pulsed field (Miura et al. 1993). Another microscopic observational animal study using PEMF for which heat was reported as a nonissue was performed on rat cremaster circulation (Smith et al. 2004). Exposing the cremaster to a pair of Helmholtz coils with approximate 3.7 kHz pulses (pulse width of 5.9 ms and duty cycle of 8.8%) resulted in arteriole vasodilation of 9% after only 2 min of exposure, although the field intensity at the target was not specified. These PEMF-related diameter changes can have substantial flow effects, but it is unclear if they are because of induced currents, electric field changes that affect membrane potentials, or other processes since it is well established that various forms of electrical stimulation result in altered skin blood flow with increases reported in skin of forearm (McDowell et al. 1999; Cramp et al. 2002) and leg (Noble et al. 2000). Further study is clearly warranted.

14.4 Conclusion

Whether or not all or any of the mechanisms are fully understood, it would appear that based on the cited materials and approaches spanning many years and the current analysis of these findings, both SMF and PEMF are modalities that have variable impacts on vascular cells, blood flow, and related hemodynamics. There is thus good reason to further explore the way in which differences in doses and field arrangements in the case of SMF and differences in parameter values in the case of PEMF affect vascular components and blood flow in humans. Further, the findings that specific SMF and PEMF are associated with specific responses in blood and vascular cells, blood circulation parameters, and aspects of tissue modification would indicate that basic science and clinical approaches and potential expected outcomes of EMF studies or treatments should take these into account. Because of the potential for such specificity, it is vitally important that studies and protocols designed and carried out under headings such as basic research or clinical evaluations or therapeutic interventions contain within such protocols and reports of findings sufficient detail regarding biological or physiological targets and the specific field parameters applied at these targets. It is via such specificity that we may facilitate further advances in the area of electromagnetic—physiological interactions and optimize the chances of discovery of mechanisms and new therapeutic applications.

References

Abraham, P., G. Leftheriotis, B. Desvaux, M. Saumet and J. L. Saumet (1994). Diameter and velocity changes in the femoral vein during thermal stress in humans. *Clin Physiol* **14**(1): 15–21.

Asano, M., K. Yoshida and K. Tatai (1965). Microphotoelectric plethysmography using a rabbit ear chamber. *J Appl Physiol* **20**(5): 1056–1062.

Bali, R. and U. Awasthi (2011). Mathematical model of blood flow in small blood vessel in the presence of magnetic field. *Appl Math* **2**: 264–269.

Bartholow, R. (1887). *Medical Electricity.* Lea Brothers & Co., Philadelphia, PA.

Berg, H. (1999). Problems of weak electromagnetic field effects in cell biology. *Bioelectrochem Bioenerg* **48**(2): 355–360.

Bor Fuh, C., Y. S. Su and H. Y. Tsai (2004). Determination of magnetic susceptibility of various ion-labeled red blood cells by means of analytical magnetapheresis. *J Chromatogr A* **1027**(1–2): 289–296.

Brix, G., S. Strieth, D. Strelczyk, M. Dellian, J. Griebel, M. E. Eichhorn, W. Andra and M. E. Bellemann (2008). Static magnetic fields affect capillary flow of red blood cells in striated skin muscle. *Microcirculation* **15**(1): 15–26.

Chen, I. H. and S. Saha (1984). Analysis of an intensive magnetic field on blood flow. *J Bioelectricity* 3(1–2): 293–298.

Ciuti, G., D. Righi, L. Forzoni, A. Fabbri and A. M. Pignone (2013). Differences between internal jugular vein and vertebral vein flow examined in real time with the use of multigate ultrasound color Doppler. *AJNR Am J Neuroradiol* 34(10): 2000–2004.

Colbert, A. P., M. S. Markov and J. S. Souder (2008). Static magnetic field therapy: Dosimetry considerations. *J Altern Complement Med* 14(5): 577–582.

Columbia, T. B. and P. Richet (1880). The magnet in medicine. *Science* 1(5): 59–60.

Cramp, F. L., G. R. McCullough, A. S. Lowe and D. M. Walsh (2002). Transcutaneous electric nerve stimulation: The effect of intensity on local and distal cutaneous blood flow and skin temperature in healthy subjects. *Arch Phys Med Rehabil* 83(1): 5–9.

Donkin, H. (1878). Remarks on metallic and magnetic therapeutics. *Br Med J* 2(930): 619–620.

El-Aragi, G. M. (2013). Effect of electrohydraulic discharge on viscosity of human blood. *Phys Res Int* 2013(Article ID 203708). doi:10.1155/2013/203708.

Erdman, J. W. (1960). Peripheral blood flow measurements during application of pulsed high frequency currents. *Am J Orthop* 2: 196–197.

Gamgee, A. (1878). An account of a demonstration on the phenomena of hystero-epilepsy given by Professor Charcot: And on the modification which they undergo under the influence of magnets and solenoids. *Br Med J* 2: 545–548.

Gmitrov, J. (2013). Static magnetic field effect on microcirculation, direct versus baroreflex-mediated approach. *Electromagn Biol Med* 32(4): 448–462.

Gmitrov, J. and C. Ohkubo (2002). Verapamil protective effect on natural and artificial magnetic field cardiovascular impact. *Bioelectromagnetics* 23(7): 531–541.

Gmitrov, J., C. Ohkubo and H. Okano (2002). Effect of 0.25 T static magnetic field on microcirculation in rabbits. *Bioelectromagnetics* 23(3): 224–229.

Guy, A. W., J. F. Lehmann and J. B. Stonebridge (1974). Therapeutic applications of electromagnetic power. *Proc IEEE* 62(1): 55–75.

Haik, Y., V. Pai and C.-J. C. Chen (2001). Apparent viscosity of human blood in a high static magnetic field. *J Magn Magn Mater* 225: 180–186.

Han, K. H. and A. B. Frazier (2004). Continuous magnetophoretic separation of blood cells in microdevice format. *J Appl Phys* 96(10): 5797–5802.

Haverkort, J. W., S. Kenjeres and C. R. Kleijn (2009). Computational simulations of magnetic particle capture in arterial flows. *Ann Biomed Eng* 37(12): 2436–2448.

Hedenius, P., E. Odeblad and L. Wahlstrom (1966). Some preliminary investigations on the therapeutic effect of pulsed short waves in intermittent claudication. *Curr Ther Res* 8(7): 317–321.

Higashi, T., N. Asjoda and T. Takeuchi (1997). Orientation of blood cells in static magnetic field. *Phys B* 237–238: 616–620.

Higashi, T., A. Yamagishi, T. Takeuchi, N. Kawaguchi, S. Sagawa, S. Onishi and M. Date (1993). Orientation of erythrocytes in a strong static magnetic field. *Blood* 82(4): 1328–1334.

Hughes, S., A. J. El Haj, J. Dobson and B. Martinac (2005). The influence of static magnetic fields on mechanosensitive ion channel activity in artificial liposomes. *Eur Biophys J* **34**(5): 461–468.

Ichioka, S., M. Iwasaka, M. Shibata, K. Harii, A. Kamiya and S. Ueno (1998). Biological effects of static magnetic fields on the microcirculatory blood flow *in vivo*: A preliminary report. *Med Biol Eng Comput* **36**(1): 91–95.

Ichioka, S., M. Minegishi, M. Iwasaka, M. Shibata, T. Nakatsuka, J. Ando and S. Ueno (2003). Skin temperature changes induced by strong static magnetic field exposure. *Bioelectromagnetics* **24**(6): 380–386.

Ichioka, S., M. Minegishi, M. Iwasaka, M. Shibata, T. Nakatsuka, K. Harii, A. Kamiya and S. Ueno (2000). High-intensity static magnetic fields modulate skin microcirculation and temperature *in vivo*. *Bioelectromagnetics* **21**(3): 183–188.

Intaglietta, M., N. R. Silverman and W. R. Tompkins (1975). Capillary flow velocity measurements *in vivo* and *in situ* by television methods. *Microvasc Res* **10**(2): 165–179.

Keltner, J. R., M. S. Roos, P. R. Brakeman and T. F. Budinger (1990). Magnetohydrodynamics of blood flow. *Magn Reson Med* **16**(1): 139–149.

Kinouchi, Y., H. Yamaguchi and T. S. Tenforde (1996). Theoretical analysis of magnetic field interactions with aortic blood flow. *Bioelectromagnetics* **17**(1): 21–32.

Kolin, A., N. Assali, G. Herrold and R. Jensen (1957). Electromagnetic determination of regional blood flow in unanesthetized animals. *Proc Natl Acad Sci USA* **43**(6): 527–540.

Kvandal, P., S. A. Landsverk, A. Bernjak, A. Stefanovska, H. D. Kvernmo and K. A. Kirkeboen (2006). Low-frequency oscillations of the laser Doppler perfusion signal in human skin. *Microvasc Res* **72**(3): 120–127.

Kvernmo, H. D., A. Stefanovska, M. Bracic, K. A. Kirkeboen and K. Kvernebo (1998). Spectral analysis of the laser Doppler perfusion signal in human skin before and after exercise. *Microvasc Res* **56**(3): 173–182.

Kwan, R. L., W. C. Wong, S. L. Yip, K. L. Chan, Y. P. Zheng and G. L. Cheing (2015). Pulsed electromagnetic field therapy promotes healing and microcirculation of chronic diabetic foot ulcers: A pilot study. *Adv Skin Wound Care* **28**(5): 212–219.

Li, F., Y. Yuan, Y. Guo, N. Liu, D. Jing, H. Wang and W. Guo (2015). Pulsed magnetic field accelerate proliferation and migration of cardiac microvascular endothelial cells. *Bioelectromagnetics* **36**(1): 1–9.

Li, Z., E. W. Tam, A. F. Mak and R. Y. Lau (2007). Wavelet analysis of the effects of static magnetic field on skin blood flowmotion: Investigation using an *in vivo* rat model. *In Vivo* **21**(1): 61–68.

Liebesny, P. (1938). Athermic short wave therapy. *Arch Phys Ther* **19**: 736–740.

Markov, M. S. (2007). Expanding use of pulsed electromagnetic field therapies. *Electromagn Biol Med* **26**(3): 257–274.

Markov, M. S. and A. P. Colbert (2000). Magnetic and electromagnetic field therapy. *J Back Musculoskelet Rehabil* **15**(1): 17–29.

Martel, G. F., C. Andrews and C. G. Roseboom (2002). Comparison of static and placebo magnets on resting forearm blood flow in young, healthy men. *J Orthop Sports Phys Therapy* **32**: 518–524.

Martino, C. F. (2011). Static magnetic field sensitivity of endothelial cells. *Bioelectromagnetics* **32**(6): 506–508.

Martino, C. F., H. Perea, U. Hopfner, V. L. Ferguson and E. Wintermantel (2010). Effects of weak static magnetic fields on endothelial cells. *Bioelectromagnetics* **31**(4): 296–301.

Mayrovitz, H. N. (1992). Skin capillary metrics and hemodynamics in the hairless mouse. *Microvasc Res* **43**(1): 46–59.

Mayrovitz, H. N. (2015). Electromagnetic fields for soft tissue wound healing. In Markov, M. S. (ed.) *Electromagnetic Fields in Biology and Medicine*. CRC Press, Boca Raton, FL, 231–251.

Mayrovitz, H. N. and S. G. Carta (1996). Laser-Doppler imaging assessment of skin hyperemia as an indicator of trauma after adhesive strip removal. *Adv Wound Care* **9**(4): 38–42.

Mayrovitz, H. N. and E. E. Groseclose (2002). Neurovascular responses to sequential deep inspirations assessed via laser-Doppler perfusion changes in dorsal finger skin. *Clin Physiol Funct Imaging* **22**(1): 49–54.

Mayrovitz, H. N. and E. E. Groseclose (2005). Effects of a static magnetic field of either polarity on skin microcirculation. *Microvasc Res* **69**(1–2): 24–27.

Mayrovitz, H. N., E. E. Groseclose and D. King (2005). No effect of 85 mT permanent magnets on laser-Doppler measured blood flow response to inspiratory gasps. *Bioelectromagnetics* **26**(4): 331–335.

Mayrovitz, H. N., E. E. Groseclose, M. Markov and A. A. Pilla (2001). Effects of permanent magnets on resting skin blood perfusion in healthy persons assessed by laser Doppler flowmetry and imaging. *Bioelectromagnetics* **22**(7): 494–502.

Mayrovitz, H. N., E. E. Groseclose and N. Sims (2002). Assessment of the short-term effects of a permanent magnet on normal skin blood circulation via laser-Doppler flowmetry. *Sci Rev Altern Med* **6**(1): 9–12.

Mayrovitz, H. N., S. J. Kang, B. Herscovici and R. N. Sampsell (1987). Leukocyte adherence initiation in skeletal muscle capillaries and venules. *Microvasc Res* **33**(1): 22–34.

Mayrovitz, H. N. and P. B. Larsen (1992). Effects of pulsed electromagnetic fields on skin microvascular blood perfusion. *Wounds* **4**: 197–202.

Mayrovitz, H. N. and P. B. Larsen (1995). A preliminary study to evaluate the effect of pulsed radio frequency field treatment on lower extremity peri-ulcer skin microcirculation of diabetic patients. *Wounds* **7**: 90–93.

Mayrovitz, H. N. and J. A. Leedham (2001). Laser-Doppler imaging of forearm skin: Perfusion features and dependence of the biological zero on heat-induced hyperemia. *Microvasc Res* **62**(1): 74–78.

Mayrovitz, H. N., J. Moore and E. Sorrentino (1990). A model of regional microvascular ischemia in intact skin. *Microvasc Res* **39**(3): 390–394.

Mayrovitz, H. N., N. Sims and J. M. Macdonald (2002). Effects of pulsed radio frequency diathermy on postmastectomy arm lymphedema and skin blood flow: A pilot investigation. *Lymphology* **35**(suppl): 353–356.

Mayrovitz, H. N., M. P. Wiedeman and A. Noordergraaf (1975). Microvascular hemodynamic variations accompanying microvessel dimensional changes. *Microvasc Res* **10**(3): 322–329.

Mayrovitz, H. N., M. P. Wiedeman and A. Noordergraaf (1976). Analytical characterization of microvascular resistance distribution. *Bull Math Biol* **38**(1): 71–82.

McDowell, B. C., C. McElduff, A. S. Lowe, D. M. Walsh and G. D. Baxter (1999). The effect of high- and low-frequency H-wave therapy upon skin blood perfusion: Evidence of frequency-specific effects. *Clin Physiol* **19**(6): 450–457.

McKay, J. C., F. S. Prato and A. W. Thomas (2007). A literature review: The effects of magnetic field exposure on blood flow and blood vessels in the microvasculature. *Bioelectromagnetics* **28**: 81–98.

McNamee, D. A., M. Corbacio, J. K. Weller, S. Brown, R. Z. Stodilka, F. S. Prato, Y. Bureau, A. W. Thomas and A. G. Legros (2011). The response of the human circulatory system to an acute 200-μT, 60-Hz magnetic field exposure. *Int Arch Occup Environ Health* **84**(3): 267–277.

Miura, M. and J. Okada (1988). Burst-type radio-frequency electromagnetic radiation antagonizes vasoconstriction. *Kitakanto Med J* **38**: 389–396.

Miura, M. and J. Okada (1991). Non-thermal vasodilatation by radio frequency burst-type electromagnetic field radiation in the frog. *J Physiol* **435**: 257–273.

Miura, M., K. Takayama and J. Okada (1993). Increase in nitric oxide and cyclic GMP of rat cerebellum by radio frequency burst-type electromagnetic field radiation. *J Physiol* **461**: 513–524.

Morris, C. and T. Skalak (2005). Static magnetic fields alter arteriolar tone *in vivo*. *Bioelectromagnetics* **26**(1): 1–9.

Nakano, N., J. Otsuka and A. Tasaki (1971). Fine structure of iron ion in deoxymyoglobin and deoxyhemoglobin. *Biochim Biophys Acta* **236**(1): 222–233.

Noble, J. G., G. Henderson, A. F. Cramp, D. M. Walsh and A. S. Lowe (2000). The effect of interferential therapy upon cutaneous blood flow in humans. *Clin Physiol* **20**(1): 2–7.

Ohkubo, C. and H. Okano (2015). Magnetic field influences on the microcirculation. In Markov, M. S. (ed.) *Electromagnetic Fields in Biology and Medicine*. CRC Press, Boca Raton, FL, 103–128.

Ohkubo, C., H. Okano, A. Ushiyama and H. Masuda (2007). EMF effects on microcirculatory system. *Environmentalist* **27**: 395–402.

Ohkubo, C. and S. Xu (1997). Acute effects of static magnetic fields on cutaneous microcirculation in rabbits. *In Vivo* **11**(3): 221–225.

Okada, J. and M. Miura (1990). EMF activates soluble guanylate cyclase from rat lung *in vitro*. *Jap J Physiol* **40**(S45).

Okano, H., J. Gmitrov and C. Ohkubo (1999). Biphasic effects of static magnetic fields on cutaneous microcirculation in rabbits. *Bioelectromagnetics* **20**(3): 161–171.

Okano, H., H. Ino, Y. Osawa, T. Osuga and H. Tatsuoka (2012). The effects of moderate-intensity gradient static magnetic fields on nerve conduction. *Bioelectromagnetics* **33**(6): 518–526.

Okano, H., H. Masuda and C. Ohkubo (2005). Decreased plasma levels of nitric oxide metabolites, angiotensin II, and aldosterone in spontaneously hypertensive rats exposed to 5 mT static magnetic field. *Bioelectromagnetics* **26**(3): 161–172.

Okano, H. and C. Ohkubo (2003). Effects of static magnetic fields on plasma levels of angiotensin II and aldosterone associated with arterial blood pressure in genetically hypertensive rats. *Bioelectromagnetics* **24**(6): 403–412.

Okano, H. and C. Ohkubo (2006). Elevated plasma nitric oxide metabolites in hypertension: Synergistic vasodepressor effects of a static magnetic field and nicardipine in spontaneously hypertensive rats. *Clin Hemorheol Microcirc* **34**(1–2): 303–308.

Okano, H., N. Tomita and Y. Ikada (2007). Effects of 120 mT static magnetic field on TGF-β-inhibited endothelial tubular formation *in vitro*. *Bioelectromagnetics* **28**(6): 497–499.

Okano, H., N. Tomita and Y. Ikada (2008). Spatial gradient effects of 120 mT static magnetic field on endothelial tubular formation *in vitro*. *Bioelectromagnetics* **29**(3): 233–236.

Okazaki, M., N. Maeda and T. Shiga (1987). Effects of an inhomogeneous magnetic field on flowing erythrocytes. *Eur Biophys J* **14**(3): 139–145.

Peterson, F. and A. E. Kennelly (1892). Some physiological experiments with magnets at the Edison laboratory. *N Y Acad Med* **56**: 729–732.

Pilla, A. A. (2012). Electromagnetic fields instantaneously modulate nitric oxide signaling in challenged biological systems. *Biochem Biophys Res Commun* **426**(3): 330–333.

Pilla, A. (2015). Pulsed electromagnetic fields. In Markov, M. S. (ed.) *Electromagnetic Fields in Biology and Medicine*. CRC Press, Boca Raton, FL, 29–47.

Shercliff, J. A. (1953). Steady motion of conducting fluids in pipes under transverse magnetic fields. *Proc Camb Phil Soc* **49**: 136–144.

Silverman, D. R. (1964). A comparison of the effects of continuous and pulsed short wave diathermy: Resistance to bacterial infection in mice. *Arch Phys Med Rehabil* **45**: 491–499.

Silverman, D. R. and L. Pendleton (1968). A comparison of the effects of continuous and pulsed short-wave diathermy on peripheral circulation. *Arch Phys Med Rehabil* **49**(8): 429–436.

Sirmatel, O., C. Sert, C. Tumer, A. Ozturk, M. Bilgin and Z. Ziylan (2007). Change of nitric oxide concentration in men exposed to a 1.5 T constant magnetic field. *Bioelectromagnetics* **28**(2): 152–154.

Smith, T. L., D. Wong-Gibbons and J. Maultsby (2004). Microcirculatory effects of pulsed electromagnetic fields. *J Orthop Res* **22**(1): 80–84.

St. Pierre, T. G. and J. Dobson (2000). Theoretical evaluation of cell membrane ion channel activation by applied magnetic fields. *Eur Biophys J* **29**(6): 455–456.

Sud, V. K. and G. S. Sekhon (1989). Blood flow through the human arterial system in the presence of a steady magnetic field. *Phys Med Biol* **34**(7): 795–805.

Suda, T. and S. Ueno (1994). Magnetic orientation of red blood cell membranes. *IEEE Trans Magn* **30**(6): 4713–4715.

Suda, T. and S. Ueno (1996). Microscopic observation of the behaviors of red blood cells with plasma proteins under strong magnetic fields. *IEEE Trans Magn* **32**(5): 5136–5138.

Takeuchi, T., T. Mizuno, T. Higashi, A. Yamagishi and M. Date (1995). Orientation of red blood cells in high magnetic field. *J Magn Magn Mater* **140–144**: 1462–1463.

Ueno, S., P. Lovsund and P. A. Oberg (1986). Effects of alternating magnetic fields and low-frequency electric currents on human skin blood flow. *Med Biol Eng Comput* **24**(1): 57–61.

Valtonen, E. J., H. G. Lilius and U. Svinhufvud (1973). Effects of three modes of application of short wave diathermy on the cutaneous temperature of the legs. *Eur Medicophys* **9**(2): 49–52.

Watarai, H. and M. Namba (2002). Capillary magnetophoresis of human blood cells and their magnetophoretic trapping in a flow system. *J Chromatogr A* **961**(1): 3–8.

Wenzel, F., J. Reissenweber and E. David (2005). Cutaneous microcirculation is not altered by a weak 50 Hz magnetic field. *Biomed Tech (Berl)* **50**(1–2): 14–18.

Wexler, L., D. H. Bergel, I. T. Gabe, G. S. Makin and C. J. Mills (1968). Velocity of blood flow in normal human venae cavae. *Circ Res* **23**(3): 349–359.

Willenberg, T., R. Clemens, L. M. Haegeli, B. Amann-Vesti, I. Baumgartner and M. Husmann (2011). The influence of abdominal pressure on lower extremity venous pressure and hemodynamics: A human *in-vivo* model simulating the effect of abdominal obesity. *Eur J Vasc Endovasc Surg* **41**(6): 849–855.

Xu, S., H. Okano, M. Nakajima, N. Hatano, N. Tomita and Y. Ikada (2013). Static magnetic field effects on impaired peripheral vasomotion in conscious rats. *Evid Based Complement Alternat Med* **2013**(Article ID 746968). doi:10.1155/2013/746968.

Xu, S., H. Okano and C. Ohkubo (1998). Subchronic effects of static magnetic fields on cutaneous microcirculation in rabbits. *In Vivo* **12**(4): 383–389.

Yamamoto, T., Y. Nagayama and M. Tamura (2004). A blood-oxygenation-dependent increase in blood viscosity due to a static magnetic field. *Phys Med Biol* **49**(14): 3267–3277.

Yan, Y., G. Shen, K. Xie, C. Tang, X. Wu, Q. Xu, J. Liu, J. Song, X. Jiang and E. Luo (2011). Wavelet analysis of acute effects of static magnetic field on resting skin blood flow at the nail wall in young men. *Microvasc Res* **82**(3): 277–283.

Zborowski, M., G. R. Ostera, L. R. Moore, S. Milliron, J. J. Chalmers and A. N. Schechter (2003). Red blood cell magnetophoresis. *Biophys J* **84**(4): 2638–2645.

15

Methodology of Standards Development for EMF RF in Russia and by International Commissions: Distinctions in Approaches

Yury G. Grigoriev
Russian National
Committee on Non-Ionizing
Radiation Protection

15.1 Introduction

A standard is a general term that incorporates both regulations and guidelines and can be defined as a set of specifications or rules to promote the safety of an individual or the population. The ultimate goal of electromagnetic field (EMF) standards is to protect human health. Exposure limits are intended to protect against adverse health effects of EMF exposure across the entire frequency range and modulation. Naturally, it is an axiom that appropriate standards evaluating harmful factors in the environment must be developed and that the scientific community must understand this as a necessity.

The Russian standard for base stations has already been in existence for more than 30 years and is more rigid than the maximum level recommended by the International Commission of Non-Ionizing Radiation Protection (ICNIRP). This distinction has been discussed at scientific meetings for many years—unfortunately, without result.

The second EMF source of mobile communication—the mobile phone—has no sufficient substantiation on exposure limits. The irradiation of a brain is not limited and is not supervised. The children using mobile phones are especially at high risk.

15.2 Methodology for the Risk Assessment of RF EMF for the Population: History of Development of Standardization in the USSR (and Later in Russia)

About 60 years ago, in 1958, the maximum permissible levels (MPLs) of radiofrequency EMF (EMF RF) for professionals had been regulated at the state level for the first time in the USSR and in the world. These standards were based on the results of clinical research of professionals and on the results of experimental studies with animals.

With the development of the tele-radio networks in the USSR, there was a need to ensure the safety of the population in conditions when relatively powerful sources of RF EMF were placed at the border or within the boundaries of a residential building. This task was entrusted to the Kiev Research Institute of General and Communal Hygiene Ministry of Health of Ukraine.

The first RF EMF standard for the population, SanPiN 848-70, was approved by the Ministry of Health of the USSR in 1970 and was considered for the population exposure limit of $1\,\mu W/cm^2$ in the microwave band of 300 MHz to 300 GHz. In 1978, the USSR Ministry of Health approved the next SanPiN No. 1823-78. In this document, MPL for the population in the frequency range of 300 MHz to 300 GHz was set as $5\,\mu W/cm^2$.

The methodological bases for the development of MPLs were "guidelines on the assessment of the biological effect of low-intensity microwave radiation for hygienic at ambient conditions" (Ministry of Health of USSR, Kiev, 1981) and "methodological approaches to the study of the effect of low levels of microwave energy on human health in terms of residential areas" (USSR, Ministry of Health, 1981).

Scientific development of hygienic standards was coordinated by Federal Commission on Scientific Problems of Environment of the USSR Academy of Medical Sciences. The main co-developers were the Kiev Institute of Communal Hygiene and the Institute of Biophysics, USSR Ministry of Health. Among the leaders who studied the problem of research on the biological effects of EMF RF and the development of normative documents were academics Shandala and Ilyin and professors Savin, Palzev, Dumanskiy, Kholodov, Akoev, Shihodyrov, and Y. Grigoriev.

Standardization strategy in the USSR and Russia is discussed in detail in a series of publications: Grigoriev (1998), Grigoriev et al. (2002, 2003a, 2004), and Grigoriev and Grigoriev (2013, 2016).

Currently, three questions remain relevant for standardization:

1. Are there nonthermal biological effects of low levels of RF EMF?
2. Is it possible that the irradiation of the population with RF EMF throughout human life leads to increased adverse biological effects?
3. Is there a "threshold" level of exposure to RF EMF, and if so how do we define it?

15.2.1 Nonthermal Biological Effects

A number of Russian scientists in the 1960s–1970s have pointed to the possibility of informational or the nonthermal action of EMF RF (Presman, 1968; Ivanov-Muromsky, 1977). Numerous animal experiments have shown that clear bioeffects appear under the influence of very low RF EMF with nonthermal intensity (Frey et al., 1975; Kholodov, 1975, 1996; Adey, 1980; Chizhenkova, 1988, 2000, 2004; Lukyanova et al., 1996, 2010; Grigoriev, 1998, 2015; Lukyanova, 1999, 2015; Belyaev and Grigoriev, 2007; Belyaev, 2015). Specific possible mechanisms of thermal effects are considered (Barnes and Greenebaum, 2016).

More than 100 articles have been published on monitoring volunteers who used mobile phones. In most cases, the different functional responses of the body's systems under the influence of nonthermal EMF levels from mobile phone use have been published (Reiser et al., 1995; Dec et al., 1997; Thuroczy et al., 1999; Krause, 2002; Huber et al., 2003; Croft et al., 2008).

In our laboratory, we tested 10 volunteers aged 23–47 years. Exposure of volunteer's brain to EMF from mobile phones (MNT standards GSM-450, GSM-900, and GSM-1800) was done for a period of 5, 10, and 20 min (Grigoriev et al., 1999). There were changes in EEG brain—increasing power of the alpha—rhythm in the spectrum of the EEG, an increase in the number of spindle variations in the alpha- and beta-bands (Lukyanova et al., 1996; Lukyanova, 2002, 2015). The changes persisted for 2h after the end of irradiation.

Another study of 29 volunteers (25–40 years) exposed to EMF nonthermal intensity (up to $200\,\mu W/cm^2$) has shown the dependence of the changes in cognitive function based on the typological features of EEG (Lukyanova et al., 2010).

The role of the modulation of the carrier frequency in the formation of biological effects of RF EMF at low intensity confirms our view about the presence of nonthermal mechanism of interaction of EMF with biological tissue (Blackman, 1984; Grigoriev, 1996, 2004; Belyev and Grigoriev, 2007; Grigoriev and Grigoriev, 2013, 2016).

These results together with numerous studies conducted by scientists from many countries provide direct evidence that RF EMF intensity of up to $10\,mW/cm^2$ may have a nonthermal mechanism of action.

The problem of a possible nonthermal mechanism of action of EMF RF with low-intensity EMF has been one of the reasons for the WHO program on harmonization of the existing international standards, which unfortunately turned to be unsuccessful. One of the main reasons is that engineering committees continue to affirm that biological effects are not possible without heat development. Let us point that the biological systems are nonlinear thermodynamic systems with a number of targets within single organs or tissues. As it was shown, even a simple conformational change in the protein or a change in the ion binding is capable of initiating signal transduction modifications (Markov, 2006).

15.3 Cumulative Effect in Terms of EMF RF Repeated or Extended Long-Term Exposure

Solving this problem requires a large number of experimental and epidemiological studies. Radiobiological studies of cumulative effects occurring after repeated or chronic effects of RF EMF in terms of accumulation of negative effects and their recovery rate, the residual lesions, identification of effective compensatory processes, and the estimation of the optimal time intervals for use of a mobile phone have not been performed.

It took about 20 years of active use of mobile phones to confidently say that the accumulation process is present; we now have understood the necessity of monitoring the health of the population exposed to this new environmental factor.

In 2003, at the International Conference in Budapest "Mobile Communications and the Brain," Hardell and Mild (2003) reported the results of multiple years of epidemiological research (1997–2000). The 1617 patients aged 20–80 years were divided into 5 groups with a difference of 10 years. An increased risk of brain tumors was observed in the age group of 20–29 years, whereas for other age categories such dependence was not found. Further analysis of these data showed that persons in the age group of 20–29 years began using cell phones in their childhood. These results indicate that people who started using mobile phones from childhood and adolescence may be at increased risk than those who started using mobile phones at a later age, which directly indicates a cumulative effect.

There is evidence that RF EMF can cause development of tumors in the brain of mobile phone users after a 10–12 year "waiting period" (Hardell and Calberg, 2009). The term "heavy users" that appeared in some publications linked the unfavorable bioeffects of the prolonged mobile phone use to accumulative processes of adverse biological effects.

It has been shown that after a single exposure to low-intensity RF EMF, certain changes in the brain EEG occur (Lukyanova, 1999, 2015). During the first hours after exposure, there is a restoration of bioelectrical activity of the brain, which indicates the insinuation of compensatory processes. Naturally, in these conditions, a repeated exposure might weaken compensatory processes and lead to development of the process of accumulation (Lukyanov et al., 2015). A number of factors such as the reactivity of the organism, the increasing of the user's age, and factors of environment may influence the compensation processes.

The decision of IARC (2011), in which RF EMF is classified as a possible carcinogen by the radiation group 2B, is another confirmation of the presence of the cumulative effect of the repeated impacts of low-intensity EMF RF.

Thus, the accumulation of adverse biological effects in the conditions of repeated or long-term chronic exposure is one of the most important criteria for risk assessment of the impact of mobile phone EMF RF on the population to develop appropriate standards.

15.3.1 Threshold Levels

The threshold level is the lowest level of exposure of the physical factor (EMF RF), below which the risk to public health does not exist, is introduced in analogy with the principles of ionizing radiation.

Given the complexity of this problem, we propose to determine the threshold level as a criterion for the body's response to RF EMF exposure, but on the condition that this response should not be pathological. This reaction may be compensatory/adaptive and should exist within the physiological range. In this case, the threshold level is "conditional action level." This term emphasizes that the EMF, acting with a certain intensity and mode of irradiation on the biological object, causes certain reactions in the body, but these reactions should not be pathological. The MPL, i.e., already the so-called inactive level, is set using the "safety factor." That is, the value of the current level will be reduced by the value of the safety factor.

Summing up on this issue, it can be concluded that the Russian standards for the population had previously been installed on the basis of the threshold value that has been set on the basis of compensatory/adaptive response within the physiological norm, not on the basis of pathological effects.

In 1984, the first practical result of the scientific program of the USSR was the development of the interim sanitary norms and rules to protect the public from exposure to EMFs generated by radio facilities (VSN No. 2963-84). The MPL for the population was established at $10\,\mu W/cm^2$ for the frequency range of 300 MHz to 30 GHz.

It should be noted that despite the "temporary" status adopted in the 1984 document, its basic rules of MAL $10\,\mu W/cm^2$ also apply to the present time for risk assessment, especially for base stations. The next standard approved by the Ministry of Health in 1996 fully retained the MPL for the population, bearing in mind the base station, as $10\,\mu W/cm^2$ for the frequency range of 300 MHz to 300 GHz.

We emphasize that the studies of the biological effect of RF EMF and their evaluation in the USSR were carried out in accordance with the Ministry of Health recommendations, "Methodological approaches to the study of the effect of low levels of microwave energy on the health of human beings in populated areas" and "Guidelines on the assessment of the biological effect of low-intensity microwave radiation for hygienic at ambient conditions." Results of acute, protracted, and chronic experiments with laboratory animals were used, as well as observations of humans in a real production environment. Studies were conducted on volunteers under simulated situations.

It is important that individual fragments of these comprehensive studies were analyzed, synthesized, and presented in more than 100 master's and doctoral theses, which had previously passed very thorough testing of the control in the Higher Attestation Commission of the USSR.

15.4 Development of Standards for Cellular Communications in Russia

First of all, let us consider the physical features of each of the two RF EMF sources of mobile communication, which are currently the main sources of environmental pollution and pose a real threat to the entire population.

15.4.1 Base Stations

EMF RFs are creating a constant electromagnetic background in the environment. The exposure is at a low and nonthermal level in the presence of various frequencies and round-the-clock exposure of the population during all stages of life. There is an exposure of all groups of the population, including children, patients and radiosensitive people. This chronic exposure of the population raises the issue of probable danger and negatively perceived growth of electromagnetic pollution by base stations of environment.

Critical targets: nervous system, immune system, and brain.

15.4.2 Mobile Phones

For mobile phone users, EMF RF exposure affects various groups of the population, including children 3 years of age. It is an open source of radiation, without protection, is easy to approach, and cannot be controlled. Expanding opportunities for communication and information is perceived by the population as a factor of increased comfort and is consequently superfluous, without restrictions.

15.4.2.1 Exposure

Daily fractional exposures are repeated throughout life. Generally, the total mobile phone usage is estimated to be 30–40 min a day. Most of the population uses mobile communications without any restrictions 2–4 h a day. It is necessary to note that this daily EMF exposure of the brain occurs for the first time in the history of civilization.

Critical target: brain and its function.

Thus, these data suggest that sources of EMF RF are quite different for the risk assessment and require an individual approach to the development of standards and exposure limits.

In 1994, development of standards for mobile phone users started for the first time in Russia. Based on the hygienic standard GN 2.1.8/2.2.4.019-94, the temporary permissible levels (TPLs) of mobile phones are $100\,\mu W/cm^2$ for EMF 450, 900, and 1800 Hz, with the limitation that mobile phones cannot be used for more than 40 min a day.

Then, based on SanPiN 2.1.8/2.2.4.1383-03 (hygienic requirements for the placement and operation of radio transmitting facilities) and SanPiN 2.1.8/2.2.4.1190-03 (hygienic requirements to placement and exploitation of the mobile communication) introduced in 2003, MPL for users of mobile phones remains $100\,\mu W/cm^2$.

When determining the limit values for base stations, the RNCNIRP decided to leave the limit value for the general public of $10\,\mu W/cm^2$ unchanged, as it was set in 1984. This value was well justified by previous research, and so there was no need for changing it (Vinogradov and Dumanskiy, 1974, 1975; Shandala and Vinogradov, 1982; Shandala et al., 1983, 1985; Vinogradov and Naumenko, 1986; Vinogradov et al., 1999).

It is important to note that the MPL of $10\,\mu W/cm^2$ for the population has remained intact for more than 30 years. Previously, the standard was used only in Russia and the countries formerly in coalition with the Soviet Union. Now, MPLs of $10\,\mu W/cm^2$ or less are used as RF legal exposure limits or nonbinding recommendations for national, regional, urban, or sensitive areas for at least 20 countries worldwide (Figure 15.1).

The adoption of the standard in 2003 for the mobile phone in terms of formalizing requirements for methods of measuring the near field and for the establishment of a threshold for the evaluation of RF EMF exposure on brain function as a critical organ was not optimal. Paltsev and Rubtsova carried out comprehensive studies on establishing hygienic standards for the users of mobile phones (experiments with 110 rats, GSM 900 and 1800 MHz for 1 h/day for 40 days at 0.5 and $2\,mW/cm^2$) (Rubsova and Paltsev, 2006). In addition, the authors conducted a study on 25 volunteers and found no significant changes in the nervous and cardiovascular systems after a single 30-min exposure to EMF of different mobile phones (Suvorov et al., 2002).

They reported that data about the changes in the immune status of animals after exposure to $500\,\mu W/cm^2$ were in agreement with earlier studies, indicating that the effects of RF EMF with power density $500\,\mu W/cm^2$ cause immune changes that can be regarded as pathological (Vinogradov and Dumanskiy, 1974, 1975; Shandala and Vinogradov, 1982; Shandala et al., 1983, 1985; Vinogradov and Naumenko, 1986; Vinogradov et al., 1999).

There was a proposal to use a safety factor of 5 and set to the cell phone MPL at $100\,\mu W/cm^2$ (Russian Standard, 2003—SanPiN 2.1.8/2.2.4.1190-03). It should be emphasized that SanPiN 2.1.8/2.2.4.1190-03, for the first time, introduced the recommendation to limit cell phone use for persons younger than 18 years as well as pregnant women.

The following results of our research were used as the base for the development of the Russian standard for mobile phones (model studies on isolated hearts of frogs, experiment with imprinting, physiological studies of children). The studies conducted in our laboratory have shown that the nonthermal RF EMF effects can be significant. Experiments on isolated frog hearts were conducted by Afrikanova and Grigoriev (2005). The low intensity frequency modulation was changed over time with constant frequency, preset from 1 to 100 Hz. Irradiation was carried out at 9.3 GHz. Dimensions of the frog hearts were comparable with the wavelength of the radiation. RF EMF exposure was carried out under conditions of maximum absorption of energy.

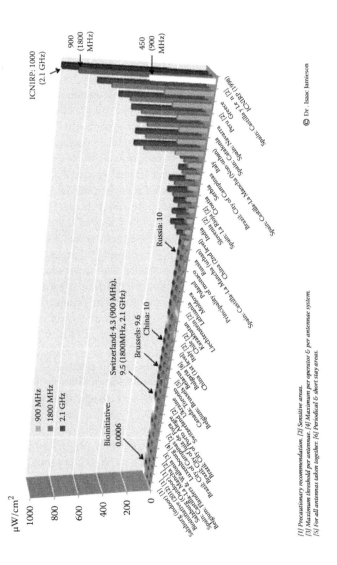

FIGURE 15.1 RF EMF legal exposure limits and nonbinding recommendations in different countries of the world. (Adapted from Jamieson, I., Changing perspectives—Improving lives. European Economic and Social Committee, Brussels, Belgium, November 4, 2014.)

The modulated signal was characterized by EMI modulation frequency ranging from 1 to 100 Hz. The pulse shape was rectangular ($S = 0.016\,\text{mW/cm}^2$). The study was conducted during exposure and after the impact of electromagnetic irradiation for 24 h. A total of 180 frog hearts were used.

A heart rate was estimated during each 30 min within 6 h from the moment of preparation of the isolated heart samples, during exposure, and also within the day after an irradiation. Simultaneously, a control "sham exposure" was run. It is important for assessing the response to irradiation that hearts in Ringer's solution can beat for 2 days. Besides, the morphological criterion of a state of excited heart tissues was studied (after a vital staining). Intact uncolored hearts on 24 h of observations showed slowing the rate of the heart on the average of 7%; the cardiac standstill was not present (Figure 15.2). The presence of the heart in stain solution during half-hour has resulted in the modification of its function. The number of the heart contractions was decreased by 30%, and 14% of hearts ceased contraction (Figure 15.2). The reaction of hearts irradiated in a continuous mode (the CW) was low and did not differ from reaction of the "colored hearts."

There was a sharp decrease in the heart rate and an increase in the number of hearts that stopped beating when the exposure was in modulated mode (Figure 15.2). The greatest effect was obtained at a frequency drift of modulation in a band of 6–10 Hz and time of exposure of 5 min. Under these conditions of exposure, slowing of the heart rate and stopping of the hearts in 85% of samples were observed. These effects were not reversible. Thus, in all, the series of the modulation frequency ranging from 1 to 100 Hz has a great influence on the function of the heart than in CW mode.

For the assessment of a role of the modulation, the procedure detailed in the general physiology of imprinting was used (Grigoriev, 1996; Grigoriev and Stepanov, 1998,

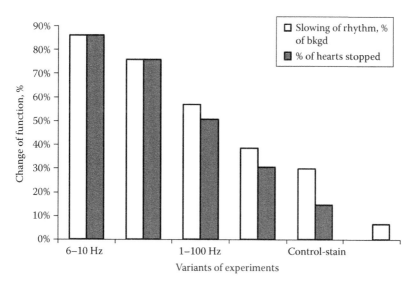

FIGURE 15.2 Change of the cardiac rhythm and cardiac standstill. The continuous regime (CW) and regime of the various modulations from 1 to 100 Hz.

2000). Imprinting is an original aspect of memorizing: The organism at birth fixes in the memory what it has seen for the first time.

One hundred and twenty-nine embryos of chickens were irradiated on day 16 of an incubation to EMF 9.3 GHz for 5 min ($S = 40\,\mu W/cm^2$), with a quantization of 10 and 40 Hz, meander, pulse duration of 2.5 ms. Besides, there was a series with a continuous irradiation (CW) and "sham" exposure.

The imprinting suppression (up to 50%) was found in newborn chicken only for the series of EMF exposure at 10 and 40 Hz (Table 15.1). In case of CW exposure ($S = 40\,\mu W/cm^2$) and in control group, the imprinting disturbance was not found.

In one experiment, we studied the possibility of fixing specific modulation mode EMF by brain (Grigoriev, 1996). The imprinting model was used in this experiment, and the signal of the EMF quantization was used as imprint stimulant. Irradiation of chickens was performed on the 16th day of incubation in a non-echo chamber: EMF at 9.3 GHz with the quantization of 1, 2, 3, 7, 9, or 10 Hz ($S = 0.04\,mW/cm^2$, duration of each irradiation was 5 min).

The possibility for the development of temporal communications in 15-days old embryos was earlier shown through an electric current and sound (Hunt, 1949). Taking into account these results, we assumed that the electromagnetic modulation waveform can be fixed by the brain, and thus gain the value of an imprint signal.

The experiment protocol was as follows: on the 16th day of incubation, embryos were irradiated with EMF with modulations of 1, 2, 3, 7, 9, and 10 Hz. After the birth of chickens, the next stage was the imprinting period (24 h after birth), and in this period, no chorionic irritator was applied to the chickens. After 48 h, strobe lights with the same frequency were shown to a chicken as imprint stimulant, from which the embryo was subjected to an electromagnetic irradiation for Day 16 of incubation. The difference between alleged light stimulant and differentiation stimulant was equal to 8 Hz. The experiments were conducted on 127 chicken embryos. A possibility of imprinting is shown in Figure 15.3. The analysis of the obtained data has allowed the author to make a deduction that the embryo brain at Day 16 of incubation can fix electromagnetic stimulant with modulation of 9 or 10 Hz and can store this information for a particular time after birth.

The analysis of 28 biological experiments conducted in the laboratories in the Soviet Union and then in Russia with use of modulated RF EMF (Grigoriev, 1996, 1999a, 2001, 2004b) leads to the following conclusions:

TABLE 15.1 Imprinting in Chickens after an EMF Irradiation of Embryos for Continuous and Modulated Regimens

Series No.	Series Name	PFD, $\mu W/cm^2$	Exposure Time	Number of Embryos	Number of Chickens with Imprinting
1	Control—sham exposure	—	—	83	81 (97%)
2	Continuous exposure	40	5	27	23 (89%)
3	10 or 40 Hz modulated exposure	40	5	19	9 (50%)

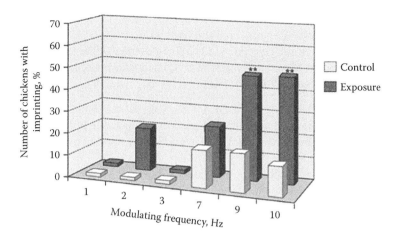

FIGURE 15.3 The number of chickens that recorded the brain electromagnetic signal (fixed imprinting).

- The exposure of biosystems to EMF with composite regimens of modulation leads to the development of bioeffects, both physiological and unfavorable, which are distinct from bioeffects induced by nonmodulated (CW) EMF.
- The acute exposure to modulated low-intensity EMF at nonthermal levels can result in the development of pathological effects.
- There is a dependence of development of a reciprocal biological response on the intensity and directness of the concrete regimen of EMF modulation; this dependence was found at all levels of biological systems—*in vitro*, *in situ*, and *in vivo*.
- As a rule, modulated EMF has invoked more expressed bioeffects than CW.
- The effects of EMF RF modulation are better pronounced at lower levels of intensity.

At the same time, we have our own experience of long-term monitoring of children who are mobile phone users. In 2006, we began long-term psychophysiological studies of primary school children who are mobile phone users and reported our results in 2014 (Grigoriev and Khorseva, 2014). It should be noted that apart from the main group in which children use mobile phones, a control group of children who did not have mobile phones and did not use was formed. The study was performed at the Lyceum in Chimki city.

Complex diagnostics of neurodynamic characteristics of children included psychophysiological indicators, assessment of neuropsychological status, and level of language development, as well as analysis of anamnesis and biographical data.

The following functional changes (preliminary results) were obtained in the first 6 years of observation:

- Fatigue (39.7%)
- Reduced ability to work in school and at home (50.7%)
- Decrease in the stability of voluntary attention (production—14.3%, accuracy—19.4%)

- Weakening of semantic memory (decrease of accuracy—19.4%, increasing time on the job—30.1%, changing the speed audiometer reaction—55.5%)
- Violation of phonemic perception (all children users). Similar results on phonemic perception of abuse were observed in India, but for the adult users of mobile phones (Panda et al., 2007, 2010, 2011).

In a recent study, Calvente et al. (2016), investigating the cognitive and behavioral function in 10-year-old children mobile phone users, concluded that "the impact of EMF RF low levels in the environment can have a negative impact on cognitive and/or behavioral development of children."

Thus, the preliminary results of the observations show that MP EMF may have a negative impact on the mental and physical health of children. The above effects reflected on the negative success of a child at school. The observed increase in the number of phonemic perception disorders increased the likelihood of errors in speech and writing and also reduced the effectiveness of the speech therapist. Despite the fact that in most cases, the change was within the age norm, stable values were below the normal limits.

The following factors allow us to conclude that the potential risk to the health of children who use mobile phones is very high:

- Absorption of electromagnetic energy by the head of a child is much higher than in the head of adults (children's brain tissue has a higher conductivity, the size of the child's head is smaller, and the skull bone of the child is thin).
- The distance from the antenna to the brain is short, because the child's ear shell is very soft and has almost no layer of the cartilage.
- The child's body is more sensitive to EMFs than adults.
- The child's brain is more vulnerable to the effects of EMF.
- The brains of children have a greater propensity to accumulation of adverse reactions in the context of repeated exposures to EMF.
- EMF RF may have an adverse effect on cognitive functions.
- Today's children use mobile phones at an early age and will continue to use them during their lifespan, and so the duration of the exposure of children to electromagnetic radiation will be substantially larger than that of modern adult users.

According to the members of the Russian National Committee of Non-Ionizing Radiation Protection (RNCNIRP, 2008), some possible disorders that might originate in children who use mobile phones include weakened memory, decline of attention, reduction of mental and cognitive abilities, irritability, sleep disturbance, tendency to stress reactions, and increased epileptic readiness.

It is also possible to expect the development of the adverse effects in older age as the result of the accumulation of adverse effects both in cells and in various functional systems of the body: brain tumors, tumors of the auditory and vestibular nerves (at age 25–30 years), Alzheimer's disease, "dementia," depressive syndrome, and other manifestations of degeneration of the nervous structures of the brain (at age 50–60 years).

Children users of mobile phones are not able to know that their brains are subjected to EMF, risking their health. This is a significant factor in moral ethics for parents.

Also important is that the risk of EMF RF exposure is not less than the risk for children's health from tobacco or alcohol.

Considering the above, we believe that children should be assigned to the high-risk group, and in fact considering a full range of circumstances, they can be equated to the "professionals." As a result, there is a need to develop specific standards for children age. Justification of these conclusions will be found in our monographs published in 2014 and 2015: Grigoriev and Khorseva (2014) and Grigoriev and Grigoriev (2013, 2016).

Our proposal for the development of specific standards for children may have additional justification. Indeed, in Russia, the only country that approved SanPin in 2003, it was recommended to restrict the use of mobile phones in children below 18 years. However, this document is still accessible only to professionals working in this field. The population does not have information about the existence of such recommendations, and children, with the permission of parents, use mobile communication and spend a maximum amount of time on cell phones, without supervision. In fact, there are two independent processes: on one hand, there is a formal recommendation on limiting the use of mobile phones by children and adolescents, and on the other hand, they use WiFi without control, even with the support of their parents, exceeding all reasonable limits on the use of mobile communications. The existence of a separate corresponding normative instrument for children will allow to draw attention of the public and government authorities to the problem and effectively implement preventive measures.

15.5 Principles for Development of International Standards/Recommendations

Currently, international standards are developed by ICNIRP, IEEE, CENELEC, and other international and national commissions. Their methodology uses only the results of experimental animal studies obtained under the conditions of acute effects and thermal-level EMF RF (Bernhard, 1999).

Any standard safety margin depends on the predetermined threshold. Outside Russia, the threshold level is determined on the basis of "stable pathological reactions" in the conditions of acute exposure to RF EMF heat level (WHO Handbook, 2002).

Taking into account the methodology of this standardization—*risk assessment of RF EMF on the basis of thermal effects in acute single exposure and determining the threshold criterion of stable pathological response*—a series of regulations was published, ICNIRP Guidelines (1998), IEEE S95.1-2005, and CENELEC EN 50166-2.2000.

In the WHO publication "Establishing a dialogue on risks from electromagnetic fields" (2002), it is stated that "Exposure limits are based on effects related to short-term acute exposure rather than long-term exposure, because the available scientific information on the long-term low level effects of exposure to EMF fields is considered to be insufficient to establish quantitative limits."

Moreover, in this edition, WHO considers it possible that the indicative or threshold level of exposure shall be established on the basis of data on early adverse biological effects of acute exposure to EMF, but it needs to sct a high safety factor of 50 for preventing adverse effects on human health caused by local or general increase in body temperature. It must be remembered that in this case, it is a one-time acute exposure, but in real life, EMF exposure will be permanent and lifelong. In this case, we can expect the development of other adverse effects that are not associated with the heating of the entire body or individual organs. It is necessary to note that Western standards are based on the settlement data.

Thus, in our opinion, the WHO has controversial and perhaps erroneous ideas about the principles of hygienic regulation. The author regrets that these principles have been used in the ICNIRP and IEEE guidelines.

Unfortunately, the development of international standards has ignored the view of many scientists about the possibility of nonthermal mechanism of the implementation of the biological action of low levels of EMF (Presman, 1968; Frey et al., 1975; Kholodov, 1975, 1996; Ivanov-Muramsky, 1977). Furthermore, it was not taken into account that at low levels of exposure to RF EMF, a significant role in shaping the biological effects can have a major effect (Grigoriev, 1996, 1999b, 2001, 2004, 2006; Adey, 2002; Markov, 2006; Belyaev and Grigoriev, 2007; Belyaev, 2015).

Our long experience with ionizing and non-ionizing radiations led us to formulate the following postulate: "The development of hygiene standards for the population should take into account the actual conditions of EMF RF exposure of the population—local or total exposure, acute single exposure or chronic, constant, or repeated exposure; the functional importance of 'critical organ' or 'critical body systems'; and effect on all population groups or only on certain limited groups of the population" (Grigoriev, 1997, 2008a).

Taking into account this postulate, we can make a clear conclusion that the Western standards do not meet the basic hygienic requirements. We make this conclusion based on the evaluation of the modern electromagnetic situation in the habitat of the population. In reality, the population has never met with high (thermal) RF EMF levels, especially in the context of an acute exposure. The current population is subjected to round-the-clock chronic exposure to RF EMF throughout the life span, and intermittent irradiation of the brain occurs. In all these cases, the impact is observed at nonthermal RF EMF levels. In addition, Western regulations do not take into account events that occurred for the first time during the life of our civilization. *Children who use mobile phones voluntarily irradiate their brains. This EMF RF exposure of the brain occurs every day, and the fractional exposure is projected for many years.*

We criticized the Western standards because they do not correspond to the actual conditions of RF EMF exposure on the population (report in 2003 at an international seminar in China, Grigoriev et al., 2003b). The criticism of these standards was continued in 2008 at an international conference in London on "EMF and Health—a global assessment" (Grigoriev, 2008). However, until now, we do not have enough clarification from ICNIRP as to how their standards and guidelines can be used in relation to very different conditions of RF EMF exposure on the population.

We further wish to note that there is no scientific background to extrapolate international standards to the real conditions of the habitat of the population. Completely absent are studies and papers on the methodology of transition to a rationing of acute exposure levels of thermal EMF RF to chronic long-term effects of low levels of radiation.

As a result of the existence of different methodologies, rationing RF EMF led to large differences in the recommended MPL for the people who are directly related to the risk assessment of the EMF of base stations; while MPL (Ministry of Health) established in Russia is $10\,\mu W/cm^2$, ICNIRP level is 100 times higher ($1000\,\mu W/cm^2$).

This analysis of the methodology of RF EMF regulation abroad allows us to conclude that the current so-called International Recommendations/Guidelines (ICNIRP, 1998) and the IEEE Standards (S95.1-2005), CENELEC (EN 50166-2.2000) do not correspond to existing conditions of RF EMF exposure on the population and cannot guarantee the safety of the public health.

Interestingly, this view was confirmed by the European Parliament in 2009 (in Paragraph 22 of the EMF resolution):

> 22. Notes that the limits on exposure to electromagnetic fields which have been set for the general public are obsolete, since they have not been adjusted in the wake of Council Recommendation 1999/519/EC of 12 July 1999 on the limitation of exposure of the general public to electromagnetic fields (0 Hz to 30 GHz), obviously take no account of developments in information and communication technologies, of the recommendations issued by the European Environment Agency or of the stricter emission standards adopted, for example, by Belgium, Italy and Austria, and do not address the issue of vulnerable groups, such as pregnant women, newborn babies and children. (The European Parliament approved the EMF resolution, April 12, 2009. The votes by the MEPs were: 559 yes, 22 against and 8 abstentions)

15.6 Children and the Possible Standardization of EMF RF

We believe that it is necessary within the framework of the development problems of the methodology of EMF RF standards to specifically consider additional criteria for risk assessment related to the exposure of children to RF EMF who became active users of mobile phones.

Western experts working on new standards, completely ignoring the problem of childhood cell phone use do not take into account the WHO opinion on the higher sensitivity of children to environmental factors in the International standards: "children are different from adults."

Children have a unique vulnerability. As they grow and develop, there are "windows of susceptibility": periods when their organs and systems may be particularly sensitive to the effect of certain environmental threats (WHO, 2003).

Previously, we paid attention to this problem (Grigoriev, 2004a, 2005, 2007, 2008, 2012, 2013; Belyev and Grigoriev, 2007; Fragopoulou et al., 2010; Markov and Grigoriev, 2013, 2015). The Russian National Committee on Non-Ionizing Radiation Protection adopted six resolutions on the need to protect children from the EMF of mobile phones (2001, 2004, 2008, 2009, 2011, 2012). Some of these decisions have been translated into English, and at our request, the WHO Secretariat has sent the decision to the members of the WHO "International EMF Project" advisory committee.

In Russian SanPiN 2.1.8/2.2.4.1190-03, Item 6.9 (2003) recommended to limit the possibility of the use of mobile phones by children.

Finally, in 2014–2016, several books and several articles were published confirming the need to address this problem (Grigoriev and Khorseva, 2014; Gandhi, 2015; Markov and Grigoriev, 2015; Grigoriev and Grigoriev, 2016). In 2012, American Academy of Pediatrics appealed to the Federal Communication Commission (FCC) of USA and urgently requested the revision of the methodological approach and standards to protect the health of children from mobile phone EMF (Kucinich, 2012).

15.7 The Real Picture of Today

The electromagnetic burden on the population is growing daily. At the same time, over the last 20 years, debates are still continuing on the following topic: Is the health of the population at risk because of increasing pollution due to RF EMF from the base stations and mobile phones?

The brains of almost all people on earth are exposed to EMF radiation. However, practically, there are no restrictions for the use of mobile communications. Having the advantages and convenience of mobile communication, the population is ignoring the information about the possible risks to their health. This threat affects everybody, including children aged 3–4 years. Pregnant women do not protect their fetuses from exposure to EMF.

The scientific community is watching this picture and is waiting for the results of this uncontrolled global experiment (Markov and Grigoriev, 2013). We saw similar hazards during the Victorian period in Britain (wallpaper with mercury and toys with lead).

15.8 What to Do?

In 2013, we were invited to Brussels for the EMF Workshop on Risk Communication of the European Commission for Health and Consumer Protection (DG SANCO) to present our viewpoint on the possible risk of mobile phone EMF to public health. The main theme of our report was that there are four postulates that show the risk to public health from mobile communication (Grigoriev, 2013). It is necessary to convince the population and to create an environment of reasonable restrictions on the use of this communication.

The first postulate: "EMF—harmful type of radiation." Mobile communication uses RF EMF. This type of electromagnetic radiation is considered harmful. Exceeding the permissible levels can cause disease; therefore, it requires hygienic control. *This is the absolute truth.*

The second postulate: "The brain and EMF." The mobile phone is an open source of EMF, and there is no protection for valuable human organs. EMFs affect the brain during mobile phone use. Nerve structures inside the internal ear (the vestibular and the auditory apparatus) are located directly under the beam of EMF. *This is the absolute truth.*

The third postulate: "Children and EMF." For the first time, in history the child's brain is subjected to RF EMF. There are no results of the study of chronic local RF EMF exposure on the brain. Children are more vulnerable to external environmental factors. This opinion was expressed by WHO (2003) and in the Parma Declaration (WHO European Region, 2010). *This is the absolute truth.*

Fourth postulate: "The lack of adequate recommendations/standards." There is no agreement on the methodology for determining the EMF RF remote control and for the development of international standards, and there are no results from 20 years of debate on this issue. *This is a real fact.*

15.9 My Suggestion

I have extensive experience in research on issues related to "ionizing radiation and health" (over 60 years) and "non-ionizing radiation and health" (about 40 years).

I believe that the time has come to provide the public with full information on the possible dangers of mobile communication for their health. The abovementioned four postulates allow the public to comprehend the likely risks to their health from uncontrolled use of mobile communication. Given the global situation, which occurred due to the use of mobile communications for the past 25 years, I believe that at this stage, only the people themselves can take more effective measures to limit the effects of RF EMF on their own health.

As a temporary measure of limiting exposure to EMF on the population, it is necessary to introduce the concept of *"voluntary risk"*; that is, the mobile phone use should be a product of self-selection on the background of the official public information about possible health hazards.

I appeal to colleagues: Do not sin against the truth!

15.10 Conclusion

Of course, new sources of electromagnetic radiation are creating additional problems in the development of standards. Public health protection issues in connection with the use of mobile communications have become completely different. The use of mobile phones has led to the local long-term RF EMF exposure to the brain. The normative level is not considered a permanent RF EMF exposure on the brain of the user. Existing regulations do not address to the real hazard RF EMF exposure. Given these circumstances, standards cannot currently guarantee the well-being of adults and children.

Children mobile phone users were included in the group of high risk. In this regard, there is a need to develop more appropriate stringent standards to ensure absolute security for growing children. Existing standards should take into consideration the vulnerable group of people hypersensitive to RF EMF.

> Given that the current regulations are outdated, it is necessary to carry out complex research into possible biological effects on conditions of chronic exposure to low-intensity EMF RF, bearing in mind, above all, long-term exposure on the brain at all levels of development.

As a temporary measure of limiting exposure to EMF on the population, it is necessary to introduce the concept of "voluntary risk"; that is, mobile telephony should be a product of self-selection on the background of the official public information about possible health hazards.

References

Adey WR (1980) Frequency and power windowing in tissue interactions with weak electromagnetic. *Proceedings of the IEEE* 68(1), 47–55.

Afrikanova EA, Grigoriev YG (2005) Influence of electromagnetic radiation of different modes on heart activity (in the experiment). *Radiation Biology, Radioecology* 36(5), 691–699 (in Russian).

Barnes F, Greenenbaum B (2016). Some effects of weak magnetic fields on biological systems: RF fields can change radical concentrations and cancer cell growth rates. *IEEE Power Electronics Magazine* 3(1), 60–68.

Belyaev I (2015) Biophysical mechanisms for nonthermal microwave effects. In: Markov M (Ed.) *Electromagnetic Fields in Biology and Medicine*, CRC Press, Boca Raton, FL, 49–67.

Belyev IY, Grigoriev YG (2007) Problems in assessment of risks from exposures to microwaves of mobile communication. *Radiation Biology, Radioecology* 47(6), 727–732.

Bernhard J-H (1999) Criteria for the development of international standards. Electromagnetic fields: Biological effects a hygienic standardization. In: *Proceedings of International Meeting*, Moscow, USSR, May 18–22, 1998. WHO, Geneva, Switzerland, 19–30.

Blackman C (1984) Biological effects of low frequency modulation of RF radiation. In: Cahill P (Ed.) *Biological Effects of Radiofrequency Radiation*, U.S. Environmental Protection Agency, Durham, NC, 588–592.

Calvente I, Pérez-Lobato R, Núñez MI, et al. (2016) Does exposure to environmental radiofrequency electromagnetic fields cause cognitive and behavioral effects in 10-year-old boys? *Bioelectromagnetics* 37, 25–36.

Chizhenkova RA (1988) Slow potential and spike unit activity of the cerebral cortex of rabbits exposed to microwaves. *Bioelectromagnetics* 9(40), 337–345.

Chizhenkova RA (2000) Burst activity in pulse flows of cortex neuron populations under low-intensity microwaves. In: *Millennium Workshop on Biological Effects of Electromagnetic Fields*, University of Ioannina, Ioannina, Greece, 104–108.

Chizhenkova RA (2004) Pulse activity of populations of cortical neurons under microwave exposures of different intensity. *Bioelectrochemistry* 63(12), 343–346.

Croft R, Hamblin D, Spong J, et al. (2008) The effect of mobile phone electromagnetic fields on the alpha rhythm of human electroencephalogram. *Bioelectromagnetics* 29, 1–10.

Dec S, Cieslak E, Miszczak J (1997) Abstract book. In: *Second World Congress for Electricity and Magnetism in Biology and Medicine*, Bologna, Italy, 273.

Fragopoulou A, Grigoriev Y, Johansson O, et al. (2010) Scientific panel on electromagnetic field health risks: Consensus points, recommendations, and rationales. Scientific meeting: Seletun, Norway, November 17–21, 2009. *Reviews on Environmental Health* 25(4), 1–11.

Frey A, Feld S, Frey B (1975) Neural function and behavior: Defining the relationship. *Annals of New York Academy of Sciences* 247, 433–439.

Gandhi OP (2015) Yes the children are more exposed to radio-frequency energy from mobile telephones than adults. *IEEE Access* 3, 985–988.

Grigoriev Y (1998) The Russian standards and the opinion about international harmonization of electromagnetic standards. In: *International Seminar on Electromagnetic Fields. Global Need for Standards Harmonization*, Ljubljana, Slovenia, 1–6.

Grigoriev Y (1999a) Modulation and EMF somatic effects. In: *Proceedings of 1st International Medical Scientific Congress*, November 29–30, Rome, Italy, 78–86.

Grigoriev Y (2001) Role of modulation in development of EMF somatic effects. In: *Proceedings of WHO Meeting on EMF Biological Effects and Standards Harmonization in Asia and Oceania*, October 22–24, Seoul, Korea, 77–80.

Grigoriev Y (2004a) Mobile phones and children: Is precaution warranted? *Bioelectromagnetics* 25(5), 322–323.

Grigoriev Y (2006) Development of electromagnetic field somatic effects: Role of modulation. *Proceedings of the Latvian Academy of Sciences. Section B, Natural, Extract and Applied Sciences* 60(10), 11–15.

Grigoriev Y (2008a) Russian NCNIRP and standards. New condition of EMF RF exposure and guarantee of population health. In: *Proceedings of International Conference "the Global Issue of EMF and Health,"* London.

Grigoriev Y (2013) Four indisputable postulate/truth to the risk assessment of mobile communications for public health (our opinion). In: *EMF Workshops on Risk Communication*, SANCO, Brussels, Belgium, 6–7.

Grigoriev Y, Lukyanova S, Rynskov V, et al. (1999) Proceedings of International Meeting, Moscow, USSR, May 18–22. WHO, Geneva, Switzerland, 501–512.

Grigoriev Y, Shafirkin A, Vasin A (2002) Standardization of RF EMF for Russian population. A retrospective study and a modern point of view. In: *Electromagnetic Fields and Human Health*, 93–123 (in Russian).

Grigoriev Y, Shafirkin A, Vasin A (2003a) Biological effects of chronic exposure to electromagnetic fields, radio frequency low-intensity (valuation strategy). *Radiation Biology, Radioecology* 43(5), 501–511 (in Russian).

Grigoriev Y, Shafirkin A, Vasin A (2004) To improve the methodology for the valuation of radio frequencies of EMF. In: *Annual Book. The Russian National Committee on Non-ionizing Radiation Protection*, Moscow, USSR, 108–150 (in Russian).

Grigoriev Y, Stepanov V (2000) Microwave effect on embryo brain: Dose dependence and the effect of modulation. In: Klauenberg B, Miklavic D (Eds) *Radio Frequency Radiation Dosimetry*. Kluwer Academic Publishers, Dordrecht, the Netherlands, 31–37.

Grigoriev Y, Vasin A, Grigoriev O, et al. (2003b) Harmonization options for EMF standards: Proposals of Russian National Committee on Non-ionizing Radiation Protection (RNCNIP). In: *Proceedings of 3rd International EMF Seminar in China*, October 13–17, 55–56.

Grigoriev YG (1996) Role of modulation in the biological effect of EMF. *Radiation Biology, Radioecology* 36(5), 659–670 (in Russian).

Grigoriev YG (1997) A man in an electromagnetic field (the current situation, the expected biological effects and hazards assessment). *Radiation Biology, Radiology* 37(4), 690–703 (in Russian).

Grigoriev YG (1999b) The Russian standards, EMF RF. In: *Inaugural Round Table on World EMF Standards Harmonization Minutes of Meeting*. WHO, Zagreb, Croatia/Geneva, 1998, 57–60.

Grigoriev YG (2004b) Biological effects under the influence of modulated electromagnetic fields in acute experiments (on the basis of national studies). In: *RNCNIRP, Annual Book 2003*. Moscow, USSR, 16–72 (in Russian).

Grigoriev YG (2005) The electromagnetic fields of cell phones and health of children and adolescents (a situation that requires urgent action). *Radiation Biology, Radioecology* 45(4), 442–450.

Grigoriev YG (2007) Modern problems of electromagnetic radiobiology. Immediate and long-term objectives. In: *Scientific Conference "New Directions in Radiobiology,"* RAS, Moscow, USSR, 27–33 (in Russian).

Grigoriev YG (2008b) Children risk in the evaluation of the danger of mobile EMF (forecast health of present and future generations). In: *Proceedings of the International Scientific Conference. Electromagnetic Radiation in Biology "BIO-EMI 2008,"* Kaluga, USSR, 207–209.

Grigoriev YG (2012) Mobile communications and health of population: The risk assessment, social and ethical problems. *The Environmentalist* 32(2), 193–200.

Grigoriev YG, Grigoriev OA (2013) *Cellular Communication and Health. Electromagnetic Environment. Radiobiological and Hygienic Problems. The Prognosis of Hazard.* 565 p. (in Russian).

Grigoriev YG, Grigoriev OA (2016) *Cellular Communication and Health. Electromagnetic Environment. Radiobiological and Hygienic Problems. The Prognosis of Hazard,* 2nd edition, 572 p. (in Russian).

Grigoriev YG, Khorseva NI (2014) *Mobile Communications and Children Health. Assessment of Hazard of Mobile Phone Use by Children and Teenagers. Recommendations to Children and Parents,* 229 p. (in Russian).

Grigoriev YG, Stepanov VS (1998) Formation of memory (imprinting) in chicks after prior exposure to electromagnetic fields of low levels. *Radiation Biology, Radioecology* 38(2), 223–231.

Hardell L, Carlberg M (2009) Mobile phones, cordless phones and the risk for brain tumors. *International Journal of Oncology* 35, 5–17.

Hardell L, Mild H (2003) Mobile and cordless telephones and association with brain tumors in different age groups. Abstract book. In: *5-th COST 281 MCM and Workshop "Mobile Telecommunications and the Brain,"* Budapest, Hungary, November 15–16, 13–14.

Huber R, Schuderer J, Graf T, et al. (2003) Radio frequency electromagnetic field exposure in humans: Estimation of SAR distribution in the brain, effects on sleep and heart rate. *Bioelectromagnetics* 24(4), 262–276.

Hunt EL. (1949) Establishment of conditioned responses in chick embryos. *Journal of Comparative and Physiological Psychology* 42(2), 107–117.

IARC/WHO (2011) Classifies radiofrequency electromagnetic fields as possibly carcinogenic to humans. Press Release No. 208, May 31, 3 p.

ICNIRP (1998) Guidelines for limiting exposure to time-varying electric, magnetic, and electromagnetic fields (up to 300 GHz). *Health Physics* 74(4), 494–522.

Ivanov-Muramsky KA (1977) *Electromagnetic Biology.* Naukova Dumka, Kiev, Ukraine, 77 (in Russian).

Jamieson I (2014) Changing perspectives—Improving lives. Revised version of graphic from presentation given at the European Economic and Social Committee's "Electromagnetic hypersensitivity, Public Hearing—EESC," November 4, EESC, Brussels, Belgium.

Jun S (2016) The reciprocal longitudinal relationships between mobile phone addiction and depressive symptoms among Korean adolescents. *Computers in Human Behavior* 58, 179–186.

Kholodov YA (1975) *The Reactions of the Nervous System to the EMF.* Nauka, Moscow, USSR, 284 p. (in Russian).

Kholodov YA (1996) *Influence of Electromagnetic and Magnetic Fields on the Central Nervous System.* Nauka, Moscow, USSR, 284 p. (in Russian).

Krause C (2002) EMF effects on human cognitive processes and the EEG. Abstract book. In: *24th BEMS Annual Meeting*, Quebec, Canada, 12.

Kucinich, D. (2012) American Academy of Pediatricians endorses cell phone safety legislation. AAP press release, OpEd News.com. http://www.opednews.com/articles/American-Academy-of-Pediat-by-Dennis-Kucinich-121213-724.html

Lukyanov SN (1999) The reaction of the central nervous system in the short-term low-intensity microwave radiation. In: *Proceedings of Electromagnetic Fields: Biological Action and Hygienic Regulations.* WHO, Geneva, Switzerland, 401–408.

Lukyanov SN (2002) Phenomenology and the genesis of the changes in total bioelectric activity of the brain to electromagnetic radiation. *Radiation Biology, Radioecology* 42(3), 308–314 (in Russian).

Lukyanov SN (2015) *The Electromagnetic Field of Microwave Nonthermal Intensity as a Stimulus to the Central Nervous System*, Moscow, USSR, 200 p.

Lukyanov SN, Grigoriev OA, Grigoriev YG, et al. (2010) The dependence of the biological effects of electromagnetic field radio frequency thermal intensity of the typological features of the human electroencephalogram. *Radiation Biology, Radioecology* 50(6), 1–11.

Lukyanov SN, Karpikova NI, Grigoriev OA, et al. (2015) Accumulation of neurological effects of repeated non-thermal electromagnetic radiation exposure. *Radiation Biology, Radioecology* 55(2), 169–179 (in Russian).

Lukyanov SN, Makarov VP, Rynskov VV (1996) The dependence of the total changes of bioelectrical activity of the brain to low-intensity exposure from flux density. *Radiation Biology, Radioecology* 36(5), 706–709 (in Russian).

Markov M, Grigoriev YG (2013) Wi-Fi technology—An uncontrolled global experiment on the health of mankind. *Electromagnetic Biology and Medicine* 32(2), 200–208.

Markov M, Grigoriev Y (2013) Wi-Fi technology—An uncontrolled experiment on the health of mankind. *Electromagnetic Biology and Medicine* 32, 200–208.

Markov M, Grigoriev Y (2015) Protect children from EMF. *Electromagnetic Biology and Medicine* 34(3), 251–256.

Markov MS. (2006) Therapy with magnets – myth or reality. In: Kostarakis P (Ed.) Proceedings of Forth International Workshop "Biological effects of electromagnetic fields," Crete 16–20 October 2006, ISBN# 960-233-172-0, p.10–17.

Ministry of Health of USSR (1981) *Guidelines on the Assessment of the Action of the Biological Low-Intensity Microwave Radiation for Hygienic at Ambient Conditions.* The USSR Ministry of Health, Kiev, Ukraine, 1981, 28 p.

Panda NK, Jain R, Bakshi J, et al. (2007) Audiological disturbances in long-term mobile phone users. *Journal of Otolaryngology-Head and Neck Surgery* 137(2), 131–132.

Panda NK, Jain R, Bakshi J, et al. (2010) Audiologic disturbances in long-term mobile phone users. *Journal of Otolaryngology-Head and Neck Surgery* 39(1), 5–11.

Panda NK, Modi R, Munjal S, et al. (2011) Auditory changes in mobile users: Is evidence forthcoming? *Journal of Otolaryngology-Head and Neck Surgery* 144(4), 581–585.

Presman AS (1968) *Electromagnetic Fields and Wildlife.* Science Publications, Moscow, USSR, 288 p. (in Russian).

Reister H, Dimfler W, Shober F (1995) The influence of electromagnetic fields on human brain activity. *European Journal of Medical Research* 1(1), 27–32.

RNCNIRP (2008) Decision: Children and mobile phones: The health of the following generations is in danger. In: *Annual Book-2008*, Moscow, USSR, 118–119.

Rubsova NV, Palsev YP (2006) Situation and perspectives of the safety compliance for using mobile phone. *Journal of Bezopasnost Zhiznedeyatelnosti* 2, 29–33 (in Russian).

Shandala MG, Vinogradov GI (1982) Auto-allergic effects of microwave electromagnetic energy exposure on rats with special reference to the fetus and offspring. *Vestnik AMN* 10, 13–16 (in Russian).

Shandala MG, Vinogradov GI, Rudnev MI, et al. (1985). Non-ionizing microwave radiation as an inducer of auto-allergic processes. *Gigiena i Sanitaria* 8, 32–35 (in Russian).

Shandala MG, Vinogradov GI, Rudnev MI, et al. (1983) Effects of microwave irradiation on some parameters of cell immunity under conditions of chronic exposure. *Radiobiologia* 23, 544–546 (in Russian).

Suvorov GA, Fingers YP, Rubtsova NB, et al. (2002) Questions of biological effect and hygienic regulation of EMF generated by the mobile communication means. *Journal of Medicine of Labor and Industrial Ecology* 9, 10–18.

Thuroczy G, Kubinyi G, Sinay H, et al. (1999) Human electrophysiological studies on the influence of RF exposure emitted GSM cellular phones. In: Bersani F (Ed.) *Electricity and Magnetism in Biology and Medicine.* Academic/Plenum Publishers, New York, 721–724.

USSR, Ministry of Health. (1981) *Methodological Approaches to the Study of the Effect of Low Levels of Microwave Energy on the Human Health in Terms of Residential Areas.* Hunt E. // Comp. Physiol. Psychology, 1949, V.42 P. 107–109.

Vinogradov GI, Andriyenko LG, Naumenko GM (1999) The phenomenon of adaptive immunity when exposed to non-ionizing microwave radiation. *Radiobiology* 31(5), 718–721 (in Russian).

Vinogradov GI, Dumanskiy YD (1974) Changing the antigenic properties of tissues and autoimmune processes when exposed to microwave energy. *Bulletin of Experimental Biology and Medicine* 8, 76–79 (in Russian).

Vinogradov GI, Dumanskiy YD (1975) On the sensitizing effect of electromagnetic fields of ultrahigh frequency. *Hygiene and Sanitation* 9, 31–35 (in Russian).

Vinogradov GI, Naumenko GM (1986) Experimental modeling of autoimmune reactions upon exposure to non-ionizing microwave radiation. *Radiobiology* 26(5), 705–708 (in Russian).

WHO (2003) Healthy environments for children, WHO Backgrounder No. 3.

WHO European Region (2010) *Parma Declaration of European Region of WHO*, Brussels, Belgium.

WHO Handbook (2002) *Establishing a Dialogue on Risks from Electromagnetic Fields.* WHO, Geneva, Switzerland, 66 p.

16

The Intracellular Signaling System Controlling Cell Hydration as a Biomarker for EMF Dosimetry

Sinerik Ayrapetyan
UNESCO

Electromagnetic fields (EMFs) exist in nature and have consequently always been present on earth; whereas, in recent decades, environmental exposure to man-made sources of EMFs has risen constantly, driven by demand for electricity, increasingly more specialized wireless technologies, and changes in the organization of society [European Parliament, Report on health concerns associated with electromagnetic fields (2008/2211(INI)), February 23, 2009], leading to the increase of different diseases. Thus, health control and health risk assessment become two of the essential problems in public health. The solution to these problems is extremely important for better health promotion and disease prevention. To solve these problems, it is extremely important to estimate the environmental effects on health adequately. As EMF [especially extremely low frequency of EMF (ELF EMF) and radiofrequency of EMF (RF EMF)] is one of the main components of environmental pollution in densely populated areas, with a tendency to increase as a result of technical progress, the possible hazardous effects of EMF become of great public concern.

In spite of the fact that biological effects of EMF can be considered as a proven fact, and it is established that the impact of EMF on organisms depends not only on its thermodynamic characteristics (thermal) but also on its nonthermal (NT) effects, physicochemical characteristics of the EMF exposure medium, as well as the initial functional state of the organism, The World Health Organization (WHO), International Commission on Non-Ionizing Radiation Protection (ICNIRP), and other national and international health control organizations, having mission to monitor beneficial and hazardous effects of EMF, traditionally base their instruction on thermodynamic characteristics [specific absorption rate (SAR)] of EMF. It is obvious that SAR cannot serve as a reliable marker for estimation of possible beneficial and hazardous effects of EMF on the health of organisms (in different environmental mediums and in different initial metabolic states), which is extremely important for vulnerable social groups, especially for people of different ages and sexes with different functional states. EMF risk groups have been identified by European legislation. However, a shortcoming is noticed in the risk assessment methods on how to quantify and assess the EMF-induced effects for the risk groups (children, pregnant women, and people with medical implants; Directive 2013/35/EU). Therefore, for adequate estimation of beneficial and hazardous effects of EMF on organisms, novel approaches based on a universal and extra-sensitive biomarker reflecting the effects of different components of environmental medium as well as the initial functional state of the organism are of vital importance. The discovery of such a biomarker as a parameter for detection of biological effects of EMF on an organism could open a new avenue for dosimetry of environmental pollution, including EMF. On the basis of the previous two-decade study of our group, we suggest metabolically controlled cell hydration as a biomarker for adequate estimation of biological effects of EMF on cells and organisms. In this chapter, an attempt has been made to support this statement.

16.1 Cell Aqua Medium as a Primary Target for EMF Effects on Cells and Organisms

As the biological effects of EMF can also be recorded at intensities much less than the thermal threshold (Devyatkov, 1973; Adey, 1981; Markov, 2004; Kaczmarek, 2006; Gapeyev et al., 2009; Grigoriev, 2012), classic thermodynamics theories fail to explain this phenomenon (Foster, 2006; Binhi and Rubin, 2007). Therefore, it is suggested that the target(s) of such weak signals could have a quantum mechanical nature (Belyaev, 2012; Blank and Goodman, 2012; Gapeyev et al., 2013). However, the nature of a primary target through which the NT effect of EMF on cells and organisms is realized has not been elucidated yet and serves as one of the core problems of modern electromagnetobiology.

Some researchers try to explain these effects in terms of a quantum mechanical approach, considering different biochemical reactions with participation of uncoupled electrons, such as the electrons transferring cytochrome c to cytochrome oxidize during oxidation of malonic acid, the activity of Na^+/K^+-adenosine triphosphate (ATPase), and during other target reactions (Belyaev, 2012; Binhi, 2012; Blank and Goodman, 2012). However, high sensitivity of formation and breaking processes of hydrogen bonds to EMF that occur in cell aqua medium (Klassen, 1982; Chaplin, 2006), and especially during metabolically driven collective dynamics of intracellular water molecules (Ayrapetyan and De, 2014), makes the role of individual biochemical reactions inessential in NT effect of EMF on intracellular metabolism. In this respect, the so-called "water" hypothesis seems more reliable for explaining the biological effect of EMF. As the valence angle of water molecules between O–H bonds, which determines its dissociation and physicochemical properties, is highly sensitive to different environmental factors (Szent-Gyorgyi, 1968; Klassen, 1982; Lednev, 1991; Ayrapetyan et al., 1994a, 2015; Chaplin, 2006), according to this hypothesis, EMF can modulate cell metabolism by changing physicochemical properties of water, which is a common medium for metabolic reactions in cells and organisms.

Our previous studies have shown that the effect of EMF on physicochemical properties of water solutions and formation of reactive oxygen species (ROS) in it depends on Ca^{2+} concentration in water solution (Ayrapetyan et al., 1994b), temperature, background radiation, light, and gas composition of the medium (Ayrapetyan et al., 1994a, 2015). It is known that the formation of H_2O_2 in physiological solution (PS) upon the impact of EMF is one of the main messengers for modulation of cell metabolism (Ayrapetyan et al., 2005; Gudkova et al., 2005; Gapeyev et al., 2009). Previously, it has also been shown that the effects of ELF EMF and microwave (MW) on H_2O_2 contents in PS depend on chemical and physical characteristics of the environmental medium (Baghdasaryan et al., 2012).

As can be seen in Figure 16.1, 10-min-treated PS by 8Hz-modulated MW with the intensity of $SAR = 3.8\,mW/g$ at normal background radiation (NBGR) and illumination at 18°C decreases H_2O_2 content by 13%, while at low background radiation (LBGR) and in dark conditions, the same MW radiation has no significant effect on it. At present, it is a well-documented fact that EMF-treated water solutions have modulation effects on different types of organisms (including plants, microbes, and human beings) (Ayrapetyan et al., 2015; Mayrovitz, 2015). However, the variability of water structure that is dictated by its high sensitivity to different environmental factors makes it difficult to use EMF-induced changes of water structure as a marker for estimation of beneficial and hazardous effects of EMF on organisms. This process becomes more complicated due to the fact that the sensitivity of cells to EMF-treated water also depends on the initial metabolic state of the cells.

As can be seen in Figure 16.2, water uptake by seeds in dormant state is insensitive to EMF-treated solution, while during the germination period, EMF-treated water solution strongly elevates water uptake by seeds compared with nontreated water solution.

Thus, the obtained data allow us to suggest that the metabolically driven intracellular water dynamics, which takes place during anabolic and catabolic processes, is a more sensitive target for EMF than simple thermodynamic processes such as osmotic water absorption by seeds in dormant state.

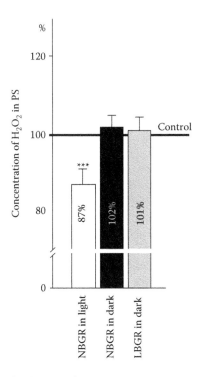

FIGURE 16.1 8-Hz–modulated MW effect on H_2O_2 content in PS at *NBGR in light, NBGR in dark*, and LBGR *in dark* mediums at 18°C temperature condition. Columns indicate the mean of 10 independent measurements, vertical bars represent the standard error of mean (SEM), ***$p < .005$ as compared to control (*NBGR in light*). (From Ayrapetyan, S., et al. In: M.S. Markov (Ed.) *Electromagnetic Fields in Biology and Medicine*, CRC Press, Boca Raton, FL, 196, 2015, Figure 13.3.) NBGR, normal background radiation; LBGR, low background radiation; PS, physiological solution.

The fact that there is a direct correlation between metabolic activity of cells and EMF sensitivity of cell hydration has also been shown by our previous study on age-dependent magnetic sensitivity of brain and heart muscle tissue hydration of rats (Heqimyan et al., 2015).

As can be seen in Figure 16.3, SMF exposure on young animals (metabolic activity is high) leads to dehydration in both brain cortex and heart muscle tissues, while in old animals (metabolic activity is depressed) it leads to overhydration of brain cortex tissue and has no effect on the hydration of heart muscle tissues.

It is known that isolated neurons and heart muscles of mollusks are convenient experimental models for the study of their general physiological properties, as in *in vitro* experimental conditions they can function in a stable manner for several hours. Therefore, in our studies, the isolated neurons and heart muscles of *Helix pomatia* have been chosen as experimental models for studying the cellular mechanism of EMF-treated PS on their functional activities. It has been shown that magnetized PS (MPS) and MW-treated PS have dehydration effects on isolated neurons of snail (Figure 16.4).

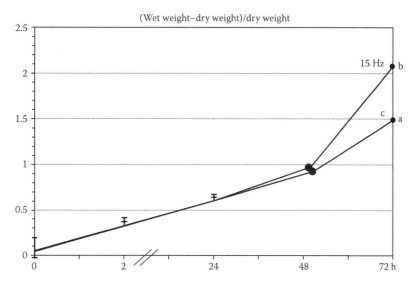

FIGURE 16.2 The effect of preliminary ELF EMF-treated distilled water (DW) on seed hydration during 72 h of incubation. (a) Seed hydration in cold (4°C) DW and (b) seed hydration in DW at room temperature (20°C). "C" indicates *control*—"sham-treated." (From Ayrapetyan, S., et al. In: M.S. Markov (Ed.) *Electromagnetic Fields in Biology and Medicine*, CRC Press, Boca Raton, FL, 198, 2015, Figure 13.5.)

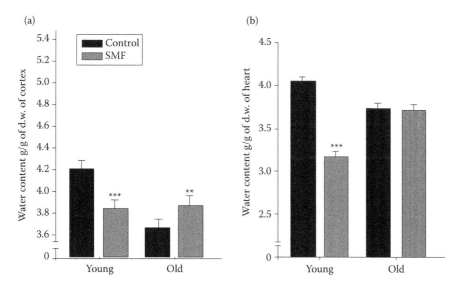

FIGURE 16.3 Change of water content in brain cortex (a) and heart muscle (b) tissues of young and old rats after SMF exposure. Ordinate indicates the mean value of tissues' water content. Error bars indicate the SEM for three independent experiments. ** and *** indicate $p < .01$ and $p < .001$, respectively. (From Heqimyan, A., et al., In: M.S. Markov (Ed.) *Electromagnetic Fields in Biology and Medicine*, CRC Press, Boca Raton, FL, 223, 2015, Figure 16.7.) SMF, static magnetic field.

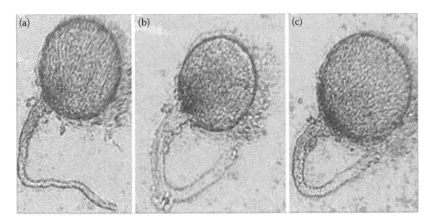

FIGURE 16.4 The effect of 9.3 GHz MW (SAR = 50 kW/kg) 5-min–pretreated PS effect on snail neuronal volume. In normal (sham-exposed) PS (a) after 2–3 min (b) and 5–7 min (c) of incubation in preliminary MW-treated PS (light microscopic picture). (From Ayrapetyan, S.N., In: Ayrapetyan, S.N., Markov M.S., (Eds.), *Bioelectromagnetics: Current Concepts*, Springer, Dordrecht, the Netherlands, 61, 2006, Figure 24.)

Figure 16.4 demonstrates the effect of 20-Hz-modulated 9.3 GHz MW (SAR = 50 kW/kg)-pretreated PS on the neuronal size of snail. As can be seen, as a result of replacement of sham PS with 5-min MW-pretreated PS (at the same sham temperature), time-dependent changes of neuronal volume are observed. During the first 2–3 min after the replacement, there is shrinkage and during 5–7 min after returning the neuron to sham solution, the neuron becomes more swollen than it was in the initial state.

It is known that the permeability of cell membrane for water molecules is much higher than the permeability of cell membrane for ions (Borgnia et al., 1999; Hoffman et al., 2015) and that intracellular osmotic pressure exceeds the extracellular one (Evans, 2009). Therefore, to keep cell volume in a steady state, the osmotic water uptake must be balanced by water efflux from the cell. As EMF-treated PS leads to neuronal shrinkage, it is predicted that before stabilization of cell volume the transient net water efflux from the cell is observed, whereas, after returning the neuron to nontreated PS, cell swelling takes place, which is followed by transit net water uptake.

Our study performed on intracellularly dialyzed squid axons and intact neurons of snails has shown that water fluxes through membrane have a crucial role in regulation of cell membrane permeability for Na^+, Ca^{2+}, and K^+: water influx and efflux through cell membrane have activation and inactivation effects on inward ionic currents [particularly, on Na^+ current (I_{Na}) and Ca^{2+} current (I_{Ca})] and opposite effects on outward K^+ currents (I_k), respectively (Kojima et al., 1984; Ayrapetyan and Rychkov, 1985; Ayrapetyan et al., 1988; Suleymanian et al., 1993). At the same time, it has been shown that there is a direct correlation between the number of functionally active ionic channels in the membrane and the surface of the cell membrane (Ayrapetyan et al., 1988). Figure 16.5 shows the mentioned effects of cell shrinkage on membrane current, in which the correlation between the time-dependent changes of cell volume and amplitudes of I_k are demonstrated.

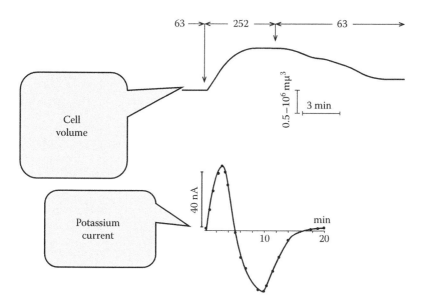

FIGURE 16.5 Time course of both cell volume (up curve) and K outward current (down curve) changes after replacing the normal isotonic solution by a hypertonic one, and the recovery of the K current after returning to the isotonic solution. The hypertonic solution was replaced by the isotonic solution after the 10th minute. (From Ayrapetyan, S.N., et al., *Comp. Biochem. Physiol.*, 89, 181, 1988, Figure 4.)

Figure 16.5 makes it obvious that after applying hypertonic solution, when water flows outward through the membrane 30 s before stabilization of cell volume, the I_k increases despite the fact that the amplitude of the command pulse (50 mV) remains constant. After 2.5 min, the I_k reaches its maximum value and then begins to decrease. Such a decrease of I_k lasts up to the 10th min, i.e., until stabilization of the neuronal volume in the hypertonic solution. It is necessary to note that the kinetics of the I_k is significantly changed, that is, the process of achievement of the peak value of the I_k is essentially accelerated. From these data, it can be concluded that EMF radiation-induced shrinkage of neurons (Figure 16.4) leads to transient water efflux through the membrane and decreases the sustaining membrane surface, with both of them having a depressing effect on membrane excitability (I_{Na}) (Ayrapetyan et al., 1988).

It has also been shown that both hypertonic PS (Ayrapetyan and Arvanov, 1979) and MPS (4 Hz, 0.2 mT) have a depression effect on acetylcholine-(Ach)-induced current in the neuronal membrane of snail (Ayrapetyan et al., 2004).

The fact that the depressing effect of MPS on Ach-induced current (Figure 16.6a) disappears in cold medium (Figure 16.6b) indicates that the observed depressing effect of MPS on membrane chemosensitivity has a metabolic nature.

The results showing that both MPS and NT MW-treated PS have modulation effect on cell hydration have been demonstrated on brain cortex and heart muscle tissues of rats in *in vivo* and *in vitro* experiments (Ayrapetyan, 2015; Heqimyan et al., 2015).

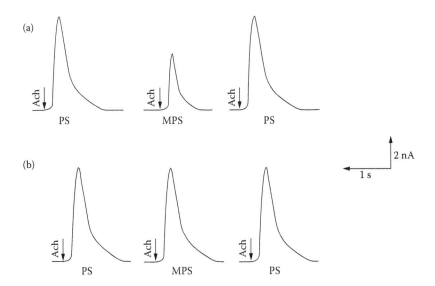

FIGURE 16.6 The effect of magnetized PS (MPS) on Ach-induced current in D-type neuron in voltage-clump experiments, where the clumping potential (E_c) is equal to the resting membrane potential ($E_r = -50\,mV$). (a) At room temperature (23°C) and (b) in cold medium (12°C). The rows show the transit (30 ms) application of $10^{-4}\,M$ Ach-containing PS. The time of preincubation of neurons in tested physiological solution (PS and MPS), before the application of Ach was 3 min. The intervals between Ach applications were 5 min. (From Ayrapetyan S.N., In: Ayrapetyan S.N., Markov M.S., (Eds.), *Bioelectromagnetics: Current Concepts*, Springer, Dordrecht, the Netherlands, 51, 2006, Figure 18.)

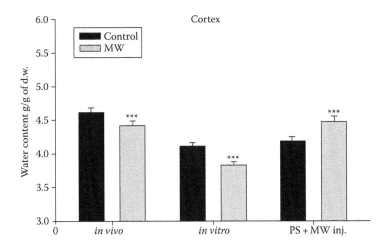

FIGURE 16.7 The effects of direct exposure on brain (*in vivo*) and brain slices bathing in PS (*in vitro*) and IP injection of MW-treated PS on brain tissue hydration (PS + MW inj.). (From Ayrapetyan, S., et al., In: M.S. Markov (Ed.) *Electromagnetic Fields in Biology and Medicine*, CRC Press, Boca Raton, FL, 210, 2015, Figure 13.20.)

Direct exposure of NT MW on rats' brain *in vivo* and NT MW exposure on brain cortex slices containing saline show dehydrating effects on cortex slices, while the intraperitoneal (IP) injection of MW-treated PS with the same intensity has a hydrating effect on them (Figure 16.7).

From these data, it can be assumed that the effect of IP injection of EMF-treated PS on brain tissue hydration and the direct radiation effect of EMF on an organism are not equivalent as water molecules are not the only target for EMF effects.

Thus, it becomes obvious that knowledge on the metabolic nature of the cellular sensor through which the EMF-treated water molecules and water dissociation products modulate cell metabolism seems extremely important for choosing an adequate biomarker for EMF dosimetry from the point of public health.

16.2 The Correlation between Cell Hydration and Na$^+$/K$^+$-ATPase α_3 Isoform

Cell hydration is one of the fundamental cell parameters determining metabolic activity. Cell hydration regulates metabolic activity of cells via (1) "folding–unfolding" mechanisms of intracellular macromolecules (Minkoff and Damadian, 1976; Parsegian et al., 2000), (2) cell-surface-dependent changes of the number of functionally active protein molecules, having enzymes (Ayrapetyan et al., 1984), receptors (Ayrapetyan et al., 1985), ionic channel-forming properties (Ayrapetyan et al., 1988), and (3) water-flux-induced changes of membrane permeability for ions. Therefore, metabolically controlled cell hydration is of vital importance for living cells.

Metabolically controlled cell hydration is realized by three pathways: (1) by anti-gradient membrane transporting mechanisms of osmotically active particles through membrane, (2) by intracellular signaling system controlling myosin-like protein contractility in cytoskeleton and absorption properties of intracellular structure, and (3) by endogenous production of water due to intracellular oxidative phosphorylation process.

It is known that, among a number of metabolic mechanisms involved in cell volume regulation, Na$^+$/K$^+$ pump has a central role, which is proven by the following facts:

1. Na$^+$/K$^+$ pump generates Na$^+$ gradient on the membrane, which serves as an energy source for a number of secondary ionic transporters in membrane, such as Na$^+$/H$^+$, Na$^+$/Ca^{2+}, Na$^+$/sugars, amino acids, and other osmolytes (Ayrapetyan and Sulejmanian, 1979; Lang, 2007; Hoffmann et al., 2009).

2. Na$^+$/K$^+$ pump working in electrogenic regime (Ayrapetyan, 1969; Thomas, 1972) inhibits I_{Na} and I_{Ca} uptake by the cell via three pathways: (1) hyperpolarization of membrane potential (Hodgkin, 1964), (2) water efflux through ionic channels (Kojima et al., 1984; Ayrapetyan et al., 1988), and (3) cell-shrinkage-induced decrease of the number of functioning ionic channels in membrane (Ayrapetyan et al., 1988).

3. Na$^+$/K$^+$ pump plays a major intracellular signaling role in regulation of absorption properties of intracellular structures (Xie and Askari, 2002).

4. Na^+/K^+ pump, being the highest metabolic energy (ATP) utilizing mechanism, stimulates the intracellular oxidative phosphorylation process that causes the release of water molecules in cytoplasm (releasing 42 water molecules per 1 molecule glucose oxidation) (Lehninger, 1970).

From the aforementioned data, it can be assumed that the Na^+/K^+ pump activation-induced water efflux through cell membrane, which is due to both its electrogenecity-induced dehydration and oxidation-induced stimulation of endogenous water release, has a strong inhibitory effect on Na^+ and Ca^{2+} uptake by the cell, while pump inactivation has the opposite effect.

As the Na^+ gradient on membrane is the highest inward going ion gradient on the membrane and serves as an energy source for a number of exchanges, such as Na^+/H^+ and Na^+/Ca^{2+} exchanges, its decrease is a common consequence of any cell pathology. *Therefore, the net water influx- and efflux-induced activation and inactivation of membrane permeability for Na^+, respectively, can be considered as a marker for estimation of potential harmful and beneficial effects of any environmental factors, including EMF.* This suggestion supports the hypothesis of Prof. Tasaki and his coworkers, according to whom neuronal swelling and shrinkage during action potentials are due to water influx and efflux, respectively, and ionic currents are just consequences of cytoskeleton contraction (Iwasa et al., 1980; Terakawa, 1985). Therefore, the attempt of some researchers to explain the effects of EMF and homeopathic concentrations of chemical substances on cells and organisms from the point of classic membrane theory seems unrealistic, as according to this theory, a couple of millivolts, or no less than 10^{-9} M agonists, are needed for the activation of the potential- or agonist-dependent ionic channels (Katz, 1966). However, at present, there is a great amount of data in the literature on the changes of intracellular ionic homeostasis upon the effects of EMF, having intensity less than thermal thresholds (see WHO-BioInitiative, 2012; Blackman, 2015; Pilla, 2015) and extremely low concentrations $<10^{-13}$ M biologically active substances (Ayrapetyan and Carpenter, 1991a,b; Azatian et al., 1998). Thus, from the these data, it can be assumed that water molecules, the quantum-mechanical properties of which are extra-sensitive to weak environmental factors, having intensity even less than the thermal threshold (such as EMF), serve as a messenger for the metabolic mechanism controlling water balance on membrane. *Therefore, it is suggested that the elucidation of the metabolic nature of the mechanism responsible for controlling water balance on membrane could allow us to perform adequate EMF dosimetry from the point of public health.*

At present, it is known that in neuronal and muscle tissues of healthy animals, three isoforms ($\alpha_1, \alpha_2, \alpha_3$) of α catalytic subunits of Na^+/K^+-ATPase (working molecules of Na^+/K^+ pump) can be identified (Juhaszova and Blaustein, 1997). They are characterized by different affinities to cardiac glycoside ouabain (specific inhibitor for Na^+/K^+-ATPase) and functional activities: α_1 (with low affinity) and α_2 (with middle affinity) isoforms are involved in transportation of Na^+ and K^+, while α_3 (with high affinity) is not directly involved in transporting Na^+ and K^+ and has only signaling function (Xie and Askari, 2002; Wu et al., 2013). Therefore, the physiological role of the high-affinity α_3 receptor is still a matter of discussion (Blaustein and Lederer, 1999; Xie and Askari, 2002; de Lores Arnaiz et al., 2014; Romanovsky et al., 2015).

Our earlier study has shown that like in mammalian neurons (Juhaszova and Blaustein, 1997) in snail neurons as well the curve of dose-dependent [³H]-ouabain binding with membrane consists of three components: two (10^{-11}–10^{-9} and 10^{-9}–10^{-7} M ouabain) saturated and one (10^{-7}–10^{-4} M) linear component (Ayrapetyan et al., 1984).

As can be seen in Figure 16.8, ouabain at concentrations of >10^{-7} M only has an inhibitory effect on $^{22}Na^+$ efflux from the neurons (in exchange with Rb^+; K^+) (Ayrapetyan et al., 1984), while low concentration of ouabain has a reverse dose-deendent activation effect on $^{22}Na^+$ efflux in exchange with Ca^{2+} (Saghian et al., 1996). It has been shown that there is a direct correlation between cell hydration (cell volume) and the number of functionally active ouabain receptors (pump units) in the membrane. The number of all three families of ouabain receptors in cell membrane also increases in response to cell swelling, which can be due to inactivation of Na^+/K^+ pump by K^+-free saline, activation of agonist- and potential-dependent ionic channels in membrane (Ayrapetyan, 2012), ionic radiation (Dvoretsky et al., 2012), as well as those cell pathologies that are accompanied by cell swelling (e.g., cancer) (Danielyan et al., 1999a,b). Furthermore, cell hydration-dependent variability of α_3 receptors is more pronounced than the variability of α_1 and α_2 receptors (Ayrapetyan, 2012).

Thus, it is suggested that Na^+/K^+ pump activation-induced cell shrinkage (Ayrapetyan and Sulejmanian, 1979; Carpenter et al., 1992), which decreases the number of pump units (α_1 and α_2) functioning in the membrane, serves as a negative feedback through which the autoregulation of Na^+/K^+ pump is realized (Ayrapetyan et al., 1984). However, the role of α_3 receptors, having only signaling function and being more dependent on cell hydration than α_1 and α_2 receptors, in autoregulation of Na^+/K^+ pump has not been elucidated yet.

Special interest for the functional significance of α_3 isoforms grew after the discovery of the anticancer properties of nanomolar concentrations of ouabain by the laboratory of Prof. Askari (Kometiani et al., 2005). At present, it is well documented that in cancer cells, which are overhydrated [water content is more than 90% as compared with normal cells (70%–73%)] (Damadian, 1971), α_3 isoforms are expressed (Shibuya et al., 2010; Ayrapetyan et al., 2012; Weidemann, 2012). Recently, we have shown that in mice

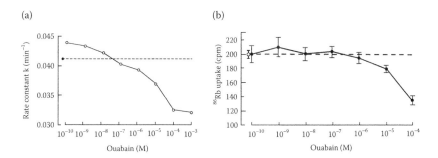

FIGURE 16.8 The dependence of the rate constant of $^{22}Na^+$ efflux from cell on the concentration of ouabain from 10^{-10} to 10^{-3} M (a). $^{86}Rb^+$ uptake (K^+ uptake) as a function of ouabain concentration in cell aqua medium (b). The dashed line indicates the rate constant in normal Ringer's solution. (From Ayrapetyan, S.N., et al., *Cell. Mol. Neurobiol.*, 4, 375, 1984, Figure 8; Saghian, A.A., et al., *Cell. Mol. Neurobiol.*, 16, 493, 1996, Figure 2.)

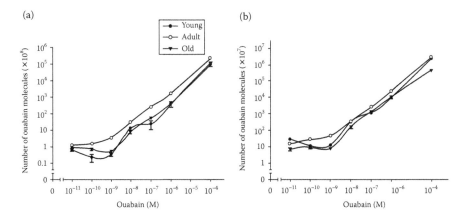

FIGURE 16.9 Number of ouabain molecules as a function of different concentrations of [³H]-ouabain in brain cortex (a) and heart muscle (b) tissues of three animal groups. Abscissas indicate [³H]-ouabain concentrations; ordinates are logarithmic and define the number of ouabain molecules in tissues. Each point in the curve is the mean ± SD for three independent experiments. Error bars of each point are not detected because they are blended with each other. (From Heqimyan, A., et al., In: M.S. Markov (Ed.) *Electromagnetic Fields in Biology and Medicine*, CRC Press, Boca Raton, FL, 218, 2015, Figure 16.1.)

carrying sarcoma 180, the increase of hydration in nonexcitable tissues is also accompanied by the expression of α_3 receptors, which are absent in healthy animals (Ayrapetyan et al., 2012). It has also been shown that the decrease in the number of α_3 receptors in the membrane of both neurons and myocytes is accompanied by age-dependent cell dehydration (Narinyan et al., 2012, 2013, 2014; Heqimyan et al., 2012, 2015).

As can be seen in Figure 16.9, all the three zones of [³H]-ouabain binding curves are more pronounced in brain cortex and heart muscle tissues of young rats compared to those in adult and old ones.

As has been described in a number of studies (Lucchesi and Sweadner, 1991; Gao et al., 2002; Blaustein et al., 2009; de Lores Arnaiz et al., 2014), the expression of high-affinity receptors reaches its maximum in adult animals, and then their number decreases with age. These data clearly indicate that α_3 receptors have a more pronounced functional-dependent character than α_1 and α_2. Thus, the highest sensitivity of cell hydration to EMF-treated PS and the direct correlation between cell hydration and the number of α_3 receptors functioning in membrane, allow us to suggest that α_3 receptor-dependent signaling system can serve as a gate for metabolic cascade through which the EMF-induced modulation effects on intracellular metabolism is realized. To check this suggestion, the role of α_3 receptor-dependent signaling system in regulation of cell hydration has been studied.

16.3 The Role of α_3 Isoform-Dependent Signaling System in Regulation of Cell Hydration

From the works of Prof. Adey and Prof. Blackman, it has been known that Ca^{2+} has a key role in realization of the biological effect of EMF (Blackman et al., 1979; Adey, 1981).

However, the role of α_3 isoform-dependent signaling system in controlling $[Ca^{2+}]_i$ has not been elucidated yet.

At present, it is well established that all three isoforms of Na^+/K^+-ATPase have a crucial role in regulation of Na^+/Ca^{2+} exchange in membrane (Juhaszova and Blaustein, 1997; Xie and Askari, 2002; Wu et al., 2013; de Lores Arnaiz et al., 2014; Romanovsky et al., 2015). Since the pioneering work of Baker et al. (1969), the role of Na^+/K^+-ATPase in regulation of Na^+/Ca^{2+} exchange is traditionally explained by the negative correlation between Na^+/K^+ pump and Na^+/Ca^{2+} exchange (Blaustein et al., 2009). However, as it has been shown by our earlier work performed on snail neurons (Ayrapetyan et al., 1984; Saghian et al., 1996) and later by other laboratory studies on mammalian cells, nanomolar concentrations of ouabain do not inhibit Na^+/K^+ pump but stimulate Na^+/Ca^{2+} exchange in reverse mode (R Na^+/Ca^+ exchange) (Xie and Askari, 2002; Wu et al., 2013; de Lores Arnaiz et al., 2014; Romanovsky et al., 2015).

Our comparative study of the effect of 10^{-9} and 10^{-4} M ouabain on R Na^+/Ca^{2+} exchange ($^{45}Ca^{2+}$ uptake) in brain and heart muscle tissues of rats has shown that the activation effect of 10^{-9} M ouabain on R Na^+/Ca^{2+} exchange is incomparably higher than the 10^{-4} M ouabain effect, having a strong inhibitory effect on Na^+/K^+ pump activity (Nikoghosyan et al., 2015). Therefore, these data make the explanation of 10^{-9} M ouabain-induced activation of R Na^+/Ca^{2+} exchange by Na^+/K^+ pump inhibition seem unrealistic (Ayrapetyan et al., 2015).

Figure 16.10 shows that 10^{-9} M ouabain-induced activation of $^{45}Ca^{2+}$ uptake has an age-dependent decreasing character, which indicates that this effect depends on the

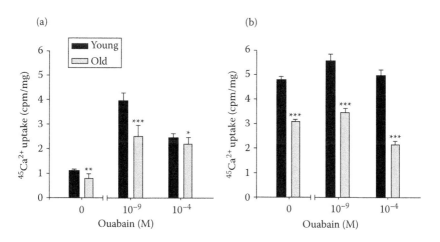

FIGURE 16.10 $^{45}Ca^{2+}$ uptake in brain cortex (a) and heart muscle (b) tissues after PS and nonradioactive ouabain (at 10^{-9} and 10^{-4} M) IP injections. Abscissas indicate ouabain doses (PS injection, i.e., 0 M ouabain, 10^{-9}, 10^{-4} M); ordinates indicate $^{45}Ca^{2+}$ uptake content counting per minute. Error bars indicate the SEM for three independent experiments. *, **, and *** indicate $p < .05$, $p < .01$, and $p < .001$, respectively. (From Heqimyan, A., et al., In: M.S. Markov (Ed.) *Electromagnetic Fields in Biology and Medicine*, CRC Press, Boca Raton, FL, 225, 2015, Figure 16.9.)

Dosimetry in Bioelectromagnetics

functional state of cells. The fact that there is an age-induced increase of $[Ca^{2+}]_i$ (Khachaturian, 1989) allows us to explain that the age-dependent impairment of activation effect of nM ouabain on $^{45}Ca^{2+}$ uptake is due to the increase of $[Ca^{2+}]_i$.

As the differences between the gradients of Na^+ and Ca^{2+} (Baker et al., 1969) serve as an energy source for Na^+/Ca^{2+} exchange, 10^{-9} M ouabain-induced activation of R Na^+/Ca^{2+} exchange, when Na^+ gradient on membrane is constant, can be explained by the increase of Ca^{2+} gradient on membrane as a result of $[Ca^{2+}]_i$ decrease.

It is known that Ca^{2+} pump, localized in the membrane of endoplasmic reticulum (ER) and transporting Ca^{2+} from cytoplasm to ER, is activated by cyclic adenosine monophosphate (cAMP) (Brini and Carafoli, 2009). On the basis of the data that in neurons nM ouabain-induced activation of R Na^+/Ca^{2+} exchange is accompanied by the increase of intracellular cAMP (Saghian et al., 1996) and that nM ouabain has elevation effect on the intracellular cAMP in different tissues (Siegel et al., 1999), nM ouabain-induced activation of R Na^+/Ca^{2+} exchange can be considered as a consequence of $[Ca^{2+}]_i$ decrease because of activation of cAMP-dependent Ca^{2+} pump in ER membrane.

Since Na^+/Ca^{2+} exchange functions in stoichiometry of 3Na:1Ca, the activation of R Na^+/Ca^{2+} exchange could cause cell dehydration. However, our previous study has shown that IP injection of 10^{-9} M ouabain-induced activation of R Na^+/Ca^{2+} exchange is accompanied by hydration in brain cortex and heart muscle tissues. Furthermore, it has been shown that 10^{-9} M ouabain-induced tissue hydration is more pronounced than 10^{-4} M ouabain-induced cell hydration (Narinyan et al., 2014; Ayrapetyan et al., 2015; Nikoghosyan et al., 2015). It is known that 10^{-4} M ouabain increases cell hydration, which is due to inhibition of Na^+/K^+ pump, having electrogenic character (Ayrapetyan and Sulejmanian, 1979; Carpenter et al., 1992). But the fact that 10^{-9} M ouabain has a more pronounced effect on cell hydration than 10^{-4} M ouabain serves as additional evidence that nM ouabain-induced cell hydration cannot be considered as the result of Na^+/K^+ pump inactivation. We came to the same conclusion by studying the activation of R Na^+/Ca^{2+} exchange as a result of decrease of Na^+ concentration in cell aqua medium ($[Na^+]_0$) (Nikoghosyan et al., 2015). It has been shown that the decrease of $[Na^+]_0$ has a double effect on cell hydration: from one side, through activation of R Na^+/Ca^{2+} exchange it brings dehydration, from the other side through activation of oxidative processes in the cell it causes hydration because of H_2O release in cytoplasm. It is worth noting that as a consequence of these two effects water efflux from the cell takes place, which in its turn inhibits inward going I_{Na} and I_{Ca} (Ayrapetyan et al., 1988). The activation of R Na^+/Ca^{2+} exchange leading to membrane hyperpolarization (Baker et al., 1969; Saghian et al., 1996) and resulting in cell shrinkage (Iwasa et al., 1980) is the next mechanism through which the decrease of membrane permeability takes place. Thus, it can be assumed that nM ouabain-induced activation of cAMP-dependent R Na^+/Ca^{2+} exchange has a multisided protective function for cells. However, this protective reaction has a transient character as its continuous activation brings an increase of $[Ca^{2+}]_i$ which poisons intracellular metabolism and leads to the impairment of endogenous H_2O release.

Taking into consideration that *in vivo* state endogenous ouabain-like hormones constantly circulate in blood, it is suggested that the activation of α_3 receptors leading to the increase of $[Ca^{2+}]_i$ can be compensated by pushing Ca^{2+} from the cells, which is realized by Ca^{2+} pump in cell membrane and Na^+/Ca^{2+} exchange in forward mode (F Na^+/Ca^{2+} exchange) (Blaustein and Lederer, 1999). At present it is established that cyclic guanosine monophosphate (cGMP) has an activation effect on these two mechanisms of pushing Ca^{2+} from the cells (Azatian et al., 1998; Brini and Carafoli, 2009). As Ca^{2+} pump has high affinity to Ca^{2+} and low capacity for Ca^{2+} transportation, cGMP-activated F Na^+/Ca^{2+} exchange has a fundamental role in compensation of cAMP-dependent increase of $[Ca^{2+}]_i$ (Brini and Carafoli, 2009). As $[Ca^{2+}]_i$ has a strong inhibitory effect on Na^+/K^+ pump activity (Skou, 1957), cGMP-dependent removal of $[Ca^{2+}]_i$ from the cell has a pivotal role in keeping Na^+/K^+ pump in active state. It is known that intracellular cGMP contents can be increased by a number of biologically active substances, even by their homeopathic concentrations (Azatyan et al., 1994; Ayrapetyan, 2012), static magnetic field (SMF), and 4-Hz pulsing magnetic field (PMF) exposures (Ayrapetyan et al., 1994a) as well as by high $[Ca^{2+}]_i$-induced activation of calmodulin-NO-cGMP cascade (Denninger and Marletta, 1999).

It is obvious that the extra-sensitivity of intracellular cGMP and cAMP contents to the effects of both EMF and homeopathic concentrations of biologically active substances cannot be explained from the point of classic membrane theory according to which the signal transduction in a cell is explained by the changes of the conductive properties of membrane (Katz, 1966). Therefore, such a high sensitivity of intracellular signaling system to weak environmental factors can be explained by high sensitivity of soluble guanylate and adenylate cyclases to the water structure changes upon the impact of weak chemical and physical factors (Klassen, 1982; Chaplin, 2006).

Previously, we have shown that the threshold for the changes of intracellular cGMP by biologically active substances, such as synaptic transmitters and ouabain, is much lower than that for the changes of intracellular cAMP contents (Azatyan et al., 1994; Ayrapetyan, 2012). It means that cGMP-activated release of $[Ca^{2+}]_i$ from the cell, which reactivates Na^+/K^+ pump, serves as a primary cell response to the changes of chemical composition of cell aqua medium, while the activation of cAMP-dependent R Na^+/Ca^{2+} exchange realizing transient protective function, which leads to the increase of $[Ca^{2+}]_i$-induced poisoning of cell metabolism (inactivation of Na^+/K^+ pump), serves as a secondary cell response.

Thus, on the basis of the presented data it can be concluded that in living cells there are a minimum of two intracellular signaling systems that control Na^+ gradient on membrane, which is sensitive to the structural changes of cell aqua medium: (1) cGMP-activated $[Ca^{2+}]_i$ efflux which activates Na^+/K^+ pump leading to water efflux, which in its turn decreases membrane permeability for ions and cell hydration and (2) cAMP-activated R Na^+/Ca^{2+} exchange increasing $[Ca^{2+}]_i$, which poisons cell metabolism (Na^+/K^+ pump inactivation) and leads to water uptake as a result of depressing metabolic generation of water efflux from the cell.

Thus, it can be concluded that the activation of α_3 isoform upon the impact of EMF-induced water structure changes leading to elevation of intracellular cGMP (causes cell hydration and decreases membrane permeability) could be considered as beneficial, while the EMF-induced activation of α_3 isoform increasing cAMP (increases both water uptake and membrane permeability) could be considered as hazardous.

16.4　The Na+/K+-ATPase α_3 Isoform-Dependent Signaling System Controlling Cell Hydration as a Marker for EMF Dosimetry from the Point of Public Health

As it has been noted earlier, the dysfunction of metabolic mechanism controlling Na+ gradient on the membrane that increases [Na+]$_i$, serves as a common consequence of any pathology. Therefore, the metabolically driven water efflux, having an inhibitory effect on membrane permeability for Na+ (P_{Na}) (Ayrapetyan et al., 1988), can be considered as a primary mechanism responsible for cell pathology. Thus, the factor-induced imbalance of water fluxes (water net fluxes) in the membrane bringing to appropriate changes of P_{Na} could serve as a cellular marker for detection of the biological effects of environmental factors (including homeopathic concentrations of chemical substances as well as EMF) on cells and organisms. *On the basis of the presented data that cGMP through decreasing [Ca²+]$_i$ stimulates water efflux, which has inhibitory effect on* P$_{Na}$, *we suggest that the factor-induced elevation of cGMP can be considered as a key protective mechanism of the cell. Whereas, the elevation of cAMP-induced increase of [Ca²+]$_i$ leading to generation of net water influx, which in its turn activates* P$_{Na}$ *as a result of metabolic poisoning, can be considered hazardous for cells and organisms.*

As noted earlier, nM ouabain-induced activation of cAMP-dependent R Na+/Ca²+ exchange has a biphasic effect on cell hydration: (1) transient protective effect (water efflux-induced decrease of P_{Na}) and (2) [Ca²+]$_i$-induced poisoning (water influx-induced increase of P_{Na}). Therefore, we suggest that the EMF sensitivity of nM ouabain-induced cell hydration could serve as a marker for estimation of beneficial and hazardous effects of EMF on cells and organisms: the EMF, having a depressing effect on nM ouabain-induced changes of cell hydration, can be considered as beneficial, whereas the EMF, having elevation effect on nM ouabain-induced changes of cell hydration, can be considered as hazardous. In this respect, the experimental data regarding the effects of SMF, PMF, and MW on neuronal and muscle cells are presented below.

16.4.1　SMF and PMF Effects on nM Ouabain-Induced Activation of the Signaling System Controlling Cell Hydration

Our previous study on snail neurons has shown that SMF and PMF exposures lead to the increase of cGMP contents, which is accompanied by the decrease of both cAMP and Ca²+ uptake (R Na+/Ca²+ exchange) (Table 16.1) (Ayrapetyan et al., 1994a, 2005).

The study performed on the tissues of different organs of adult rats (Danielyan et al., 1999a) as well as healthy and cancer tissues of women's breast (Danielyan et al., 1999b) has shown that 0.2-mT SMF has a dehydration effect on tissues, which is accompanied by the decrease of [³H]-ouabain binding with cell membrane. However, in the case of both age-induced cell dehydration (Narinyan et al., 2012; Heqimyan et al., 2015) and overhydration of cells in cancer tissues (Damadian, 1971), impairment of magnetic sensitivity of [³H]-ouabain binding with cell membrane is observed. Furthermore, in both cases the impairment of magnetic sensitivity of [^{3}H]-ouabain binding with α_3 receptors is more pronounced compared with α_1 and α_2 receptors (Figures 16.11 and 16.12).

It is known that the dysfunction of Na$^+$/K$^+$ pump (during cell pathology and aging) increases [Ca^{2+}]$_i$ (Khachaturian, 1989), (which brings to the decrease of membrane receptors' affinity to ouabain Narinyan et al., 2014). *As α_3 receptors are not directly involved in Na$^+$/K$^+$ pump activity (Xie and Askari, 2002) and their sensitivities are depressed in cell pathology (including aging) as a result of [Ca^{2+}]$_i$ increase, it is suggested that intracellular signaling system is a target for the effect of EMF.*

TABLE 16.1 MPS on Intracellular ^{45}Ca^{2+} and Cyclic Nucleotide in Heart Muscle and Neurons of Snails

Tissue Type	Exposure Medium	Intracellular ^{45}Ca^{2+} Content (cpm/mg Wet Weight) and Percentage of Their Changing Compared to the Control	Cyclic Nucleotide Contents (pM/mg Wet Weight) and Percentage of Their Changing Compared to the Control	
			cAMP	cGMP
Muscle	Sham exposed	1210 ± 110 (100%) $T = 6.399, p = .03$	0.53 ± 0.04 (100%) $T = 3.532, p = .24$	0.43 ± 0.01 (100%) $T = 3.554, p = .24$
	SMF exposed	698 ± 85 (~57%) $T = 6.399, p = .03$	0.43 ± 0.03 (82%) $T = 3.532, p = .24$	0.61 ± 0.08 (157.9%) $T = 3.554, p = .24$
Neuron	Sham exposed	100 ± 10.07 $p < .05$ comp. to control	100 ± 9.6	100 ± 11.6
	SMF exposed 4.6, 38 mT	74.33 ± 9.22 62.67 ± 9.00 $p < .05$ comp. to control	79.1 ± 9.5 $p < .02$	164.9 ± 41.6 $p < .05$

Source: Ayrapetyan et al., In: M. S. Markov (Ed.) *Electromagnetic Fields in Biology and Medicine*, CRC Press, Boca Raton, FL, 205, 2015, Table 13.1.

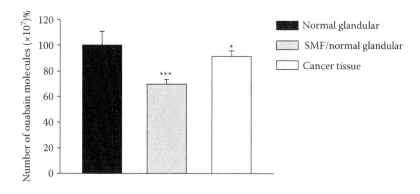

FIGURE 16.11 The effect of 0.2-mT SMF on [^{3}H]-ouabain binding with cell membrane in healthy and cancer breast tissue slices bathing in PS containing 10^{-9} M ouabain. SMF, static magnetic field. *, **, and *** indicate $p < .05$, $p < .01$, and $p < .001$, respectively

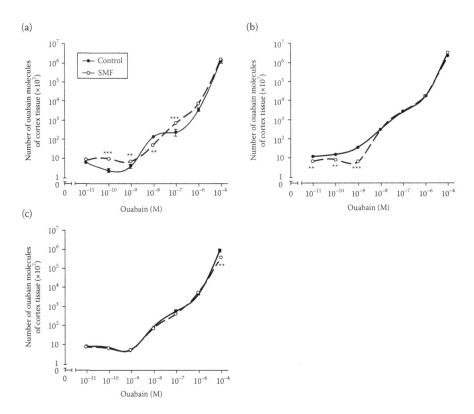

FIGURE 16.12 Curves of [³H]-ouabain binding in brain cortex tissue of young (a), adult (b), and old (c) animal groups after SMF exposure. Abscissas indicate [³H]-ouabain concentrations; ordinates are logarithmic and define the number of ouabain molecules in tissues. Each point in the curve is the mean ± SD for three independent experiments. Error bars of each point are not detected because they are blended with it. ** and *** indicate $p < .01$ and $p < .001$, respectively. (From Heqimyan, A., et al., In: M.S. Markov (Ed.) *Electromagnetic Fields in Biology and Medicine*, CRC Press, Boca Raton, FL, 225, 2015, Figure 16.2.)

As has been shown in Figure 16.3, SMF has a dehydration effect on brain cortex and heart muscle tissues of young animals, while it has a hydration effect on neurons and no effect on heart muscle of old animals. The fact that the dysfunction of Na^+/K^+ pump could be a reason for age-dependent reverse effect of SMF on tissue hydration in aging is supported by the data presented in Figure 16.13.

This Figure shows that SMF has a dehydration effect on the brain cortex of young rats when Na^+/K^+ pump is in active state (ouabain-free medium), while the same intensity of SMF has in hydration effect on brain cortex tissue when the Na^+/K^+ pump is a inactive state (10^{-4} M ouabain IP injection). These data allow us to suggest that the impairment of magnetic sensitivity in cell pathology and aging is due to the dysfunction of intracellular signaling system leading to the increase of $[Ca^{2+}]_i$, which has an inactivation effect on Na^+/K^+ pump. This suggestion is supported by the study on the effect of SMF on

$^{45}Ca^{2+}$ uptake by cells in normal PS containing K^+ (Na^+/K^+ pump is in active state) and in K^+-free PS (Na^+/K^+ pump is in inactive state) (Figure 16.14).

Previously, we have shown that α_3 isoform-dependent signaling system is a common sensor for both 4-Hz EMF and 4-Hz mechanical vibration (MV). The data presented in Figure 16.14 clearly indicate that SMF and MV have a depressing effect on $^{45}Ca^{2+}$ uptake

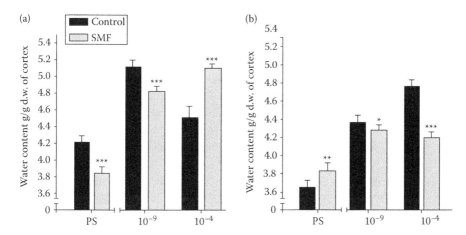

FIGURE 16.13 Age-dependent change of water content after SMF exposure in brain cortex tissues of young (a) and old animals (b) after IP injection of PS and PS with 10^{-9} and 10^{-4} M of ouabain. Abscissas indicate ouabain doses (PS injection, i.e., 0 M ouabain, 10^{-9}, 10^{-4} M). Ordinates indicate the mean value of tissues' water content. Error bars indicate the standard error of the SEM for three independent experiments. *, **, and *** indicate $p < .05$, $p < .01$, and $p < .001$, respectively. (From Heqimyan, A., et al., In: M.S. Markov (Ed.) *Electromagnetic Fields in Biology and Medicine*, CRC Press, Boca Raton, FL, 225, 2015, Figure 16.8.) SMF, static magnetic field.

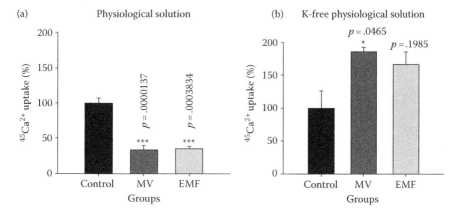

FIGURE 16.14 The effects of MV on $^{45}Ca^{2+}$ uptake by neuronal ganglia in normal (a) and K^+-free PS (b). (From Ayrapetyan, S.N., *Environmentalist*, 32, 210, 2012, Figure 6.) MV, mechanical vibration; EMF, Electromagnetic fields.

by cell in normal PS, while they have an activation effect on $^{45}Ca^{2+}$ uptake (R Na^+/Ca^{2+} exchange) by cell in K^+-free PS.

Thus, the presented data allow us to conclude that depending on the initial functional state of the organism, the same intensity of SMF could have different effects on cell hydration in neuronal and heart muscle tissues.

As can be seen in Figure 16.13, SMF having an activation effect on cGMP-dependent F Na^+/Ca^{2+} exchange depresses the effect on nM ouabain-induced cell hydration in both young and old animals, in spite of the fact that SMF has a dehydration effect in young animals and hydration effect in old animals. Therefore, the depressing effect of SMF on nM ouabain-induced cell hydration can be considered as a marker for the beneficial effect of EMF. This suggestion is supported by the study of SMF effect on nM ouabain-induced activation of R Na^+/Ca^{2+} exchange both in young and old animals.

The data show that in both young and old animals SMF has a depressing effect on nM ouabain-induced activation of R Na^+/Ca^{2+} exchange (Figure 16.15), i.e., it removes nM ouabain-induced poisoning effect on cells as a result of elevation of $[Ca^{2+}]_i$.

16.4.2 The MW Effect on nM Ouabain-Induced Activation of the Signaling System Controlling Cell Hydration

As the penetration of NT MW into biological tissue is less than 1 mm, it is suggested that the effect of MW on an organism is realized through changing physicochemical properties of water components of skin and sub-skin tissues (Ziskin, 2006; Deghoyan et al., 2013). As is noted above, there is a direct correlation between metabolic activity and EMF sensitivity of cell hydration (Ayrapetyan and De, 2014). As pacemaker

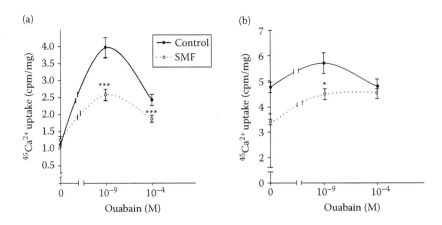

FIGURE 16.15 Curves of $^{45}Ca^{2+}$ uptake in brain cortex tissues of young (a) and old (b) animals after PS and nonradioactive ouabain (at 10^{-9} and 10^{-4} M concentrations) IP injections received after SMF exposure. Abscissas indicate ouabain doses (PS injection, i.e., 0 M ouabain, 10^{-9}, 10^{-4} M). Ordinates indicate $^{45}Ca^{2+}$ uptake content counting per minute. Error bars indicate the SEM for three independent experiments. * and *** indicate $p < .05$ and $p < .001$, respectively. (From Heqimyan, A., et al., In: M.S. Markov (Ed.) *Electromagnetic Fields in Biology and Medicine*, Boca Raton, FL, 227, 2015, Figure 16.10.)

activity of heart muscle has high metabolic activity, it is predicted that it could have high sensitivity to MW-induced water structure changes in skin and sub-skin tissues. Therefore, we have chosen heart muscle as an experimental model for the study of the effect of brain exposure by MW on intracellular signaling system controlling cell hydration.

As can be seen in Figure 16.16, IP injection of sham PS leads to muscle hydration, while MW-treated PS leads to muscle hydration, which is more pronounced than in the case of sham PS injection.

Figure 16.17 shows that dose-dependent ouabain effects on heart muscle hydration are significantly changed upon the exposure of rats' heads with NT MW. The NT MW exposure on brains of rats preliminarily injected with 10^{-10} and 10^{-9} M ouabain has more pronounced hydration effects on muscle than sham exposure. It is worthy to note that like in the case of 10^{-10} and 10^{-9} M ouabain, MW-treated PS has a hydration effect on heart muscle, while at higher concentrations MW has the opposite effect of what was recorded in the case of ouabain effect. It means that the effects of MW and nM ouabain are realized through the same signaling system, i.e., through activation of cAMP-dependent R Na^+/Ca^{2+} exchange.

To check this suggestion, the effects of NT MW-treated PS on $^{45}Ca^{2+}$ uptake (R Na^+/Ca^{2+} exchange) and muscle hydration at 10^{-9} and 10^{-4} M (Na^+/K^+-pump inhibitory) ouabain and ouabain-free mediums were studied.

As can be seen in Figure 16.18, IP injection of NT MW-treated PS compared with IP injection of sham-treated PS leads to strong activation of $^{45}Ca^{2+}$ uptake by muscle which is not accompanied by significant changes of muscle hydration (Figure 16.18b). However, IP injection of NT MW-treated PS with 10^{-4} M ouabain (Na^+/K^+ pump is in inhibited

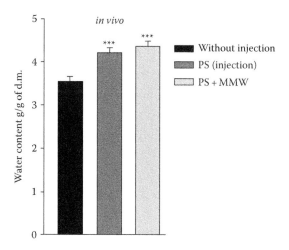

FIGURE 16.16 The effects of NT MW-treated PS on muscle hydration *in vivo* (IP injection of sham- and NT MW-treated PS). Each bar indicates the ±SEM of 30 samples. *, **, and *** indicate $p < .05$, $p < .01$ and, $p < .001$, respectively. PS, physiological solution; MW, microwave.

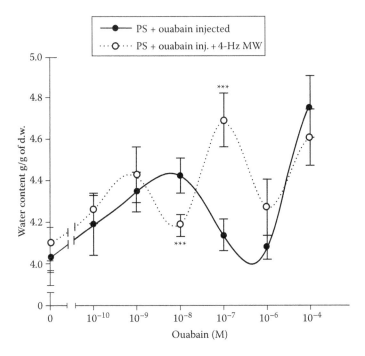

FIGURE 16.17 Dose-dependent [³H]-ouabain effect on heart muscle hydration after NT MW exposure. The kinetics of dose-dependent [³H]-ouabain effect on heart muscle hydration in sham condition (*continuous lines*) and after 15 min NT MW exposure of rat's heart muscle (*dashed lines*). Each point indicates the ±SEM of 30 samples. *** indicates $p < .001$.

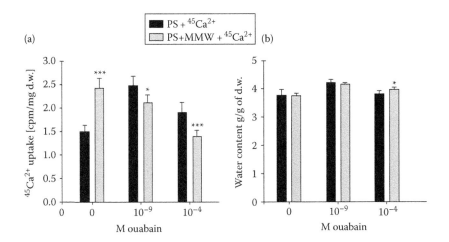

FIGURE 16.18 The effects of IP injection of sham- (*black columns*) and NT MW-treated (*gray columns*) PS containing ouabain-free, 10^{-9} and 10^{-4} M ouabain on (a) $^{45}Ca^{2+}$ uptake and (b) heart muscle hydration. Each bar indicates the ±SEM of 30 samples. * and *** indicate $p < .05$ and $p < .001$, respectively.

state) has no significant effect on $^{45}Ca^{2+}$ update by muscle, but leads to the increase of muscle hydration (Figure 16.18).

To find out whether the effect of MW-treated PS is realized through the same signaling system (cAMP-dependent R Na^+/Ca^{2+} exchange), we have studied the effect of MW-treated PS on intracellular cAMP and cGMP. The obtained data presented in Figure 16.19 indicate that MW-treated PS injection leads to the increase of cAMP and decreases cGMP content.

Thus, MW-induced activation of cAMP-dependent R Na^+/Ca^{2+} exchange, which has a poisoning effect on cell metabolism, could be considered as a potential harmful effect able to lead to cell pathology.

Thus, on the basis of the presented data, it can be concluded that:

1. Na^+/K^+-ATPase α_3 isoform-dependent signalling system controlling cell hydration is a common target for EMF.

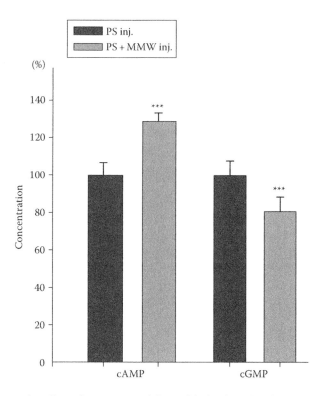

FIGURE 16.19 The effects of IP injection of sham- (*black columns*) and NT MW-treated (*gray columns*) PS on inracellular contents of cAMP (*left columns*) and cGMP (*right columns*) contents in heart muscle tissues. Each bar indicates the ±SEM of 30 samples. ** and *** indicate $p < .01$ and $p < .001$, respectively. cAMP, cyclic adenosine monophosphate; cGMP, cyclic guanosine monophosphate.

2. In young animals (with low $[Ca^{2+}]_i$), SMF and PMF with the intensity of $\leq 0.2\,mT$ have an activation effect on cGMP-dependent F Na^+/Ca^{2+} exchange leading to the decrease of $[Ca^{2+}]_i$ and reactivation of Na^+/K^+-pump, which in its turn by activation of intracellular oxidation-induced release of water molecules generates water efflux and inhibits P_{Na} of cell membrane. Therefore, short exposures of $\leq 0.2\,mT$ SMF and PMF (30 and 15 min, respectively) have beneficial effects on an organism.

3. NT MW has an activation effect on cAMP-dependent R Na^+/Ca^{2+} exchange leading to the increase of $[Ca^{2+}]_i$, which brings intracellular poisoning and imbalance-induced water influx causing an increase of P_{Na} of cell membrane. Therefore, MW has a hazardous effect on an organism and is able to generate cell pathology.

4. nM ouabain-induced activation of α_3 isoform leads to stimulation of cAMP-dependent R Na^+/Ca^{2+} exchange, which transiently activates intracellular oxidation-induced generation of water efflux having an inhibitory effect on P_{Na} of cell membrane, which is followed by $[Ca^{2+}]_i$-induced poisoning of cell metabolism leading to generation of net water influx-activated P_{Na} and P_{Ca}, as a result of depression of metabolically dependent water efflux from the cell. Therefore, nM ouabain-induced hydration changes serve as a diagnostic marker for estimation of the initial functional state of the organism. In a healthy state, nM ouabain has a hydration effect on tissue, while in a non-healthy state it leads to dehydration.

Thus, the EMF sensitivity of nM ouabain-induced tissue hydration can be used as a marker for EMF dosimetry: the EMF, which is able to counteract nM ouabain-induced effect on tissue hydration, i.e., depresses nM ouabain-induced hydration in healthy and dehydration in non-healthy organisms, can be considered as beneficial. Whereas, the EMF that has the same poisoning effect of nM ouabain can be considered as hazardous.

References

Adey WR. 1981. Tissue interactions with nonionizing electromagnetic field. *Physiol. Rev.* 61:435–514.

Ayrapetyan SN. 1969. Metabolically-dependent part of membrane potential and electrode properties of giant neuron membrane of mollusk. *Biofizika* 14(6):1027–1031.

Ayrapetyan SN. 2006. Cell aqua medium as a primary target for the effect of electromagnetic fields. In: *Bioelectromagnetics: Current Concepts.* Ayrapetyan SN, Markov MS, eds. Springer, Dordrecht, the Netherlands, pp. 31–61.

Ayrapetyan SN. 2012. Cell hydration as a universal marker for detection of environmental pollution. *Environmentalist* 32:210–221.

Ayrapetyan SN. 2015. The role of cell hydration in realization of biological effects of nonionizing radiation (NIR). *Electromagnet. Biol. Med.* 34(3):197–210.

Ayrapetyan SN, Arvanov VL. 1979. On the mechanism of the electrogenic sodium pump dependence of membrane chemosensitivity. *Comp. Biochem. Physiol. A Physiol.* 64(4):601–604.

Ayrapetyan SN, Arvanov VL, Maginyan SB, Azatyan KV. 1985. Further study of the correlation between Na-pump activity and membrane chemosensitivity. *Cell. Mol. Neurobiol.* 5(3):231–243.

Ayrapetyan SN, Avanesian AS, Avetisian TH, Majinian SB. 1994a. Physiological effects of magnetic fields may be mediated through actions on the state of calcium ions in solution. In: *Biological Effects of Electric and Magnetic Fields*, Vol. 1. Carpenter D, Ayrapetyan S, eds. Academic Press, New York, pp. 181–192.

Ayrapetyan SN, Baghdasaryan N, Mikayelyan Y, Barseghyan S, Martirosyan V, Heqimyan A, Narinyan L, Nikoghosyan A. 2015. Cell hydration as a marker for nonionizing radiation. In: *Electromagnetic Fields in Biology and Medicine*. Markov M, ed., CRC Press, Boca Raton, FL, pp. 193–215.

Ayrapetyan SN, Carpenter DO. 1991a. The modulatory effect of extremely low doses of mediators on functional activity of the neuronal membrane. *Zh. Evol. Biokhim. Fiziol.* 27(2):146–151 (in Russian).

Ayrapetyan SN, Carpenter DO. 1991b. Very low concentration of acetylcholine and GABA modulate transmitter responses. *NeuroReport* 2(10):563–565.

Ayrapetyan SN, De J. 2014. Cell hydration as a biomarker for estimation of biological effects of nonionizing radiation on cells and organisms. *Sci. World J.* 2014, Article ID 890518, 8 pages.

Ayrapetyan SN, Grigorian KV, Avanesian AS, Stamboltsian KV. 1994b. Magnetic fields alter electrical properties of solutions and their physiological effects. *Bioelectromagnetics* 15(2):133–142.

Ayrapetyan SN, Hunanian ASh, Hakobyan SN. 2004. 4Hz EMF treated physiological solution depresses Ach-induced neuromembrane current. *Bioelectromagnetics* 25(5):397–399.

Ayrapetyan GS, Papanyan AV, Hayrapetyan HV, Ayrapetyan SN. 2005. Metabolic pathway of magnetized fluid-induced relaxation effects on heart muscle. *Bioelectromagnetics* 26:624–630.

Ayrapetyan SN, Rychkov GY. 1985. The presence of reserve ion channels in membranes of giant snail neurons and giant squid axons. *Dokl. Akad. Nauk SSSR* 285(6):1464–1467 (in Russian).

Ayrapetyan SN, Rychkov GY, Suleymanyan MA. 1988. Effects of water flow on transmembrane ionic currents in neurons of *Helix pomatia* and in squid giant axon. *Comp. Biochem. Physiol.* 89:179–186.

Ayrapetyan SN, Sulejmanian MA. 1979. On the pump-induced cell volume changes. *Comp. Biochem. Physiol. A Physiol.* 64(4):571–575.

Ayrapetyan SN, Suleymanyan MA, Saghyan AA, Dadalyan SS. 1984. Autoregulation of electrogenic sodium pump. *Cell. Mol. Neurobiol.* 4(4):367–383.

Ayrapetyan SN, Yeganyan L, Bazikyan G, Muradyan R, Arsenyan F. 2012. Na^+/K^+ pump α_3-isoform-dependent cell hydration controlling signaling system dysfunction as a primary mechanism for carcinogenesis. *J. Bioequiv. Bioavailab.* 4(7):112–120.

Azatian KV, White AR, Walker RJ, Ayrapetyan SN. 1998. Cellular and molecular mechanisms of nitric oxide-induced heart muscle relaxation. *Gen. Pharmacol.* 30(4):543–553.

Azatyan KV, Karapetyan IC, Ayrapetyan SN. 1994. Effect of acetylcholine of low doses of $^{45}Ca^{2+}$ influx into the *Helix pomatia* neurons. *Biol. Mem. Harwood Academic Publishers GmbH* 7(3):301–304 (reprinted in the United States).

Baghdasaryan NS, Mikayelyan YR, Barseghyan S, Dadasyan EH, Ayrapetyan SN. 2012. The modulating impact of illumination and background radiation on 8Hz-induced infrasound effect on physicochemical properties of physiological solution. *Electromagnet. Biol. Med.* 31(4):310–319.

Baker PF, Blaustein MP, Hodgkin AL, Steinhardt SA. 1969. The influence of calcium on sodium efflux in squid axons. *J. Physiol.* 200:431–458.

Belyaev A. 2012. Evidence for disruption by modulation role of physical and biological variables in bioeffects of non-thermal microwaves for reproducibility, cancer risk and safety standards. In: *BioInitiative Working Group, Section 15. Bioinitiative Report WHO 2012*. Sage C, Carpenter D, eds., WHO Press, Switzerland pp. 1–71.

Binhi VN. 2012. Two types of magnetic biological effects. *Biophysics* 57:237–243.

Binhi VN, Rubin AB. 2007. The kT paradox and possible solutions. *Electromagnet. Biol. Med.* 26:45–62.

BioInitiative Report WHO: Working Group. 2012. *A Rationale for Biologically-Based Public Exposure Standards for Electromagnetic Radiation*, Sage C, Carpenter DO, eds., WHO Press, Switzerland.

Blackman CF. 2015. Replication and extension of Adey group's calcium efflux results. In: *Electromagnetic Fields in Biology and Medicine*. Markov MS, ed., CRC Press, Boca Raton, FL, pp. 7–14.

Blackman CF, Elder JA, Weil CM, Benane SG, Eichinger DC, House D. 1979. Induction of calcium-ion efflux from brain tissue by radiofrequency radiation: Effects of modulation-frequency and field strength. *Radio Sci.* 14(6S):93–98.

Blank M, Goodman RM. 2012. Electromagnetic fields and health: DNA-based dosimetry. *Electromagnet. Biol. Med.* 31(4):243–249.

Blaustein MP, Lederer WJ. 1999. Sodium/calcium exchange. Its physiological implications. *Physiol. Rev.* 79(3):763–854.

Blaustein MP, Mordecai P, Jin Zhang, Hamlyn JM. 2009. The pump, the exchanger, and endogenous ouabain: Signaling mechanisms that link salt retention to hypertension. *Hypertension* 53:291–298.

Borgnia M, Nielsen S, Engel A, Agre P. 1999. Cellular and molecular biology of the aquaporin water channels. *Annu. Rev. Biochem.* 68:425–458.

Brini M, Carafoli E. 2009. Calcium pumps in health and disease. *Physiol. Rev.* 89(4):1341–1378.

Carpenter DO, Fejtl M, Ayrapetyan SN, Szarowski DH, Turner JN. 1992. Dynamic changes in neuronal volume resulting from osmotic and sodium transport manipulations. *Acta Biol. Hung.* 43(1–4):39–48.

Chaplin MF. 2006. Information exchange within intracellular water. In: *Water and the Cell*. Pollack GH, Cameron IL, Wheatley DN, eds. Springer, Dordrecht, the Netherlands, pp. 113–123.

Damadian R. 1971. Tumor detection by nuclear magnetic resonance. *Science* 171:1151–1153.

Danielyan AA, Ayrapetyan SN. 1999a. Changes of hydration of rats' tissues after *in vivo* exposure to 0.2 Tesla steady magnetic field. *Bioelectromagnetics* 20(2):123–128.

Danielyan AA, Mirakyan MM, Grigoryan GY, Ayrapetyan SN. 1999b. The static magnetic field on ouabain H3 binding by cancer tissue. *Physiol. Chem. Phys. Med. NMR* 31(2):139–144.

Deghoyan A, Simonyan R, Wachtel H, Ayrapetyan S. 2013. The skeletal muscle imped-ancemetric characteristics as a marker for detection of functional state of organ-ism. *ISRN Biophys.* 2013, Article ID 948074, 7 pages.

de Lores Arnaiz GR, Ordieres MG. 2014. Brain Na(+), K(+)-ATPase activity in aging and disease. *Int. J. Biomed. Sci.* 10:85–102.

Denninger JW, Marletta MA. 1999. Guanylate cyclase and the ·NO/cGMP signaling pathway. *Biochim. Biophys. Acta* 1411(2–3):334–350.

Devyatkov ND. 1973. Effect of a SHF (mm-band) radiation on biological objects. *Uspekhi Fizicheskikh Nauk.* 110:453–454 (in Russian).

Dvoretsky AI, Ayrapetyan SN, Shainskaya AM. 2012. High-affinity ouabain receptors: Primary membrane sensors for ionizing radiation. *Environmentalist* 32(2):242–248.

Evans D. 2009. *Osmotic and Ionic Regulation: Cells and Animals.* CRC Press, New York.

Foster KR. 2006. The mechanisms paradox. In: *Bioelectromagnetics: Current Concepts.* Ayrapetyan SN, Markov MS, eds. Springer, Dordrecht, the Netherlands, pp. 17–29.

Gao J, Wymore RS, Wang Y, Gaudette GR, Krukenkamp IB, Cohen IS, Mathias RT. 2002. Isoform-specific stimulation of cardiac Na/K pumps by nanomolar concen-trations of glycosides. *J. Gen. Physiol.* 119:297–312.

Gapeyev AB, Kulagina T, Alexander V. 2013. Exposure of tumor-bearing mice to extremely high-frequency electromagnetic radiation modifies the composition of fatty acids in thymocytes and tumor tissue. *Int. J. Radiat. Biol.* 89:602–610.

Gapeyev AB, Mikhailik EN, Chemeris NK. 2009. Features of anti-inflammatory effects of modulated extremely high-frequency electromagnetic radiation. *Bioelectromagnetics* 30:454–461.

Grigoriev Y. 2012. Evidence for effects on the immune system supplement. Immune sys-tem and EMF RF. In: *BioInitiative Working Group, Section 8. Bioinitiative Report WHO 2012.* Sage C, Carpenter D, eds., WHO Press, Switzerland pp. 1–24.

Gudkova Olu, Gudkov SV, Gapeev AB, Bruskov VI, Rubanik AV, Chemeris NK. 2005. The study of the mechanisms of formation of reactive oxygen species in aqueous solutions on exposure to high peak-power pulsed electromagnetic radiation of extremely high frequencies. *Biofizika* 50:773–779 (in Russian).

Heqimyan A, Narinyan L, Nikoghosyan A, Ayrapetyan S. 2015. Age-dependent mag-netic sensitivity of brain and heart muscles. In: *Electromagnetic Fields in Biology and Medicine.* Markov MS, ed., CRC Press, Boca Raton, FL, pp. 217–230.

Heqimyan A, Narinyan L, Nikoghosyan A, Deghoyan A, Yeganyan A. 2012. Age dependency of high affinity ouabain receptors and their magneto sensitivity. *Environmentalist* 32:228–235.

Hodgkin AL. 1964. *The Conduction of the Nervous Impulse.* Liverpool University Press, Springfield, IL.

Hoffmann EK, Sørensen BH, Sauter DP, Lambert IH. 2015. Role of volume-regulated and calcium-activated anion channels in cell volume homeostasis, cancer and drug resistance. *Channels (Austin)* 9(6):380–396.

Hoffmann P, Boeld TJ, Eder R, Huehn J, Floess S, Wieczorek G, Olek S, Dietmaier W, Andreesen R, Edinger M. 2009. Loss of FOXP3 expression in natural human CD4+CD25+ regulatory T cells upon repetitive *in vitro* stimulation. *Eur. J. Immunol.* 39(4):1088–1097.

Iwasa K, Tasaki I, Gibbons RC. 1980. Mechanical changes in crab nerve fibers during action potentials. *Science* 210:338–339.

Juhaszova M, Blaustein M. 1997. Na⁺ pump low and high ouabain affinity α subunit isoforms are differently distributed in cells. *Proc. Natl. Acad. Sci. USA* 94(5):1800–1805.

Kaczmarek LK. 2006. Non-conducting functions of ion channels. *Nat. Rev. Neurosci.* 7:761–771.

Katz B. 1966. *Nerve, Muscle and Synapse.* McGraw-Hill, New York.

Khachaturian ZS. 1989. The role of calcium regulation in brain aging: Reexamination of a hypothesis. *Aging* 1:17–34.

Klassen VI. 1982. *Magnetized Water Systems.* Chemistry Press, Moscow, 296 p. (in Russian).

Kojima M, Ayrapetyan S, Koketsu K. 1984. On the membrane potential independent mechanism of sodium pump-induced inhibition of spontaneous electrical activity of Japanese land snail neurons. *Comp. Biochem. Physiol. A Physiol.* 77(3):577–583.

Kometiani P, Liu L, Askari A. 2005. Digitalis-induced signaling by Na⁺/K⁺-ATPase in human breast cancer cells. *Mol. Pharmacol.* 67(3):929–936.

Lang F. 2007. Mechanisms and significance of cell volume regulation. *J. Am. Coll. Nutr.* 26:613–623.

Lednev VV. 1991. Possible mechanism for the influence of weak magnetic field interactions with biological systems. *Bioelectromagnetics* 18:455–461.

Lehninger AL. 1970. Mitochondria and calcium ion transport. *Biochem. J.* 119:129–138.

Lucchesi PA, Sweadner KJ. 1991. Postnatal changes in and skeletal muscle. *J. Biol. Chem.* 267:769–773.

Markov MS. 2004. Myosin light chain phosphorylation modification depending on magnetic fields. I. Theoretical. *Electomag. Biol. Med.* 23:55–74.

Mayrovitz HN. 2015. Electromagnetic fields for soft tissue wound healing. In: *Electromagnetic Fields in Biology and Medicine.* Markov MS, ed., CRC Press, Boca Raton, FL, pp. 231–252.

Minkoff L, Damadian R. 1976. Biological ion exchanger resins. X. The cytotonus hypothesis: Biological contractility and the total regulation of cellular physiology through quantitative control of cell water. *Physiol. Chem. Phys.* 8(4):349–387.

Narinyan L, Ayrapetyan G, Ayrapetyan S. 2012. Age-dependent magnetosensitivity of heart muscle hydration. *Bioelectromagnetics* 33(6):452–445.

Narinyan L, Ayrapetyan G, Ayrapetyan S. 2013. Age-dependent magnetosensitivity of heart muscle ouabain receptors. *Bioelectromagnetics* 34(4):312–322.

Narinyan L, De J, Ayrapetyan S. 2014. Age-dependent increase in Ca^{2+} exchange magnetosensitivity in rat heart muscles. *Biochem. Biophys.* 2(3):39–49.

Nikoghosyan A, Heqimyan A, Ayrapetyan S. 2015. Primary mechanism responsible for age-dependent neuronal dehydration. *Int. J. Basic Appl. Sci.* 5(1):5–14.

Parsegian VA, Rand RP, Rau DC. 2000. Osmotic stress crowding, preferential hydration and binding: A comparison of perspectives. *Proc. Natl. Acad. Sci. USA* 97:3987–3992.

Pilla AA. 2015. Pulsed electromagnetic fields: From signaling to healing. In: *Electromagnetic Fields in Biology and Medicine.* Markov MS, ed., CRC Press, Boca Raton, FL, pp. 29–48.

Romanovsky D, Mrak RE, Dobretsov M. 2015. Age-dependent decline in density of human nerve and spinal ganglia neurons expressing the α_3 isoform of Na/K-ATPase. *Neuroscience* 310:342–353.

Saghian AA, Ayrapetyan SN, Carpenter DO. 1996. Low concentrations of ouabain stimulate Na/Ca exchange in neurons. *Cell. Mol. Neurobiol.* 16(4):489–498.

Shibuya K, Fukuoka J, Fujii T, Shimoda E, Shimizu T, Sakai H, Tsukada K. 2010. Increase in ouabain-sensitive K^+-ATPase activity in hepatocellular carcinoma by overexpression of Na^+, K^+-ATPase α3-isoform. *Eur. J. Pharmacol.* 638(1–3):42–46.

Siegel GJ, Agranoff BW, Albers RW, Fisher SK, Uhler MD. 1999. *Basic Neurochemistry: Molecular, Cellular and Medical Aspects*, 6th edition. Lippincott-Raven, Philadelphia, PA.

Skou J. 1957. The influence of some cations on an adenosine triphosphatase from peripheral nerves. *Biochim. Biophys. Acta* 23:394–401.

Suleymanian MA, Ayrapetyan VY, Arakelyan VB, Ayrapetyan SN. 1993. The effect of osmotic gradient on the outward potassium current in dialyzed neurons of *Helix pomatia*. *Cell. Mol. Neurobiol.* 13(2):183–190.

Szent-Gyorgyi A. 1968. *Bioelectronics: A Study in Cellular Regulations, Defense, and Cancer.* Academic Press, New York, pp. 54–56.

Terakawa S. 1985. Potential-dependent variation of the intracellular pressure in the intracellularly perfused squid axon. *J. Physiol.* 369:229–248.

Thomas RC. 1972. Electrogenic sodium pump in nerve and muscle cells. *Physiol. Rev.* 52(3):563–594.

Weidemann H. 2012. "The Lower Threshold" phenomenon in tumor cells toward endogenous digitalis-like compounds: Responsible for tumorigenesis. *J. Carcinog.* 11:2.

Wu J, Akkuratov EE, Bai Y, Gaskill Cassie M, Askari A, Liu L. 2013. Cell signaling associated with Na^+/K^+-ATPase: Activation of phosphatidylinositide 3-kinase IA/Akt by ouabain is independent of Src. *Biochemistry* 52(50):9059–9067.

Xie Z, Askari A. 2002. Na^+/K^+-ATPase as a signal transducer. *Eur. J. Biochem.* 269:2434–2439.

Ziskin MC. 2006. Physiological mechanisms underlying millimeter wave therapy. In: *Bioelectromagnetics Current Concepts.* Ayrapetyan SN, Markov MS, eds. Springer, Dordrecht, the Netherlands, pp. 241–225.

17

Clinical Dosimetry of Extremely Low-Frequency Pulsed Electromagnetic Fields

William Pawluk
Medical Editor

Dosimetry classically looks at the intensity of the magnetic field presented by a pulsed electromagnetic field (PEMF) signal at the tissue. However, from a clinical perspective and outside experimental, theoretical, or mathematical modeling scenarios, the clinical considerations for this dosimetry are much more complicated and require a modified "model" within which future research could be directed. To address this, it is recommended to think about this expanded concept of dosimetry as a clinical dosimetry model.

Clinicians are often told to understand that PEMFs drop off in intensity as you move farther away from the source of the signal. They may not know that this is subject to the inverse square law. Further addressing point, clinicians often do not understand the inverse square law and how to do their own calculations or estimations of the magnetic field from their particular therapeutic devices at the target tissue.

Additionally, it may be very instructive to have some examples from the research literature, for at least one condition, pain, to see how variable stimulation systems might be and to assess their potential benefits. Clinicians need to consider the target tissue and the putative source of the pain or at least the target dysfunction or pathology. Ideally, an awareness of the clinical research literature would be helpful to clinicians to know which treatment parameters would be most effective.

17.1 Clinical Scenario

Clinicians attempting to apply PEMFs to patients are faced with a daunting array of decisions. Not only must they consider which devices are available, but also consider their intensities, frequencies, waveforms, and applicator configurations, among others. Decision making is further complicated by the nature of patients themselves, and so clinicians must consider the therapeutic goals, the nature of the physiologic dysfunction or pathology presented, the duration of individual treatments, the time course(s) of treatments, affordability, practicality, and whether treatments are applied in the clinical setting or by the patient personally in the home or nonoffice setting.

It is not likely that clinicians or patients would have available an array of devices to choose from, whether decisions to select treatment devices are based on research evidence or other aspects of decision making. Choices are most often made by influence from those marketing commercially available devices. Therefore, clinicians typically will select one or perhaps two devices to serve their needs across a spectrum of health conditions.

A simpler scenario would be for the clinician to treat only one or two individual conditions, based on which condition would have probably the best evidence base. The treatment of pain is a very common indication for the use of PEMFs. Even this single problem encompasses a vast array of causes and areas of the body that would need to be treated (Pawluk, 2015). The evidence base certainly shows effectiveness for PEMFs in the management of pain, but there have been a wide range of different devices and device design parameters tested. But, unfortunately, many of the studies reviewed did not use commercially available devices.

Clinical experience with the use of PEMFs spanning at least two decades has shown that most commercially available devices provide at least some level of benefit for almost all conditions. Unlike the research setting, clinical application of PEMFs can extend over long periods of time. A very large percentage of patients treated by clinicians with PEMFs are for chronic, long-term conditions.

Clinical experience with individual patients dictates the amount of exposure necessary for a given device to achieve sustainable results. This is especially true when one is trying to heal an underlying physiologic imbalance or pathologic process to achieve durable benefits. While most medical therapies are not curative, but rather help to control specific symptoms (such as pain) or improve specific physiologic parameters (such as blood pressure), PEMFs provide the opportunity to alter the morphology of the tissue being treated, because of the depths to which PEMFs safely penetrate the body, and therefore move closer to the possibility of obtaining "cure." Clinical experience also shows that many patients are often able to reduce or stop other medical therapies with the continuing use of PEMFs, thereby reducing their own and the health-care system's cost and the risk of side effects.

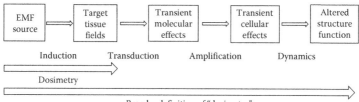

FIGURE 17.1 Clinical dosimetry model. (Adapted from Bowman, J., RF exposures to the general public: Lessons from "dosimetry" for ELF-EMF epidemiology. Joint NIOSH/DOE Workshop. EMF exposure assessment and epidemiology: Hypotheses, metrics, and measurements. National Institute for Occupational Safety and Health (NIOSH), Cincinnati, OH, 1994.)

When many different devices are compared by a clinician over an extended time, one of the most important considerations for the clinician is how quickly results will be achieved. Clinical experience has shown that low-intensity PEMF systems generally take longer to produce results acceptable to the patient or clinician. Typically, physiologic measures may improve more quickly, but outcome change measures often take significantly longer to achieve desired results.

This clinical observation was recently reinforced by a review of the effectiveness of low-intensity whole-body PEMF systems (Hug and Röösli, 2012). The studies reviewed in this paper focused on osteoarthritis (OA) of the knee or cervical spine, fibromyalgia, pain perception, skin ulcer healing, multiple sclerosis-related fatigue, or heart rate variability and well-being. The maximum study time across the studies lasted up to 12 weeks. Maximum magnetic flux density ranged from 3.4 to 200 μT. This review concluded that scientific support for benefits was insufficient. It is highly likely that for the field intensities evaluated, the studies were significantly underpowered or not conducted for a long-enough duration.

The abovementioned review and clinical experience across many different types of PEMF systems demonstrate that a different concept of magnetic field intensity needs to be considered, called the clinical dosimetry model (Figure 17.1).

17.2 Clinical Dosimetry Model

PEMFs of various configurations can produce beneficial effects for conditions as diverse as pain, wounds, and fractures. This has all been achieved with low-intensity, nonthermal, noninvasive PEMFs, with many configurations over a very broad frequency range and exposure parameters.

Dosimetry may classically be considered as a certain magnetic field dose delivered into the tissues, which can then cause a cascade of events in the tissues. Multiple mechanisms for the biologic actions of PEMFs include induction of electrical currents and ion binding, among others. Pilla (2006) likely suggested that each of the aspects of actions of PEMFs in the body have their own ranges of parameters to optimize their respective effects. It is also likely that most of the different actions of PEMFs are happening relatively simultaneously once tissues are exposed. In other words, even relatively precisely designed PEMF signals are likely to have a diverse range of actions that are not necessarily tunable. This means that while one may select a specific signal to reduce inflammation, it would be likely to have other biologic actions as well simultaneously such as improving circulation, affecting DNA, antioxidant actions, etc. Therefore, it becomes challenging to determine which specific biologic action produced the ultimate clinical outcome.

So, in this clinical dosimetry model, it cannot be assumed that effects necessarily follow in a linear fashion from induction, to transduction, to application, and finally to dynamics. In fact, it is also likely that these processes would be iterative as physiologic changes occur and different levels or stages of homeostasis are reached, ultimately reaching complete homeostasis or "normality."

In Figure 17.1, the EMF source (considering both the PEMF apparatus producing the EMF and any applicators) is used to apply the PEMF to the body, therefore creating induction in the target tissues, usually a fairly large area of stimulation, encompassing one or more organs, and sometimes a whole body simultaneously. Any electric fields induced in the tissues are transduced by various molecules in the exposed tissues to varying degrees to create molecular effects; for example, mobilization of calcium ions. Often these molecular effects are transient when the magnetic field is removed and the currents extinguished. Once these molecular effects are in motion, they may be amplified by various mechanisms in the target tissue to actually produce cellular changes. A comparable physiologic example is touching a finger to anything or with anything, which will lead to a subsequent amplification and distribution of the signal to the brain, thereby resulting in the perception of touch or even pain. If enough cellular changes accumulate from a PEMF stimulus, then altered structure and/or function may follow. Applied with a sufficient intensity and over a long enough period of time, structural change may be more or less permanent.

In terms of the clinical dosimetry model, it is quite probable that different dosimetries may be necessary to optimize changes at each of the individual steps. It cannot be assumed that what is optimal at one step in the model will be adequate for the next step. So, while the goal in the development of any therapeutic paradigm is optimization of all the steps along the way (perhaps considered bottom-up research), the ultimately important optimization happens when goal homeostasis is achieved at the structure and function level (perhaps considered top-down research).

It is this concept, clinical dosimetry, that applies to the clinical application of PEMFs. The clinician has to deduce from the marketed or stated magnetic field intensity what the "dose," that is, the maximum field intensity of the applicator is. The dose and its associated frequencies and field distribution, applied for a specified period of time and repeated as often as is clinically indicated, are expected to result in specific biologic or clinical effects. Clearly, from a scientific perspective, this is a much more complex process than just the magnetic field intensity "dose." For the clinician, the model has to be simplified to be understood and properly used.

17.3 Attenuation of Extremely Low-Frequency Magnetic Fields

Dorland's Illustrated Medical Dictionary (Dorland, 2011) defines dosimetry as the accurate and systematic determination of doses. Relative to PEMFs, clinically and simply speaking, dosimetry equates to the measurement of the field strength of a PEMF signal at various distances away from the applicator. More relevant clinically, the depth into the body the PEMF signal will penetrate and at what intensity is what matters. Extremely low-frequency (ELF) PEMFs, typically <1 MHz, do not get attenuated by passing through tissue, whether bone or soft tissue. Practically speaking, ELF flux density inside living tissues is equal to that applied externally with no intervening tissue (Polk, 1992). Magnetic permeability of tissues and cells is, to within a fraction of 1%, equal to that of free space. So, for practical purposes, ELF PEMF doses from an applicator when measured at specified distances from the applicator in free space can be essentially equivalent to those found at similar depths in the body.

Therefore, any loss of the intensity of the magnetic field has almost entirely to do with the normal loss of a magnetic field amplitude with the distance from the EMF source. Field intensity loss varies with distance—inverse square rule. It is helpful to the clinician to be able to numerically and visually see the impact of distance on magnetic field intensity.

Using Newton's inverse square rule, the following formula applies: $I_1 \times d_1^2 = I_2 \times d_2^2$, where I = intensity and d = distance. Intensity at a distance from the EMF source would be $I_2 = (I_1 \times d_1^2) / d_2^2$ (www.nde-ed.org).

On the basis of the inverse square rule, various field intensities have been calculated by the author at various depths or distances away from the magnetic applicator source (Table 17.1), assuming the selection of a single point at a selected distance from a simply configured PEMF signal. Dose calculations from EMF sources that are more complexly configured would have to have that considered, and many researchers report three-dimensional graphs of magnetic field intensities plotted at various distances.

Table 17.1 below provides the calculated field intensities with the distance from the source of the EMF. Figure 17.2 is a plot of the data from Table 17.1 for selected EMF intensities (from 1 to 100 mT) plotted at increments of 1 cm from the source of the EMF. It can be clearly shown from these calculations that magnetic field intensities drop off extraordinarily rapidly from the source.

TABLE 17.1 Calculated Magnetic Field Intensities (I_1) versus Distance from the Magnetic Field Source (d_2) with Field Intensities from 1 to 100 mT

I_1 (mT)	0	1	2	3	4	5	d_2 (cm) 6	7	8	9	10	11	12	13
1	1.0	0.3	0.1	0.1	0.04	0.03	0.02	0.02	0.01	0.01	0.01	0.01	0.01	0.01
10	10.0	2.5	1.1	0.6	0.4	0.3	0.2	0.2	0.1	0.1	0.1	0.1	0.1	0.1
20	20.0	5.0	2.2	1.3	0.8	0.6	0.4	0.3	0.2	0.2	0.2	0.1	0.1	0.1
30	30.0	7.5	3.3	1.9	1.2	0.8	0.6	0.5	0.4	0.3	0.2	0.2	0.2	0.2
40	40.0	10.0	4.4	2.5	1.6	1.1	0.8	0.6	0.5	0.4	0.3	0.3	0.2	0.2
50	50.0	12.5	5.6	3.1	2.0	1.4	1.0	0.8	0.6	0.5	0.4	0.3	0.3	0.3
60	60.0	15.0	6.7	3.8	2.4	1.7	1.2	0.9	0.7	0.6	0.5	0.4	0.4	0.3
70	70.0	17.5	7.8	4.4	2.8	1.9	1.4	1.1	0.9	0.7	0.6	0.5	0.4	0.4
80	80.0	20.0	8.9	5.0	3.2	2.2	1.6	1.3	1.0	0.8	0.7	0.6	0.5	0.4
90	90.0	22.5	10.0	5.6	3.6	2.5	1.8	1.4	1.1	0.9	0.7	0.6	0.5	0.5
100	100.0	25.0	11.1	6.3	4.0	2.8	2.0	1.6	1.2	1.0	0.8	0.7	0.6	0.5

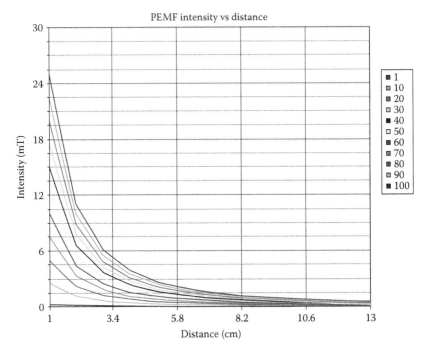

FIGURE 17.2 Calculated PEMF intensity from Table 17.1 in mT versus distance in centimeters (cm).

The data in Table 17.1 may be graphically represented to show what the rapidly declining intensities look like. The color codes for the intensities are in the legend to the right on the graph for intensities varying from 1 to 100, by 10 mT.

In Figure 17.3, at the distance of 6 cm, all of the field cluster very closely together. Chart graphed to start intensity measurements at 6 cm away from the applicator, it is easy to see how the intensities at 10 cm and more separate out based on initial intensity.

17.4 Dose Necessary to Achieve Clinical Outcomes

Because field intensities drop off this rapidly, the question becomes what is the magnetic field intensity, that is, "dose," necessary to achieve the desired clinical outcomes?

While an exhaustive evaluation of published studies is beyond the scope of this chapter, some selected studies will help to demonstrate that there are differences in the effects of magnetic field intensities for the treatment of different conditions. It would be expected that different metabolic processes, tissues, pathologies, underlying physiologic states, and health status would all be important and possibly demonstrate more stark variation in benefit. The problem is that most research does not get into the necessary detail about the nature of the fields used (Markov, 2015) to allow replication of the research. This includes, at the least, total dose delivered and even measurements of dB/dT, in general; never mind what would be expected or seen at the target tissue. There are valid reasons for this situation, usually practical, that is, lack of funding or inadequate technical support from

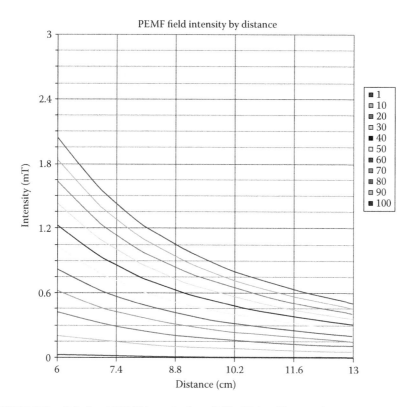

FIGURE 17.3 Calculated PEMF intensity from Table 17.1 in mT versus distance in centimeters (cm), starting at 6 cm from the PEMF source.

engineering, biology, and clinical PEMF experts. Still, that does leave the clinician with inadequate information upon which to make appropriate clinical decisions.

17.5 Rat Paw Carrageenan Dosimetry Study

The intention of this study (Dennis R, Assess the effects of ELF PEMF in a model of acute inflammation, Personal communication, 2013) was to assess the impact on inflammation of progressively increasing field intensities of a PEMF signal. Rats were experimentally evaluated using the carrageenan paw inflammation model. The rats were divided into six groups. Group 1 was a saline vehicle control. The standard treatment group was given a dose of oral dexamethasone 1 h before carrageenan challenge. Groups 3 through 6 received exposure to escalating intensity magnetic coils before the carrageenan challenge. Coils were placed in a planar array below the bottom of the rat cage. The magnetic field ran for three patterns back to back—a 5-Hz square wave for 10 min, immediately followed by 50 Hz for 10 min at a positive polarity, followed by the opposite polarity for another 10 min. The whole frequency package recycled every 30 min continuously for the 9-h study period. The gauss slew rates, for the four PEMF-treated groups were 400, 800,

1200, and 1600 kG/s. Rat paw edema was measured at 0, 1, 2, 4, and 8 h after carrageenan dose by hind limb plethysmography.

The data were supplied by the Dennis research group (Dennis R, Assess the effects of ELF PEMF in a model of acute inflammation, Personal communication, 2013).

Footpad volumes were compared to baseline, and the differences in volume were calculated. The vehicle control showed the largest differences in footpad volume from baseline. The dexamethasone standard treatment group showed the least footpad volume changes within an hour through to the end of the study. The difference in footpad volumes was evaluated for each group for each time point and treatment (Figure 17.4).

As would be expected, at time 0 there were no differences. At 1 h, differences among the four PEMF-treatment groups began to emerge but were not significant. At 2 h, three of the four PEMF-treatment groups showed a statistically significant change. At 4 h, all groups showed statistically significant changes and even greater improvement in paw edema reduction. There was an absolute gradient difference between the different magnetic field strengths used, with the highest field strength showing the greatest reduction in edema, fairly closely approximating dexamethasone. All groups showed less paw edema reduction at 8 h. Dexamethasone clearly suppressed inflammation the most (Figure 17.5).

Again, the dexamethasone group showed the least inflammation-induced paw volume inflammatory changes. The three highest intensity PEMF groups showed statistically significant differences in inflammatory volume. The 1200 kG/s group had a marginally larger reduction in inflammation than the other two best groups.

When considered from a slightly different perspective, the percentage disease suppression (calculated as the mean of the differences between the control mean area under the curve (AUC) for individual treated animals divided by the control mean AUC and multiplied by 100), is greater relative to the control group when the numbers are larger. The dexamethasone group had an 84% disease suppression, while the magnetic 800 and 1600 kG/s groups had 75% and 79% disease suppression, respectively.

At least in this classic type rat paw inflammation study, magnetic field dose intensification made clear differences in experimental animal model inflammation reduction.

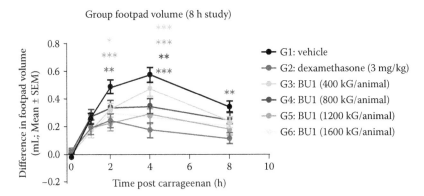

FIGURE 17.4 Group footpad volume difference between the carrageenan-injected and saline-injected paws was calculated by time and field intensity. Significance (one-way Analysis of variance (ANOVA) and post hoc Dunnett's test): $^*p \leq .05$, $^{**}p < .01$, $^{***}p < .001$, compared to Group 1.

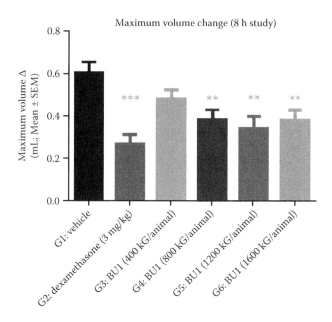

FIGURE 17.5 Footpad volume difference between maximum volume and baseline by time and field intensity. Significance (one-way ANOVA and post-hoc Dunnett's test): $^*p \leq .05$, $^{**}p < .01$, $^{***}p < .001$, compared to Group 1.

17.6 Sample of Studies on Pain Management Using ELFs by Intensity

A selected number of studies of varying field intensities were used to demonstrate effects on pain management.

Low-intensity studies

- Even weak PEMFs affect pain perception and pain-related Electroencephalogram (EEG) changes in humans, a 2 h exposure to 0.02–0.07 mT/20–70 µT. ELF MFs caused a significant positive change in pain-related EEG patterns (Sartucci et al., 1997).
- Back pain or whiplash syndrome treated with a very low-intensity (up to 30 µT) PEMF twice a day for 2 weeks along with usual pain medications relieves pain in 8 days in the PEMF group versus 12 days in the controls. Headache is halved in the PEMF group, and neck and shoulder/arm pain improved by one-third versus medications alone (Thuile and Walzl, 2002).
- 75-Hz, rectangular PEMFs after arthroscopic treatment of knee cartilage. Two groups: lower intensity control (MF at 0.05 mT/50 µT) and active (MF of 1.5 mT), for 90 days, 6 h per day. Knee score values at 45 and 90 days were higher in the active group at 90 days ($p < .05$). Non-steroidal anti-inflammatory drugs (NSAID) use was 26% in the active group and 75% in the control group ($p = .015$). At 3-year

follow-up, the percentage of those completely recovered was higher in the active group ($p < .05$) (Zorzi et al., 2007).

- 50 Hz, 105 μT PEMF was applied for 30 min, in Grade 3 knee OA for 3 weeks. Pain improved significantly in both sham and active-treatment groups relatively equally ($p < .000$). Active-treatment patients had significant improvement in morning stiffness and in activities of daily living (ADL) compared to that in controls (Ay and Evcik, 2009).
- Acute 30-min MF exposure (≤ 400 μT, <3 kHz). A significant pre–post effect in VAS-pain and anxiety ratings was seen in the fibromyalgia patients, $p < .01$ (Shupak et al., 2006).
- Small, battery-operated PEMF devices with very weak field strengths have been found to benefit musculoskeletal disorders. Because of the low strength used, treatment at the site of pain may need to last between 11 and 132 days, between 2 times per week, 4 h each, and, if needed, continuous use (Fischer G, Relieving pain in diseases of the musculoskeletal system with small apparatuses that produce magnetic fields, Personal communication, 2002).

Medium-intensity studies

- 50-Hz sinusoidal, 35-mT PEMF for 15 min, for 15 treatment sessions improved hip arthritis pain in 86% of patients. Average mobility without pain improved markedly (Rehacek et al., 1982).
- PEMFs of 35–40 mT, 20 min/day for 20–25 days gave relief or eliminated pain about 90%–95% of the time for lumbar OA (Mitbreït et al., 1986).
- PEMFs of 0.5–1.5 mT used at the site of pain and related trigger points also helped (with some remaining pain-free for 6 months after treatment) (Rauscher and Van Bise, 2001).
- Even chronic musculoskeletal pain treated with MFs for only 3 days, once per day, could eliminate and/or maintain chronic musculoskeletal pain (Stewart and Stewart, 1989).
- Diabetic neuropathy patients had 20 exposures to variable sinusoidal PEMF, 40 Hz, 15 mT, every day for 12 min. Reduction of pain and paresthesias, vibration sensation, and improved muscle strength was seen in 85% of patients, all better than sham controls (Cieslar et al., 1995).

High-intensity studies

- High-intensity pulsed magnetic stimulation (HIPMS), intensity up to 1.17 T, was studied in recovery after injury in post-traumatic/postoperative low-back pain, reflex sympathetic dystrophy (RSD), neuropathy, thoracic outlet syndrome, and endometriosis. Outcome pain VAS difference was 0.4–5.2 with sham versus 0–0.5 for active treatments; pain reduction likely due to induced eddy currents (Ellis, 1993).

Even on the basis of this abbreviated review, it can be seen that the PEMFs used would have to penetrate to various depths depending on the conditions studied. These conditions included the following organs or tissues: brain, back, spine and spinal cord, knee, hip, neck and shoulders, muscles, nerves and blood vessels in the lower extremities, and pelvic organs. When a very low-intensity PEMF was used (Zorzi) as the control 50 µT versus 1.5 mT in the active group, some benefit was still seen in the control group. It would have been better to have seen a sham control used to assess what the difference would be between a very low-intensity field and a placebo effect. As indicated by one of the authors (Fischer G, Relieving pain in diseases of the musculoskeletal system with small apparatuses that produce magnetic fields, Personal communication, 2002), using low-intensity PEMFs may require treatments to last between 11 and 132 days, between 2 times per week, 4 h each, and, if needed, based on the patient, continuous use. By contrast, lumbar and hip arthritis using 35+ mT fields for 15–20 min a day for upward of 25 days were needed to obtain symptomatic benefit (Mitbreĭt et al., 1986).

Here are some questions that a clinician might ask. For treatment of a specific spot on the hip joint, assuming a depth of penetration needed at 12 cm, per the inverse square law, a 20-mT peak intensity magnetic field would be delivering about 0.1 mT to the tissue. Obviously, given the configuration of the magnetic field and the volumetric configuration of the hip joint, the calculated dose would vary throughout the joint. Similar calculations may be done for the knee, a smaller joint, with an assumed radius of between 8 and 10 cm, for a PEMF of 20-mT peak intensity magnetic field, the delivered dose would be estimated to be between 0.2 and 0.3 mT. When considering using PEMFs between 10 and 100 µT (0.01–0.1 mT) at 8 cm, the calculated PEMF dose would be 0.00016–0.00156 mT or 160–1560 nT. How active can PEMFs be at this level of intensity in tissues? presumably they wouldn't be clinically meaningful.

Except for animal studies, such as the rat inflammation study above, it is hard to know what the actual delivered dose is at the target tissue. In fact, it is hard sometimes to determine what the actual target tissue is. In the case of a forearm fracture, tendinitis of the elbow, eye lesions, dysfunctions, temporomandibular joint dysfunctions, etc., that is, superficial tissues, dose calculations are easier.

It is suggested (Polk, 1992) that chemical reactions in tissues involving free radicals can be sensitive to PEMFs at <1 mT. On the basis of the physics, this reactivity requires the presence of a companion Static magnetic field, which the earth provides naturally. This situation would certainly apply to the clinical setting. There are likely other mechanisms involved as well.

Clinicians need to also consider that at nT or pT levels of intensity of the applicator or even internal in the body, external interferences, which are also plentiful in the clinical setting, could also affect clinical results. These external interferences could include radio frequencies emitted by cell phones, routers, nearby cell towers, smart meters, and other electrical equipment often present in the clinical and home setting. Also, external interferences can create unpredictable constructive or destructive wave patterns. So, in some cases, results may be enhanced, and in other cases, the results may be compromised, making treatment results unpredictable. This issue by itself may create an argument for using medium to higher intensity PEMFs. If low-intensity treatment systems are used in the clinical research setting, because of external interference effects, potential external interferences must be measured, accounted for, and mitigated.

Dosimetry is not just a matter of intensity; it's also dependent on the treatment time, frequency of treatments, and the length of total treatment time. Clinical experience seems to indicate that higher intensities typically require less treatment time or at least less time to see dependable results.

17.7 Summary

Clinical dosimetry is extraordinarily complex and clinicians typically lack sufficient data to determine optimal treatment parameters. There are few adequate clinical guidelines, based on empirical evidence, that describe the specifications that a PEMF system should have to produce optimal results for the condition being addressed. Elements of clinical decision making often constrain the clinician in providing the optimal system or treatment protocol for any given condition or person. Conditions being treated cover the range or spectrum of problems involving all aspects of the body. Optimal parameters have not been determined for whole-body treatment whether for conditions such as osteoporosis, hypertension, widespread vascular disease, or health maintenance, among others. Fortunately, from a risk benefit perspective, the risk for clinicians appears to be less one of overtreatment than it is undertreatment. Clinical office treatments are practically limited by time available and cost, making undertreatment more likely. Home based self treatment may be less likely to under treat.

Risk is not only clinical but also financial. As shown by the Hug review, whole body systems, which can often be very expensive, may not deliver expected results for specific health conditions, at least not in the desired or studied time frames.

Certainly, considerably more research needs to be done on the clinical aspects of the application of PEMFs and appropriate clinical dosimetry.

References

Ay S, Evcik D. (2009) The effects of pulsed electromagnetic fields in the treatment of knee osteoarthritis: A randomized, placebo-controlled trial. *Rheumatol Int* 29(6): 663–666.

Bowman J. (1994) RF exposures to the general public: Lessons from "dosimetry" for ELF-EMF epidemiology. Joint NIOSH/DOE Workshop. EMF exposure assessment and epidemiology: Hypotheses, metrics, and measurements. National Institute for Occupational Safety and Health (NIOSH), Cincinnati, OH.

Cieslar G, Sieron A, Radelli J. (1995) The estimation of therapeutic effect of variable magnetic fields in patients with diabetic neuropathy including vibratory sensibility. *Balneol Pol* 37(1): 23–27.

Dorland WAN. (2011) *Dorland's Illustrated Medical Dictionary*, 32nd edition. Elsevier Health Sciences, Philadelphia, PA.

Ellis WV. (1993) Pain control using high-intensity pulsed magnetic stimulation. *Bioelectromagnetics* 14(6): 553–556.

Hug K, Röösli M. (2012) Therapeutic effects of whole-body devices applying pulsed electromagnetic fields (PEMF): A systematic literature review. *Bioelectromagnetics* 33(2): 95–105.

Markov M. (2015) XXIst century magnetotherapy. *Electromagn Biol Med* 34(3): 190–196.

Mitbreĭt IM, Savchenko AG, Volkova LP, Proskurova GI, Shubina AV. (1986) Low-frequency magnetic field in the complex treatment of patients with lumbar osteochondrosis. *Ortop Travmatol Protez* (10): 24–27.

NDE-Ed. Calculating intensity with the inverse square law. https://www.nde-ed.org/ Education resources/Communitycollege/radiationsafety/safelife/distance.php.

Pawluk W. (2015) Magnetic fields for pain control, chapter 17. In: *Electromagnetic Fields in Biology and Medicine.* Markov MS, ed., CRC Press, Boca Raton, FL.

Pilla AA. (2006) Mechanisms and therapeutic applications of time-varying and static magnetic fields. In: *Handbook of Biological Effects of Electromagnetic Fields*, 3rd edition. Barnes F, Greenebaum B, eds, CRC Press, Boca Raton, FL.

Polk C. (1992) Dosimetry of extremely-low-frequency magnetic fields. *Bioelectromagnetics* 13(Suppl 1): 209–235.

Rauscher E, Van Bise WL. (2001) Pulsed magnetic field treatment of chronic back pain. In: *23rd Annual Meeting of Bioelectromagnetics Society*, Saint Paul, MN. Abstract 6–3: 38.

Rehacek J, Straub J, Benova H. (1982) The effect of magnetic fields on coxarthroses. *Fysiatr Revmatol Vestn* 60(2): 66–68.

Sartucci F, Bonfiglio L, Del Seppia C, Luschib P, Ghionec S, Murria L, Papi F. (1997) Changes in pain perception and pain-related somatosensory evoked potentials in humans produced by exposure to oscillating magnetic fields. *Brain Res* 769(2): 362–366.

Shupak NM, McKay JC, Nielson WR, Rollman GB, Prato FS, PhD, Thomas AW. (2006) Exposure to a specific pulsed low-frequency magnetic field: A double-blind placebo-controlled study of effects on pain ratings in rheumatoid arthritis and fibromyalgia patients. *Pain Res Manag* 11(2): 85–90.

Stewart DJ, Stewart JE. (1989) The destabilization of an abnormal physiological balanced situation, chronic musculoskeletal pain, utilizing magnetic biological device. *Acta Med Hung* 46(4): 323–337.

Thuile C, Walzl M. (2002) Evaluation of electromagnetic fields in the treatment of pain in patients with lumbar radiculopathy or the VAS-pain and anxiety ratings pain and anxiety ratings whiplash syndrome. *Neuro Rehabil* 17: 63–67.

Zorzi C, Dall'Oca C, Cadossi R, Setti S. (2007) Effects of pulsed electromagnetic fields on patients' recovery after arthroscopic surgery: Prospective, randomized and double-blind study. *Knee Surg Sports Traumatol Arthrosc* 15(7): 830–834.

18

Enhancement of Nerve Regeneration by Selected Electromagnetic Signals

Betty F. Sisken
University of Kentucky

18.1 Introduction

Injuries to the nervous system are a major clinical problem. Over 20 million Americans are victims of accidents and medical disorders affecting nerves. Peripheral nerves are particularly sensitive to being crushed or transected by blunt trauma, especially those located just under the skin. Although peripheral nerves can regenerate, they do so slowly and incompletely, and any treatment that increases the rate of regeneration is of great importance since the quicker it occurs the less deterioration of the end organ. Many different techniques to increase the rate of regeneration have been tested and adopted, involving at times chemical agents, growth factors, and electric or electromagnetic fields. Data obtained from *in vitro* and *in vivo* studies indicate that specific waveforms and current density delivered by pulsed electromagnetic fields (PEMFs) can enhance the rate of regeneration of injured peripheral nerves (Sisken et al., 1990; Walker et al., 2007; Gordon et al., 2009).

It is clear now that both basic science and clinical applications need establishment of an accurate protocol for the use of electromagnetic fields in this area. The problem is that nearly every study utilizes different EMF signals and/or different targets. In my laboratory, over the years, the model of chick dorsal root ganglia (DRG) has been used consistently, and I believe that this is a good approach for dosimetric evaluation in nerve regeneration.

In the following pages, data will be presented from my research in cooperation with Dr Stephen Smith and Dr Arthur Pilla who spent most of his scientific work devoted to developing PEMF signals (Pilla, 2015). The techniques addressed nerve regeneration, so that they could be developed to be not only noninvasive and portable but also be easily adaptable for use on humans and animals (Lundborg, 1988).

It should be noted that these data were predated by investigators who worked on *in vitro* preparations of nerve tissue. They include Ingvar (1920), Karssen and Sager (1934), Weiss (1934), Peterfi and Williams (1934), and Marsh and Beams (1946).

Since the chick embryo model is well known and nerve tissue dissected from it is easy to be obtained, we began a series of experiments (Sisken and Smith, 1975) designed to explore the effects of electric current on growth of the embryonic sensory ganglia *in vitro*. In our first series, we used chick trigeminal ganglia (TG) that was cultured in plastic dishes with metal electrodes to allow delivery of current levels of up to 11.5 nA to the ganglia. Such low levels were found to have at least three major effects on this culture system:

1. The outgrowth of nerve fibers from the explant was more abundant, and these fibers were longer and more highly branched.
2. The survival of neurons within the explanted ganglion was enhanced by the electric current, as determined on sections of the ganglia. The neurons had rounded cell bodies containing a central nucleus, cytoplasmic Nissl substance, and surrounding glia; these were found in 93.1% of the direct current-treated cultures and in 53.5% of untreated cultures after 4 days in culture.
3. Using time-lapse cinematography, neurons, fibers, and nonneuronal cells were stimulated to grow in the direction of the cathode. The rate of migration to the cathode was calculated to be 2.4 mm/day in a 12-day-old chick embryo ganglia explant.

After establishing a good model culture system, and obtaining positive effects of direct current on neuronal growth, it was important to search for a noninvasive technique that eventually could be used clinically. The advent of using pulsed electromagnetic technology for arthritis and orthopedics (Bassett et al., 1977; Fukuda and Kameyana, 1979; Bassett, 1989) and nerve studies (Ito and Bassett, 1983; Parker et al., 1983; Orgel et al., 1984) made it the obvious choice for such a study. This chick ganglia model makes it very easy to investigate different electromagnetic signals; therefore, we began testing a series of pulsed electromagnetic signals in our culture model (Sisken and Lafferty, 1978). After these experiments were finished, we used the same model to test whether static magnetic fields alone could affect neurite outgrowth. Static magnets have a long history for use against pain, dating back to early Christian times, although there is still controversy as to their effectiveness. Therefore, it was interesting to test the effects of magnetic fields since the embryonic ganglia we used were part of the sensory system affecting pain. To this end, we tested static magnetic fields (SMFs) from 22.5 to 90 mT (225–900 G; Sisken et al., 2007).

The long-term goal of our research has been to determine the most effective way to use low-level electromagnetic fields to stimulate nerve regeneration *in vitro*, and to apply this information to the *in vivo* situation. Data from our *in vitro* system reveals a "window of current" that yields an enhanced response. It is similar to the current window found to be effective for stimulating nerve growth and regeneration *in vivo* (Sisken et al., 1989).

18.2 The *In Vitro* Model

This *in vitro* model uses chick embryo sensory ganglia [either TG or DRG] to test the effects of applied direct current (DC), PEMF, and SMFs on nerve regeneration using the criteria of regrowth of axonal processes (neurite outgrowth).

In most cases, we compared our results to nontreated control cultures and to cultures treated with a well-characterized growth factor, nerve growth factor (NGF). Initial studies were performed on 12-day chick embryo TG but were later switched to 8–9-day chick embryo DRG for all future studies since we could obtain more ganglia per embryo. DRG are composed of sensory neurons whose distal axons innervate end organs like skin and muscle and communicate mechano-sensory information like pain, temperature, and vibration to the spinal cord and brain.

As stated previously, embryonic ganglia are easy to obtain, and grow well in culture dishes, which make them ideal for studying growth, survival, and tolerances to specific agents. An example of regeneration of axons and dendrites (neurites) from the centrally located neuronal cell bodies can be seen in Figure 18.1. Unless specified, the neurites are always accompanied by nonneuronal cells such as glia and fibroblasts, which help produce important growth factors. All culture dishes containing ganglia and culture medium were placed on plastic shelves in a 37°C incubator to eliminate interference of metallic shelves with the testing of any electromagnetic fields being used in the incubator.

18.3 Quantifying Growth and Regeneration

The method used initially for determining growth in explanted ganglia was designed by Levi-Montalcini (1966) for assessing the effects of NGF on regeneration of nerve fibers. Growth was determined visually by assessing the area of neurite outgrowth. Other investigators devised more quantitative methods for measuring neurite lengths and numbers (Bilsland et al., 1999; Hynds and Snow, 2002). Shah et al. (2004) produced an interactive processing program to measure six parameters (neurite area, neurite length, neurite number, center area, neurite density, and total area) using a single method. Explants were not fixed or stained to eliminate those artifacts but photographed under a microscope in culture dishes containing saline. We used this method to produce growth curves of chick embryo DRG in culture media containing various concentrations of NGF (as seen in Figure 18.2).

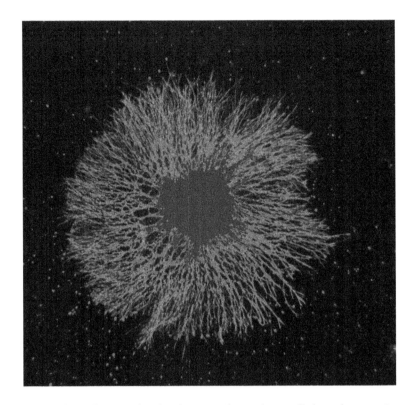

FIGURE 18.1   (See color insert.) A dorsal root ganglion with centrally-located neurons in orange are surrounded by their yellow-colored neurites (axons and dendrites) at 3 days *in vitro* in medium containing 10 ng/mL NGF.

Figure 18.2 shows the sequence of steps used for measuring the six parameters. The central area containing the nerve cell bodies are outlined, followed by drawing the neurites in a different color, thereby determining their length and number. Elimination of the central area allows the measurement of neurite area, which when combined with the central area gives the total area of the growing explant (Shah et al., 2004).

NGF was initially found in tumors of mouse sarcomas and used by Levi-Montalcini (1966) to test for its growth-promoting activity in embryonic dorsal root and sympathetic ganglia. The activity of NGF upregulated after trauma stimulates the early and late stages of the regenerating nerve and is produced *in vivo* by the Schwann cells of the distal stump. Not only does NGF promote an increase in the number of neurites from ganglion neurons *in vitro* and *in vivo* but it also acts to increase the survival of the neurons. It has been and still is used as a standard for comparison with other chemical and physical agents in assessing growth parameters. Figure 18.4 demonstrates the quantitative effect of different concentrations of NGF on the measured parameters (Shah et al., 2004).

Figure 18.3 illustrates the stimulation of specific parameters as a result of NGF concentration.

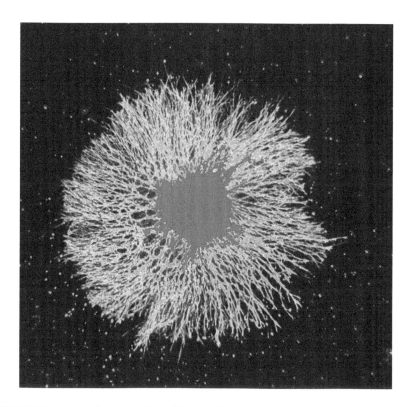

FIGURE 18.1 A dorsal root ganglion with centrally-located neurons in orange are surrounded by their yellow-colored neurites (axons and dendrites) at 3 days *in vitro* in medium containing 10 ng/mL NGF.

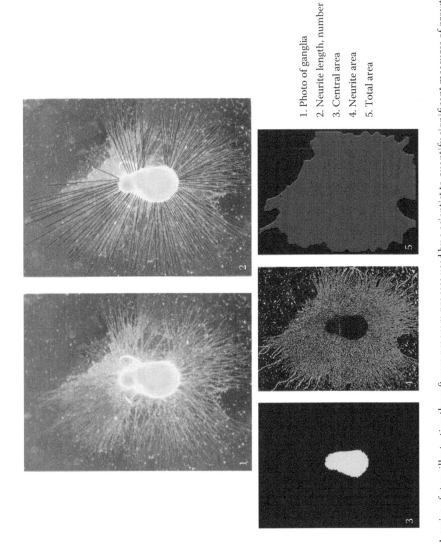

1. Photo of ganglia
2. Neurite length, number
3. Central area
4. Neurite area
5. Total area

FIGURE 18.2 A series of steps illustrating the software program process used by a scientist to quantify significant measures of growth.

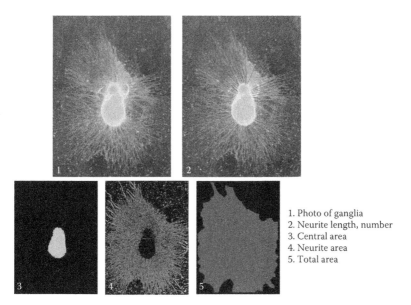

1. Photo of ganglia
2. Neurite length, number
3. Central area
4. Neurite area
5. Total area

FIGURE 18.2 **(See color insert.)** A series of steps illustrating the software program process used by a scientist to quantify significant measures of growth. (From Shah, A., et al., *J. Neurosci. Methods*, 136, 123–131, 2004. With permission.)

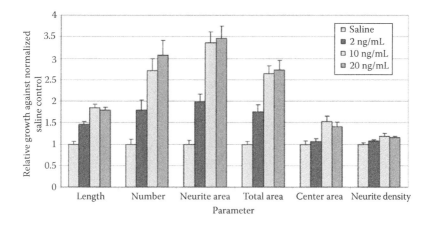

FIGURE 18.3 This graph illustrates the data obtained using the software program described in Figure 18.2 as a function of NGF concentration. The data is expressed as a function of normalized saline control. (From Shah, A., et al., *J. Neurosci. Methods*, 136, 123–131, 2004. With permission.)

18.4 Direct Current Studies

With the establishment of this chick embryo model, we decided to begin investigating different electromagnetic signals. In 1975, using a platinum electrode system and much lower current densities, we confirmed the observations of earlier reports of cathodal attraction by demonstrating that fibers from chick embryo TG stimulated with DC levels of 0.0011–11.5 nA/mm² were oriented to the cathode (Sisken and Smith, 1975). We used a time-lapse system for measuring growth rate by culturing the ganglia in a Rose chamber with electrodes inserted into it. We determined the rate of growth of the nerve fibers to the cathode to be 100 μ/h (2.4 mm/day), which is comparable to the rate found in regenerating nerves in humans.

To address the problem of nerve regeneration *in vitro* on a long-term basis, we changed our system so that 60-mm culture dishes could be used instead of the original 35-mm ones, allowing us to culture more DRG per dish. The configuration of the electrodes was therefore modified so that the nonuniform field was maintained to allow us to keep the cultures alive for longer periods of time (Sechaud and Sisken, 1981).

To substantiate our findings, we tested the protein synthetic ability of some of the nerve ganglia after exposure to DC or NGF as an indicator of growth of new neurites (Figure 18.4). DC and NGF both stimulated incorporation of 3H-leucine/ganglia by a factor of 2, although the sympathetic ganglia response was less remarkable. These data mimic the results obtained in our morphological observations that DC acts like NGF in stimulating enhanced growth but is not as effective as adding NGF exogenously to these explants (Figure 18.4).

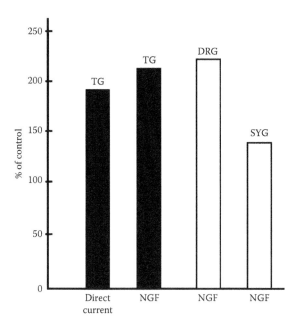

FIGURE 18.4   Protein synthesis determined by the incorporation of 3H-leucine into trigeminal ganglia (TG), dorsal root ganglia (DRG), and sympathetic ganglia (SYG) as a function of direct current (DC) or nerve growth factor (NGF). (From Sisken B. and Lafferty J.F., *Bioelectrochemistry and Bioenergetics* 5, 459, 1978.)

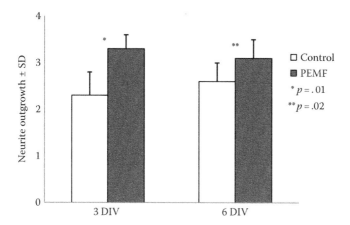

FIGURE 18.5   Pulsed electromagnetic field stimulation of cultures of DRG after 3 and 6 days *in vitro*. The PEMF signal was 0.5 G at 2 Hz, and the cultures were exposed to these fields for 2 h/day for 2 days. Although this stimulation was performed for both time periods, the greatest effect was seen in the 3 day *in vitro* culture.

18.5 Pulsed Electromagnetic Fields

Our chick embryo ganglia model was then used to test specific electromagnetic pulse signals (PEMFS) designed by Bietic Research, Inc. (New York) or by Electro-Biology, Inc. (Parsippany, NJ). Both clinical and animal studies had been run to assess PEMF action on bone unification. Results from one of the first PEMF signals tested in our *in vitro* system are presented in Figure 18.5 and enforces the positive decision of using noninvasive PEMF rather than the invasive DC electrode system we had started with. The PEMF used in Figure 18.5 significantly stimulated growth of neurites at 3 and 6 DIV (DIV = days *in vitro*).

18.6 Importance of PEMF Orientation

Our next focus was to assess the possible correlation of neurite growth with current density and field geometry; we ran two types of experiments, each with their own controls. The first was placing the Helmholtz coil in the incubator parallel to the plastic shelf (H position) generating vertical magnetic field pulses. The second was to turn the coils perpendicular to the shelf, thereby generating horizontal pulses (V position). In each case, the cultured ganglia in plastic dishes were placed between the coils and incubated for 2 days (12 h on/off). The waveform, shown in Figure 18.6, changed at a rate of 5.3 T/s, 325 μs pos., 20 μs neg., 72 Hz, 15 mV (Electro-Biology, Inc.). Rings were drawn on a piece of transparent plastic and used to define the boundaries of "inner" vs. "outer" areas. The inner ring had a radius of 14 mm, and the outer ring radius was 14 mm. This plastic sheet

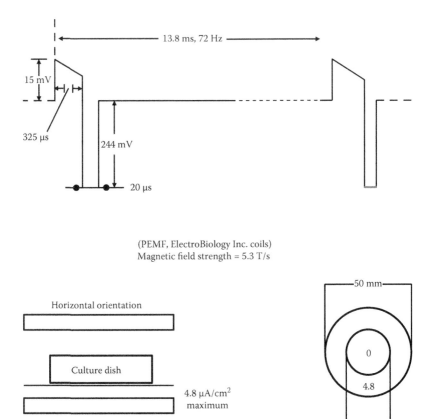

FIGURE 18.6 PEMF signal and dish measurements.

with circles on it was placed under each dish as the ganglia were identified as to ring location and neurite outgrowth determined.

18.6.1 Horizontal Orientation

In the H orientation, the current density in the inner ring of the dish was 0 and the maximum was 4.8 µA/cm^2 in the outer ring. In the V orientation, the current density in the inner ring is maximal at 0.48 µA/cm^2 and lower in the outer ring (Sisken et al., 1984).

Neurite outgrowth was measured in explants in both rings of each dish. After 3, 4, and 7 days of incubation, growth of the neurites was scored on a scale of 0–5. The results indicated that neurite growth in general was stimulated with these fields (Figure 18.7) but did not depend on the coil orientation and current density.

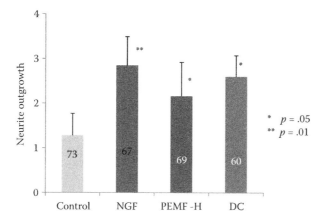

FIGURE 18.7 Neurite outgrowth in cultures of DRG treated with NGF, horizontally placed coils (PEMF-H), and DC were compared to saline controls. The number of ganglia tested are given inside each bar. These data demonstrate results similar to those found in the first study (1975) indicating that NGF consistently produces a higher score than DC, and in this experiment, PEMF-H as well.

Results in Figure 18.7 show the neurite outgrowth assessed in comparable numbers of ganglia in four groups. The highest score was for the NGF group followed by the DC and PEMF-H groups relative to the sham control.

In Figure 18.8, comparisons of neurite outgrowth in inner and outer rings of both PEMF-H and sham control can be seen, indicating that current density does not play a major role in the stimulation by these fields.

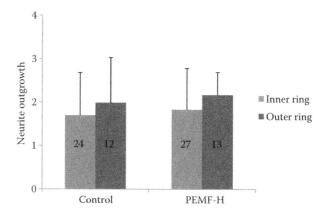

FIGURE 18.8 Comparisons of neurite outgrowth in inner and outer rings of both PEMF-H and sham control indicate that current density does not play a major role in the stimulation by these fields. (Adapted from Sisken B.F., et al., *J. Bioelectricity*, 3, 81–101, 1984.)

18.6.2 Vertical Orientation

In the V orientation when the coils are standing upright perpendicular to the plastic shelf, the current density in the inner ring is $0.48\,\mu A/cm^2$ and almost 0 in the outer ring.

Figure 18.9 shows neurite outgrowth in ganglia exposed to PEMF when the coils were placed vertically (V), were compared to sham controls, NGF, and DC. NGF produced the significantly largest increase in area ($p = .001$) with DC ($p = .001$) and PEMF-V ($p = .05$) following. Dunnett's Multiple Comparison Test.

In Figure 18.10, neurite outgrowth in ganglia exposed to PEMF when the coils were placed vertically (V), were compared to sham controls, NGF, and DC. Ganglia

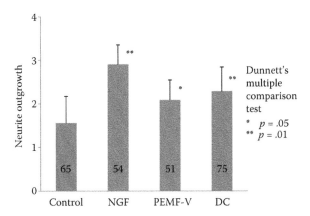

FIGURE 18.9 Neurite outgrowth in ganglia exposed to vertically placed coils (PEMF-V), were compared to sham controls, NGF, and direct current (DC). NGF produced the significantly largest increase in outgrowth ($p = .001$) followed by DC ($p = .001$) and PEMF-V ($p = .05$) when assessed using Dunnett's Multiple Comparison Test.

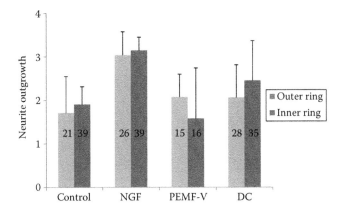

FIGURE 18.10 Comparisons of neurite outgrowth in inner and outer rings of control, NGF, PEMF-V, and DC indicate that current density does not play a major role in the stimulation by these fields.

in the inner circle of the dish were compared to those in the outer circle of the dish. The number of ganglia analyzed is printed on each bar in the figure. Again, it appears that current density is not a main component of the stimulation of neurite outgrowth.

18.7 Static Magnetic Fields

This sensory ganglia *in vitro* model was used to test if SMFs with and without NGF affected neurite outgrowth. The permanent magnet (100 mT) was placed on the bottom, and culture dishes containing DRG in culture medium were stacked vertically above it.

Magnets of 100 mT, as measured by the manufacturer, were placed on plastic shelves and culture dishes containing DRG explants were set above the surface of the magnet so that we could test the effects of static magnetic fields of 90, 45, and 22 mT on neurite outgrowth. Some dishes were fed culture medium with added NGF, or saline as a sham control. The system for obtaining data on neurite area for the explants exposed to 90 and 45 mT and different concentrations of NGF or saline are shown in Figure 18.11.

The results of exposure of DRG explants to 90 and 45 mT are illustrated in Figure 18.12. There was no stimulation of growth unless NGF was present at any magnetic field level, and higher field strength levels were necessary to promote significant neurite area even at low concentrations of NGF. It is interesting to note that 90 and 45 mT significantly suppressed neurite outgrowth with the highest concentration of NGF.

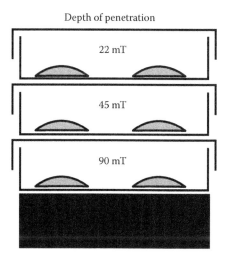

FIGURE 18.11 The depth of penetration of various strengths of static magnetic fields.

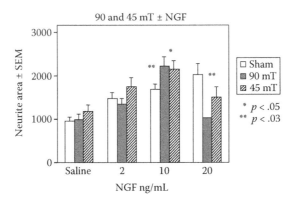

FIGURE 18.12 The effects of exposure to different magnetic fields indicate that the fields alone with saline had no effect, but when combined with NGF at 10 ng/mL showed a good response of 20 ng/mL. At this high concentration of NGF, either with 90 or 45 mT magnetic fields, neurite outgrowth was suppressed.

18.8 Discussion

This study clearly demonstrates that using the chick DRG as a model allowed us to test a variety of electromagnetic signals. The data presented support the claim of previous investigators that *in vitro* growth of sensory ganglia is significantly stimulated by DC and that this growth orients to the cathode. Importantly, this stimulation occurs when explants are treated with currents of 10 nA. The enhancement of growth (1–2 mm/day) is consistent with that of regenerating peripheral nerves in humans. These *in vitro* results have been confirmed by Jaffee and Poo (1979), Greenebaum et al. (1996), Subramanian et al. (1991), and Sutton et al. (1993). Additionally, selected noninvasive PEMFs are capable of mimicking the effects of DC, although neither are capable of producing the maximal effects found with NGF. Coil orientation and current density in the PEMF experiments were not correlated with enhanced growth, a conclusion also reached by Greenebaum et al. (1996).

Following the effects of PEMF technology in *in vitro* models and *in vivo* studies on nerve regeneration (Wilson and Jagadeesh, 1976; Apesos and Sisken, 1983; Ito and Bassett, 1983; Raji and Bowden, 1983; Orgel et al., 1984), we began our studies on nerve regeneration in crushed rat sciatic nerve using a 3G, 2Hz single pulse EMF, similar to the first signal tested in our early *in vitro* studies, only with a higher magnetic field. With these Helmholtz coils, we performed our first experiments in the laboratory of Martin Kanje and Goran Lundborg to test our PEMF signal on their crushed rat sciatic nerve model. Initially, we assessed the length of nerve regenerated in a 6-day period after injury and found a 22% increase in the regeneration rate as compared to sham controls (Sisken et al., 1989). The same stimulation response was obtained when rats were exposed to these fields before lesioning the nerve (Kanje et al., 1993). Since these fields had no effect on fast axonal transport (Sisken et al., 1995), it appeared that only slow axonal transport was affected. Additionally, new protein synthesized 2 weeks after exposure to PEMF was found to shift these polypeptides to a smaller size (Sisken et al., 1990). We also recorded walking patterns to determine if this PEMF signal had long-lasting effects and found that functional recovery was significantly enhanced 43 days after crush lesion (Sisken, Walker, Orgel, 1993; Walker et al., 1994).

These *in vitro* and *in vivo* experiments indicate that specific electromagnetic signals which are noninvasive and nonpainful techniques may well have useful clinical applications. Importantly, the use of the *in vitro* chick embryo ganglia model helped to establish dosimetric parameters (signal strength, frequency, mode of application, and exposure time) that led to successful *in vivo* experiments on nerve regeneration in a mammal. This is even more important because for specific types of nerve injury different signals might be necessary and the thorough dosimetry of the 3-D distribution of the applied field is of great importance.

Acknowledgments

This chapter is dedicated to Dr. Arthur A. Pilla who passed away in the fall of 2015. Without his help I could not have accomplished most of my work related to electromagnetic fields. I also wish to thank Dr. Ben Greenebaum for many helpful discussions and Jonathan Bullard-Sisken for his help with the figures.

References

Bassett CAL. (1989) Fundamental and practical aspects of therapeutic uses of electromagnetic fields (PEMFs). *Critical Review Biomedical Engineering* 17:451–529.

Bassett CAL, Pilla AA, Pawluk R. (1977) A non-surgical salvage of surgically-resistant pseudoarthroses and non-unions by pulsing electromagnetic fields. *Clinical Orthopaedics and Related Research* 124:117–131.

Bilsland J, Rigby M, Young L, Harper, S. (1999) A rapid method for semi-quantitative analysis of neurite outgrowth from chick DRG explants using image analysis. *Journal of Neuroscience Methods* 92:75–85.

Fukuda J, Kameyana M. (1979) Enhancement of Ca spikes in nerve cells of adult mammals during neurite growth in tissue culture. *Nature* 279:546–548.

Gordon T, Sulaiman OA, Ladak A. (2009) Electrical stimulation for improving nerve regeneration: Where do we stand? *International Review of Neurobiology* 87:433–444.

Greenebaum B, Sutton CH, Subramanian Vadula M, Battocletti JH, Swiontek T, DeKeyser J, Sisken BF. (1996) Effects of pulsed magnetic fields on neurite outgrowth from chick embryo ganglia. *Bioelectromagnetics* 17:293–302.

Hynds DL, Snow DM. (2002) A semi-automated image analysis method to quantify neurite preference/axon guidance on a patterned substratum. *Journal of Neuroscience Methods* 121(1):53–64.

Ingvar S. (1920) Reaction of cells to the galvanic current in tissue cultures. *Proceedings of the Society for Experimental Biology and Medicine* 17:198–199.

Ito H, Bassett CAL. (1983) Effect of weak pulsing electromagnetic fields on neural regeneration in rats. *Clinical Orthopaedic and Related Research* 181:283–290.

Jaffee L, Poo MM. (1979) Neurites grow faster toward the cathode than the anode in a steady field. *Journal of Experimental Zoology Part A: Ecological Genetics and Physiology* 209:115–125.

Kanje M, Rusovan A, Sisken BF, Lundborg G. (1993) Pretreatment of rats with pulsed electromagnetic fields enhance regeneration of the rat sciatic nerve. *Bioelectromagnetics* 14: 353–360.

Karssen A, Sager B. (1934) Sur l'influence de courant electrique sur la croissance des neuroblasts in vitro. *Arch Exp Zellforsch* 16:255–259.

Levi-Montalcini R. (1966) The nerve growth factor: Its mode of action on sensory and sympathetic nerve cells. *Harvey Lectures* 60:217.

Lundborg G. (1988) *Nerve Injury and Repair,* Churchill Livingstone, New York.

Marsh G, Beams HW. (1946) In vitro control of growing chick nerve fibers by applied electric currents. *Journal of Cellular Physiology* 27:139–157.

Orgel M, O'Brian W, Murray H. (1984) Pulsing electromagnetic field therapy in nerve regeneration. An experimental study in the cat. *Plastic and Reconstructive Surgery* 73:173–182.

Parker B, Bryant C, Apesos J, Sisken BF, Nickell T. (1983) The effects of pulsed electromagnetic fields (PEMF) on rat sciatic nerve regeneration. *Transactions of the Bioelectrical Repair and Growth Society* 2:19.

Peterfi T, Williams SC. (1934) Elektrische Reizversuche an gezuchteten Gewebezellen. II. Versuche an verschiedene Gewebekulturen. *Arch Exp Zellforsch* 16:230–240.

Pilla AA. (2015) Pulsed electromagnetic fields from signaling to healing. In Markov MS *(ed.) Electromagnetic Fields in Biology and Medicine.* CRC Press, Boca Raton, FL, 29–48.

Sechaud P, Sisken BF. (1981) Electrochemical characteristics of tantalum electrodes used in culture medium. *Bioelectrochemistry and Bioenergetics* 8:633.

Shah A, Fischer C, Knapp CF, Sisken BF. (2004) Quantitation of neurite growth parameters in explant cultures using a new image processing program. *Journal of Neuroscience Methods* 136:123–131.

Sisken BF, Jacob J, Walker J. (1995) Acute treatment with pulsed electromagnetic fields (PEMF) and its effect on fast axonal transport in normal and regenerating nerve. *Journal of Neuroscience Research* 42:692–699.

Sisken BF, Kanje M, Lundborg G, Herbst E, Kurtz W. (1989) Stimulation of rat nerve regeneration with pulsed electromagnetic fields. *Brain Research* 485:309–316.

Sisken BF, Kanje M, Lundborg G, Kurtz W. (1990) Pulsed electromagnetic fields stimulate nerve regeneration in vitro and in vitro. *Restorative Neurology and Neuroscience* 1:303–30.

Sisken B, Lafferty JF. (1978) Comparison of the effects of direct current and nerve growth factor on the regeneration of neurites in vitro. *Bioelectrochemistry and Bioenergetics* 5:459.

Sisken BF, McLeod B, Pilla AA. (1984) PEMF, direct current and neuronal regeneration: effect of field geometry and current density. *Journal of Bioelectricity* 3:81–102.

Sisken BF, Midkiff P, Tweheus A, Markov M. (2007) Influence of static magnetic fields on nerve regeneration *in vitro. Environmentalist* 27:477–481.

Sisken B, Smith S. (1975) The effects of minute direct electrical currents on cultured chick embryo trigeminal ganglia. *Journal of Embryology and Experimental Morphology* 33:29.

Sisken BF, Walker J, Orgel M. (1993) Prospect on clinical applications of electrical stimulation for nerve regeneration. *Journal of Cellular Biochemistry* 52:404–409.

Subramanian M, Sutton CH, Greenebaum B, Sisken BF. (1991) Interaction of electromagnetic fields and nerve growth factor on nerve regeneration *in vitro,* Chapter 25. In Brighton CT, Pollack SR (Eds) *Electromagnetics in Biology and Medicine.* San Francisco Press, San Francisco, CA, 145–151.

Sutton CH, Vadula MS, Greenebaum B, Sisken BF. (1993) Random directional distribution of asymmetric neurite outgrowth in chick dorsal root ganglia exposed to pulsed electromagnetic fields and nerve growth factor. In Blank M (ed.) *Electricity and Magnetism in Biology and Medicine.* San Francisco Press, San Francisco, CA, 351–53.

Walker J, Evans JM, Resig P, Guarnieri S, Meade P, Sisken BF. (1994) Enhancement of functional recovery following crush lesion to the rat sciatic nerve by exposure to pulsed electromagnetic fields. *Experimental Neurology* 125:302–305.

Walker J, Kryscio R, Smith J, Pilla AA, Sisken BF. (2007) Electromagnetic field treatment of nerve crush injury in a rat model: Effect of signal configuration on functional recovery. *Bioelectromagnetics* 28:256–263.

Weiss P. (1934) In vitro experiments on the factors determining the course of the outgrowing nerve fiber. *Journal of Experimental Zoology Part A: Ecological Genetics and Physiology* 68:393–448.

19

The Conundrum of Dosimetry: Its Applications to Pharmacology and Biophysics Are Distinct

Wayne Miller

Nura Health, SPC

19.1 Context Matters

Kittens raised from birth in a physical environment that includes only vertical stripes cannot "see" horizontal. If you take them out of a vertical stripes environment and put them on a normal tabletop—they will walk right off the end of it. We do not know what they actually "see." What we do know is that the integration of various developmental neurobiological processes that are required to "see" and respond to "horizontal" have not been operationalized. This inability to process "horizontal" information is actually counterproductive to their survival. Namely, they walk off the end of the table. Though they are born with an innate biological capacity to grasp "horizontal," the cognitive impairment initially imposed by environmental constraints is then reinforced by all subsequent learning. In effect, their capacity to process information from the outside world is skewed.

Can exogenous factors, such as the teachings we are indoctrinated with, influence and limit the development of human cognitive abilities in an *analogous* manner to the information processing deficit experienced by these "vertical only" cats? Asked somewhat differently, is the process of learning in a classroom environment, for which concrete rewards and punishments exist, *materially different* from experiential learning for which the rewards and punishments are more reflective of individual circumstances?

Psychologists and most laymen intuitively suspect that the answer to this question is—no. After all, the notion of imposed foundational teachings describes the intended impact of religious indoctrination—the establishment of a belief system that will serve as a filter for all information processing thereafter. Dr. Wiley Souba, MD and MBA from Dartmouth Medical School, provided some insight into this question when he examined the cognitive process at the heart of scientific inquiry (Souby 2011). He said: "It is an implicit assumption of the scientific method that there is a right answer to the questions posed by science. But answers are only as good as the language within which they and the questions that generate them are framed." Linguists know that the development of language frames subsequent cognitive ability. Souba continues: "Description implies differentiation. The concepts we (scientists) use require contrast. However, nothing except God can exist without a context."

In other words, *context* matters. The history of scientific examination illustrates that we do not often take that into consideration. Perhaps we are missing something and that something is an essential characteristic of the thing (noun) or process (verb) under examination. In theory, scientific exploration is supposed to be substantively different from religious indoctrination. In practice, the distinction may be much more gray than we would like to admit. There is a plethora of evidence that this is the case.

19.2 In Scientific Inquiry, the Ontological Problem Begins with How We Formulate a Question

In the discipline of psychology there is a well-known process called "anchoring." The term describes a type of information processing bias in which a certain element of information is taken as a "given"—a kind of first premise. Once automatically accepted, that information serves as a foundational filter through which all subsequent information is processed. Therefore, all subsequent information appears to reinforce that which is already known—the premise. With anchoring, a kind of cognitive tautology gets created. One could characterize the slow and unevenly distributed paradigm shifts of scientific progress as examples of how powerful the influence of anchoring is at maintaining the status quo. This statement is consistent with the conclusions of Thomas Kuhn who studied the history of how science materially advances and published a seminal work, *The Structure of Scientific Revolutions*, on the topic (Kuhn 1962).

What does this have to do with the concept of dosimetry? Much more than one might think.

19.2.1 The Prevailing Mental Model of Dosimetry Is Anchored in Reductionism

Werner Heisenberg, whose Uncertainty Principle won him a Nobel Prize in Physics, once said: "What we observe is not nature itself, but nature exposed to our method of questioning." It is the hypothesis of the author that those scientists and clinicians involved in conventional medicine have not sufficiently wrestled with the implications of Heisenberg's statement. The topic of dosimetry, grounded in its pharmaceutical roots, offers fertile ground on which to conduct that wresting process. An examination of dosimetry, as applied to the context of biophysics and biophysical therapies will stimulate the reader to think about what is required to match observations to hypothesis. That process will reveal that dosimetry is a natural window by which one can view systems thinking.

When examined under bright lights, reductionism will be seen to have occupied a central role in the construction of the prevailing mental model from which our notion of health and health care have been built. A vigorous examination will unveil that a lack of reconciliation exists between this foundational framework of clinical medicine and the results produced by those who practice it. For example, the "evidence" at the heart of most clinical services offered by the best clinical talent, the specialists from the academic centers, is built from what the scientific literature refers to as Class "B" or Class "C" evidence, a.k.a. "expert opinion" (Tricoci et al. 2009; Feuerstein et al. 2014). This is very different from the commonly held belief about clinical medicine—that it is grounded in established scientific fact. The scientific substance behind the notion of "expert opinion," it turns out, is very weak. One is compelled to ask: just what kind of expertise are we relying upon when, in the United States, 80% of health-care expenditures (the cost of chronic disease), including those services driven by virtue of "expert opinion," do not result in the patient's problems being resolved?

This chasm can be explained—though the explanation is all too often vigorously resisted. Scientific progress is not promoted when a hypothesis is shot down by virtue of cognitive bias, be it anchoring or otherwise. Indeed, a hypothesis that seeks to explain the lack of reconciliation between a foundational mental model and the results produced from it ought not to be the target of ridicule. *The target of ridicule ought to be the lack of reconciliation itself.*

A path to capturing dramatic efficiencies and new realms of scientific discovery will unfold when the focus of conversation is on closing the reconciliation gap between the foundational principles of reductionism and the real-world results derived when those principles are assumed to be "settled" science. Those who bear responsibility for operating population health-care systems will do well to take note of this distinction.

19.2.1.1 What Is an Alternative Mental Model?

Though the historical record is filled with pioneers in science, medicine, and philosophy, who pointed out the limitations of reductionism, over the past 300 years virtually all talent and information used to build our institutions of science, government, and economy were educated, trained, and grounded in this framework. By contrast, "systems thinking" is said to be a "holistic" framework for understanding something. As a first premise it posits that all elements within an ecosystem (be it macro, micro, or nano) are relevant to understanding how the ecosystem (or any element of it) functions. In short, context matters, and nothing can be understood without examining the environment in which something develops, lives, and thrives. Consistent with this epistemological orientation is the old adage that "the whole is indeed greater than the sum of its parts."

Scientists grounded in classical reductionism may roll their eyes at such a statement. By contrast, scientists who have developed the skill of engaging in systems thinking have learned that an attribute of structure and function, consistent with the principle quoted earlier, is that of "emergent properties." Emergent properties are those functional characteristics of a system that are only apparent when that system is operationally intact—in the environment for which it was designed. If one disintegrates the components of a system into their isolated parts, one cannot anticipate (measure) the structure or functionality of a phenomenon that is an emergent property. This makes the property a systems phenomenon. This distinction is very relevant to the topics of health and health care, as the former is thought to be an emergent property that reflects organizational homeostasis. By contrast, the latter has been established as a political, social, and financial phenomenon in response to a reductionist mental model of illness and disease.

We often call this political, social, and financial structure a health-care "system," but it really isn't a "system" at all. Its parts were never engineered to work together for a common purpose one might characterize as optimizing a patient's return to homeostasis. Indeed, it is now broadly recognized that the health-care "system" (but for acute health-care services) operates for the purpose of managing sickness and the production of revenue that comes from that activity. The pursuit of patient homeostasis as a goal of health-care services and the pursuit of revenue production for delivering those services are not related, though all too often we act, and speak, as though they are.

19.2.1.1.1 *Why Does This Distinction about Mental Models Matter?*

The study of physics offers many insights into this issue. The scientific discipline of physics came to systems thinking more than 100 years ago. Physicists find systems thinking to be a natural format for examining the real world. For example, they are comfortable with the notion of "energy fields" as essential components of all natural phenomena—including life itself. They are comfortable with the fact that even though energy fields cannot be directly seen, they can be measured (albeit sometimes with great difficulty). They are comfortable that different types and forms of matter respond to the influence of energy fields in distinct ways. Indeed, the field of material sciences, on which all technological advances are dependent, recognizes that this is a true phenomenon. Physicists are comfortable that all life (all movement) is dependent on the energy use, storage, and conservation of the system that is in motion. By contrast, the discipline of clinical medicine has only barely begun to grapple with the implications of what physicists have known for a century.

A long time ago, Einstein warned us that: "We cannot solve our problems with the same thinking we used when we created them." Given that warning, when it comes to closing the gap between foundational principles and the results produced when they are followed, one is compelled to ask: what is it about how we think of the problem that doesn't allow us to solve it?

We will know that the question above has been answered when some new way of thinking about the patient's chronic problem leads to therapeutic interventions that resolve the chronic condition(s). In an age of scarce budgetary resources, the adoption of a systems-centric paradigm holds great promise for patients. Why? As the exploration of the "-omic" sciences over the last decade has irrevocably demonstrated—human beings are complex adaptive systems. Interactivity of complex subsystems is itself complex and may not be linear in its construction or operation. At the very least, we have irrefutable data on the inadequacy of the results produced from following the conventional reductionist framework. Thus, simple engineering tells us that if that doesn't work—it is time to try something else. A systems approach to thinking about human health and disease is likely to be fertile ground for both discovery and treatment. Souba cites additional authors who have examined this topic and concluded: "Transformative learning requires shifting an entrenched frame of reference, a process that involved critically self-reflecting on the assumptions upon which our interpretations, beliefs, and habits of mind are based" (Souba 2011).

The protestations of learned men and women from prestigious institutions notwithstanding, the concept of dosimetry is an excellent window that helps to illuminate the chasm that exists between a reductionist and a systems framework of scientific inquiry and clinical medicine. Capturing data on this chasm will be a great benefit to any society or institution that takes on the challenge. First, if properly constructed, gathering such data will allow a contrast between the outcomes of conventional patient-care methods and those generated from a systems approach to diagnosis and treatment. That contrast, for any given diagnostic code, will be highly instructive to policymakers and those engaged in allocating capital to pay for health-care benefits. In finance terms, such a contrasting mechanism establishes a kind of economic value added (EVA) analysis of health-care capital allocation.

Adoption of an EVA analytical framework would represent a material advance to the creation of a pay-for-value-based health-care finance structure. Indeed, the application of such an EVA contrasting mechanism, well known for identifying value in business operations, will shed much light on assumptions that are now taken for granted. It will naturally lead to the adoption and promotion of cost-effectiveness as a foundational principle for health-care systems, be they public or private. Historically, cost-effectiveness has never enjoyed such an elevated status in the design, operation, regulation, or commercialization of practices, protocols, or products in health care. In finance terms, the entire development cycle that has driven innovation into health-care services has been oriented to commercializing those products or services that enhanced the revenue of the institutions and professionals delivering the care. Establishing an informatics system that enables a determination of "cost-effectiveness" of health-care services moves patient care into the core of the structure and function of health-care policy. That is where it belongs.

19.3 Dosimetry: The Historical Construct

On the one hand, the notion of dosimetry seems so simple. Interestingly, the explanation one finds in most reference dictionaries speaks to the notion as it applies to the measurement of ionizing radiation. Of six reference texts consulted, only one included the definition of dosimetry as a pharmacologist would think about it (*Mosby's Medical Dictionary*, 9th edition): "The accurate determination of medicinal doses based on body size, sex, age and other factors." That definition is instructive in that the words "accurate" and "other factors" are used with such aplomb. It is important to recognize that a sanguine acceptance of such a definition belies the clinical complexity of the concept and the evolution of that complexity. Further, when it comes to the use of that concept in the context of applying electromagnetic fields (EMFs) in medicinal therapies, the complexity one must take into account is orders of magnitude greater.

To engage in the examination of these distinctions, we begin by adopting a somewhat different and more clinically oriented definition than the one given earlier. Dosimetry in biology and medicine can be defined as: "The amount of a particular exogenous influence that must be introduced into a biological system to produce a particular result." Of course, from a scientist's point of view, a corollary to this question must be asked: Can the notion of dosimetry lead to reliable measurements that can be consistently reproduced?

When the "substance" is an experimental compound, for example, a drug in the middle of the regulatory approval process, and the biological system is a human being, we think we know how to identify dosimetry. After all, over the last 50 years, the government regulatory apparatus has standardized the examination and approval processes of such compounds for the purposes of promoting safety and defining efficacy of a commercial product. In reality, such processes are very slow to adapt new vistas of information. Some will recognize the slow rate of adaptation to new knowledge as a necessary but appropriate attribute of scientific progress. They will argue such adoption should be slow and methodical. Such individuals might take comfort that a flexible apparatus *even* exists and *can* be expanded as the evolution of knowledge dictates. Others will note that in the elapsed time before new knowledge is driven into clinical health-care services (NIH's Institute of Medicine estimates this at 17 years), human suffering continues unaided by the new knowledge. Furthermore, during that 17-year period, precious financial capital, dedicated to providing patient-care services, will be "stuck" paying for *less effective* but better-known patient interventions.

The utility of the reductionist paradigm was a real achievement of the 18th century. It led to the development of the gold standard in research. Once one examines systems thinking and applies that framework to scientific and clinical research, it becomes clear that the gold standard of scientific inquiry is far more limited than we were led to believe from our training. It was just the best our instructors knew at the time. Or was it? Did various academic, cognitive, and financial biases built into the cultures of our institutions discourage adoption of the more expansive systems framework? The work of Burr at Yale in the 1930s and Becker at the College of Physicians and Surgeons at Columbia are examples of early pioneers who articulated high-quality proof that the influence of bioelectromagnetic fields were profound in living systems. Their efforts were dismissed by virtue of factors other than their intellectual merit (Burr and Hovland 1937; Becker and Selden 1985).

It is now clear, that in the context of examining a complex adaptive system or systems, the notion of isolating dependent and independent variables in an experimental protocol is far more cumbersome than has been acknowledged to date. This recognition has driven the rise and acceptance of Bayesian statistics as a tool for measurement (https://en.wikipedia.org/wiki/Bayesian_statistics). Driven primarily by the limits of our knowledge, the presence of confounding factors in the development of methodological practices has been a constant influence on the evolution of the pharmaceutical construct for dosimetry. Indeed, historically, that construct has served as a mental model to which the life sciences community has been collectively anchored. That anchoring compels us to examine information through the familiar filter of conducting scientific inquiry in which variables are isolated from context rather than through the inclusive filter of systems thinking.

19.3.1 Confounding Variables

There are many poignant examples where the presence of confounding variables in research protocols were ignored, sometimes in deference to commercial interests and other times as expression of ignorance. As a causal matter either influence produced the same result—the advancement of knowledge was impeded. Before we look at such examples, an examination of the evolutionary process that dosimetry experienced as our knowledge base expanded in pharmacology is instructive. It will help illuminate the challenge of conducting research on complex systems.

The development of dosimetry in a pharmaceutical context had its own wild evolutionary ride that, to the chagrin of some, hasn't reached an end stage. Now part of the historical record, it wasn't always as multidimensional as it is now. It started with a simple metric—mg/kg of weight for the experimental compound. Each evolutionary step beyond that was greeted with skepticism and derision about whether it was really necessary or even important at all. That's the incremental nature of evolutionary developments. For example, the delivery of a therapeutic agent intramuscularly, intravenously, or through the digestive system established subsets of dosimetry measures, and each method of administration gave us new and useful insights into patient care. Eventually, the questions arose as to whether a therapeutic dosage varied with age, sex, or sexual maturity? Every time the complexity of dosimetry expanded, we acquired more knowledge and more tools were developed to refine the quality of research.

Witness what has been learned in the domain of pharmacogenomics or the intersection of pharmacokinetics and the human biome. In the context of pharmaceutical development, the complexity of dosimetry, as illustrated by the study of drug interactions with the human biome, now reflects methodological confounding factors we didn't think much about 10 years ago. However, we now have insight that the use of a simplistic set of methodological practices like mg/kg was a mistake. Nature was more complicated, and those complications mattered. Such methodological mistakes often provide windows into future domains of discovery. That is one reason why the sooner they are articulated, the more we can accelerate the discovery process.

Consider the \$3+ billion dollars spent identifying the human genome. A worthwhile investment no doubt, but let's recall how it happened. When the genomic pioneers were pursuing government funding for the exercise, they told a compelling story. It went something like this. There were about 150,000 proteins and there must be a gene for each one. Once we identified the sequencing of each gene, humanity could get to work building a pharmacopeia of 150,000 remedies (accounting for a dysfunction in each protein). Voilà, human disease would come to an end.

It turns out the genome wasn't a static blueprint. It was a dynamic one. The dynamism was dramatically different from the conceptual framework (mental model) envisioned when the notion of a static blueprint dominated the best thinking in academia. It was the best we could conjure up at the time—and it was wrong. Mental models eventually get broken, and when they do, our knowledge expands dramatically. In every age, scientists tend to forget this. They think of their own skepticism as an inherent property of high-quality scientific thought. It may be so long as it is not offered at the altar of defending the predominant mental model.

In a very short period of time, we recognized that genes are expressed and that the process of expression was itself dependent on a whole host of endogenous and exogenous factors that described a complex biochemical, biomechanical, and biophysical dance. The process of expression, which was once thought of as a step-rated linear phenomenon, gave way to the notion of personalized medicine.

The new dimension of personalized medicine notwithstanding, the unending revelation about the presence of confounding factors in research calls into question the capacity for inquiry to legitimately isolate dependent and independent variables from one another. In retrospect, unraveling physiological complexity is an unending process that has revealed time and again that confounding variables rendered prior conclusions suspect. For example, as the context of environmental considerations such as toxic burden or subclinical pathogenic infections become more visible, their respective consequences continue to call into question the notion of studying isolated elements of a complex adaptive system. When one considers the impact of confounding factors (known as well as measurable but not known anomalies), the methodological integrity of the scientific literature takes on a certain question mark that makes the veracity of conventional inquiry more challenging to accept.

In the context of the existing scientific literature, consider the impact on something much more pedestrian than epigenetics. Some of the more simple constructs of our scientific examination were also confounding factors. Take for example the biome of lab animals. The food pellet ingredients fed to rats by different suppliers of laboratory animals varies. Consequently, when the exact same methodological investigation in multiple labs attempted to define common biochemical pathways in rats, the results were inconsistent. It turns out that the biome of the various laboratory species, raised on distinct diets, were different from one another. Multiple labs were running identical study methodologies but had violated the cardinal rule of the scientific examination—the testing methodology must isolate dependent and independent variables or its conclusions will be tainted. The following article illustrates the challenge of this often overlooked variable (www.the-scientist.com/?articles.view/articleNo/44600/title/Inside-a-Lab-Mouse-s-High-Fat-Diet/). As an aside, when was the last time you read a disclosure about the animal feed given to the rats or mice subjects in a study?

All of the examples mentioned earlier demonstrate what has always been true and will continue to be. There is a tendency to explain our understanding through that which we already know. Thomas Kuhn, in his seminal classic *The Structure of Scientific Revolutions*, warned us of this cognitive trap (Kuhn 1962, p. 92). Having examined the evolution of scientific thought over the last several centuries, Kuhn, who taught at Columbia and MIT, concluded the following:

> Though logical inclusiveness remains a permissible view of the relationship between successive scientific theories – it is a historical implausibility.

The author cannot stress enough the importance of digesting this statement—especially for those involved in deciphering scientific inquiry for the purpose of formulating policy development and resource allocation planning. Historically speaking, following the logic of what is known does not lead us to a new domain of understanding. There is a companion notion to this conclusion that comes from a discipline within finance known as behavioral finance. It is called a heuristic. The term describes a kind of mental shortcut taken by using an extension of known information to interpret information that is not known. There are different types of heuristics. Strictly speaking, a heuristic is not an error in cognitive processing. Rather it is way of thinking about information in which the thought process itself is driven by the comfort (albeit unconscious) that comes from using that which is familiar. Like the process of anchoring, heuristics introduce a bias into information processing, and scientific as well as clinical research is subject to these biases with profound consequences.

When a real scientific breakthrough occurs and one looks back on how the transition from the old understanding to new understanding has progressed, it is always considered a brilliant "discovery." In one sense, it is less a discovery than it is a reconfiguration of information that was already known into a new structure of understanding. This was certainly exemplified when we moved from a static view of the genome to a dynamic one. It was also the case when Einstein transformed the construct called "time" from a static to a dynamic one. In both cases, the transition moved to a realization that reflected the operation of a larger more complex system *at work*. Is human homeostasis any different?

The acceptance and promulgation of heuristics tends to retard the process of examining existing information in a new way that enables a breakthrough in understanding to be achieved. In the context of health care, what does this mean? It has meant that the institutionalization of the reductionist conceptual framework had led to the development of clinical services as we know them. It means that the development of the prevailing clinical toolbox has been retarded and has led to a predominance of two tools—drugs and surgical intervention. Meanwhile, a more expansive toolbox sits on the sidelines because we are stuck in the grips of mental models that discourage and thwart the exploration and discovery of the benefits that systems thinking could offer. This discouragement of development has also brought about a political and social acceptance for the wasting of precious social and fiscal resources that could otherwise be allocated in a manner that reflects other paradigmatic shifts that have come before.

There is irony here. Every time we attempt to follow our existing understanding into the domain of the unknown, we run the very real risk of missing fundamental components of how nature has organized itself. The risk of stretching what we know across the chasm of the unknown encourages us to engage in activities that reinforce our conditioning and make us more like the kittens raised in the vertically striped environment. We think what we "see" is complete—but it isn't—it is just familiar. It turns out, as it always has been in scientific inquiry, a new way of connecting the dots is required.

Scientific research is not the only place in which this discouragement and dismissal occurs. Finance, a topic in which information is, in theory, more rapidly disseminated and accepted amongst the community, has Warren Buffett, whose commentary on the straightjacket of familiar frameworks is legendary. Buffett, the famous American investor and one of the world's wealthiest people, has been quoted as saying:

> What the human being is best at doing is interpreting all new information so
> that their prior conclusions remain intact.

Given the two spectacular scientific advances cited [the genome and time (relativity)] one might think that the notion of systems thinking, the processing of information that recognizes the primacy of the dynamic interaction of any phenomenon with the ecosystem in which it operates, would be readily embraced in health care. Alas, this has not been the case.

19.3.2 Linearity: Complexity Is a Quantitative and a Qualitative Phenomenon

One of the great distinctions about the notion of dosimetry as historically developed in pharmacology versus its application to biophysics is that the math of the former tends to be linear, whereas the math of the latter tends to be nonlinear. What is the difference and why is it important? Figure 19.1 is illustrative.

Contrast these simple constructions with what happens when the complexity of the system's organization is orders of magnitude greater. The exact biochemical pathways of glycan metabolism in Figure 19.2 are not important. What is important is that one can readily imagine that a set of equally complex relationships touching the two-dimensional plane on which these transactions are illustrated might intersect this illustration at various points. In imagining that three-dimensional map it becomes abundantly clear that

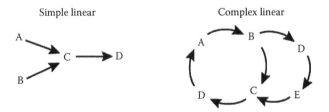

FIGURE 19.1   On the left, conventional linear analysis appears causal. On the right, with a one order of magnitude difference from the left graph, linear analysis is now more complicated. "C" is influenced by A, B, D, and E. The examination of causality is clearly influenced by multiple factors in a differential manner that can be examined and described though by math that is a bit more complicated. Such is the nature of a multivariant system. One can imagine that a three-dimensional representation of three or four or ten feedback loops has complexities built in that do not lend themselves to identification in a conventional testing environment.

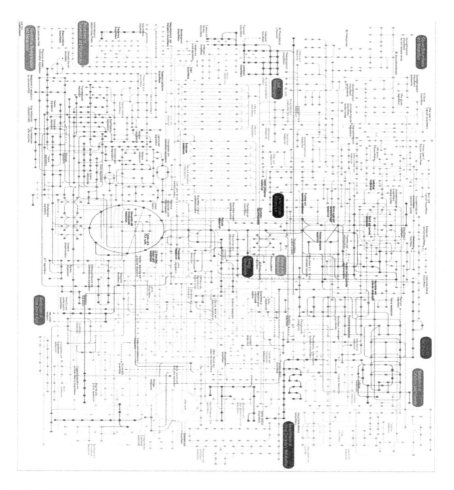

FIGURE 19.2 The pathways of glycan metabolism. The graphic illustrates that the notion of causality is far more challenging to assess and quantify than is implied from the results of a randomized double-blinded controlled study. The multitude of variables that confound experimental results may or may not be isolated from one another. A three-dimensional graph that would intersect this graph would more accurately represent the plethora of pathways and interactions that must occur for a system to be operationalized.

the notion of causality is quite complicated. Indeed, strict causality can be said to be a relative phenomenon—dependent on a whole host of variables operating in a matrix whose components are very challenging to isolate. An attempt to identify causality will produce some kind of statistical result but that result will only reflect the operational status of a matrix when the data snapshot is taken. During that moment, many components of the system are themselves influenced by multiple variables acting within the larger system. In this regard data snapshots of complex systems can be problematic. They appear to reflect the operation of a system in a natural state but that is, as Heisenberg referred to earlier, an artifact of measurement—not necessarily a glimpse into reality.

The example of moonlighting proteins should be noted. These are proteins whose three-dimensional confirmations stimulate functional changes. One can identify the presence of the protein but isolating its three-dimensional confirmation in the context of an unrelated experimental protocol adds a layer of complexity to the study (Constance 2011).

As the above example illustrates, the simplicity of how we have thought about dosimetry simply doesn't take into account the kinds of things one must address when one is examining a complex system. The construct of dosimetry that seeks a simplistic endpoint must adapt to the lessons of history, which illustrate that a multitude of confounding factors are present—whether we test for them or not.

The math of very complex linear systems and nonlinear systems (those whose operations may contain the feature of turbulence) is illustrated in Figure 19.3. Here is another example of the challenge of isolating dependent and independent variables in a meaningful way. Being able to isolate such variables, core to the concept of the gold standard of biomedical research, cannot be assumed—yet all too often it is. This is an example of how and where the "gold standard" may be insufficient as a tool for scientific examination of a complex adaptive system. One would never expect the pharmaceutical examination of dosimetry to generate such a graph. The physics involved is measuring wave number, which is related to wavelength, and frequency, and was borrowed from Naval research on sonar. It is a two-variable complex linear system. Indeed, if one presented such information, meant to characterize the relationship between dosage and biological effect, to drug regulatory authorities one might expect them to shake their head and tell the principal investigator to go back and start over. The manner in which this principle applies to biological systems is beyond the scope of this article. The author notes, however, that there is no corollary in the pharmaceutical construct of dosimetry.

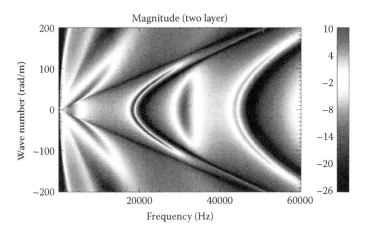

FIGURE 19.3 The image represents the dynamic elastic response of a thick-walled fluid-filled cylinder to an excitation on the outside surface of the cylinder. Although not extracted from a biological system, the graph illustrates the complexity of a complex two-variable linear system. The image is an exemplar of the challenge of applying a simple lens to the interpretation of results that occur in a complex environment.

We use linear analysis to fill in the gaps of data because we have accepted the notion, which flows from a reductionist framework, that the relationship between two variables might have complexities—but nothing of this magnitude. One can imagine that 20 randomized double-blinded clinical trials attempting to document the nature of such two-variable complex linear system might be conducted with methodological purity but still reach no conclusion based on each representing a different measurement on this graph. How would we ever know if we were attempting to measure a complex linear system or a nonlinear system or gibberish? Someone attempting to construct a meta-analysis of these imagined 20 studies would almost certainly conclude that: "the research to date is inconclusive." Again, Heisenberg's commentary about measurement appears prescient.

19.3.2.1 Applying the Pharmaceutical Construct to Complex Systems Has Consequences

This challenge of measurement in complex systems may explain the pronouncement made by Sharon Bagley, a science writer for Reuters, who penned an article printed on March 28th 2012 in which she disclosed that Glenn Bagley (not related), who spent a decade as head of research at Amgen, directed Amgen scientists to replicate 53 seminal oncology papers. The Amgen scientists could only replicate 7. According to Bagley's article, Bayer Healthcare, a unit of the German pharmaceutical giant, had a similar experience in 2011 as reported by Bayer's Vice President and head of Target Discovery. According to Begley and Ellis (2012) they could only replicate 25% of 47 foundational papers.

An article that appeared in *The Economist* magazine entitled: *Trouble at the lab: Scientists like to think of science as self-correcting. To an alarming degree, it is not* published on October 19, 2013 summarized much of this challenge of scientific inquiry. While the article points to challenges of defining and articulating the right conclusions on the basis of the statistical analysis performed, the author points to the possibility that the more basic problem of how we begin to frame a problem and how we carry the familiarity of that framing to the next set of unknowns may be inadequate to the task. This is not proof positive that the culprit of the replication failure is due to the inadequacy of the measurement constructs to correctly assess the structure or operation of a complex system. However, it does support the hypothesis.

When one attempts to apply the pharmaceutical mental model of dosimetry on to the realm of biophysics, that transposition does not fit well. Why should it? There are numerous characteristics of biophysical interdictions and effects that have no close pharmaceutical corollary. For example, with regard to cardiovascular disease the notion of linearity and nonlinearity is very relevant in the characterization of heart disease. Indeed, Kensey and Cho (2007) illustrated that nonlinear (non-Newtonian) behavior of a circulatory system in the process of developing arteriosclerosis is fundamental to the progress of the disease process. They point to the importance of whole blood viscosity, whose non-Newtonian behavior reflects something more complex than the mechanical laminar fluid dynamics that all cardiologists are trained to recognize. Physicists know that non-Newtonian phenomenon exists in biological systems. At the extreme end is the phenomenon of turbulence—also a nonlinear phenomenon. At the very least this means that the math we know in statistical analysis cannot be used to describe it—or more rightly said, the consequences of its presence, which can be measured but is not understood. This notion is reviewed in the classic text, *Chaos* by James Gleick (1987).

Another element that is very relevant in physics but has not been grounded in the reductionist construction of clinical medicine is the Right-Hand Rule (RHR). The RHR states that where there exists an electrical charge that moves in a certain direction, a magnetic field accompanies that charge and travels at a right angle to its flow. This is a very interesting notion—one that is missing from the foundational teaching of physiology and also in all medical education in the United States.

Given that the brain, the heart, all muscles, and even cell membranes operate on the basis of electric charges or gradients that move, what is the likelihood that the magnetic field (albeit small) that accompanies such charges or gradients has no impact on the structure or functionality of the system? As components of a system must communicate with one another and that communication must contain or reflect a certain operational intelligence, what is the influence of the magnetic field on an organism's homeostasis? Given that cell signaling mechanisms often rely upon changes to electrical gradients or charge inside and/or outside the cell or organelles, the likelihood seems very remote.

Until a few years ago the notion of cell signaling would have been taken as one of those topics in science that is at the fringe. The framework of understanding physiology was classic biochemistry, and the notion of WHAT made biochemistry the "core" science required to understand the workings of human physiology (and therefore, clinical medicine) was that it seemed like it could or would explain most everything. Though we didn't know it at the time, by memorizing the Krebs cycle we were being introduced to systems thinking. However, because we were so intent on memorizing the cycle for the purposes of passing an exam, we likely didn't digest that the molecular activity we memorized reflected the functional operation of the organism—its energy system. Cellular organization reflected the operational integrity of energy production and the influence of charge that moved around the cycle components. We thought of charge movement like taking electrons and moving them from place to place. Of course, we didn't know at the time what nuclear physicists now know. A clinician not trained in any physics of biophysics would not have that information. What might you suppose he/she is missing?

The author's wake-up call to systems thinking came when he realized that pharmacology was the study of the carrier of the signal, but *it was the signal* that was important. Signal processing contained language and included syntax, grammar, and punctuation. In short, signal processing provided a framework for intelligent communication and organization—a foundational requirement of running a complex adaptive system that is designed to be responsive to organization, structure, and function (interactivity with environment). Signal processing contained the capacity for intelligence—something that never made sense—even when one pondered the most sophisticated Brownian movement.

19.4 Signal Processing without the Biochemistry

For those who might resist the conclusion about the relevance of the signal not its carrier, the author suggests that the reader examine the technology developed by a company located in Seattle, Washington. The company developed a technology that can record the vibration of molecules. As their first preclinical study they recorded the molecular vibration of a form of Taxol, a common chemotherapeutic agent and fed the recording of the molecular vibration directly into certain tumors on dogs. Almost all of the dogs had no measurable tumor left after these weeks of feeding the recording directly into the tumor site. Further research into the utility of using signals rather than the biochemical carriers of the signals appears very promising (Butters et al. 2014). Thus, we may have yet another glimpse into the consequences of adopting systems thinking rather than reductionism. The notion that biochemistry may not be required and only the signals of the molecules may be needed to produce the desired therapeutic effect has been given stunning reinforcement from this work (see www.nativis.com).

What does such a development mean to our sense of dosimetry? Given that a type of signal input into the body has to be absorbed and doesn't stay at the target site—this looks much like a pharmaceutical construct. But is it? Signals reflect the physics that explains them. One of the key characteristics of signals is the math that describes them. There are two principles in particular that apply to signal processing that do not have a clear corollary in the pharmaceutical world. One is resonance and the other is coherence. Both of these attributes are relevant to the contemplation of systems thinking and to the practical consideration of understanding biophysics and the clinical treatments that can result from its introduction into patient care.

19.4.1 Resonance

The principles and properties involved in energetic field interactions are qualitatively different in character from molecular interactions with membranes and proteins. The more one knows about energy fields the more one is compelled to view dosimetry as a far more complex phenomenon than the classic definition portends. The implications of that complexity are dramatic. For example, the math of resonance is logarithmic not arithmetic. This is known to anyone who plays music. The simple rule of resonance is derived from the observation that all matter vibrates. That vibration is a direct reflection of the material that makes up the matter. This is true whether the matter is represented on the periodic table or is made from multiple molecules with a unique three-dimensional configuration. The natural frequency of that vibration is unique to the material and to its organization. When an external source of energy is applied to any matter and that source vibrates at the same frequency as the matter to which it is applied, the increase in the energy of the entire system (resident matter plus exogenous source) is increased logarithmically. This is very interesting and represents yet another example of how the pharmaceutical construct of dosimetry differs from the construct in the context of biophysics.

We would expect that an energetic pulse introduced into the body will create resonant harmonics among a multitude of different tissues. Given what we know from material sciences, those harmonics would be expected to impact the local environment. However, given current methodological limitations, the characterization of such harmonics is impossible to isolate and therefore measure with precision. That said, the impact of some energetic input may be measurable on the operation of the entire complex adaptive system as has been demonstrated with the use of the heart rate variability (HRV) technology. The body of data with regard to HRV is especially interesting as it provides a window into the clinical construct of homeostasis in the form of providing a quantitative measure of the operational balance between the sympathetic and parasympathetic systems. This is a particularly interesting example of how the dosimetry construct can be thought about in a manner that is distinct from how it is thought about in pharmacology. When a patient is diagnosed with dysautonomia, there is typically a component of their autonomic nervous system that is sufficiently dysfunctional to cause a clinical diagnosis. However, the pharmaceutical construct of dosimetry gives us no insight or tools, analogous to HRV with which to assess homeostasis in a broader systemic context (McCraty 2015).

19.4.2 Coherence

Like resonance, coherence is a property of physics that has no clear corollary in the pharmaceutical realm. The coherence of two waves expresses how well correlated the waves are in time and in motion. Coherence describes several characteristics of waves, some of which are relevant for understanding quantum physics and the others are relevant for constructing high-quality sound systems. Coherence is a relevant topic in biophysics that can be used as a theoretical construct for explaining physiological phenomena that the application of biochemistry methods might not otherwise identify. For example, Loppini demonstrated that the Islets cells of Langerhans, well studied for their role in the development and progression of diabetes, are structurally and functionally constructed so that the electrical activity of the cells are synchronized over the Islet (Loppini et al. 2014). This has led to the identification of a coherent intercellular correlation. Loppini and his team concluded:

> Our numerical simulations agree with the observed possibility of formation of coherent molecular domains and how this depends on glucose concentration and on Islet size. In particular, we find that the power density spectrum of membrane voltage exhibits scale-free self-similarity features .... as an evidence of coherent molecular dynamics.

In their conclusions, Loppini et al. also cite various studies in which clusters of brain cells with phase-locked amplitude modulated oscillations have been documented. The physiological functionality induced by coherence appears to be an emergent property of organizational architecture. In other words, context matters. Their presence in complex adaptive systems implies characteristics that are materially different from how we have thought of system components in which functionality reflects a more mechanical Newtonian architecture. Without a grasp of biophysical principles one would expect such functionality and structural design to be invisible to the investigator.

19.5 Physics: The Mental Model for the 21st Century

One cannot argue that the application of physics to diagnostics has not made a stunning contribution to clinical medicine since Roentgen took the first X-ray of his wife's hand in 1895. In the operation of a magnetic resonance imaging (MRI) devise, a patient is placed into a large magnetic field whereupon smaller magnetic fields are then used to calibrate the impact of the larger field until such time as the change in the structure of water molecules in the patient's body emits a radiofrequency signal to which an algorithm is then applied and a three-dimensional picture is constructed from that signal. However, it is interesting that if you ask a conventionally trained MD how an X-ray works—they will often laugh and respond with humor—"by pressing the button." Rather, the physics of having excited (high energy) particles travel through biological material at different rates is the basis for forming a "picture." MRIs took the value of physics as a platform for diagnostic applications, to a whole new level. It is the author's prediction that the application of physics to therapeutics (biophysical therapies) will make an equivalent contribution in coming decades.

19.6 More on the Biophysical Construct of Dosimetry

Dosimetry in biophysics is complicated. There are simply too many confounding variables to make the assumption that dependent and independent variables can be isolated from one another. Macromolecular crowding is the term that has been applied to the study of just how crowded the inside of most cells actually are. There could be concentrations of proteins on the order of ~500 g/L and as many as 4300 different types of molecules in a very small space (Minton 2006). Given the notion of macromolecular crowding, the issue of generating variable effects on the components of a system may be relevant to operation of the components or to the whole system in some differential manner. How to study such impact, however, has yet to be resolved. Mathematical modeling has begun to be used as a core strategy in research; however, we must not dismiss that examining crowded conditions is challenging because their complexity is still a proverbial black box. In the context of applying an energetic stimulus that will impact directly or indirectly 4300 different types of molecules begs the issue that biochemistry has ignored for far too long. Brownian movement is not a cause in the operation and organization of the complexity.

The consideration of crowded conditions is not a singular phenomenon. It contains various processes, each of which is complex in and of its own right. Processes such as the unfolding of proteins so that they can do what they were designed to do, the shifting concentration gradients in various intracellular sites such as organelles, and finally how the operations are calibrated so that the larger cellular collections (tissue) can coordinate their activity and the organism's homeostasis can still can be managed. In the context of biophysical therapies, dosimetry in a macromolecular crowded world cannot go down the path of conventional interpretation. There is not enough time or money left to wait until every biochemical transaction can be isolated in the context of its contribution to the status of the system.

Every once in a while, the extraordinary construction of life is simply breathtaking. A solution to the conundrum begins with examining nature as it is. As a practical matter, if we want to advance our understanding and construct a practical system of delivering services that promote health, we have to come to terms with fundamental principles of the construction of the phenomenon called health. It helps to think about the very nature of evolution. Molecularly speaking, we humans evolved in cooperation with the molecular environment in which our ancestors developed. Cooperation with the exogenous molecular environment had to be a foundational principle of that evolutionary activity or we wouldn't be here. That cooperation reflects the influence of multiple systems which are in part connected by one common bridge—food. Food is that resource found in the external environment which, when ingested and digested, becomes the raw material from which our human system survives and/or thrives. Dosimetry in the context of food as medicine is complex because we know that food has direct and indirect epigenetic impacts and such impacts will vary in a manner that makes personalized medicine make sense. We are at the beginning stages of this effort and anyone engaged in the activity will tell you—it is an exercise in unraveling a complex matrix (Ferguson 2014).

It is an interesting phenomenon that pursuit of our understanding of complexity has, in one sense, all been worked out through evolution. Nature knows what it is doing. We aren't smart enough yet to get it because we have insisted on studying her in a manner that does not reflect how she is built or how she de facto operates. In documenting natural phenomena, we would do well to remember that foundational principles are essential to the construction and operation of a complex adaptive system. In addition to food, water is one of those foundational elements of our "system" that is routinely dismissed in our pursuit of life sciences research. Yet, talk with a PhD in physics about water and their eyes light up with delight at the curious nature of the substance. They will tell you that water is not well understood.

How could that be? It is a curious thing indeed that the very notion of a patient's hydration is typically not documented in pharmaceutical research. Thus, a discussion about the impact of a patient's hydration on dosimetry of pharmacologic agents is substantially unexplored. When one looks at the topic of patient hydration in PubMed, one finds some papers that examine the topic with regard to a single condition—for example, end-stage renal failure. However, on the broader topic of hydration it is a fair statement to say that "adequate" hydration is assumed except in the most desperate of cases. In a manner of speaking, no one really knows whether the matter of hydration is relevant or not to assess pharmaceutical dosimetry.

By contrast, the author has interviewed dozens of clinicians who have employed biophysical therapeutic devices in patient care. These clinicians will often comment on the need to hydrate the patient before treatment or the treatment will be less effective. They assume that this observation is related to water's nature of being a conductor. While that makes sense from an engineering perspective, it is not well documented with regard to biological systems and their homeostasis. One might imagine that with 99% of all molecules in the body being water molecules and water by volume accounting for ~75% of the weight of a healthy adult, the notion of hydration and its role in health and disease would be well established. Surely, someone must have done this already. However, that wouldn't be accurate. Indeed, water is often dismissed as an almost irrelevant element of any experimental protocol.

In this regard, the work of Dr. Gerald Pollack from the University of Washington on the interactions between water and membranes, what has come to be called "exclusion zones," seems to imply that water is very relevant to the operational integrity of human physiology (Pollack 2013). From Pollack's work and others one can adopt a biophysical point of view that the behavior of water in exclusion zones may reflect a kind of biological amplifier function, a mechanism for calibrating a membrane's responsiveness to extracellular signals. One can imagine, from an evolutionary perspective, the value such a signal calibration mechanism would offer an organism by enhancing its range of responsiveness to its environment. By contrast, a binary on–off switch that limited the range of responsiveness would not be as beneficial a construct. After all, the notion that adaptation enhances resiliency is a foundational principle of evolution. A mechanism that enhanced adaptive capacity would likely have a complex functionality in its construction or operation or both.

The cycle of anchoring and breakthrough continues while the wisdom of Kuhn's work shines a light on the path forward. By promoting the study of systems biology—the multidisciplinary field of study that arises when one applies systems thinking to clinical medicine—we can accelerate our learning. This is not a theoretical notion. The activities of the Institute for Functional Medicine (IFM), the Personalized Medicine Lifestyle Institute and the Institute for Systems Biology have already documented the value of applying systems thinking to clinical medicine.

The implied lesson of this chapter has to do with how we think about thinking about a problem. This is not a new phenomenon of course. This issue has always been at the heart of scientific inquiry. The history of science is replete with attempts to examine new phenomenon through the lens of what was already "known." It is often the application of a new lens to old data that provides the opportunity for breakthrough thinking and "discovery." The distinction between static and dynamic is simple—isn't it? But the implications are profound in more ways than could have been anticipated before the breakthrough was recognized. This is the fertile ground in which dosimetry and biophysical therapies live. It is the stuff young scientists dreamed of when they contemplated their dedication to such a career.

References

Becker RO, Selden G (1985) *The Body Electric*. HarperCollins, New York.

Begley CG, Ellis LM (2012) Drug development: Raise standards for preclinical cancer research. *Nature* 483:531–533.

Burr HS, Hovland CI (1937) Bio-electric potential gradients in the chick. *Yale Journal of Biology and Medicine* 9:247–258.

Butters J, Figueroa X, Butters B (2014) Non-thermal radio frequency stimulation of tubulin polymerization *in vitro*: A potential therapy for cancer treatment. *Open Journal of Biophysics* 4(4):147–168.

Ferguson LR (2014) *Nutrigenomics and Nutrigenetics in Functional Foods and Personalized Nutrition*. CRC Press, Boca Raton, FL.

Feuerstein JD, Akbari M, Gifford AE, Hurley CM, Leffler DA, Sheth SG, Cheifetz AS (2014) Systematic analysis underlying the quality of the scientific evidence and conflicts of interest in interventional medicine subspecialty guidelines. *Mayo Clinic Proceedings* 89(1):16–24.

Gleick J (1987) *Chaos: Making a New Science*. Penguin Books, New York.

Jeffery CJ (2011) Proteins with neomorphic moonlighting functions in disease. *IUBMB Life* 63(7):489–494.

Kensey KR, Cho YI (2007) *The Origin of Atherosclerosis: What Really Initiates the Inflammatory Process*, 2nd ed. SegMedica, Inc., Summersville, WV.

Kuhn TS (1962) *The Structure of Scientific Revolutions*. University of Chicago Press, Chicago, IL.

Loppini A, Capolupo A, Cherubini C, Gizzi A, Bertolaso M, Filippi S, Vitiello G (2014) On the coherent behavior of pancreatic beta cell clusters. *Physics Letters A* 378(44):3210–3217.

McCraty R (2015) *Science of the Heart: Exploring the Role of the Heart in Human Performance*, Volume 2. HeartMath Institute, Boulder Creek, CA.

Minton AP (2006) How can biochemical reactions within cells differ from those in test tubes? *Journal of Cell Science* 119:2863–2869.

Pollack GH (2013) *The Fourth Phase of Water: Beyond Solid, Liquid, and Vapor*. Ebner & Sons, Seattle, WA.

Souby W (2011) The language of discovery. *Journal of Biomedical Discovery and Collaboration* 6:53–59.

Tricoci P, Allen JM, Kramer JM, Califf RM, Smith SC (2009) Scientific evidence underlying the ACC/AHA clinical practice guidelines. *JAMA* 301(8):831–841.

Index

Printed and bound by CPI Group (UK) Ltd, Croydon, CR0 4YY

01/11/2024

01782619-0016